T0405765

THE ROUTLEDGE HANDBOOK OF THE PHILOSOPHY OF PATERNALISM

While paternalism has been a long-standing philosophical issue, it has recently received renewed attention among scholars and the general public. *The Routledge Handbook of the Philosophy of Paternalism* is an outstanding reference source to the key topics, problems, and debates in this exciting subject and is the first collection of its kind. Comprising twenty-seven chapters by a team of international contributors, the handbook is divided into five parts:

- What Is Paternalism?
- Paternalism and Ethical Theory
- Paternalism and Political Philosophy
- Paternalism without Coercion
- Paternalism in Practice

Within these sections central debates, issues, and questions are examined, including: How should paternalism be defined or characterized? How is paternalism related to such moral notions as rights, well-being and autonomy? When is paternalism morally objectionable? What are the legitimate limits of government benevolence? To what extent should medical practice be paternalistic?

The Routledge Handbook of the Philosophy of Paternalism is essential reading for students and researchers in applied ethics and political philosophy. The handbook will also be very useful for those in related fields, such as law, medicine, sociology, and political science.

Kalle Grill is Associate Professor of Philosophy at Umeå University, Sweden. He has published extensively on the concept of paternalism and its normative content. His research also covers related issues, such as nudging, respect for preferences, state neutrality, and various issues in public health policy.

Jason Hanna is Associate Professor of Philosophy at Northern Illinois University, USA. His research addresses issues in normative and applied ethics, including the permissibility of paternalism and the defensibility of deontological constraints on harming. He has previously held a Faculty Fellowship at Tulane University's Murphy Institute.

ROUTLEDGE HANDBOOKS IN APPLIED ETHICS

For more information on this series, please visit
www.routledge.com/Routledge-Handbooks-in-Applied-Ethics/book-series/RHAE

Applied ethics is one of the largest and most diverse fields in philosophy and is closely related to many other disciplines across the humanities, sciences, and social sciences. *Routledge Handbooks in Applied Ethics* are state-of-the-art surveys of important and emerging topics in applied ethics, providing accessible yet thorough assessments of key fields, themes, thinkers, and recent developments in research.

All chapters for each volume are specially commissioned, and written by leading scholars in the field. Carefully edited and organized, *Routledge Handbooks in Applied Ethics* provide indispensable reference tools for students and researchers seeking a comprehensive overview of new and exciting topics in applied ethics and related disciplines. They are also valuable teaching resources as accompaniments to textbooks, anthologies, and research-orientated publications.

AVAILABLE:

THE ROUTLEDGE HANDBOOK OF GLOBAL ETHICS
Edited by Darrel Moellendorf and Heather Widdows

THE ROUTLEDGE HANDBOOK OF FOOD ETHICS
Edited by Mary Rawlinson

THE ROUTLEDGE HANDBOOK OF NEUROETHICS
Edited by Syd Johnson and Karen S. Rommelfanger

FORTHCOMING:

THE ROUTLEDGE HANDBOOK OF THE ETHICS OF DISCRIMINATION
Edited by Kasper Lippert-Rasmussen

THE ROUTLEDGE HANDBOOK OF THE ETHICS OF CONSENT
Edited by Peter Schaber

THE ROUTLEDGE HANDBOOK OF ETHICS AND PUBLIC POLICY
Edited by Annabelle Lever and Andrei Poama

THE ROUTLEDGE HANDBOOK OF THE PHILOSOPHY OF PATERNALISM

Edited by Kalle Grill and Jason Hanna

LONDON AND NEW YORK

First published 2018
by Routledge
2 Park Square, Milton Park, Abingdon, Oxon OX14 4RN

and by Routledge
711 Third Avenue, New York, NY 10017

Routledge is an imprint of the Taylor & Francis Group, an informa business

British Library Cataloguing-in-Publication Data
A catalogue record for this book is available from the British Library

Library of Congress Cataloging-in-Publication Data
Names: Grill, Kalle, editor.
Title: The Routledge handbook of the philosophy of paternalism / [edited by] Kalle Grill and Jason Hanna.
Other titles: Handbook of the philosophy of paternalism
Description: New York : Routledge, 2018. | Series: Routledge handbooks in philosophy | Includes bibliographical references and index.
Identifiers: LCCN 2017038976 | ISBN 9781138956100 (hardback : alk. paper) | ISBN 9781315657080 (e-book)
Subjects: LCSH: Paternalism—Moral and ethical aspects. | Paternalism—Political aspects.
Classification: LCC JC571 .R7725 2018 | DDC 172—dc23
LC record available at https://lccn.loc.gov/2017038976

ISBN: 978-1-138-95610-0 (hbk)
ISBN: 978-1-315-65708-0 (ebk)

Typeset in Bembo
by Apex CoVantage, LLC

CONTENTS

NOTES ON CONTRIBUTORS

Kristoffer Ahlstrom-Vij is Reader in Philosophy at Birkbeck College, University of London. He works mainly in social epistemology and on the foundations of epistemic value and virtue. His book *Epistemic Paternalism: A Defence* was published by Palgrave Macmillan in 2013.

Chrisoula Andreou is Professor of Philosophy at the University of Utah and Executive Editor of the *Canadian Journal of Philosophy*. Her current research projects lie in the areas of practical reasoning, action theory, ethical theory, and applied ethics.

Richard Arneson has taught at the University of California, San Diego, since 1973. He holds the Valtz Family Chair in Philosophy there. He writes mainly on social justice issues and especially on theories of distributive justice. He also works on applied ethics topics.

Emma C. Bullock is Assistant Professor of Philosophy and Research Associate at the Centre for Ethics and Law in Biomedicine (CELAB) at Central European University, Budapest, Hungary. Her primary research interests are in medical ethics, normative ethics, and epistemology, especially in relation to issues surrounding the value of autonomy and justified paternalism.

Michael Cholbi is Professor of Philosophy at California State Polytechnic University, Pomona. He has written on a diversity of topics in ethics, including suicide, punishment, paternalism, and Kantian ethics. He is currently completing a book manuscript on philosophical issues related to grief.

Sarah Conly is Associate Professor of Philosophy at Bowdoin College. She is the author of *Against Autonomy: Justifying Coercive Paternalism* (Cambridge University Press, 2013) and *One Child: Do We Have a Right to More?* (Oxford University Press, 2016).

Péter Cserne is Senior Lecturer in Law at the University of Hull, UK. His research focuses on legal reasoning, private law theory, and the philosophical foundations of law and economics. He is the author of *Freedom of Contract and Paternalism: Prospects and Limits of an Economic Approach* (Palgrave Macmillan, 2012).

Melissa Seymour Fahmy is Associate Professor of Philosophy at the University of Georgia. She is the author of several articles on Kant's ethics including "Kantian Practical Love," "Love, Respect, and Interfering with Others," and "Understanding Kant's Duty of Respect as a Duty of Virtue."

Moti Gorin is Assistant Professor of Philosophy at Colorado State University. He writes and teaches in moral and political philosophy and in bioethics. His articles have appeared in *American Philosophical Quarterly*, *The American Journal of Bioethics*, *Hastings Center Report*, *Public Health Ethics*, and *Philosophia.*

Kalle Grill is Associate Professor of Philosophy at Umeå University, Sweden. His research mainly concerns paternalism and related issues, such as nudging, respect for preferences, and various issues in public health policy. He also works on population ethics.

Daniel Groll is Associate Professor of Philosophy at Carleton College in Northfield, Minnesota. His main area of research is paternalism, although he also does work on moral disagreement, moral testimony, and issues in clinical medical ethics.

Jason Hanna is Associate Professor of Philosophy at Northern Illinois University. His research addresses issues in normative and applied ethics, including the permissibility of paternalism and the defensibility of deontological constraints on harming.

Dana Howard is a philosopher interested in ethics, political philosophy, and bioethics. She received a PhD in Philosophy in 2014 from Brown University and was a postdoctoral fellow in the National Institute of Health Clinical Center Department of Bioethics, and currently teaches at the Ohio State University in the Department of Philosophy and the College of Medicine.

Heidi M. Hurd is the David C. Baum Professor of Law and Philosophy at the University of Illinois. She has published extensively in the fields of criminal law, torts, and jurisprudence, lectured and taught around the globe, and testified before Congress on aspects of the American criminal justice system.

Serene J. Khader is Jay Newman Chair in Philosophy of Culture at Brooklyn College and Associate Professor of Philosophy at the CUNY Graduate Center. She works in global justice, feminist philosophy, and normative ethics, and much of her work considers autonomy under conditions of oppression. She is the author of *Adaptive Preferences and Women's Empowerment* (Oxford, 2011) and *Decolonizing Universalism: Transnational Feminist Ethics* (Oxford, 2018), and a co-editor of the *Routledge Companion to Feminist Philosophy* (2017).

Suzy Killmister is Lecturer in Philosophy at Monash University, Australia. Her first book, *Taking the Measure of Autonomy*, is due to come out with Routledge in 2017.

Peter de Marneffe is Professor of Philosophy at Arizona State University. He is the author of *Liberalism and Prostitution* (Oxford University Press, 2010) and *The Legalization of Drugs* with Doug Husak (Cambridge University Press, 2005).

Muireann Quigley holds the Chair in Law, Medicine, and Technology at Birmingham Law School in the UK. Her research centers on the philosophical analysis of law and policy, often with a focus on issues in the biosciences and medicine.

Jason Raibley is Associate Professor of Philosophy and Director of the Applied Ethics Forum at California State University–Long Beach. His main research areas are the theory of personal well-being, metaethics, moral psychology, and bioethics.

Jonathan Riley is Professor of Philosophy and Political Economy, Tulane University, and a founding editor of the Sage journal *Politics, Philosophy and Economics*. He has published extensively on Mill's philosophy. His most recent publications are *Mill's On Liberty* (Routledge, 2015), "Freedom of Speech" in the online *Oxford Research Encyclopedia on Politics*, and *Mill's Radical Liberalism: A Study in Retrieval* (Routledge, 2018).

Gina Schouten is Assistant Professor of Philosophy at Harvard University. Her research interests include gender justice, educational justice, and political legitimacy, including especially questions about whether political liberalism can constitute an adequate theory of legitimacy.

Danny Scoccia is Professor Emeritus at New Mexico State University, having taught in its philosophy department for 28 years and retired in 2016. He continues to write on various topics in social/political philosophy and biomedical ethics.

Michael Slote is UST Professor of Ethics at the University of Miami and Honorary Professor at Hubei University in China. A member of the Royal Irish Academy and former Tanner Lecturer, he has recently been working on a project that seeks to integrate Chinese and Western philosophy.

Andreas Stokke is Pro Futura Scientia Fellow at the Swedish Collegium for Advanced Studies and Senior Lecturer at the Department of Philosophy at Uppsala University, Sweden. He works mainly on topics in the philosophy of language and epistemology, and has published papers on semantics, pragmatics, lying, assertion, and the nature of insincere speech.

George Tsai is Associate Professor of Philosophy at the University of Hawaii. He has published articles in venues including *Philosophy and Public Affairs*, *Journal of Political Philosophy*, *Pacific Philosophical Quarterly*, *Social Theory and Practice*, and *Oxford Studies in Agency and Responsibility*.

Peter Vallentyne is Florence G. Kline Professor of Philosophy at the University of Missouri, USA. He writes on issues of liberty and equality in the theory of justice (and left-libertarianism in particular) and, more recently, on enforcement rights (rights to protect primary rights).

Steven Wall is Professor of Philosophy at the University of Arizona, where he is a member of both the Center for the Philosophy of Freedom and the Politics, Philosophy, Economics and Law Program. He co-edits *Oxford Studies in Political Philosophy*.

INTRODUCTION

Jason Hanna and Kalle Grill

The moral debate over paternalism concerns the conditions under which it is permissible to intervene in a person's affairs for that person's own good. This debate raises a set of issues – for instance, about the significance of personal autonomy, the norms appropriate to various interpersonal relationships, and the limits of government authority – that have been widely debated among both scholars and the general public. Philosophical discussion of such issues goes back at least to Plato's *Republic*, which argued that the ideal society would be ruled by a class of benevolent philosopher-kings. The greatest influence on contemporary discussions of paternalism, however, has come from J.S. Mill. In *On Liberty*, Mill argued for his famous principle of liberty, which holds that society can legitimately coerce a person only in order to prevent harm to others. In particular, Mill rejected the view that society can legitimately coerce a person for that person's own good. As Mill put it, "the only purpose for which power can be rightfully exercised over any member of a civilized community, against his will, is to prevent harm to others. His own good, either physical or moral, is not a sufficient warrant" (1977 [1859]: 223).

While paternalism has been a long-standing philosophical issue, it continues to attract attention and provoke debate. Current interest in the moral debate over paternalism appears to be driven by the interplay of at least three factors. First, despite many local setbacks, the past several decades have seen an acceleration of diversity in lifestyles coupled with a commitment to universal freedom and self-expression. In the larger perspective, this trend may be seen as a continuation of enlightenment commitments to liberty and equality, and the massive democratization of Western countries since the 17th century. The appeal of liberalism has gradually strengthened the idea of universal rights, including rights to liberty. Benevolent authority is no longer a generally accepted relationship between rulers and citizens, employers and workers, or doctors and patients. While liberals understand that autonomy or self-rule requires cooperation with, or even assistance from, others, paternalism has become more controversial as freedom of choice and lifestyle has been achieved more universally. In terms of particular political debates, the prohibition of homosexual interactions came under scrutiny in the UK in the late 1950s and instigated the debate between H.L.A. Hart and Patrick Devlin on society's interests in such apparently private matters (Devlin 1959; Hart 1963). Though most contemporary liberals would be quick to champion the values of freedom and individuality in this particular case, other cases may be more controversial: recent lifestyle debates concern the medical diagnosis of transgender people and the widespread legal practice of requiring sterilization as a condition of sex change (Winter et al. 2016), the legal prohibition of incest between consenting adults (Hörnle 2014; Deutscher Ethikrat 2016), and the legal prohibition of non-therapeutic self-mutilation and extreme body modification (Schramme 2008).

Second, the recent uptick of interest in paternalism may be partly explained by contemporary research in the social sciences. The "heuristics and biases" literature has extensively documented some of the ways in which human decision-making commonly seems to go wrong (e.g., Kahneman 2011). To be sure, descriptive empirical findings in the social sciences cannot by themselves tell us when, if ever, we ought to involve ourselves in another person's affairs. But this research might give us good reason to revisit a common argument against paternalism. One of Mill's antipaternalist arguments appeals to the idea that each individual is likely to be the best judge of his or her own interests: "with respect to his own feelings and circumstances, the most ordinary man or woman has means of knowledge immeasurably surpassing those that can be possessed by any one else" (1977 [1859]: 277). Mill's critics have long objected that he paints an overly optimistic picture of human deliberative abilities (Hart 1963: 32–34). The contemporary social sciences may thus merely confirm a widespread perception. Still, the costs of a Millian commitment to freedom may become more apparent once we recognize precisely how, and how often, people fall into common deliberative errors. For instance, people are often unrealistically optimistic about the likelihood that risky decisions will turn out well. They are sometimes influenced by the way in which options are presented, and they tend to severely discount costs and benefits that will come to fruition only in the distant future. Psychological research on these deliberative errors has led some to conclude that paternalistic intervention, used appropriately, can often be an effective way to advance individual well-being (e.g., Conly 2013). Others have argued that because the frailty of human deliberation also affects the decisions of professionals and politicians, we ought to remain suspicious of liberty-limiting intervention in purely personal choices (e.g., Glaeser 2006). Seen in this light, an awareness of human limitations may complement the trend toward greater individuality and freedom of expression.

Third, the moral debate over paternalism has high practical stakes. There is now greater awareness of the ways in which the global burden of disease is affected by lifestyle issues such as diet and sedentary behavior, and the debate over paternalism has taken on new urgency in light of several policy disputes in public health. The great health risk of tobacco smoke was widely acknowledged only in the 1950s, and since then many other links between health and lifestyle choices have been identified. Just as many governments have imposed increased taxes on tobacco, some have proposed or adopted similar measures in order to reduce the consumption of sugary beverages. Public health experts around the world have endorsed bans or restrictions on the sale of certain unhealthy foods, such as those containing trans fat. In the United States, there has been disagreement about the permissibility of laws requiring people to purchase health insurance. Anti-drug laws continue to be hotly debated. These debates often pit those who claim that the relevant policy is objectionably paternalistic, on the one hand, against those who claim either that the policy is not paternalistic or that the relevant type of paternalism is not objectionable, on the other. The justifiability of a range of public policies would appear to depend on whether, and when, paternalism is objectionable.

The three factors discussed above – the liberal commitment to diversity and individuality, the increased awareness of our deliberative deficiencies, and the high practical stakes – help to explain why paternalism is a topic of growing interest and concern. The battle lines in the debate are not always clear, however, and continue to be redrawn. While many philosophers agree that there is something at least presumptively objectionable about paternalism, they disagree about what this is – and, indeed, about how we ought to understand paternalism. And the details here are obviously important: some objections to paternalism may be more easily overcome than others, and some forms of paternalism may be more problematic than others.

The remainder of this brief introduction provides an entry point into philosophical debate over paternalism by briefly describing some of the issues taken up in the chapters to follow. We refer to specific chapters as necessary, in order to indicate where readers may turn for a fuller discussion. Obviously it would be difficult for any collection of papers to address *every* issue related to the philosophy of paternalism (or any other philosophical issue, for that matter). Nonetheless, we hope that this volume gives a thorough and accurate accounting of where the discussion stands and also indicates avenues for further inquiry.

1 Conceptual questions

At least on the face of it, a fruitful discussion of the normative status of paternalism would seem to require an account of what paternalism is, or of what the debate over paternalism is ultimately about. Gerald Dworkin once defined paternalism as "interference with a person's liberty of action justified by reasons referring exclusively to the welfare, good, happiness, needs, interests or values of the person being coerced" (1971: 108). This rough definition has provoked considerable disagreement, as Danny Scoccia discusses in his contribution to this volume (Chapter 1). Consider just two possible complications for Dworkin's rough definition. First, many critics have argued that paternalism need not involve "interference" with liberty. It may seem paternalistic for the government to offer certain forms of aid "in kind" – for instance, to provide needy citizens with vouchers to buy groceries, but prohibit them from using these vouchers to buy junk food. Although measures of this sort may be paternalistic, they do not obviously limit anyone's liberty. Second, it may be unclear how we ought to understand the claim that the paternalist involves himself in another person's affairs *for that person's own good*. According to one way of understanding this claim, the paternalist agent is *motivated* by the aim of securing the subject's own good. According to an alternative understanding, the relevant question is instead whether intervention is most plausibly justified on the grounds that it serves the subject's own good, or, perhaps, whether intervention could be justified if, and only if, the subject's interests are allowed to count in its favor (for related discussion, see Chapter 4).

It seems fair to say that there is no consensus account of what makes acts and policies paternalistic. In one sense, however, this result is hardly surprising. Ideally, a characterization of paternalism should latch onto features that we regard as morally interesting, or that indicate something about why paternalism is morally controversial. Yet philosophers disagree about why paternalism is presumptively objectionable, and even about whether it is presumptively objectionable. Disputes about how to characterize paternalism may reflect these deeper disagreements. In this regard, at least, the debate over paternalism is unlike other debates in applied ethics, such as those over abortion or capital punishment. There is very little disagreement about what abortion or capital punishment is, and very little disagreement about whether any particular act ought to be classified as an instance of abortion or capital punishment. By contrast, there is often disagreement about whether some particular act or policy is paternalistic. (Are social insurance programs paternalistic? Indoor smoking bans?) And even if we were to agree about whether an act or policy can be appropriately described as paternalistic, we might disagree about whether it is paternalistic *in the way that matters morally*.

Perhaps, then, moral and political philosophers should simply focus on characterizing their favored principles of paternalism. A principle of paternalism might represent a view about why it matters, and how it matters, that intervention in a person's affairs would be good for her. To develop a complete view of this sort, one would need to delineate the sorts of intervention or behavior that one is concerned about, and one would then need to explain the conditions under which this sort of behavior is, or is not, justified. This approach is at least arguably exemplified by some prominent philosophical discussions of paternalism. According to Joel Feinberg's influential theory, for instance, we ought to distinguish (at least) four different principles regarding the limits of the criminal law: a "harm principle," which holds that it is a good reason in favor of criminal legislation that it would prevent some people from imposing non-consensual harm on others; an "offense principle," which holds that it is a good reason in favor of criminal legislation that it is necessary to prevent serious offense to others; a principle of legal paternalism, which holds that it is a good reason in favor of criminal legislation that it would prevent people from (voluntarily) behaving in ways likely to harm themselves; and a principle of legal moralism, which holds that it is a good reason in favor of criminal legislation that it prevents ethically undesirable behavior even if this behavior does not cause harm or offense to anyone (1984: 26–27; for related discussion, see Chapter 22). On Feinberg's view, we ought to reject the principles of legal paternalism and legal moralism. (The relationship between these two principles is discussed in Chapter 3.) But of course a critic might disagree with Feinberg's moral view while nonetheless accepting the distinctions among the principles he describes.

Any attempt to defend, or reject, a single unified principle of paternalism is likely to confront a number of complications, however. As Feinberg himself observes, virtually no one rejects *all* intervention designed to protect people from their own self-regarding behavior. After all, paternalism as directed toward children often seems innocuous (see Chapter 26). Most people would also accept intervention that protects adults from foolish choices they might make while drunk or otherwise impaired. Those who believe that there is a morally important distinction between intervention on behalf of incompetent decision-makers and intervention on behalf of competent decision-makers must explain what this difference consists in (on this issue, see Chapter 2). Ideally, such an explanation should also offer some guidance in our dealings with those judged to be incompetent (see Chapter 25).

Moreover, there may be important differences between paternalism in the legal-institutional setting and paternalism in the interpersonal setting – and, indeed, among different areas of the law (or different interpersonal relationships). One aspect of the legal-institutional setting is that several people are involved both on the delivering and on the receiving end of paternalism, even though paternalism is standardly defined in terms of one-on-one interactions (see Chapter 4). As a result, it may sometimes be more difficult to evaluate putatively paternalistic laws and policies than it is to evaluate one-off interpersonal interventions. Since instances of legal paternalism sometimes raise questions that are arguably extraneous to the core normative debate – for instance, questions about whether some policy really represents the most prudent investment of tax dollars – reflection on interpersonal interactions may sometimes help to isolate whatever we might find morally salient about paternalism. Even here, however, we might ask whether paternalism is always usefully modeled on the behavior of one person who interferes in another person's behavior without that person's consent or against her will. In some cases, people hoping to overcome their own weakness of will or other rational failings may positively welcome the involvement of other agents or indeed the state. In this way, it may be possible for someone to treat herself paternalistically (see Chapter 5).

Once we appreciate the different contexts in which questions about paternalism arise, it may become more difficult to say, simply, that one is "for" or "against" paternalism. One might favor apparently paternalistic behavior when it takes place in certain contexts or involves the use of certain means, but not when it takes place in other contexts or involves the use of other means. We should also take care to ensure that our normative responses to a practice or policy are not influenced merely by the label attached to it – for instance, whether it is commonly classified as "paternalistic."

2 Moral and political theory

Regardless of how paternalism is ultimately characterized, it may seem to represent a fault line along which various moral and political theories differ. Some theories, such as consequentialism, would seem quite friendly toward paternalism: if the best overall consequences could be produced by limiting a person's liberty or interfering with her decision-making, then it appears a consequentialist would have no basis on which to oppose such intervention (see Chapter 7). In contrast, theories that emphasize the importance of individual choice or personal autonomy may seem much more hostile to paternalism (see Chapters 8 and 10).

Nonetheless, reaching a view on the permissibility of paternalism is not likely to be a simple matter of "applying" one's favorite moral or political theory, in part because reflection on the issue of paternalism may inform one's choice among theories and also indicate how competing theories might be most plausibly developed. Indeed, in some cases it may be no easy task to determine what a particular normative theory implies about paternalism. While Mill is commonly taken to be a staunch antipaternalist, he was also a utilitarian, and he seemed to argue against paternalism at least partly on utilitarian grounds (see Chapter 13). Likewise, although the view that people can have duties to themselves may seem to be a natural ally of paternalism, Michael Cholbi argues in his contribution that this appearance is misleading (Chapter 9). To take one further example, although libertarianism is commonly thought to provide a firm bulwark against paternalism, many libertarians would accept intervention when a person is about to harm himself as a result of temporary incompetence or when it is otherwise impossible to ask for consent to benevolent intervention. Libertarian or choice-prioritizing accounts of rights should arguably be formulated so as to accommodate this result (see Chapter 15). Ultimately, then, our views on paternalism are likely to influence our more general moral commitments, just as these commitments are likely to influence our views on paternalism.

In examining the ethics of paternalism, it may sometimes be sensible to leave aside debates about comprehensive moral or political theories to focus on the specific values and principles that seem most pertinent. Perhaps two of the most salient such principles or values are *beneficence*, which may often speak in favor of paternalism, and *respect*, which may often speak against it. It might seem that an adequate view of paternalism must resolve the apparent conflict between beneficence and respect, either by specifying the conditions under which one of these values takes priority over the other, or by indicating how they ought to be balanced against each other in cases of conflict. Yet while this thought captures something important, it also risks oversimplifying the debate. Partly this is because people have different views about how to understand well-being and accordingly about how to understand beneficence. On some views, for instance, it may be intrinsically good for a person that she makes her own choices (for relevant discussion, see Chapter 6).

But perhaps more importantly, people have different views about which forms of behavior are, and are not, appropriately respectful. The operative notion of respect might be situated within a broadly Kantian moral theory (see Chapter 8). It might also be claimed that, Kantianism aside, paternalistic intervention is disrespectful insofar as it involves a failure to treat others as equals (on this issue, see Chapter 16). Critics of such proposals, however, may argue that appropriate respect involves respect not for what people happen to want but for what they have reason to want (see Chapter 14). In a similar vein, Joseph Raz (1986) has argued that autonomy is valuable only when it is cultivated and exercised in the pursuit of valuable and worthwhile activities. Intervention in imprudent and self-destructive behavior may not always show a lack of respect. There may also be important differences between the sorts of respect that we can legitimately demand from other individuals and the sorts of respect that we can legitimately demand from governments: perhaps a government shows appropriate respect for its citizens by remaining neutral among conceptions of the good life, and a commitment to such neutrality may have important implications about the conditions under which paternalistic intervention is likely to be justified (for relevant discussion, see Chapters 14 and 17).

Disagreement about what respect involves or entails can sometimes be traced to disagreement about precisely what ought to be respected. It has been claimed that paternalism is inconsistent with respect for autonomy, that it is inconsistent with respect for individual rights, and that it is inconsistent with respect for persons. These claims are not obviously equivalent. Acts and policies that do not obviously limit liberty or autonomy may nonetheless violate individual rights, depending on which rights we have (see Chapter 10). Moreover, it is not clear that the best way to respect persons is to respect their autonomy. Indeed, some critics have argued that because many reasonable people deny that autonomy is valuable, it would be disrespectful for the government to reject paternalism by appealing to this controversial value, Arneson 1989: 435; (Quong 2011: 99–100). Moreover, even if we agree that autonomy ought to be respected, we may disagree about what autonomy is. As Suzy Killmister observes in her contribution to this volume, there are a number of different accounts of autonomy, which may oppose paternalism to differing degrees, or not at all (Chapter 12).

If we characterize the core normative issue raised by paternalism as a potential conflict between a principle of beneficence and a principle of respect, then we need to clarify the notion of respect at issue. Moreover, we should be careful not to overlook other important values. Paternalistic intervention may sometimes serve to promote equality (see Chapter 16). If less prudent people are disproportionately represented among the worse-off members of society, and if a complete ban on paternalism would give such people greater freedom to make imprudent decisions, then egalitarians may have special reason to permit paternalism under at least some conditions. Traditionally overlooked values, such as trust and care, may also have a bearing on the justifiability of intervention, especially in interpersonal settings. Paternalism may sometimes be acceptable not on the grounds that it promotes the impartial good, but rather on the grounds that it is an appropriate way of showing concern for a person in the context of a particular close relationship (see Chapters 11 and 27).

3 Paternalism in practice

While theories of paternalism are inextricably linked to broader moral and political theories, paternalism remains interesting for many people largely because of its practical dimensions. Questions about paternalism have been especially salient in medical practice (see Chapters 24 and 25). At least traditionally, medical professionals have enjoyed very broad discretion to act in their patients' interests, even without the patients' consent or approval. Nowadays, however, it is widely accepted that, when possible, medical practitioners ought to obtain a patient's informed consent before performing any medical procedure. The requirements of informed consent generally require that physicians be truthful with their patients (for related discussion of paternalistic deception, see Chapter 20). It might appear, then, that medical paternalism has given way to a regime that more thoroughly reflects the importance of patient autonomy.

It is not entirely clear, however, how much we ought to read into this. Some critics of supposedly paternalistic behavior by medical professionals condemn such behavior precisely on the grounds that it is likely to do more harm than good (Buchanan 1978). If this is the central problem that critics have identified with "medical paternalism," however, it may be that medical paternalism is in fact justified when it would serve the patient's own good, even if medical professionals are not always able to determine whether or not their behavior will do so.

Moreover, even if medical paternalism is *generally* wrong, it is surely sometimes permissible to treat patients without their consent. Most of us believe, for instance, that it is sometimes permissible to provide medical treatment to incompetent patients when they are incapable of providing informed consent, and perhaps even when they actually oppose or resent treatment (Chapter 25). In such cases, however, it may make a difference whether the patient has *never* been competent to make his own decisions (say, because he is a child or has been incompetent for his whole life) or whether the patient instead *was* competent but is no longer (say, because he developed dementia in old age).

While some ethicists have been interested in theories of paternalism largely because of their implications for medical practice, others have focused more broadly on issues in law and public policy, including public health policy. Some policies, such as laws requiring the use of seatbelts, strike many people as both reasonable and paternalistic. Yet it is sometimes claimed that if we endorse paternalism in principle, we will have to accept a wide range of policies that most people would regard as clearly objectionable, if not draconian.

In evaluating various policies, however, we should be careful to observe that it is not always entirely clear *why* a policy is supported, or *how* it might best be supported. Many people seem to support drug laws because they believe that drugs are harmful to users. Yet at the same time, some people support drug laws because they believe that drug use harms *non-users*, while others support drug laws because they believe that drug use is itself immoral, and still others support drug laws for some combination of these reasons. We might here attempt to isolate supposedly paternalistic reasons or rationales and ask whether, and when, they can justify or contribute to the justification of public policies, especially public policies that limit individual liberty.

In this regard, many liberal political philosophers have been especially suspicious of paternalistic rationales for *criminal* legislation. Yet even if such suspicion is warranted, this would not necessarily mean that criminal laws can never be justified on paternalistic grounds, and in her contribution to this volume, Heidi M. Hurd considers several distinctions that may be relevant to the justifiability of paternalistic criminal laws (Chapter 22). Moreover, we should bear in mind that arguments against paternalistic criminal laws may not impugn paternalism in other areas of the law. For instance, some restrictions on freedom of contract may best be justified on paternalistic grounds (see Chapter 23). And even if we conclude that the government should not prohibit the sale or use of dangerous products such as tobacco, it might nonetheless discourage such behavior through less coercive means. For instance, the government might impose higher taxes on tobacco products. It might also "nudge" people to quit smoking by placing graphic warning labels on cigarette packaging or disseminating information about the health risks of smoking in the way most likely to encourage healthy behavior. Though the concept of "nudging" is controversial, the practice of shaping the environment in order to affect choice outcomes is established practice in public health policy, for instance (for relevant discussion, see Chapter 18). Even if such influence is non-coercive, it may be paternalistic because it uses non-rational means (as discussed in Chapter 1) and problematic because it is manipulative or deceptive (as discussed in Chapters 19 and 20). However, paternalistic manipulation may sometimes be justified, and may not even be *pro tanto* problematic when directed at an agent who is unable to recognize or respond to reasons that, were she to recognize and respond to them, would promote her good. A somewhat similar point may be made in defense of epistemic paternalism: in certain institutional settings, such as criminal trials or scientific experiments, limitations on individuals' freedom to conduct inquiry in whatever way they see fit may help them to form more reliable beliefs (see Chapter 21).

Finally, even if we set aside medical and policy contexts, virtually all of us confront questions about paternalism in our personal lives. How much control should we exercise over our adolescent children (see Chapter 26)? How should we care for our aging parents? How involved should we be in the lives of our friends? How should we react when someone we know makes a bad decision (see Chapter 27)? While these questions may call for very different answers than the more commonly discussed questions about paternalism in public policy, they suggest that the philosophical issues raised by paternalism are pervasive.

4 Conclusion

Paternalism is a live philosophical issue, and debate over it is likely to continue. In particular, philosophers are likely to focus on what, if anything, is objectionable about paternalism – for instance, about whether it is a distinctive moral category that involves or constitutes a distinctive moral wrong – and about the conditions under which it permissible to paternalistically interfere (on a certain understanding of paternalism) with another person's choices. We hope that this volume fosters an appreciation of the complexity and significance of these questions.

References

Arneson, R. (1989) "Paternalism, Utility and Fairness," *Revue Internationale de Philosophie* 43: 409–437.

Buchanan, A. (1978) "Medical Paternalism," *Philosophy & Public Affairs* 7: 371–390.

Conly, S. (2013) *Against Autonomy: Justifying Coercive Paternalism*, Cambridge: Cambridge University Press.

Deutscher Ethikrat. (2016) "Incest Prohibition," Berlin.

Devlin, P. (1959) *The Enforcement of Morals*, Oxford: Oxford University Press.

Dworkin, G. (1971) "Paternalism," in R. A. Wasserstrom (ed.) *Morality and the Law*, Belmont, CA: Wadsworth, pp. 107–126.

Feinberg, J. (1984) *The Moral Limits of the Criminal Law, Volume 1: Harm to Others*, New York: Oxford University Press.

Glaeser, E.L. (2006) "Paternalism and Psychology," *University of Chicago of Law Review* 73: 133–156.

Hart, H. L. A. (1963) *Law, Liberty, and Morality*, Stanford: Stanford University Press.

Hörnle, T. (2014) "Consensual Adult Incest," *New Criminal Law Review: An International and Interdisciplinary Journal* 17: 76–102.

Kahneman, D. (2011) *Thinking, Fast and Slow*, New York: Farrar, Straus and Giroux.

Mill, J. S. (1977 [1859]) "On Liberty," in J. Robson (ed.) *The Collected Works of J.S. Mill*, Vol. 18, Toronto and London: University of Toronto Press and Routledge, pp. 213–310.

Quong, J. (2011) *Liberalism Without Perfection*, New York: Oxford University Press.

Raz, J. (1986) *The Morality of Freedom*, New York: Oxford University Press.

Schramme, T. (2008) "Should We Prevent Non-Therapeutic Mutilation and Extreme Body Modification?" *Bioethics* 22: 8–15.

Winter, S., Diamond, M., Green, J., Karasic, D., Reed, T., Whittle, S. and Wylie, K. (2016) "Transgender People: Health at the Margins of Society," *The Lancet* 388: 390–400.

PART I

What is paternalism?

1
THE CONCEPT OF PATERNALISM

Danny Scoccia

Before we can determine under what circumstances, if any, paternalism is morally justified, we need to have some sense of what it is. We can answer the normative question only after answering (or assuming an answer to) the conceptual one.

1 Normative concepts; the standard definition of paternalism; motivational vs. justificatory reasons

One of the key questions about the concept of paternalism is whether it is normative.[1] A concept is normative just in case some basic or non-basic normative term (e.g., good/bad, right/wrong, justified/unjustified) is needed to state its definition. "Paternalistic" is often used pejoratively, as part of an accusation ("you're being paternalistic!"). This might lead one to think that a negative evaluation is built into the concept, as one is in the cases of "thief," "cruel," and many other "thick" evaluative terms.

A large number – perhaps the majority – of moral and political theorists assume a definition of paternalism according to which:

P acts paternalistically toward Q (her "target") just in case:

1 P limits Q's liberty or interferes with Q's decision-making;
2 against Q's will, without his consent, or contrary to his preferences; and
3 for Q's own good.

Let's call this the "standard" definition.[2] If it included the explicit requirement that the means by which the paternalist interferes with choice must violate some *prima facie* moral rule, then it would certainly be normative (Gert and Culver 1976). But the standard definition does not include such a condition.

There is no question that some paternalism is all things considered justified. A mother's forbidding her small children to play with matches, sharp knives, etc. for their own good is the very paradigm of a paternalistic action, and it is clearly morally permissible.[3] Paternalism is *prima facie* wrong, however, if all of it stands *in need of* justification. Deception too seems to be *prima facie* wrong, but it consists in (something like) intentionally causing what one thinks are false beliefs in others, making the concept/definition of it non-normative. The same may be true of paternalism on the standard definition. On the assumption that it is *prima facie* wrong to "interfere" with other's choices without their consent, then conditions (1) and (2) imply that paternalism is *prima facie* wrong. Paternalism can be *prima facie* wrong without that wrongness being guaranteed by the concept itself.

Still, the standard definition might yield a normative concept because of (3). (3) implies that paternalism is undertaken for a certain reason, but "reason" here is ambiguous, referring either to a motivation or a justification. This ambiguity gives rise to two competing conceptions of what paternalism is. On an act/motivating reason conception of it, paternalism is a type of action (or practice, law, policy, etc.) that is motivated by altruism, friendship, belief in a duty of beneficence, or the like. The paternalist *believes* that her interference benefits her target and that this benefit justifies the interference. She need not believe that it by itself justifies it, or even that it is the most important or weightiest reason in support of it. She need only believe that her interference would not be justified if it didn't prevent harm-to-self. Since it is a non-normative question whether this motivation is present in the paternalist, the motivational reading of (3) yields a non-normative version of the standard definition. On the reading of (3) where a reason is a justification, one needs to know whether an interference really would benefit its target to know whether there are good paternalistic reasons that support it, and thus, whether it is paternalism. Since it is a normative question whether interference would benefit or harm someone, the justifying reason conception makes paternalism a normative concept.

Whether interference with choice that prevents catastrophic harm from befalling the target is for that reason all things considered justified is a normative question that this essay does not address. It addresses only the conceptual question and proceeds by offering clarifications and criticisms of the standard definition. Toward the end it proposes a broader definition.

2 Paternalistic reasons; paternalism vs. moralism; de Marneffe's "hybrid" characterization

On the justificatory reading of (3), paternalism is, first and foremost, a normative view like retributivism ("desert is a good reason to impose suffering or deprivation on another"). "Anti-paternalism" is the view that promoting the good of another person is *not* a good reason to interfere with his acts or choices, so paternalism, being the opposite of this, must be the view that promoting another's good is a good and sometimes by itself sufficient reason to do so. But on this reading of (3) paternalism may also refer to certain types of act (law, policy, etc.) whose justification requires an appeal to paternalistic reasons. If thirsty Kurt asks me to share my canteen of water with him and I do so, the fact that he benefits is a good justifying reason for my action, but it is not a paternalistic one. My action doesn't "interfere" with any of his choices, so there can't be paternalistic reasons for it. The justifying reason conception should make paternalism a "compound" of an act type (an "interference") and a justification, as Kalle Grill (2007) has argued.

Justification refers to the good of the person whose choices are interfered with rather than the good of other persons. This distinguishes a paternalistic reason from a "prevention of harm to others" rationale for interference. While that distinction is well understood, the distinction between paternalistic reasons and moralistic ones is often blurred.[4] Consider laws forbidding prostitution, "dwarf tossing," or "same-sex sodomy," defended on the grounds that while those who freely engage in the activities may do no harm to themselves or others, they degrade themselves, thus violating a moral duty to self to avoid self-degradation. This defense of the laws appeals to moralistic rather than paternalistic reasons.[5]

The main difference between the act/motivating reason and the justifying reason conceptions of paternalism becomes apparent if we suppose, first, that a law banning meth use was enacted by legislators all of whom believed that the ban is needed to prevent many people from ruining their own lives, and second, that there is a cogent "prevention of harm to others" rationale for the ban that the legislators wrongly rejected. The act/motivating reason conception implies that this ban is paternalistic, while the justifying reason conception implies that it isn't.

Peter de Marneffe (2006) has noted that the act/motivating reason conception does not sit well with the project of "reconciliation" that many antipaternalists pursue. "Reconciliation" consists in identifying a nonpaternalistic justification for a policy that one wishes to defend, but is usually defended by others on paternalistic grounds. Richard Arneson (1980: 271–272) pursues this reconciliation strategy with anti-dueling laws, Seana Shiffrin (2000) with the unconscionability doctrine in contract law, and Elizabeth Anderson (1999: 300–301) with the welfare state's requirement that everyone contribute to social insurance schemes like Medicare and Social Security. Upon identifying a plausible nonpaternalistic rationale for such policies, the "reconciler" claims that the policies are "not really" paternalistic no matter what motivated those who enacted them. De Marneffe suggests that reconcilers operate with a conception of paternalism based on motivating *and* justifying reasons. According to this "hybrid characterization," P acts paternalistically toward Q only if P intends to promote Q's good *and* there is no good nonpaternalistic rationale sufficient to justify P's action. To see the attraction of this view, suppose that the members of a religious cult believe (perhaps somewhat inconsistently) both that murder is a sin the commission of which dooms one to eternal damnation and that all murder victims go straight to heaven and suffer no net harm from the premature ending of their earthly lives. The cult establishes a community on a remote island and enforces a criminal ban on murder; the motivation for the ban is to save the souls of would-be murderers. Should we think that this ban on murder is paternalistic? The "hybrid" view explains the appeal of a "no" answer here.

3 "Mixed" paternalism

Let's say that an act, law, or policy is "mixed" paternalism on the justifying reason conception if paternalistic reasons are necessary but not by themselves sufficient to justify it; they are sufficient only in conjunction with other, nonpaternalistic reasons. An act, law, or policy is "mixed" paternalism on the act/motivating reason conception if the actor in question believes that paternalistic reasons are necessary but not sufficient to justify it. What are some plausible examples of paternalism that's "mixed" in this sense?

De Marneffe complains that the antipaternalism that motivates "reconcilers" often distorts their thinking about how the laws they support are best justified. He claims that this occurs with the defense of anti-prostitution laws given by some feminists. Instead of defending such laws on the grounds that prostitution is prudentially bad for most prostitutes and johns, these feminists defend them on the grounds that they help prevent the rights violations of women coerced into prostitution. De Marneffe objects that this flies in the face of the evidence, that "the testimony of prostitutes suggests that prostitution is generally voluntary" (2006: 93). Note that even if de Marneffe is right, "generally" is compatible with a sizable minority of prostitutes being coercively exploited. Those of us who think that anti-prostitution legislation of some sort is justified might "split the difference" between de Marneffe and these feminists. Perhaps the paternalistic and prevention of harm to others rationales are each necessary but only jointly sufficient to justify it. If so, then this is an example of "mixed" paternalism on the justifying reason conception.

Consider next Prohibition enacted in the United States in the 1920s. What motivated Congress and state legislatures that enacted it to do so? Surely some legislators were motivated primarily by concern for the welfare of heavy drinkers, some by prevention of the harm to others that such drinkers cause, others by moralism, and still others by a combination of these rationales. Paternalistic motivation may well have been causally necessary but not sufficient to produce this ban on liquor. If so, then it's an example of "mixed" paternalism on the act/motivating reason conception of what paternalism is.

4 Broad vs. narrow conceptions of paternalism; the soft/hard distinction

Suppose that while very drunk, you accept a dare to perform an exceedingly dangerous stunt, but I intercede and restrain you for your own good. Or suppose a paranoid schizophrenic during a psychotic episode claims that small extraterrestrials have invaded his body and to kill them he must ingest a can of drain cleaner; relatives restrain him and have him committed to a psychiatric hospital for treatment. Are these examples of paternalism? On a broad conception of paternalism, like the standard one with which we began, the answer is "yes." The interference in these two cases is clearly morally justified, and on a broad conception of paternalism there are many circumstances in which it is justified. On this conception paternalism is more like homicide than murder. Those who operate with a narrow conception of paternalism treat it as more analogous to murder (i.e., as seldom if ever justified). They will object that the standard definition is overinclusive: the range of Q ought to be limited to competent adults, and the range of acts and choices in condition (1) ought to be limited to unimpaired and well-informed ones. Interfering with the choices of drunks or psychotics for their own good isn't "really" paternalism at all. "Real" paternalism, on the narrow conception, is limited to cases like forcing a life-saving blood transfusion on an adult Jehovah's Witness who has refused to consent to one, or the state's requiring that all motorcyclists, including those well informed of the risks of helmetless riding, wear a helmet.

Joel Feinberg's discussion of paternalism shows ambivalence about which of these conceptions of paternalism to adopt. On the one hand he draws a distinction between "hard" and "soft" paternalism – the former targets the substantially voluntary choices of competent adults, while the latter targets substantially nonvoluntary choices (Feinberg 1986: 12).[6] Thus, interference with the drunken daredevil or the psychotic is soft paternalism, which implies Feinberg's acceptance of the broad conception. On the other hand Feinberg admits to not being entirely comfortable with that conception. "It is not clear that 'soft paternalism' is 'paternalistic' at all, in any clear sense," he says (Feinberg 1986: 12). A reply to Feinberg is that the narrow conception of paternalism is clearly underinclusive. Again, a mother's forbidding her small children to play with matches, sharp knives, etc. for their own good is the very paradigm of paternalism.

Soft paternalism includes not just cases involving children, drunks, or other incompetents, but ones involving ignorance or mistaken belief about the consequences of one's choice, such as John Stuart Mill's bridge-crosser, who is about to cross a bridge that unbeknownst to him is dangerously unstable, and there is no time to warn him (1977 [1859]: 294). It might be objected that intervention in such cases isn't paternalism even on the standard definition, because condition (2) is not satisfied. After all, the bridge-crosser has no will or wish to commit suicide or run dangerous risks.[7]

To see why condition (2) (or something like it) is necessary, consider a couple of examples. Dworkin cited a law prohibiting dueling as an example of paternalism, but suppose, as does Arneson in his discussion of this case, that while the vast majority of the upper-class males in the society to which the law applies prefer to accept a duel challenge if one is made over losing honor by declining one, they also prefer even more never receiving such challenges in the first place. The law solves a collective action problem and, under these assumptions, gives its targets the outcome that they themselves judge best. Another relevant example is:

> *Extortionate Demand*: Samantha is starving, but she has $50. The only grocer in town would happily sell her a loaf of bread for $5, but knowing that Samantha is desperate, the greedy grocer is ready to exploit her and demand $50, which Samantha would have to pay. However, the state intervenes, blocking the sale at that price with laws that forbid "price gouging."

Neither of these examples involves paternalism, because in both the interference enables its targets to realize their own present actual goals or wishes. For interference with someone's choices to count as paternalism, the paternalist must override or ignore those goals or wishes.[8]

Mill's bridge-crosser is different from these cases, however, because he *does* will or wish to cross the bridge. That is, his actual *de re* intention is to cross *this* bridge *now*. The paternalist who detains him thwarts that intention. It's true that the bridge-crosser would not have the intention if he knew how dangerous it would be to try crossing the bridge. But that's irrelevant. There remains a clear sense in which interference overrides his will but not Samantha's or the upper-class males assumed to have the preferences stipulated by Arneson. It satisfies all of the conditions laid down by the standard definition and thus counts as an example of paternalism under that definition – albeit of the softest, most easily justified kind. Since cases like it count as paternalism only on a broader conception of paternalism, they give us another reason to reject the narrow conception for being underinclusive.

5 Different senses of "his own good" and different types of paternalism

The requirement that paternalistic interference with another's actions or choices be for that very person's "own good" can be cashed out in different ways. So far we've been assuming (as do most discussions of paternalism) that it refers to one's *prudential* good (well-being, one's interests in life, health, happiness, and so on). But while prudential paternalism is the most common and important type of paternalism, it is not the only one.

The purpose of prudential paternalism is to get its targets to act *prudently*. Prudent choice should not be confused with economically *rational* choice. The most rational choice is the one that maximizes the satisfaction of one's presently held preferences (values being a kind of preference) weighted according to their strength, given one's beliefs about the world. Economic rationality imposes no limits on the objects of preference, no requirement that one value one's own prudential good rightly. If one truly valued the life of a housefly more than one's limb, and if sacrificing the limb were the only way to save the fly, then it would be rational to make the sacrifice. Economic rationality also does not require that one value one's prudential good in a temporally neutral way. Prudence, by contrast, requires both that one give one's own prudential good its proper due (neither over- nor under-valuing it) and that one value future benefits/harms as strongly as present ones (except insofar as their lying in the future makes them less certain). Thus, the choice of a smoker who could (with effort) quit but never tries because he values "the pleasures of smoking" now more than longevity or his future health, is probably imprudent but not irrational. Finally, while the rationality of risk seeking or aversion depends entirely on one's attitude toward risk, prudence imposes limits on how much of either is "reasonable," perhaps even permitting none (i.e., requiring *expected* welfare maximization).

The goods that prudential paternalism aims to augment needn't be limited to the mundane ones of health, longevity, financial security, and the like. There are a number of competing theories – hedonism, desire-satisfactionism, and objective list theories – about what one's prudential good consists in, and prudential paternalism is compatible with all of them. An objective list theory might include autonomy on its list, claiming that it has intrinsic prudential value even if one does not value, derive enjoyment from, or have any pro-attitude toward it. A prudential paternalism based on this theory might support a ban on highly addictive euphoric drugs even if it reduced the long-term happiness of users, because the increase in their long-term autonomy outweighs the reduced happiness.

There are other types of paternalism besides the prudential kind, because it is possible for interference with another's choices to be for that person's "good" without being calculated to provide him with prudential benefits or protect him from prudential harm. There are a few possibilities here. Paternalism might seek to prevent its targets from making choices contrary to their own considered "conception of the good." The choice required by one's currently held conception of the good just is the choice one would make if one were economically rational and had complete and accurate information about one's options. Suppose you are committed to ensuring the preservation of an endangered bird species and that requires using all of your vast wealth to purchase a huge tract of land for a wildlife sanctuary. Because of your weak will, you hesitate to do it when the opportunity presents itself, so I threaten or maybe "guilt trip" you into doing it. Since I've intervened "for your own good" (i.e., your own conception of the good), it seems appropriate to describe my intervention as "paternalistic" despite the fact that it leaves you prudentially worse off. Gerald Dworkin (2017) called this "weak" paternalism.[9] When economists with libertarian sympathies criticize paternalism for being "inefficient," they are criticizing "weak" rather than prudential paternalism (though many of them fail to recognize the difference between the two).

Another type of non-prudential paternalism is concerned with whether its targets will form true and/or reasonable beliefs about matters unrelated to their own prudential good. Thus, if it is likely that certain evidence, though relevant to the question whether the defendant in a criminal trial is guilty, is likely to have a "prejudicial" effect on jurors that exceeds its probative value, perhaps it should be excluded to increase the likelihood of the jury's reaching the correct verdict. The ultimate goal here would be justice (conviction of the guilty, exoneration of the innocent), but promoting the "epistemic good" of the jurors is necessary to achieve that end, and excluding the prejudicial evidence promotes their epistemic good. It seems proper to describe the exclusion of the evidence for this reason as paternalistic – Alvin Goldman (1991) dubbed it "epistemic paternalism" – despite its not being justified by the fact that (or motivated by the belief that) it will benefit the jurors prudentially.[10]

Is there another kind of paternalism – "moral" paternalism – that is neither prudential paternalism nor moralism? Moral paternalism seeks to prevent its targets from doing "moral harm" to themselves and their character (Feinberg 1984: xiii; Dworkin 2005). A moral paternalist might claim that sexist pornography promotes the moral vice of disrespect for women in the males who consume it, that it is not good *for them* to have that vice, and thus restricting their access to it is justified "for their own good." But the claim that disrespect for women is a moral vice that is "bad for" the men who have it sounds prudential, and if it is, would seem to presuppose the old (going back at least to Plato's *Republic*) but questionable idea that possessing all the moral virtues, including the primarily other-regarding ones of justice and benevolence, is a key part of everyone's prudential good. Joel Feinberg rejects the possibility of a moral paternalism for this reason, claiming that it conflates moral and prudential goodness.[11] A possible reply to Feinberg's objection is that an objective list theory of prudential value is correct and all of the moral virtues belong on the objective list.[12] That is, "a person can be better off just because they are morally better, just as they can be better off just because they are happier" (Dworkin 2005: 308). This reply to Feinberg's objection agrees with Feinberg that there is no coherent conception of moral paternalism as something distinct from both moralism and prudential paternalism.

Seana Shiffrin has denied that condition (3) – "for his own good" in any of the senses distinguished in the last few paragraphs – is a necessary condition of paternalism. She offers an example of a park ranger who refuses to allow an inexperienced climber to attempt an especially dangerous climb not for the climber's own good but because his death would leave his wife grief-stricken. According to Shiffrin the ranger has acted paternalistically by substituting her own judgment for the climber's about how he should treat his wife. The ranger "intrudes into and insults" the climber's "range of agency," implying that "her judgment is superior" (Shiffrin 2000: 216–218). For Shiffrin the essence of a paternalistic act (or omission) is that it substitutes the paternalist's judgment or agency for that of her target on some matter that is within the target's "legitimate domain" of action, and it is motivated by the belief that the paternalist's capacity for practical rationality exceeds her target's. (Thus, it assumes the act/motivation rather than justifying reason conception of paternalism). If the target is a child or other incompetent, such a belief is true and acting on it is unobjectionable. If he is not an incompetent, then acting on the belief is morally objectionable even if it is true.

Shiffrin's account reflects the widely held view that any paternalism that targets competent adults disrespects them by treating them "like children." While it rejects the standard definition's requirement that paternalism must interfere with liberty or autonomy, it shares with it the implication that all paternalism is *prima facie* wrong. Whereas the standard definition traces that *prima facie* wrongness to the means that it supposes paternalism must use, Shiffrin's definition traces it to the elitist motivation that it supposes is present in the paternalist.

We'll return to the question of whether any sort of elitist motivation is necessary for paternalism that targets competent adults shortly. For now I want to defend the inclusion of a "for his own good" condition in any definition of paternalism. That condition distinguishes prudential paternalism from moralism, and distinguishing them allows us to raise the important question whether liberals ought to find hard prudential paternalism by the state more objectionable than moralism, as Feinberg (1988: 6–7) thought, or moralism more objectionable, as utilitarians and H.L.A. Hart (1963: 32–34) suppose. If the park ranger's action is motivated (or justified) by the view that the climber's self-endangerment wrongly disrespects his wife even if she knows and consents to it, then it is a kind of moralism. If the idea is that people have rather strong veto rights over any choice by a spouse to endanger life or limb, and the man doesn't have his wife's consent to attempt the climb, then the ranger's action is based on an implausible form of the "prevention of wrongful harm to others" principle. In neither case is it paternalism.

6 Limits liberty or autonomy?

A possible objection to condition (1) in the standard definition is that paternalism needn't interfere with choice or action at all. Suppose that an elderly woman on her deathbed asks to see her son and grandchildren one last time and, to spare her pointless grief, her doctors lie, telling her that their flight was delayed rather than revealing the horrible truth that they all just died in a traffic accident. Gert and Culver claim that such deathbed deception is paternalistic even though its purpose is to control the dying woman's feelings or emotions rather than her choices or actions (1976: 46–47).

In defense of condition (1), it might be said that the choices or actions referred to needn't be actual; they can be hypothetical. Thus, giving an unconscious Jehovah's Witness a blood transfusion that we know he would reject if conscious is paternalistic because we're pre-empting for his own good the choice we know he *would* make if he were in a position to make one. The deathbed deception is paternalistic on the assumption that the woman would, if given the choice, elect to hear the sad truth rather than a consoling falsehood. If she wouldn't, then the deception is not paternalistic because condition (2) isn't satisfied.

Assuming that paternalism aims to influence choice or action, actual or hypothetical, we can now ask which means of influencing it belong in the paternalist's toolbox. Condition (1) in the standard definition can be given a stronger reading that limits the means to restrictions of liberty, or a weaker one that adds any interference with autonomous choice or action. On the stronger reading the paternalist may thwart already made choices *via* physical compulsion, or she may influence prospective choices by attaching disincentives (e.g., threats, taxes) to the options she wants to discourage. An objection to a definition based on this reading of the condition is that it is underinclusive. If I intentionally exaggerate to you the dangers of alpine skiing to get you to quit it for your own good, I act paternalistically but do not limit your liberty. Nobody forced you to believe me, and you remain free to ski or not. Deception is surely one of the means by which a paternalist can "interfere" with another's decision-making.

The "nudges" defended by Cass Sunstein and Richard Thaler can likewise be paternalistic despite the fact that they do not limit liberty or do so only slightly.[13] A nudge is an influence that affects the decision-making of ordinary humans with their cognitive and volitional shortcomings but not that of an ideally "rational" person – where "rational" is cashed out epistemically (properly weighing all of the information or evidence in one's possession) and "economically" (as explained earlier) (Thaler and Sunstein 2008: 9). If an oncologist gets her loss-aversive cancer patient to choose a treatment regimen by giving him the positive survival rate rather than the equivalent negative mortality rate, she has nudged him. The oncologist's use of this "framing" nudge for her patient's own good is surely an example of paternalism despite the fact that it is neither coercive nor deceptive. Perhaps it is "manipulative."[14]

The weaker reading of condition (1) – interference with liberty *or* autonomy – yields a better, less underinclusive definition of paternalism. A person acts autonomously when she acts on reasons that she would regard as justifying if certain conditions were satisfied. Which conditions? There are a few possibilities, giving rise to progressively stronger conceptions of autonomy. These include: (a) if she were well informed of relevant facts; (b) if she were well informed and economically and epistemically rational; and (c) if she were well informed, fully rational, and fully "reasonable" in some more demanding sense (e.g., fully prudent and moral).[15] If your choice is suboptimal (relative to your conception of the good) because I've deceived or withheld information from you, then I've interfered with your autonomy on all of these conceptions. I also interfere with your autonomy on all of them if, knowing that you are prone to weakness of will, I induce you to make a choice contrary to your own better judgment by creating a temptation you are likely to succumb to. (If your weakness of will is predictable but still voluntary, I interfere with your autonomy but do not deprive you of "free will" or reduce your responsibility for your choice.)

The oncologist's nudge does not interfere with autonomy on either (b) or (c), since the patient will choose irrationally (or so it is assumed) whether or not he is nudged; his choice will reflect either risk aversion or an irrelevant framing effect. Indeed, some nudges, such as mandatory "cooling off" periods for major financial or medical care decisions, may actually enhance autonomy on all three conceptions of it. Forcing the drunk to sober up before he is allowed to decide whether to accept the dare to perform the dangerous stunt enhances rather than interferes with his autonomy.[16] If the oncologist's nudge and soft paternalism cases like that one involve paternalism despite the fact that they do not limit liberty (or do so only minimally) and do not interfere with autonomy, then a definition of paternalism that includes condition (1) is underinclusive even on the weaker reading of it.

Other examples seem to support this conclusion. Suppose that Maria's broke and hungry gambling-addicted brother asks her for a gift of cash. She believes that while he is sincere in his promise not to use the money to gamble, he can't be trusted to keep his promise. So instead of giving him cash, she offers to buy his groceries and pay his rent. Maria's "in kind" aid gives him fewer options than cash, but that's the wrong comparison if she owes him nothing; *whatever* she gives him expands his options. Her *selective* expansion of his options seems paternalistic. Or suppose that Laticia's talented adult son is a lazy layabout and her pleas that he do something with his life fall on deaf ears, so she offers him large monetary incentives to go to college and earn a degree. Her offer is intended to make the choice that (she believes) is in his long-term best interests coincide with the one that is in his more immediate self-interest, and thus more attractive to someone as short-sighted as he currently is. Laticia's "bribe" seems paternalistic. These examples provide further evidence that paternalism need not interfere with either liberty or autonomy.[17]

7 Non-rational means vs. rational persuasion; a broader definition

What do the means by which the paternalist can influence the choices or actions of others have in common? It would seem that they are all "non-rational" in the bare sense of not being "rational persuasion." One engages another in rational persuasion when one directs his attention to what one believes are good, justifying moral, prudential, or epistemic reasons, with the intention that he make his choices or form his beliefs on the basis of them. Informing others that some of their options have costs or benefits that they were unaware of is rational persuasion and thus not paternalistic.[18]

Suppose that Mill's bridge-crosser is emotionally agitated. If I warn him of the bridge's dangerous condition in my normal voice, his emotional state will prevent him from fully appreciating my warning; but if I use an especially soothing tone, it will calm him, enabling him to appreciate better the import of what I'm saying. My warning is not paternalistic, despite its use of a framing nudge, because the framing is intended to and does facilitate rational persuasion. In the oncologist example, the nudge does not have that purpose.

Putting together some of the points made in the last few paragraphs against condition (1) of the standard definition, I propose a broader definition based on the following necessary and sufficient conditions:

1 Aim – the paternalist's aim is to get another to engage in a better act, make a better choice, or form a better belief "for his own good."[19] Usually the "good" here is prudential, but it can also refer to the target's "conception of the good," or his alethic or epistemic good (forming true and/or justified beliefs). Paternalism can be "moral" only if one's "moral good" is a species of one's prudential good.
2 "Non-rational" means – the means available to the paternalist include not just restrictions on liberty and other interferences with autonomy (deception, withholding information that one had a duty to pass along, some but not all nudges, "manipulation"), but also forms that don't, such as the selective expansion of options and the offering of incentives to make better choices. The means exclude rational persuasion.
3 Belief – the paternalist believes that her target's judgment of where his good lies is either mistaken or will fail to guide his choices due to weakness of will, psychological compulsion, or some other volitional shortcoming, and further, that the means in #2 are the best, most effective ones to achieve the goal in #1. They are best because there is no opportunity

for rational persuasion; or there was an opportunity for it, the paternalist attempted it, but it failed; or there is opportunity for it but the paternalist does not attempt it because she believes it would fail or be too costly.

The reference to the paternalist's "aim" in #1 and "belief" in #3 implies the act/motivation conception of paternalism, but the definition seems to me compatible with the justifying reason conception as well (#2 specifying the type of act and #1 and #3 together the type of reason in the act/reason compound that is paternalism). #3 is really just Shiffrin's "substitution of judgment or agency" condition, which seems to do the same work as the "against his will, without his consent, or contrary to his preferences" condition in the standard definition. These conditions exclude the earlier Extortionate Demand and dueling examples from counting as paternalism. But #2 in the above definition rejects the standard definition's requirement that paternalism must interfere with liberty or autonomy, and so allows Laticia's bribe of her son and Maria's offer of in-kind aid to her brother to count as paternalism. Maria's expansion of her brother's options doesn't hinder his ability to act autonomously on any of the conceptions of autonomy distinguished earlier, but it is paternalistic because it is not rational persuasion, and it is motivated (or justified) by the judgment that he can't be trusted to refrain from using any cash he receives for gambling.

An objection to this definition is that it is *over*inclusive and too "broad" in the sense distinguished in Section 4 above. Suppose that a man asks a friend for help to complete a simple task. She tries to convince him that he relies on others too often and ought to be more self-reliant. He is unpersuaded and renews his request for help, which she then refuses for his own good (Shiffrin 2000: 213). The proposed definition implies that her refusal is paternalistic. It might be objected that this isn't an example of paternalism at all, because paternalism is *prima facie* wrong and there is nothing *prima facie* wrong in refusing to assist others when one believes that assisting them would not benefit or might even harm them, especially after one has tried to convince them of this but failed.

My reply to this objection is simply to deny that a definition of paternalism is acceptable only if it implies that *all* of it is *prima facie* wrong. It is enough if the definition implies that the most common types of it have that moral status. The issue here boils down to whether we are or should be more confident that all paternalism has this moral status, or that the examples of Maria, Laticia, and the denial of the overly needy man's request for help are instances of paternalism. I judge the broader definition to be better because I am more confident of the latter (though I'm a bit more confident about the first two examples than the last one). If you are more confident of the former, then you should opt for a definition that includes something like the "limits liberty or interferes with autonomy" condition.

A different objection to the proposed definition is that it is *under*inclusive. Having correctly observed that one may attempt rational persuasion while believing that one's capacity for practical rationality exceeds one's target's, George Tsai (2014) infers that rational persuasion too can be paternalistic. The problem with this objection is that it assumes the correctness of Shiffrin's view that this motive is essential to paternalism. It isn't. To be sure it or something like it is present in many of the examples of paternalism provided so far, including Laticia's bribe of her son and Maria's refusal to give cash to her brother. But it need not be present at all in other cases of paternalism aimed at competent adults. Consider the public health nudges defended by Thaler and Sunstein, such as painted lines across the highway that are spaced closer together as drivers approach a dangerous curve (2008: 37–38). What motivates and justifies this particular nudge is not "we traffic engineers/choice architects are better, safer, more rational drivers than everyone else" but rather "*all* human beings, including us, are predictably prone to certain types of irrationality." Jonathan Quong might claim that even this belief manifests wrongful disrespect of a "noncomparative" rather than "comparative" type (2011: 101). But while some instances of "noncomparative" disrespect are certainly wrong (e.g., one woman's insisting to another that women in general have less capacity for rationality than men), the belief that motivates this nudge hardly seems of that sort. It surely reveals no culpable lack of respect for one's self or our species to recognize that we humans often get distracted when we drive and fail to notice posted warnings to slow down, or if we do notice them, fail to appreciate them. Indeed, attempting to persuade another to make a different choice for his own good is not paternalistic, even if it is overtly condescending, manifesting the comparative "I am superior to you" type of disrespect. Of course rational persuasion done in that manner is usually morally wrong, as is most paternalism done in that manner. But that only shows that some paternalism is wrong for the same reason that some rational persuasion is. It does not obviate the distinction between paternalism and rational persuasion.

8 Conclusion

A good definition of paternalism should not exclude any clear examples of paternalism or include any clear examples of nonpaternalism. But as the examples of offering incentives to the short-sighted to choose prudently, declining a request for assistance when one believes the assistance would harm the person making it, and others mentioned above make abundantly clear, there is disagreement among moral and political theorists about which cases clearly are paternalism and which aren't. At this point we do well to remind ourselves that debates about the meanings of terms carried too far become tedious and profitless. I believe that the definition proposed in the previous section does a better job than the standard definition of getting the extension of the concept right, but I concede that others may reasonably disagree (the concept's extension has *some* indeterminacy), and I grant that in the final analysis the question of which of these two definitions is better may not be all that important. After all, whether cases like "bribing" the imprudent to choose prudently are or are not paternalism does not get us any closer to settling the more important normative question of whether or not the actions in question are morally wrong. For that task we need the correct moral theory.

Related topics

Epistemic Paternalism; Hard and Soft Paternalism; Libertarian Perspectives on Paternalism; Moralism and Moral Paternalism; Paternalism and Autonomy; Paternalism by and towards Groups; Paternalism and Well-Being.

Notes

1 See Grill (2013) for a summary of this debate.
2 Variants of it may be found in Dworkin (2017); Feinberg (1986); Arneson (1980); and VanDeVeer (1986: 22). The scope of P and Q in the definition is not limited to individuals. Q also ranges over groups (e.g., smokers) and P over corporate entities (e.g., the state).
3 In many of the examples used in this essay, the paternalist will be a female and her target male. This is done intentionally to combat the too common confusion of the meaning of the term with its etymology, or the identification of paternalism with patriarchalism.
4 In his first, seminal paper on paternalism, Gerald Dworkin (1972) includes "laws regulating certain kinds of sexual conduct, for example, homosexuality among consenting adults in private" on his list of examples of paternalism. But on both the act/motivation and justifying reason conceptions, they are

better seen as moralism. In his later "Moral Paternalism" (2005), Dworkin clearly distinguishes moralism and paternalism.

5 Another even stronger kind of moralism holds that there are immoralities that *neither harm nor wrong* anyone (what Feinberg called "free-floating evils") but are still egregious enough that they warrant efforts to prevent or prohibit them.

6 Note that for Feinberg the hard/soft distinction is based on which choices are being interfered with and why. In many discussions of Thaler and Sunstein's "libertarian paternalism," the terms are used differently, to refer to the means by which the interference is effected ("hard" meaning "uses coercive means" and "soft" meaning "uses non-coercive means").

7 See Arneson (1980: 471). See too his more recent discussion (2015: esp. 673–674).

8 Another example that condition (2) excludes is giving emergency medical treatment to an unconscious patient whom we know would consent to it if conscious. Simon Clarke (2002: 84) defends a definition that counts this as paternalism.

9 For Dworkin "weak" paternalism interferes only with those actions that are not effective as means to the target's own ends, while "strong" paternalism interferes with actions aimed at ends the paternalist judges to be misguided. Note that his weak/strong distinction is different from Feinberg's soft/hard distinction.

10 See the essay by Kristoffer Ahlstrom-Vij in this volume.

11 In particular, Feinberg (1984: 65–70) objects that identifying the two goods presupposes the dubious teleological metaphysics of the Stoics and natural law theory.

12 See the essay by Peter de Marneffe in this volume.

13 See the essay on nudging and public policy in this volume by Muireann Quigley.

14 See the essay on manipulation in this volume by Moti Gorin.

15 Whether (c) is a plausible conception of autonomy is controversial. Isaiah Berlin objected that it is "monstrous impersonation" of autonomy (1969: 134).

16 Jason Hanna (2015) provides other examples of framing nudges that enhance autonomy.

17 Shiffrin denies that paternalism must restrict liberty or interfere with autonomy. So do Archard (1990) and Quong (2011: chap. 3).

18 This does not mean that providing information cannot be paternalism. If I have chosen to remain ignorant of some matter but you ignore my choice and inform me of it for my own good (e.g., a doctor disregarding a patient's request not to be told if he is terminally ill and likely to die within months), you've engaged in paternalism, because you have *not* attempted rationally to persuade me that my "ignorance is bliss" choice is misguided.

19 Perhaps it does not have to be "another" but can be a temporally later stage of the paternalist's own self. Note that it is quite possible for one to have the belief in #3 below toward one's own future self.

References

Anderson, E. (1999) "What Is the Point of Equality?" *Ethics* 109: 287–337.

Archard, D. (1990) "Paternalism Defined," *Analysis* 50: 36–42.

Arneson, R. (1980) "Mill vs. Paternalism," *Ethics* 90: 470–489.

———. (2015) "Nudge and Shove," *Social Theory and Practice* 41: 668–691.

Berlin, I. (1969) "Two Concepts of Liberty," in *Four Essays on Liberty*, New York: Oxford University Press.

Clarke, S. (2002) "A Definition of Paternalism," *Critical Review of International Social and Political Philosophy* 5: 81–91.

de Marneffe, P. (2006) "Avoiding Paternalism," *Philosophy & Public Affairs* 34: 68–94.

Dworkin, G. (1972) "Paternalism," *The Monist* 56: 64–84.

———. (2005) "Moral Paternalism," *Law and Philosophy* 24: 305–319.

———. (2017) "Paternalism," in E. N. Zalta (ed.) *The Stanford Encyclopedia of Philosophy*, http://plato.stanford.edu/archives/sum2010/entries/paternalism/.

Feinberg, J. (1984) *The Moral Limits of the Criminal Law, Volume 1: Harm to Others*, New York: Oxford University Press.

———. (1986) *The Moral Limits of the Criminal Law, Volume 3: Harm to Self*, New York: Oxford University Press.

———. (1988) *The Moral Limits of the Criminal Law, Volume 4: Harmless Wrongdoing*, New York: Oxford University Press.

Gert, B. and Culver, C. (1976) "Paternalistic Behavior," *Philosophy & Public Affairs* 6: 45–57.
Goldman, A. (1991) "Epistemic Paternalism: Communication Control in Law and Society," *Journal of Philosophy* 88: 113–131.
Grill, K. (2007) "The Normative Core of Paternalism," *Res Publica* 13: 441–458.
———. (2013) "Normative and Non-Normative Concepts: Paternalism and Libertarian Paternalism," in D. Strech, I. Hirschberg and Georg Marckmann (eds.) *Ethics in Public Health and Health Policy: Concepts, Methods, and Case Studies*, New York: Springer.
Hanna, J. (2015) "Libertarian Paternalism, Manipulation, and the Shaping of Preferences," *Social Theory and Practice* 41: 618–643.
Hart, H. L. A. (1963) *Law, Liberty, and Morality*, Stanford, CA: Stanford University Press.
Mill, J. S. (1977 [1859]) "On Liberty," in J. Robson (ed.) *The Collected Works of John Stuart Mill*, Vol. 18, Toronto and London: University of Toronto Press and Routledge.
Quong, J. (2011) *Liberalism Without Perfection*, New York: Oxford University Press.
Shiffrin, S. (2000) "Paternalism, Unconscionability Doctrine, and Accommodation," *Philosophy & Public Affairs* 29: 205–250.
Thaler, R. and Sunstein, C. (2008) *Nudge: Improving Decisions About Health, Wealth, and Happiness*, New York: Penguin Books.
Tsai, G. (2014) "Rational Persuasion as Paternalism," *Philosophy & Public Affairs* 42: 78–112.
VanDeVeer, D. (1986) *Paternalistic Intervention*, Princeton, NJ: Princeton University Press.

2

HARD AND SOFT PATERNALISM

Jason Hanna

Many people believe that paternalism is generally wrong. Yet virtually everyone believes that it is at least sometimes permissible to intervene in a person's choices for that person's own good. J.S. Mill, a strong critic of paternalism, famously supports intervention in the following sort of case:

> If either a public officer or any one else saw a person attempting to cross a bridge which had been ascertained to be unsafe, and there were no time to warn him of his danger, they might seize him and turn him back, without any real infringement of his liberty; for liberty consists in doing what one desires, and he does not desire to fall into the river. Nevertheless, when there is not a certainty, but only a danger of mischief, no one but the person himself can judge of the sufficiency of the motive which may prompt him to incur the risk: in this case, therefore (*unless he is a child, or delirious, or in some state of excitement or absorption incompatible with the full use of the reflecting faculty*) he ought, I conceive, to be only warned of the danger; not forcibly prevented from exposing himself to it.
>
> (Mill 1977 [1859]: 294, emphasis added)

Mill points to at least two reasons why it may be permissible to intervene on behalf of the bridge-crosser. First, the bridge-crosser may simply be ill-informed: he may not know that the bridge he intends to cross is dangerous. Second, although informed, the bridge-crosser may be immature, or insane, or intoxicated, or in some other "state of excitement or absorption." Many people who share Mill's antipaternalist orientation would agree that exceptions should be made in such cases.

To accommodate this point, many philosophers distinguish between "hard" and "soft" paternalism. Joel Feinberg, the most prominent defender of this distinction, maintains that paternalistic intervention is soft when it prevents (or aims to prevent) *nonvoluntary* self-harmful choices or behavior. Intervention is hard, on this view, when it prevents (or aims to prevent) *voluntary* self-harmful choices or behavior (Feinberg 1986: 12–16). Some philosophers reject the details of Feinberg's account but agree that intervention in a person's affairs does not violate her autonomy – or is in some way easier to justify – when that person is subject to deliberative impairments that jeopardize her ability to realize her own values. I will assume that any view of this sort relies on an account of the hard/soft distinction. Though philosophers disagree about precisely how the hard/soft distinction should be drawn, any plausible account would imply that intervention on behalf of Mill's ill-informed bridge-crosser is a paradigm example of soft paternalism.

Unfortunately, the "hard/soft" terminology has been used in a further way that is likely to invite confusion. According to some writers, paternalistic interference is hard if it is coercive or imposes "material costs" on the choices it influences, while interference is soft if it does not impose such costs (Sunstein 2014: 58). According to this terminology, it might be best to think of the hard/soft distinction as a continuum: the lower the material costs imposed on the decision-maker, the softer the paternalism. Thus, printing graphic warning labels on cigarette packaging would count as a relatively soft form of paternalism, while fining people who smoke cigarettes would count as a relatively hard form of paternalism. Notice, however, that this use of the "hard/soft" terminology is indeed distinct from that favored by Feinberg and others. There may be some cases in which the only way to prevent substantially nonvoluntary behavior is through means that involve significant material costs. To return to Mill's example, suppose that the bridge-crosser has bipolar disorder and decides to cross the bridge during a manic episode. Intervention to prevent him from crossing the bridge would presumably be soft, in Feinberg's sense. But such intervention may involve significant restrictions on liberty: for instance, the only way to save the bridge-crosser may be to place him in preventative detention until he has regained his "normal" capacities.

This entry will focus on the first, broadly Feinbergian, account of the hard/soft distinction. I begin by briefly setting out Feinberg's view. I then consider several possible difficulties confronting this view before turning to some further attempts to specify the distinction. I suggest that although the hard/soft distinction appears to be a crucially important component of most antipaternalist views, and although it has been developed in promising directions, it is surprisingly difficult to draw in an intuitively satisfying way. I briefly explore some possible implications of this conclusion in the final section.

1 Feinberg's sliding scale conception of voluntariness

Feinberg is concerned with paternalism in the law, and especially the criminal law. According to the view he defends, "the state has the right to prevent self-regarding harmful conduct . . . *when but only when* that conduct is substantially nonvoluntary, or when temporary intervention is necessary to establish whether it is voluntary or not" (1986: 12, emphasis in original; see also 1971: 113). Feinberg holds that voluntary self-regarding choices are protected by the decision-maker's right of autonomy or self-sovereignty. On his view, such autonomy rights are absolute. If autonomy rights are absolute, then hard paternalism is *always* impermissible. On a closely related but more moderate view, autonomy rights are weighty but not absolute: although all voluntary self-regarding choices are protected by autonomy, it is at least sometimes permissible to intervene in such choices, perhaps because doing so would prevent exceptionally grave self-harm. I will follow Feinberg in assuming that a viable account of the hard/soft distinction will imply that hard paternalism violates autonomy, construed as a right of self-sovereignty, though I take no stand on whether such autonomy rights are best understood in absolutist terms.

By contrast, soft paternalistic intervention, on Feinberg's account, involves intervention in nonvoluntary self-regarding choices (or in self-regarding choices that there is sufficiently good reason to believe *may* be nonvoluntary).[1] Such soft paternalistic intervention supposedly does not violate autonomy, though it may sometimes be wrong for other reasons (for instance, because it would do more harm than good). Soft paternalistic intervention, when genuinely beneficial, seems relatively benign. Indeed, preventing a person from *unwittingly* harming herself may not even be genuinely paternalistic. Protecting people from nonvoluntary self-harm appears to be relevantly like protecting them from non-consensual harm at the hands of others (Beauchamp 1977: 67; Feinberg 1986: 15–16).

But how can we tell whether a choice is voluntary? Feinberg discusses a number of conditions that supposedly detract from voluntariness. These *voluntariness-reducing factors* (henceforth, VRFs) include the following: general incompetence, such as that characteristic of infants, young children, and people with severe cognitive impairments; coercion or duress that forces a choice between the lesser of two evils, whether this forced choice comes from a natural source, a threat, or an offer; manipulation, such as subliminal messaging; ignorance or mistaken belief; and temporary distortions, such as fatigue, excessive excitement, nervousness, intoxication, pain, and so forth (Feinberg 1986: 115). Of course, as Feinberg recognizes, virtually no choice is *perfectly* voluntary by these criteria. Even generally competent adults under reasonably propitious circumstances are often ignorant of some relevant facts or subject to distorting influences such as fatigue or passion. On Feinberg's view, however, the relevant question is not whether a choice is perfectly voluntary, but instead whether it meets some threshold of voluntariness, or whether it is voluntary enough. The threshold varies from case to case and depends on contextual features such as the level of risk posed by the agent's behavior. In general, Feinberg claims that the more harmful and irrevocable the agent's self-regarding choice, the higher the standard of voluntariness (1986: 117–124).

This sliding scale account of voluntariness is a crucial component of Feinberg's view. Without it, his view would appear to confront a sort of "Goldilocks problem": there may be no single standard of voluntariness that seems to be "just right" for the full range of potentially risky choices. Let me explain. On the face of it, a healthy adult's choice to commit suicide would seem to be protected by autonomy only if he is thinking exceptionally clearly and is virtually immune to various distorting influences. Many of us think that it would be permissible – and indeed wholly consistent with respect for autonomy – to prevent someone from committing suicide impulsively, or on a whim, or while in the grip of temporary depression. Here, a high standard of voluntariness seems appropriate. But of course if such a standard were applied across the board, it may turn out that relatively few of our choices are voluntary. Feinbergian antipaternalists would presumably claim that a person's choice to get an ugly tattoo would be protected by autonomy, even if it were made impulsively or on a whim. Any invariant standard of voluntariness is thus likely to seem too low when applied to some choices and too high when applied to others. The sliding scale account enables Feinberg's view to avoid this difficulty.

Yet the sliding scale account raises problems of its own. In particular, this account itself is arguably paternalistic (Buchanan and Brock 1990: 41–47; Grill 2010: 11). To see why, compare two decision-makers, each of whom must choose among the same three options. Option 1, let us suppose, is prudent and would not result in any self-regarding harm; Option 2 is likely to result in moderate, but not irreparable, self-regarding harm; and Option 3 is likely to result in very severe and irreparable self-regarding harm. Suppose that Alison is about to choose Option 2, while Betty is about to choose Option 3. Assume further that Alison and Betty are in comparable states of mind, so that each is influenced to the same extent by the various conditions on Feinberg's list of VRFs. The sliding scale account would presumably imply that there are circumstances under which Alison's choice is voluntary enough while Betty's is not. But if we conclude that it is easier to justify intervention on behalf of Betty than it is to justify intervention on behalf of Alison, *despite the fact that they are in comparable states of mind*, it seems misleading to claim that intervention in Betty's case is not "really" paternalistic. As Feinberg characterizes it, soft paternalism aims to "help implement [the subject's] real choice, not to protect [her] from harm as such" (1986: 12). In our pair of examples, however, it is far from clear what basis there could be for claiming that Alison's choice represents her "real" will, while Betty's choice does not. This is especially so if we accept Feinberg's plausible view that a choice's voluntariness is wholly distinct from its prudence (1986: 106–113).

Feinberg acknowledges the concern that his sliding scale account represents a "compromise with hard paternalism" (1986: 119). He denies that it falls prey to this objection, however. In doing so, he appears to defend the sliding scale account on epistemic grounds. Determinations of voluntariness are fallible. Some nonvoluntary choices might be mistakenly classified as voluntary ("false positives"), and some voluntary choices might be mistakenly classified as nonvoluntary ("false negatives"). There is greater reason to be on guard against "false positives" when the self-regarding risks posed by an individual's behavior are especially great or the harm risked is irrevocable. In such cases, we may thus have good reason to apply especially strict standards. It is unclear, however, that this epistemic argument supports the sliding scale account of voluntariness as Feinberg develops it. For there is a difference between claiming (a) that the riskier the agent's conduct, the more evidence we must have that it meets some fixed standard of voluntariness, and (b) that the riskier the agent's conduct, the higher the standard of voluntariness (DeMarco 2002: 241). Purely epistemic considerations would seem to support (a) rather than (b), since purely epistemic considerations would seem relevant only to evidential standards and not to the standard of voluntariness itself. Suppose, for instance, that there is overwhelmingly good evidence that a decision meets a relatively low threshold of voluntariness. Suppose also that there is overwhelmingly good evidence that the decision does *not* meet a much more demanding threshold of voluntariness. Feinberg's view seems to imply that the riskiness of the agent's choice partly determines which standard of voluntariness (the high or the low) is most appropriate. Yet epistemic considerations do not appear to support this verdict, since (*ex hypothesi*) we have overwhelmingly good evidence concerning the agent's state of mind. And in the absence of an epistemic argument for the sliding scale account of voluntariness, Feinberg's view again confronts the concern that it involves a compromise with (hard) paternalism.

2 The voluntariness-reducing factors

Let us now set aside these concerns about the sliding scale account of voluntariness to focus more directly on what this scale is supposed to measure. Recall that Feinberg's list of VRFs includes (i) general incompetence, (ii) influence by coercion, duress, or manipulation, (iii) ignorance, and (iv) temporarily "distorting" conditions, including fatigue, nervousness, time pressures, "powerful passions" such as rage, and "gripping moods" such as depression (1986: 115). It is easy to explain the relevance of some items on Feinberg's list. If a person's risky behavior is prompted by a coercive threat, then obviously it is permissible to intervene and eliminate the threat.[2] And there is a strong case for classifying ill-informed choices as nonvoluntary, at least when the decision-maker is not culpable or responsible for her ignorance (though for an argument that cases of *culpable* ignorance pose problems for Feinberg's view, see Hanna (2012)). It is less clear, however, that all of the "distorting" conditions included in category (iv) compromise voluntariness.

We might sensibly ask what these various distortions – for instance, fatigue, nervousness, powerful passions, temporary depression, and so forth – have in common. One obvious similarity is that they tend to result in imprudent decisions. They also tend to result in decisions that people later come to regret (Hodson 1977: 66). But these features fail to distinguish the deliberative "distortions" in category (iv) from other conditions, such as short-sightedness, that do not appear to detract from voluntariness. What, then, marks these distortions out as VRFs?

Feinberg suggests that whether an agent's behavior is voluntary enough to be protected by his right of self-sovereignty depends on whether "it represents him faithfully in an important way" (1986: 113). Perhaps, then, the idea is that distortions such as fatigue, powerful passions, and gripping moods compromise voluntariness because they detract from the extent to which an agent's choice faithfully represents him.

This line of argument is initially promising. Yet there is a complication. There appear to be some cases in which intuitively distorted or impaired decisions do faithfully represent the agent. For instance, some people wait to make important decisions until the last minute, when they are likely to feel the full weight of time pressures. Although the choices such a person makes are liable to be impetuous, these choices, and the means by which they are reached, may nonetheless represent him in an important way. Likewise, some people are especially prone to emotional distress and gripping moods, and some of these people do an especially poor job managing their emotions. Do the choices such people make while enraged or anxious represent them in a faithful way? Feinberg at one point claims that there may be no reason to deny the voluntariness of "rash" behavior that is nonetheless "in character" for the agent in question (1986: 112). We might wonder, however, how much further this point can be pushed. Sometimes people make ill-informed and imprudent decisions because they carelessly choose not to take the time to inform themselves (Hanna 2012). In such cases, an ill-informed choice might be "in character" for the agent, even though there would appear to be a strong case in favor of soft paternalistic intervention. Likewise, those sympathetic to Feinberg's account would probably want to argue that it is sometimes permissible, and only softly paternalistic, to intervene on behalf of someone who acts imprudently while in the grip of a towering rage, even if such behavior is hardly unexpected given his character and personality.

Perhaps the overarching concern here is that Feinberg appears to provide two accounts of what unifies the VRFs, and it is not clear that they are equivalent. On the one hand, Feinberg argues that the VRFs tend to result in "clouded judgment" and "impaired reasoning" (1986: 104). It seems clear enough how the distortions he cites, such as powerful passions and gripping moods, might fit this description.[3] On the other hand, however, Feinberg seems to think that a common feature of the VRFs is that they tend to result in choices that are out of character for the decision-maker or do not faithfully represent him (1986: 113). But this criterion appears to be different from the first, since, as we have just seen, a person's choice may result from clouded judgment but nonetheless be in character. In the end, then, a defender of the voluntariness-based hard/soft distinction may need to indicate which of the two criteria is most fundamental. And each alternative may have its costs.

Suppose first that the important question in assessing voluntariness is whether a choice results from clouded judgment or impaired reasoning. Since such a choice may nonetheless be in character for the agent, it may be difficult to claim that it is alien to her will. And without the claim that the choice is alien to her will, it may be difficult to explain why putatively soft paternalistic intervention is consistent with respect for autonomy.

Suppose then that the important question in assessing voluntariness is whether a choice is in character or faithfully represents the agent. In this case, it may be easier to explain why non-voluntary choices are alien to the agent's will. But it may be more difficult to justify some of the intervention that soft paternalists have traditionally supported – for instance, intervention on behalf of generally competent people who are especially prone to the influence of distorting emotions, or intervention on behalf of ill-informed people who voluntarily chose not to gather relevant information.

Despite the potential difficulties confronting Feinberg's view, the motivation for it seems both clear and compelling. Intuitively, it often does seem relatively easy to justify intervention in choices that are ill-informed or impaired by conditions such as emotional distress, time pressures, and so forth. Perhaps, however, the relevance of these factors can be explained in terms other than voluntariness. The next section takes up this question.

3 Instrumental rationality, desires, and values

Sometimes, people behave in ways that are irrational even by their own lights. This may be because they fall prey to various forms of "reasoning failure" (Le Grand and New 2015: 82–101). For instance, they may make calculative errors in determining how to achieve their ends. They may lack imagination or foresight. Having decided on a course of action, they may lack the willpower to carry through on it. The result may be that they act in ways that fail to accurately reflect their own basic desires and values.

Taking this thought as their cue, some philosophers have held that whether intervention counts as hard (or soft) depends on how it fits with the target's own desires, values, and aims. According to one promising view of this sort, "interference does not violate one's autonomy if one would consent to it, given one's *current* preferences and values, *if* one were *instrumentally rational* and *well informed* about relevant, empirically ascertainable causal or means-ends matters" (Scoccia 2013: 80, emphasis in original). One is instrumentally rational in the sense at issue if one is disposed to take the most efficient means to one's given ends, whatever these ends happen to be. This sort of view may be what Donald VanDeVeer has in mind when he claims that paternalistic interference is autonomy-respecting provided that the target would validly consent to it if she "were aware of the relevant circumstances" and her "normal capacities for deliberation and choice were not substantially impaired" (1986: 75). Perhaps a person's "normal capacities" are impaired when she makes choices that are instrumentally irrational given her desires, aims, and values.

To be clear, the view I now want to consider does not imply that interference in instrumentally irrational choices *always* involves soft paternalism. After all, it might be true both that a certain choice is instrumentally irrational *and* that the decision-maker would strongly oppose others' interference even in her irrational choices. Scoccia, VanDeVeer, and others thus hold that whether intervention is autonomy-respecting depends on whether the target would consent to it if she were informed and instrumentally rational but otherwise possessed of her current preferences and values – including any that she might have regarding others' intervention in her affairs.

Following Scoccia, I have suggested that instrumental rationality operates on one's "current preferences and values." Yet there appears to be a difference between mere preferences (or desires), on the one hand, and values, on the other. To say that one *values* one's marriage, for instance, is not *merely* to say that one desires that it continue. One who values a marriage is likely to nurture one's affection for a spouse, to identify with one's (positive) attitudes toward the relationship, and so forth (Raibley 2010: 608). Moreover, when one values something – such as a relationship, a career, a project, an activity, or an ideal – one is typically committed to it, and this commitment is typically stable over time. At least in ordinary cases, one who values a marriage will not cease to do so overnight, without thereby calling into question the extent to which one valued the marriage in the first place. Mere desires are different. If one now desires to drink a cup of coffee, one might not take any steps to nurture or maintain this desire. Desires that are not associated with value commitments are often susceptible to waxing and waning over fairly short periods of time: someone might desire a cup of coffee now, but not an hour from now. To be sure, the contrast here should not be overstated. While value commitments are typically fairly stable, they do change; and while desires often wax and wane, some are fairly stable. Moreover, our desires are often aligned with our values, and our values may sometimes direct us to satisfy certain desires. (Indeed, someone who values "living in the moment" may be able to realize this value only by catering to whatever whims she happens to have.) Nonetheless, it seems that we have a different sort of regard for the things we value than we do for the objects of mere desire.

This point raises a complication for the instrumental rationality-based account of the hard/soft distinction. The strength of a person's desire for something may be out of proportion to the extent to which she values it, as often happens when we act against our better judgment. Thus, someone might have difficulty quitting smoking, even if she values longevity more than she values the pleasures of smoking, and even if she is aware of the health effects of her habit. Someone might have difficulty saving money for her child's college education, even if she values her child's education more than she values the superfluous luxuries she routinely buys. And so forth. In these sorts of cases, it may not be all that helpful to claim that a person's choice is autonomous if it reflects her current desires and values, for the desires on which she acts point in a different direction than her values; what she most wants to do does not always correspond to what she most values. In cases of this sort, we can ask whether the target's desires, or rather the target's values, ought to be privileged by an account of the hard/soft distinction.

Consider first the view that the target's current desires, rather than her current values, ought to be privileged. This proposal confronts two problems. First, it seems arbitrary. The idea animating the instrumental rationality account is that hard paternalism imposes values on people, while soft paternalism does not (Scoccia 1990: 320). But if someone intervenes in a person's affairs on behalf of that person's values, it does not seem that values have been "imposed" on her at all. Second, it is not clear how the proposal could be applied to decision-makers who are impaired by conditions other than simple ignorance. When a person is ignorant of relevant facts, he may act in ways that are unlikely to achieve his own ends. (Consider someone who sprinkles poison in his coffee because he mistakenly believes it to be sugar.) By contrast, if someone in the midst of an extreme depressive episode acts on a desire to do something imprudent or self-destructive, it is not obvious that his behavior is *instrumentally* irrational; after all, such behavior may be supported by his strongest current desire.[4]

Perhaps, then, the hard/soft distinction ought to privilege the target's values. On this view, intervention counts as soft provided that the target would consent to it if she were to reason correctly from her own values, where her commitment to a certain value is not always reflected by the strength of her corresponding desires. One problem for this proposal is that it may not be extensionally adequate, since it may classify as "soft" some intervention that most antipaternalists would regard as "hard." Consider the following example. Suppose that your roommate has the goal of becoming a professional musician. Unfortunately, however, she too often succumbs to the temptation of television and neglects to practice. You reasonably judge that your roommate has a good chance of achieving a musical career if, but only if, she practices much more. You thus remove the television, without your roommate's consent, and she predictably spends more time practicing. At least on the face of it, your behavior seems to be of the sort that most antipaternalists would oppose. After all, we can suppose that your roommate is a competent adult who is fully informed about the possible consequences of her behavior. (She is presumably *aware* that her failure to practice will make a musical career less likely.) Nonetheless, your behavior may help your roommate to realize her values, since (we can suppose) she *values* a career in music above her television habit, as you are aware.[5] Under these sorts of circumstances, it may well be that your roommate would consent to your intervention if she were choosing so as to promote the realization of her *values*. If she resents your attempt to encourage her to practice by removing the television without her permission, that may be largely because she is swayed by desires that do not correspond (at least in their strength) to her value commitments. The proposal we are considering would then seem to imply that intervention would be of the soft, autonomy-respecting variety. But that seems to be the wrong result.

Finally, it might be suggested that an account of the hard/soft distinction should privilege neither desires nor values. One might hold that both desires and values count: perhaps paternalistic intervention is soft when it increases some weighted sum of desire fulfillment and value realization.[6] The obvious challenge confronting this proposal is that it is unclear how we are to trade desire-satisfaction against value realization in cases of conflict. One concern here is that any attempt to show, in a given case, that a person's values are more important than her desires may threaten to show that her values always ought to take priority in cases of conflict. For instance, if one were to claim that a person's values merit more respect than her desires because her values are more constitutive of her practical identity, then it would appear that values always ought to take priority, and we would be led back to the problem discussed in the previous paragraph. A further concern is that it may be difficult to provide any criterion that specifies the relative importance of values and desires without appealing to some overarching value that itself is likely to determine the permissibility of interference. One might argue that when a person desires to act in ways that conflict with her values, as in our musician example, we should honor her desires instead of her values (or vice versa) if doing so would serve her best interest. But in this case, it would appear that the rationale for intervention ultimately appeals, in hard paternalist fashion, to the subject's interests.

4 A hard case

Perhaps the various considerations described above are not decisive against either Feinberg's voluntariness-based view or the instrumental rationality account of the hard/soft distinction. Nonetheless, I now want to turn to a case that can be used to illustrate some of the complexities raised by the various views we have discussed, not necessarily to show that all of these views ought to be rejected, but rather to cast in sharper relief the moral puzzles raised by the hard/soft distinction.

Consider a devout and lifelong Jehovah's Witness who has long accepted her religion's prohibition on blood transfusions. Suppose that she always carries a "no blood" card in her wallet indicating that she should not be given blood in an emergency, that she has written and distributed pamphlets critical of transfusions, and so forth. Now imagine that this Witness has been in an accident and is told that she will die unless she receives a transfusion very soon. (Although the accident has left her physically incapacitated, she is still conscious and able to make decisions for herself.) Feeling frightened, and acting under severe time pressures, she consents to the transfusion. Is her choice to consent to a transfusion protected by her right to autonomy? We can make this question a bit more vivid. Imagine that the doctor who is assigned to treat the Jehovah's Witness is a long-time friend of hers and is thus fully aware of her history of religious devotion. Should the doctor perform the transfusion? Would performing the transfusion violate the Witness' right to autonomy? And if the doctor does decide to perform the transfusion, would intervention to prevent it – say, by a legal authority or hospital administrator – violate the woman's right to autonomy?

Of course, under the circumstances, a transfusion would almost certainly benefit the Jehovah's Witness. For this reason, some might argue that the case raises no interesting moral questions about paternalism: after all, the Witness has actually consented to a procedure that is likely to benefit her. Even if we agree that the doctor ought to perform the transfusion, however, there is an important theoretical question about *why* this is: is it only because the transfusion would serve the Witness' interests, or is it also because the transfusion respects, or is consistent with, the Witness' (putative) right to direct the course of her own life? Since the hard/soft distinction aims to (at least partially) characterize this right, a complete account of the distinction should inform our answer to this question.

Consider first what Feinberg's account would imply about our case. On the one hand, fear, time pressures, and other features of the Witness' circumstances appear on Feinberg's list of VRFs. If these features influence the Witness' choice, then there might appear to be a strong case that her consent is substantially nonvoluntary. On the other hand, it might be argued that the fear and pressure direct the Witness' mind to what she really values. Perhaps her commitment to the tenets of her religion was not, after all, as strong as it appeared to be, or even as strong as she believed it to be. It may thus be unclear whether fear compromises the voluntariness of her choice.

A similar question arises for other accounts of the hard/soft distinction. Recall VanDeVeer's claim that intervention does not violate autonomy provided that the subject would consent if she were informed and her "normal capacities" were unimpaired. We might sensibly ask which attitude is a product of the Witness' "normal capacities": the opposition to blood transfusions that she has harbored for her whole adult life up until now, or her willingness to receive a transfusion now that she fears her own death. Likewise, in considering whether intervention to prevent the transfusion would fulfill or realize the Witness' desires and values, we must decide which desires and values (or rankings of desires and values) are relevant: those she has held for most of her life, or those she holds now.

One concern here is that the judgment that the Witness' choice is (or is not) impaired may depend on further value judgments about the prudence or reasonableness of this choice. For instance, fellow Jehovah's Witnesses may be especially likely to prefer the description of our example according to which the woman is impaired by fear and simply "chickens out" when she consents to a transfusion. They might claim that the woman's fear leads her to betray her own values. On the other hand, those who regard the traditional Witness doctrine as manifestly unreasonable are probably more likely to prefer the alternative description, according to which fear clarifies the woman's mind and helps her to see more clearly what she wants. They might claim that her true values are best revealed by the choices she makes when the chips are down, so to speak. Yet the extent to which fear, time pressures, and other factors result in impaired decision-making had better not depend on the extent to which these factors result in imprudent or unreasonable decision-making; Feinberg, among others, repeatedly claims that whether a choice is protected by rights of self-sovereignty does not depend on how wise or prudent it is (1986: 61–62, 106–113).

To be sure, the issues raised by the example we have been considering are difficult, and perhaps these issues can be resolved by a suitable account of the hard/soft distinction. The example nonetheless merits further consideration, since it highlights some of the difficulties discussed in the preceding sections of this entry.

5 Conclusion: weakening or forgoing the distinction?

Accounts of the hard/soft distinction purport to identify the difference between intervention that violates autonomy rights and intervention that does not. This entry has considered two accounts of the hard/soft distinction, one that appeals to the notion of voluntariness and one that appeals to the notion of instrumental rationality. I have argued that each account confronts apparent difficulties. To be sure, these difficulties may be merely apparent, and it is certainly possible that, with further work, they can be resolved by one of the accounts we have considered.

If they cannot, however, we might be left with a very *thin* conception of the hard/soft distinction. On such a thin conception, intervention is soft only when it prevents behavior that is induced by conditions that uncontroversially cancel voluntariness, such as (non-culpable) ignorance, or certain extreme forms of mental illness, or (non-consensual) hypnosis. Nothing that I have said above would challenge this thin view. Instead, the difficulties I have discussed arise when we turn to the common cases in which a decision-maker is affected by less severe forms of impairment, such as time pressures, gripping moods, and powerful passions. Thus, for all that I have said, one could easily hold on to a thin account of the hard/soft distinction and argue that paternalistic intervention violates autonomy unless the target is wholly incapable of voluntary choice as a result of (non-culpable) ignorance or extreme incapacity. Such a thin account of the hard/soft distinction, however, could not easily be combined with an absolutist view on which hard paternalistic intervention is always wrong. After all, most of us think that it would be permissible to prevent a person from seriously harming himself as a result of significant, but not incapacitating, emotional distress. An absolutist view could support this judgment only if combined with a "thicker" account of the hard/soft distinction.

If such a thick account proves elusive, it may be worth considering two alternative views about the significance of ignorance, powerful passions, and other VRFs. According to the first, the fact that someone is ill-informed or in the midst of a gripping mood is relevant to the justification of paternalism simply because the presence of such conditions often provides at least some evidence that she is unlikely to act prudently. J.S. Mill forcefully argued that each individual is generally the best judge of her own interests (1977 [1859]: 277). Yet when an individual is in a temporary state of depression or extreme emotional excitement, the general presumption that she is the best judge of her own interests may be defeated or called into question. Perhaps this is all there is to the hard/soft distinction.[7]

Second, it may be that people have important interests in having certain sorts of liberty and that these interests must be balanced, on a case-by-case basis, against the various other interests that would be served by intervention (Shafer-Landau 2005: 186–188). The relative weight of a person's interest in being free to make a certain choice is likely to depend on the nature of that choice – for instance, on whether it is important or trivial. Perhaps more interesting for present purposes, it is at least possible that the weight of a person's interest in being free to make a certain choice depends in a direct way on the circumstances in which she would be making it. For instance, it may be good for an individual to develop or cultivate her own character or to chart her own course through life, and the freedom to make sober choices may be more conducive to this end than the freedom to make intoxicated choices. While a view of this sort may accommodate some of the intuitions that have led philosophers to endorse soft paternalism, it may not rely on a binary account of the hard/soft distinction that would neatly categorize some interventions as autonomy-violating and others as autonomy-respecting.

Whatever we might think of these alternative views, the hard/soft distinction appears to be critically important for robust antipaternalist positions. Virtually everyone believes that intervention in ill-informed or impaired self-regarding choices is often justified. An adequate antipaternalist view must explain why this is. Because some of the most prominent explanations appeal to the putative difference between hard and soft paternalism, discussion of this distinction will continue to play an important role in the normative debate over paternalism.[8]

Related topics

Paternalism and Autonomy; Paternalism and Contract Law; Paternalism and the Criminal Law; The Concept of Paternalism.

Notes

1 For the sake of brevity, I generally set aside the parenthetical qualification in what follows, although it is an important component of Feinberg's view.
2 It may be more difficult to explain how supposedly coercive *offers* can compromise voluntariness (Callahan 1986).
3 To be sure, powerful passions and gripping moods may not *always* compromise voluntariness, since they may not always impair reasoning (Feinberg 1986: 117).
4 Someone might argue that a person in the grip of a powerful passion is unlikely to be informed in the relevant sense: although she might be aware of the relevant facts, she may fail to appreciate them. This point, however, would not show that she is ignorant in a way that would prevent her from satisfying her current (non-instrumental) desires. Instead, it would show that there is something amiss about the desires themselves.
5 We might explain the roommate's behavior in terms of the common distinction between first- and second-order desires: perhaps your roommate often desires to watch television more than she desires to practice, but nonetheless desires to desire to practice more.
6 A suggestion along these lines is defended by Grill (2015), who argues that we have reason to respect a person's actual choices, actual desires, and true or idealized desires.
7 This view may be attractive to those convinced by the arguments I offered in Hanna (2012).
8 I am grateful to Kalle Grill for helpful feedback on a draft of this chapter.

References

Beauchamp, T. (1977) "Paternalism and Biobehavioral Control," *The Monist* 60: 62–80.
Buchanan, A. and Brock, D. (1990) *Deciding for Others: The Ethics of Surrogate Decision Making*, New York: Cambridge University Press.
Callahan, J. (1986) "Paternalism and Voluntariness," *Canadian Journal of Philosophy* 16: 199–220.
DeMarco, J. P. (2002) "Competence and Paternalism," *Bioethics* 16: 231–245.
Feinberg, J. (1971) "Legal Paternalism," *Canadian Journal of Philosophy* 1: 105–124.
———. (1986) *The Moral Limits of the Criminal Law, Volume 3: Harm to Self*, New York: Oxford University Press.
Grill, K. (2010) "Anti-Paternalism and the Invalidation of Reasons," *Public Reason* 2: 3–20.
———. (2015) "Respect for What? Choices, Actual Preferences, and True Preferences," *Social Theory and Practice* 41: 692–715.
Hanna, J. (2012) "Paternalism and the Ill-Informed Agent," *Journal of Ethics* 16: 421–439.
Hodson, J. (1977) "The Principle of Paternalism," *American Philosophical Quarterly* 14: 61–69.
Le Grand, J. and New, B. (2015) *Government Paternalism: Nanny State or Helpful Friend?* Princeton, NJ: Princeton University Press.
Mill, J. S. (1977 [1859]) "On Liberty," in J. Robson (ed.) *The Collected Works of John Stuart Mill*, Vol. 18, Toronto and London: University of Toronto Press and Routledge.
Raibley, J. (2010) "Welfare and the Priority of Values," *Social Theory and Practice* 36: 593–620.
Scoccia, D. (1990) "Paternalism and Respect for Autonomy," *Ethics* 100: 318–334.
———. (2013) "The Right to Autonomy and the Justification of Hard Paternalism," in C. Coons and M. Weber (eds.) *Paternalism: Theory and Practice*, New York: Cambridge University Press, pp. 74–92.
Shafer-Landau, R. (2005) "Liberalism and Paternalism," *Legal Theory* 11: 169–191.
Sunstein, C. (2014) *Why Nudge?* New Haven, CT: Yale University Press.
VanDeVeer, D. (1986) *Paternalistic Intervention: The Moral Bounds on Benevolence*, Princeton, NJ: Princeton University Press.

Further reading

I further explore many of the issues raised above in my "Paternalism and Impairment," *Social Theory and Practice* 37(2011): 434–460. An incisive response to the views of Feinberg and VanDeVeer can be found in D. Brock, "Paternalism and Autonomy," *Ethics* 98(1988): 550–565. Further criticism of Feinberg's view is provided in R. Arneson, "Joel Feinberg and the Justification of Hard Paternalism," *Legal Theory* 11(2005): 259–284. For a view that attempts to situate something resembling the hard/soft distinction within a Kantian framework, see M. Cholbi, "Kantian Paternalism and Suicide Intervention," in C. Coons and M. Weber (eds.) *Paternalism: Theory and Practice*, New York: Oxford University Press, 2013.

3

MORALISM AND MORAL PATERNALISM

Peter de Marneffe

Moralism and moral paternalism are discussed here as views about the eligibility of certain reasons for government policy.

> *Moralism*: If an action will make the world a morally worse place, apart from any harm to individuals, this is a good reason for government policies that reduce such actions (although this reason might be outweighed by countervailing reasons).
>
> *Moral Paternalism*: If an action will cause moral harm to the agent, this is a good reason for government policies that reduce such actions (although this reason might be outweighed by countervailing reasons).

To illustrate the difference, consider two different kinds of reasons for drug laws.

> *Moralistic Reason*: It is bad for people to use mind-altering drugs when doing so is medically unnecessary, because a person misuses his mind in doing so. This misuse of the mind is bad independent of any harm to the drug-user or others, simply because the world is a morally worse place when minds are misused in this way.

Moralism implies that, if constituted by true propositions, reasons of this kind ought to be counted in favor of policies that reduce recreational drug use.

> *Moral Paternalistic Reason*: It is bad for people to have the vice of intemperance because in having this vice they have a morally worse character. This is bad for them independent of any other harm, simply because it is non-instrumentally bad for a person to have a morally worse character.

Moral paternalism implies that, if constituted by true propositions, reasons of this kind ought to be counted in favor of policies that reduce intemperance. If drug laws do this, this is a good reason for them.

The terms *moralistic* and *paternalistic* are sometimes used to characterize government policies by reference to the motives of those who support them. For example, it might be said that drug laws are paternalistic because they are motivated by a desire to protect drug users from their own self-destructive conduct. And it might be said that sodomy laws are moralistic because they are motivated by the belief that sodomy is sinful, quite apart from any harm it might cause. Here I set aside the motivational use of these terms, and focus exclusively on the eligibility of reasons. If moralistic or paternalistic reasons should never be counted, this would explain what is wrong with moralistic or paternalistic motives. If, however, moralistic or paternalistic reasons should be counted, it is hard to see what is wrong with moralistic or paternalistic motives. Furthermore, when we evaluate a government policy our primary concern is with whether there is sufficient reason for it, something on which the eligibility of reasons bears directly. In contrast, it is not clear how the motives of those who support a policy are relevant to whether there is sufficient reason for it. If there is decisive reason for a government policy, the government should adopt it, regardless of the motives of those who support it.

In this chapter, I consider moralism and moral paternalism as views about the eligibility of reasons for government policy, and, specifically, for legal prohibitions. They might also be considered as views about the eligibility of reasons in practical deliberation more generally. For example, they might be considered as views about the reasons that doctors should count in deliberating about their treatment of patients, and they might be considered as views about the reasons that family members should count in deliberating about their treatment of other family members. The most influential theoretical discussions of moralism and paternalism, however, have been about the eligibility of reasons for government policy (for example, Mill 1978 [1859]; Hart 1963; and Feinberg 1984). Furthermore, the idea that certain considerations, although good reasons in themselves, should not be counted in favor of government policy has played a central role in liberal political theory, at least since Mill's *On Liberty*, whereas this idea has not played such a central role in the theory of practical reasoning more generally. Finally, the primary function of this chapter is analytical, not prescriptive. The primary goal is to clarify the difference between moralism and moral paternalism. If we understand the difference as applied to reasons for government policy, we will also understand how they differ as applied to personal and professional relationships.

The next section introduces Feinberg's influential formulations of legal moralism and moralistic legal paternalism. Most of this chapter is then devoted to understanding the content of these two principles. Toward the end, some questions are raised about the normative validity of legal moralism.

1 Feinberg's formulations

In *The Moral Limits of the Criminal Law*, Joel Feinberg distinguishes two views:

> *Legal Moralism* (in the broad sense): It can be morally legitimate for the state to prohibit certain types of action that cause neither harm nor offense to anyone, on the grounds that such actions constitute or cause evils of other ("free-floating") kinds.
>
> (Feinberg 1984: 27)

> *Moralistic Legal Paternalism*: It is always a good reason in support of a proposed prohibition that it is probably necessary to prevent *moral harm* (as opposed to physical, psychological, or economic harm) to the actor himself. (Moral harm is "harm to one's character," "becoming a worse person," as opposed to harm to one's body, psyche, or purse).
>
> (Feinberg 1984: 27)

Understanding the difference between these two views requires understanding the difference between *moral harm*, mentioned in the second principle, and "evils of other kinds," mentioned in the first. Because moral harm is a kind of harm, we need to identify what kind of harm it is. Because evils of other kinds are not harms, we need to identify what kind of evils they are.

Although Feinberg uses the phrase "free-floating evils" in first stating the (broad) principle of legal moralism, he later distinguishes free-floating evils from another kind of harmless wrongdoing, which he calls "welfare-connected non-grievance evils" (Feinberg 1988: 19). "Evils," as he uses the term, are regrettable states of affairs (Feinberg 1988: 18). Because they are regrettable, we ought, other things equal, avoid bringing them about, if we can.

Feinberg does not define the notion of a free-floating evil, but illustrates it with a list, which includes "the wanton, capricious squashing of a beetle . . . in the wild" and "the extinction of a species" (Feinberg 1988: 20–25). He illustrates the notion of a welfare-connected non-grievance evil with an example from Derek Parfit which illustrates Parfit's *non-identity problem* (Feinberg 1988: 27–28). Parfit introduces this problem in *Reasons and Persons* with the following example:

> *The 14-Year-Old Girl.* This girl chooses to have a child. Because she is so young, she gives her child a bad start in life. Though this will have bad effects throughout this child's life, his life will, predictably, be worth living. If this girl had waited for several years, she would have had a different child, to whom she would have given a better start in life.
>
> (Parfit 1984: 358)

Parfit believes there is a good reason for the girl to wait to have a child, which is that any child she has now will have a worse start in life than a child she has later. For this reason, he thinks, there is something wrong with the girl's choice. But this wrong does not consist in harm to the child. Harming someone requires making that person worse off than he otherwise would have been and no one (the example assumes) is worse off as a result of the girl's choice. The child she has now is not worse off because his life is worth living, and he never would have come into existence had she waited to have a child. Nor are the children worse off that she would have had later had she not had a child now, because they do not exist. Consequently, if the girl's choice is wrong, it is a kind of harmless wrongdoing.

Feinberg recognizes harmless wrongdoing of this kind (Feinberg 1988: 25–33), and he distinguishes welfare-connected non-grievance evils of this kind from free-floating evils. This kind of evil is *welfare-connected* because it consists in the fact that there is less welfare in the world than there could otherwise be (on the assumption that a child the girl had later would be better off than the child she has now). Feinberg thinks many liberals would regard welfare-connected non-grievance evils of this kind as legitimate considerations in evaluating and justifying government policy (Feinberg 1988: 32–33). Furthermore, it seems that reducing evils of this kind might warrant legal prohibitions of some kind. If so, then the (broad) principle of legal moralism is valid.

2 Personal and impersonal reasons

To grasp the content of legal moralism, we must understand the kinds of evils that are produced by harmless wrongdoing, as opposed to harmful wrongdoing. To grasp the content of moralistic legal paternalism, we must understand moral harm as distinguished from other kinds of harm. And to understand the difference between legal moralism and moralistic legal paternalism, we must understand the difference between harmless wrongdoing and moral harm. I propose that we understand the general distinction between harm and harmless wrongdoing in terms of T.M. Scanlon's distinction between personal and impersonal reasons, and then understand moral harm in terms of a subset of personal reasons.

Personal reasons, as Scanlon understands them, are reasons "grounded in the moral claims or the well-being of individuals" (Scanlon 1998: 219). Impersonal reasons are reasons of other kinds – not grounded in the moral claims or well-being of individuals. Illustrating the notion of an impersonal reason, Scanlon writes:

> Many people, for example, believe that we have reason not to flood the Grand Canyon, or destroy the rain forests, or to act in a way that threatens the survival of a species (our own or some other), simply because these things are valuable and ought to be preserved and respected, and not just because acting in these ways would be contrary to the claims or interests of individuals.
>
> (Scanlon 1998: 219)

Applying this distinction to reasons for government policy – and going beyond what Scanlon actually writes here – I propose that we understand the distinction between personal and impersonal reasons in the following way. Personal reasons for the government to adopt a policy are reasons for a person to prefer his or her own situation when the government adopts this policy. Impersonal reasons for the government to adopt a policy are reasons for someone to prefer the world that is likely to result when the government adopts this policy, as distinct from the reasons there are for anyone to prefer his or her own situation in that world.

To illustrate this distinction, consider Ronald Dworkin's discussion of the morality of abortion. In *Life's Dominion*, Dworkin distinguishes two different objections to abortion, which he calls the *derivative* objection and the *detached* objection (Dworkin 1994: 11). The derivative objection is that abortion is wrong because it violates the rights of the fetus, which, Dworkin assumes, must be grounded on interests the fetus has in continuing to live. The detached objection is that abortion is wrong "because it disregards and insults the intrinsic value . . . of human life" (Dworkin 1994: 11). By "intrinsic value" Dworkin means value that is not reducible to interests (Dworkin 1994: 69). Because a first-trimester fetus has no mental life, is not conscious, and has never been conscious, there is no reason for it to care about continuing to live, and so there is no interest in continuing to live which abortion laws might protect. Dworkin believes that it nonetheless makes sense to think that first-trimester fetuses have intrinsic value, and he believes this can provide a rational basis for abortion laws, on the general principle that the government may limit individual liberty to protect things of intrinsic value (Dworkin 1994: 149).

To say, as Dworkin does, that a first-trimester fetus has no interest in continuing to live that could justify the government in prohibiting abortion is to say that there is no reason for a fetus to prefer its own situation when abortion is prohibited. Having no mental life and never having been conscious, there is no reason for a fetus to care whether or not it is aborted, and so no reason to prefer its own situation when abortion is made less likely by legal restrictions. To say, as Dworkin also does, that a first-trimester fetus has intrinsic value is to say that, other things being equal, there is good reason to prefer a world in which first-trimester fetuses are not intentionally destroyed for no good reason or destroyed thoughtlessly or carelessly. If legal restrictions reduce the number of fetuses that are destroyed in these ways, there is consequently a reason to prefer a world in which the government adopts these restrictions.

Considered as a reason for government policy, the reduction of harm, including moral harm, can be reduced to personal reasons. If a government policy will reduce an action that inflicts moral harm on someone, there is a reason for this person to prefer his own situation when this policy is adopted. In contrast, the reduction of harmless wrongdoing – considered as a reason for government policy – can be reduced only to impersonal reasons. If a government policy will reduce an action that causes harmless wrongdoing, there is a reason to prefer the world that is likely to result when this policy is adopted, but it is not a reason for anyone to prefer his own situation in that world. Formally, this is what distinguishes moral harms from non-grievance evils, considered as reasons for government policy.

3 Moral harm

Considered as reasons for government policy, all harms, including moral harms, can be reduced to personal reasons. But what distinguishes moral harms from other kinds of harm, and what kinds of moral harm are there?

Feinberg characterizes moralistic legal paternalism as follows:

> It is always a good reason in support of a proposed prohibition that it is probably necessary to prevent *moral harm* (as opposed to physical, psychological, or economic harm) to the actor himself.
>
> (Feinberg 1984: 27)

It is doubtful, though, that Feinberg would count as moral harm every harm that is not physical, psychological, or economic.

Suppose a young person will not suffer economic harm if he does not attend high school because he has an ample trust fund, and that he will also not suffer psychologically as a result (for whatever reason). In not attending high school he still loses valuable opportunities: to develop intellectually, to develop understanding and appreciation of the arts and sciences, to develop mentoring relationships with educated and caring adults, to participate in high school athletics and other organized activities such as high school plays, to socialize with other young people in an educational setting, and so on. The loss of these opportunities can be bad for a person, even if they do not result in psychological or economic harm as normally understood, but I doubt Feinberg would conclude from this that the loss of these opportunities is therefore a kind of *moral* harm. If the government were to require all students to attend high school so as to ensure that they have these opportunities, I doubt he would regard this as an instance of *moralistic* paternalism. So although we can use the idea of harm that is not physical, psychological, or economic as a guide to identifying moral harm, we should not *define* it this way.

Feinberg and other influential commentators also identify moral harm with harm to one's character. Feinberg writes:

> Moral harm is "harm to one's character," "becoming a worse person," as opposed to harm to one's body, psyche or purse.
>
> (Feinberg 1984: 27)

Danny Scoccia writes:

> By moral paternalism I understand the principle that it is right for the state to restrict people's liberty in order to prevent a deterioration in their character that is bad for them or contrary to their best interests, or in order to produce an improvement in their character that is to their benefit.
>
> (Scoccia 2000: 53)

Gerald Dworkin characterizes moral paternalism as follows:

> It is a distinct claim, and one made by Plato and Epictetus among others, that improvement in one's moral character is a good for the person. It is this last claim that differentiates MP [moral paternalism] from LM [legal moralism].
>
> (Dworkin 2005: 308)

Below I consider reasons to identify moralistic paternalism with protecting a person's character. First I identify two other things that might count as moral harm to show that the identification of moralistic paternalism with protecting a person's character is intellectually optional, not forced on us by the concept.

Moral harm, we might say, is the harm we do to ourselves in doing something morally wrong. If so, there are at least two things other than harm to one's character that might count as moral harm. First, there is the harm we do to ourselves in failing in our obligations to others. Second, there is the harm we do to ourselves in failing to treat what is precious with due care or respect.

There is good reason for each of us to meet our obligations to others, to want this for ourselves, to want it for its own sake and not only as a means to other goods, and to regret it if we fail. When we violate our obligations to others, we alter and impair our relationships with them by treating them with disrespect. Insofar as it is non-instrumentally good for us to maintain respectful relations with others, we harm ourselves in failing to meet our obligations to them. We might not be harmed physically, psychologically, or economically. We might not lose valuable opportunities as a result. But this does not mean we are not harmed by it.

There is also good reason for each of us to treat what is precious with due respect, to want this for ourselves, to want it for its own sake and not only as a means to other goods, and to regret it if we fail. When we treat what is precious carelessly, thoughtlessly, with contempt, "like trash," we are contemptible ourselves. Insofar as it is non-instrumentally good for us to treat what is precious with respect, we harm ourselves in failing to do so. We might not be harmed physically, psychologically, or economically. We might not lose valuable opportunities as a result. But this does not mean we are not harmed by it.

The two categories of moral harm just identified show that, if a person's character is understood to consist in his settled psychological dispositions to act, things might count as moral harm other than harm to one's character. Why, then, have commentators identified moralistic paternalism with the goal of protecting a person's character?

4 Why identify moral harm as harm to character?

The reason Feinberg does this is connected to his endorsement of a subjective conception of harm. He defines harm in terms of interests (Feinberg 1984: 65), and he understands interests in terms of what people actually want (Feinberg 1984: 42, 67). If we want something for its own sake, then it is in our interest to have it. If having x makes it more likely that we will have y, which we want for its own sake, then it is also in our interest to have x, even if we do not actually want x. Feinberg identifies moral harm with harm to character because he recognizes that some philosophers, such as Plato, have held that we can harm ourselves by acting viciously even if we do not want to act virtuously, and even if acting viciously does not prevent us from getting anything we want for its own sake. He writes:

> No doctrine was more central to the teaching of Socrates, Plato, and the Stoics than the thesis that a morally degraded character is itself a harm quite independently of its

> effect on its possessor's [desire-based] interests. On this issue, the implications of our own analysis of harm in terms of set-back interest are clear. If a person has no ulterior interest in having a good character [because he does not want this for its own sake], and if such a character is not *in* his (other) interests [because it does not help him achieve things that are useful to achieving what he does want for its own sake], then his depraved character is no harm to him (*pace* Plato et al.), and even if he becomes worse [in some sense], he does not necessarily become worse off.
>
> (Feinberg 1984: 66)

If, however, we reject Feinberg's subjective conception of interests, it is possible to make sense of moral harm without identifying it with harm to character.

According to a subjective conception of interests, which Feinberg endorses, a person has an interest in something if and only if (a) she actually wants it non-instrumentally for herself, or (b) it makes it more likely that she will have something she wants non-instrumentally for herself. According to an objective conception of interests, a person has an interest in something if and only if there is good reason for her to want it for herself. This conception of interests is objective in two ways. First, someone can have an interest in something they do not actually want, even if it does not bring about something they want. Second, if there is no good reason for someone to want something, they do not have an interest in it, even if they want it.

If we accept an objective conception of interests and recognize the two categories of moral harm identified in the previous section, we can explicate the notion of moral harm without identifying it with harm to a person's character. We can say that a person is harmed in failing to meet his obligations to others and in treating what is precious with disrespect, even if meeting his obligations to others and treating what is precious with respect is not something that he wants for its own sake, and even if this does not help him achieve something he wants for its own sake. This is because there is good reason for him to want to meet his obligations to others and to treat what is precious with respect, and to want this non-instrumentally for himself.

Feinberg recognizes that an objective conception of interests is possible, and he contrasts his own subjective conception of interests with an objective conception that he calls "the ideal-regarding theory":

> The ideal-regarding theory holds that it is in a person's interest ultimately not only to have his wants and goals fulfilled, but also (and often this is held to be more important) to have his tastes elevated, his sensibilities refined, his judgment sharpened, his integrity strengthened: in short to become a better person. On this view, a person can be harmed not only in his health, his purse, his worldly ambition, and the like, but also in his character. One's ultimate good is not only to *have* the things one wants, but (perhaps more importantly) to *be* an excellent person, whatever one may want. We not only degrade and corrupt a person by making him a worse person than he would otherwise be; on this view, we inflict serious harm on him, even though all his interests flourish. Socrates and the Stoics even went so far as to hold that this "moral harm" is the *only* genuine harm. Epictetus was so impressed with the harm that consists simply in having a poor character that he thought it redundant to punish a morally depraved person for his crimes. Such a person is punished enough, he thought, just by being the sort of person he is.
>
> (Feinberg 1984: 67–68)

This passage and the previous one suggest the following interpretative hypothesis: Feinberg identifies moral harm with harm to character because the only non-subjective conception of harm that occurred to him when he wrote was one held by the ancient Greeks, who also held that one could be harmed by becoming a vicious person as a result of performing vicious acts. Had he not focused so much on ancient Greek thought, he might not have glossed "moral harm" as "harm to one's character."

Scoccia suggests another reason for identifying moral paternalism with the prevention of harm to character, which is that defenders of "morals laws," such as James Fitzjames Stephen and Robert George, have relied on the following assumptions:

1 Intemperance, lack of self-control, increased weakness of will, laziness, recklessness – all are vices that are bad for or harmful to the person who has them.
2 Prostitution, homosexuality, and the consumption of pornography, marijuana, or hard drugs promote some or all of these "self-regarding" vices in whoever engages in them.
3 A criminal ban on these activities will discourage a significant number of the people tempted to engage in them from doing so.
4 By doing that, a ban will elevate the character of these people, or at least prevent its further deterioration, thereby conferring a valuable benefit on them (Scoccia 2000: 53).

Even if defenders of moral laws have relied on these assumptions, however, there is no reason, in explicating the nature of moral paternalism, to equate moral harm entirely with harm to a person's character.

5 How relevant is moralistic legal paternalism?

Although moral harm is a kind of harm, and the good of reducing harm seems like a good reason in favor of a law that reduces it, it is possible that the following is also true: for every legal prohibition, either it can be fully justified as reducing non-moral harm alone or it is not justifiable. If so, then although the truth of moral paternalism might remain an interesting theoretical issue, it would be practically irrelevant to the evaluation and justification of government policy.

Consider two scenarios, suggested by Scoccia's discussion:

Harmless Intemperance: Smith's habitual drug use leads to intemperance in the following sense: he uses drugs more than he believes he ought to and more than he actually ought to. No serious consequences follow from this. If he used drugs less often he would read more ancient Roman history. If he read more ancient Roman history, his general understanding of the world would be better. This would be a good thing in itself, but aside from this his habitual drug use has no bad effects.

Harmful Intemperance: Jones's habitual drug use leads to intemperance in the following sense: he uses drugs more than he believes he ought to and more than he actually ought to. As a result of this intemperance he fails to show up for work, fails to hold a steady job, fails to make enough money to live on, fails to take proper care of his health, develops feelings of contempt for himself and unwarranted feelings of resentment toward the world, all of which interfere with his emotional development, his friendships, and his family relationships, and prevent him from developing into a mature, self-reliant person.

Suppose the benefit of reading more ancient Roman history cannot on its own justify the government in prohibiting Smith from using drugs, and that adding the good of reducing his intemperance in using drugs does not tip the balance of reasons in favor of prohibition. Suppose, in contrast, that the good of preventing the bad consequences of Jones's habitual drug use can justify the government in prohibiting him from using drugs, and that this policy can be justified without counting in addition the good to him of reducing his intemperance in using drugs. Then although reducing the vice of intemperance would be a good thing in itself, it would be practically irrelevant in these cases to justifying prohibition. If all the relevant cases were like this one, then the truth of moralistic legal paternalism would be practically irrelevant.

6 The validity of legal moralism

The primary goal of this chapter is to explicate the content of legal moralism and moralistic legal paternalism and to identify the key differences between them. In this section I consider the normative validity of legal moralism. Many leading theorists of rights ground rights on interests, in one way or another (Rawls 1971: 11–16; Scanlon 2003: 26, 99; Raz 1986: 166, 180; Dworkin 1994: 11, 23–24). If rights are grounded on interests, it would not be surprising if the principle of legal moralism were invalid.

Suppose that rights are valid moral rules, and that a system of moral rules is valid only if it optimally protects individuals' interests. Suppose a system of rules optimally protects people's interests only if the following is true: if any other system of rules protects someone's interests better, it also protects someone's interests worse, and this loss in interest-protection to the second person is at least as great as the gain in interest-protection to the first person. Suppose, finally, that interests are constituted only by personal reasons: reasons to prefer one's own situation under one system of rules or another.

If a theory of rights like this one is correct, it is open to doubt that legal moralism is valid. Legal moralism allows impersonal reasons to count in favor of legal prohibitions. Legal prohibitions always threaten interests, and impersonal reasons do not constitute interests. So, it seems, no system of rules that optimally protects individuals' interests will allow impersonal reasons to count in favor of legal prohibitions. If they are not to count, legal moralism is invalid.

One possible response to this line of reasoning is to accept the conclusion: legal moralism is invalid. Another possible response is to reject the view that rights are grounded on interests. Another possible response is to argue that although rights are grounded on interests, our interests are optimally protected by a system of rules permitting the government to adopt policies that can be justified only by counting impersonal reasons. The upshot is that legal moralism might be valid and it might be invalid. Its validity depends on whether our rights are grounded on interests and, if so, on what systems of rules optimally protect our interests.

Scoccia has recently identified another kind of reason for thinking legal moralism is invalid. He argues that if the government may prohibit harmless wrongdoing to prevent free-floating evils, then the justification for punishing violations of these prohibitions must rest on either a retributivist or a moral education theory of punishment (Scoccia 2013: 523, 525). If these theories are false, as many believe, this would seem to show that the government may not impose penalties on harmless wrongdoing to reduce free-floating evils.

Scoccia supposes that what is bad about free-floating evils consists largely in the evil will of the person who intentionally causes them. If, for example, the careless or thoughtless destruction of a fetus is bad, much of the badness consists in the carelessness or the thoughtlessness. This leads Scoccia to conclude that the function of moralistic penalties is to punish people for their evil states of mind, but this does not follow. Retributivism, as Scoccia understands it, is the view that the primary reason for punishment is to give evil-doers their "just deserts" (Scoccia 2013: 524). One could consistently reject this view while holding that moralistic penalties are justified. By *moralistic penalties* I mean penalties that are justified as reducing free-floating evils. We can think of free-floating evils as those whose presence in the world makes it a worse place, apart from any negative effect on human or animal welfare. The absence of these evils is therefore desirable in making the world a better place. The desirability of a better world might provide sufficient reason for penalties that reduce these evils by deterring actions that produce them, provided these penalties are not too burdensome. No reference to the goal of giving evil-doers their just deserts is necessary here.

7 The scope of legal moralism

Suppose the government is permitted to limit individual liberty to prevent harmless wrongdoing, either because rights are not grounded on interests or because a system of rules that optimally protects our interests would permit the government to limit individual liberty to prevent harmless wrongdoing. Is there any reason to think that, although the government is permitted to limit individual liberty to reduce welfare-connected non-grievance evils, it is not permitted to limit individual liberty to reduce free-floating evils? Feinberg thinks many liberals would regard welfare-connected non-grievance evils as legitimate considerations in evaluating and justifying government policy (Feinberg 1988: 32–33), but he suggests there is something inherently illiberal about limiting individual liberty to prevent free-floating evils (Feinberg 1988: 20). In contrast, Ronald Dworkin, another influential liberal theorist, holds that the reduction of free-floating evils can justify some legal restrictions.

Dworkin does not offer a theoretical defense of this principle. Instead he argues that rejecting it would conflict too much with settled constitutional practice (Dworkin 1994: 149). The US Supreme Court, he says, recognizes the validity of laws that protect works of art, animal species, and the quality of life of future generations, and these laws are justified, at least partly, as protecting things of intrinsic value.

This argument is inconclusive. For one thing, it is possible that all the laws that Dworkin has in mind can be fully justified as protecting individuals' interests. It is possible, for example, that the protection of endangered animal species can be fully justified as protecting the interests of scientists in biological research or the interests of nature lovers in discovering and observing different species. Moreover, even if it is settled constitutional practice to recognize the protection of something of intrinsic value as a rational basis for limiting liberty, this sociological claim about current legal practice does not show that it is morally permissible for the government to enact laws solely to protect things of intrinsic value.

Suppose, though, that this is permissible. How does Dworkin address the challenge that if the government may limit individual liberty to prevent free-floating evils, it may limit individual liberty in all sorts of objectionable ways? Consider the traditional Christian belief that non-procreative sex acts are "intrinsic evils." By *non-procreative sex acts* I mean those that cannot possibly result in procreation, even when engaged in by persons with fully functioning reproductive organs. If such acts are intrinsic evils, and the government may limit individual liberty to prevent intrinsic evils, then, it seems, the government may legitimately enact sodomy laws. Few policies seem more contrary to liberal principles than this one. How, then, does a liberal, like Dworkin, address this objection?

First, he would deny that non-procreative sex acts are, in fact, intrinsic evils. Second, he would argue that sexual freedom is a fundamental liberty, and the prevention of intrinsic evils – even genuine ones – cannot justify restriction of a fundamental liberty. This second claim is Dworkin's position on abortion laws.

The freedom to have an abortion, Dworkin holds, is a fundamental liberty. Fundamental liberties may be limited only to protect rights and important interests of individuals. Criminal prohibitions of abortion in the first trimester are not necessary to protect rights and important interests (because a first-trimester fetus has no interests and therefore no rights). Dworkin does not infer from this that there is no rational basis for abortion laws. To the contrary, he thinks that first-trimester fetuses have intrinsic value, and that the protection of things of intrinsic value can provide a good reason for government policy. His defense of a right to abortion is that fundamental liberties may be limited only in ways that are necessary to advance a compelling state interest, and the protection of things of intrinsic value does not constitute a compelling interest of this kind. So Dworkin's endorsement of the principle of legal moralism, as interpreted here, does not commit him to allowing unjustifiable restrictions of sexual and reproductive freedom.

The previous section identified some theoretical reasons to think that the prevention of harmless wrongdoing cannot justify the government in restricting individual liberty. These general theoretical considerations, however, apply with equal force to both welfare-connected non-grievance evils and to free-floating evils. If there is no valid theoretical reason to exclude welfare-connected non-grievance evils from consideration, it is hard to see a valid justification for excluding free-floating evils. On this point, then, it seems that Dworkin is right. If the government may limit individual liberty to prevent harmless wrongdoing, then it may limit individual liberty to prevent free-floating evils as well as to prevent welfare-connected non-grievance evils.

Related topics

Paternalism and Duties to Self; Paternalism and the Criminal Law; Paternalism and Well-Being; The Concept of Paternalism.

References

Dworkin, G. (2005) "Moral Paternalism," *Law and Philosophy* 24: 305–319.

Dworkin, R. (1994) *Life's Dominion*, New York: Vintage Books.

Feinberg, J. (1984) *The Moral Limits of the Criminal Law, Volume 1: Harm to Others*, New York: Oxford University Press.

———. (1988) *The Moral Limits of the Criminal Law, Volume 4: Harmless Wrongdoing*, New York: Oxford University Press.

Hart, H. L. A. (1963) *Law, Liberty, and Morality*, Stanford, CA: Stanford University Press.

Mill, J. S. (1978 [1859]) *On Liberty*, ed. E. Rapaport, Indianapolis: Hackett.

Parfit, D. (1984) *Reasons and Persons*, New York: Oxford University Press.

Rawls, J. (1971) *A Theory of Justice*, Cambridge, MA: Harvard University Press.

Raz, J. (1986) *The Morality of Freedom*, New York: Oxford University Press.

Scanlon, T. M. (1998) *What We Owe to Each Other*, Cambridge, MA: Harvard University Press.

———. (2003) *The Difficulty of Tolerance*, New York: Cambridge University Press.

Scoccia, D. (2000) "Moral Paternalism, Virtue, and Autonomy," *Australasian Journal of Philosophy* 78: 53–71.

———. (2013) "In Defense of 'Pure' Legal Moralism," *Criminal Law and Philosophy* 7: 513–530.

4

PATERNALISM BY AND TOWARDS GROUPS

Kalle Grill

In the conceptual debate on paternalism, most proposed definitions and characterizations have this approximate form: "An agent A behaves/acts paternalistically towards a person B, if (and only if). . . ." The controversy then concerns what comes after this phrase, in terms of what sort of action can be paternalistic and what is the role of, respectively, A's motives, the possible justifications for A's behavior, B's consent or lack thereof, B's general competence and current degree of voluntariness, etc. In other words, the discussion presumes, for the most part, that paternalism is something that is done by one agent towards one other agent.

There are certainly cases of one-on-one paternalism, but in many cases, either more than one person acts paternalistically, or more than one person is treated paternalistically. Examples include paternalism by governments and organizations, paternalism by physicians towards groups of patients, and paternalism by parents or teachers towards groups of children. This chapter is focused on such group cases and how they differ from one-on-one cases. By a "group" I simply mean any collection of more than one person, with no assumptions regarding possible shared interests, intentions, or other group properties. Groups of individuals in this loose sense sometimes together perform an action or are jointly affected by some action.

If groups can be understood as normative agents and patients in their own right, over and above the individuals they consist of, it seems to me these entities can be paternalists and can be paternalized. It would be interesting to explore paternalism by and towards groups in this stronger sense of collective agency and patiency, including the relationship between the individual and the group in cases where the group is a paternalist or a target of paternalism, but some members are not, or vice versa. However, I will not discuss these issues here. Nor will I discuss interesting and important issues around discrimination and prejudice towards groups – communities, classes, nations, etc. – that can lead to their being paternalized. Many issues around group paternalism are underexplored. This chapter is aimed at addressing some of these.

The chapter has three main parts. The first concerns groups as agents of paternalism in relation to a general controversy around whether paternalism resides in actions or rather in reasons for actions. The second and third parts concern groups as patients of paternalism. The second is focused on prevention of consensual interactions that are harmful to at least one party and discusses when such prevention is paternalism. The third discusses more generally how one action that affects many can have different effects on different people and what this means for our analysis of paternalism.

1 Group paternalists: different people have different reasons

Paternalism essentially involves some sort of interference or at least involvement with its target (merely thinking about someone cannot be paternalistic) and some sort of benevolent rationale (affecting someone for purely self-interested reasons is not paternalism). There is, in other words, an action component and a reason component to paternalism. I will for the most part refer to the action component as "interference" while leaving it open what this should mean exactly. On some accounts, quite mild influences are sufficient for actions to be potentially paternalistic, and so interfering on my use of the term. I will refer to the reason component as "benevolence" or the "benevolent" or simply "paternalism-making" rationale.[1]

In this section, I will first discuss a general controversy around defining paternalism and then turn to how this controversy is relevant for groups as paternalists. The controversy concerns how the two core components of paternalism – interference and benevolence – are related. Interference is a property of actions, and benevolence is a property of reasons. Their interrelation hinges on what sort of thing can be paternalistic. The standard and quite dominant view is that this is actions and, perhaps in a derived sense, policies, laws, etc., that are produced by actions (influential proponents include Dworkin 1972; Kleinig 1983; VanDeVeer 1986; Shiffrin 2000; de Marneffe 2006). On this *action-focused view*, whether or not an action is paternalistic is partly determined by its rationale. There are competing accounts of *what kind* of rationale it is that can make actions paternalistic. The most common idea is that *motives* are paternalism-makers (e.g., VanDeVeer 1986; Shiffrin 2000), but it is also rather common to point to *justification* in some sense (e.g., Dworkin 1972). Both types of accounts have some intuitive support. It seems paternalistic to force a person into rehab with the motive that this will cure her drug addiction, whether or not this is justified and whatever is the justification for it, if there is one. On the other hand, it also seems paternalistic to force a person into rehab with some other motive and to justify this by invoking the benefit to her. John Kleinig, noting this ambiguity in the concept, proposes at one point that we should avoid talking about either motives or justifications in this context and instead stick with "'having as its rationale', with its explanation/justification ambivalence" (1983: 10). This may indicate a disjunctive definition such that an action is paternalistic if it is *either* motivated *or* justified by benevolence. Peter de Marneffe has proposed, to the contrary, a hybrid definition where *both* motive and justification are required (2006: 73–74).

"Motive" and "justification" are rather vague terms. There are arguably many kinds of motives, some conscious and occurrent, some more subtle, nonoccurrent, and less accessible to the agent. Joel Feinberg distinguishes between "conscious reasons" and "deep rationales" (1986: 16). Different psychological theories will divide up the terrain differently. When it comes to justifications, it is problematic to define paternalism in terms of the normative reasons that in fact count in favor of an interference, since whether there are any such reasons depends on the moral status of the phenomena the definition is supposed to capture. Antipaternalists will typically hold that we usually have no good or valid reasons for benevolent interference. It would be strange if this should cause them to hold that such interference is nonpaternalistic (cf. Husak 2003: 392). It is also problematic to bypass justification and invoke actual outcomes, since an interference may result in a benefit quite unexpectedly, which seems insufficient for making it paternalistic (e.g., you force me into rehab, I suffer terribly and return to addiction, but 20 years later the experience inspires me to write a bestseller, which makes me very happy). Objectively expected outcomes may be more plausible, but this notion needs spelling out and I am not aware that anyone has proposed this solution. It makes more sense, therefore, to invoke what the paternalist *takes to be* the normative reasons for her action, perhaps mistakenly, which may differ from her motives. There are also, however, the normative reasons that agents officially cite, perhaps only rhetorically. Note that actual normative reasons can be indirectly relevant since if such reasons are identified, agents can then be motivated by them or believe in them or officially point to them. When I speak of "justification" in the following, however, I will only refer, jointly, to the "taking to be" and the "citing as reason" sense.

An alternative to the action-focused view is the *reason-focused view*. On this view, actions cannot be paternalistic, which may seem unintuitive. Instead, what is paternalistic is the combination of some reasons and some actions (this view is indicated by Kleinig 1983: 12 and by Husak 2003: 390; it is endorsed and defended in Grill 2007). This view is motivated by the fact that antipaternalism is typically directed at reasons for action rather than at actions themselves. J.S. Mill's Liberty Principle, the locus classicus in this context, is not directed at any class of actions, but instead rejects benevolence as an unacceptable "end," "purpose," or "warrant" for interference (1859: I.9). An advantage of the reason-focused view is that, unlike the action-focused view, it need not distinguish between different kinds of rationales. Both the action-focused and the reason-focused view must specify what actions count as interfering and what contents of reasons count as benevolent or paternalism-making. However, only the action-focused view must then go on to say, for any combination of benevolent rationales, whether this combination makes the action paternalistic. For example, assume A forces B into rehab motivated by benefits to B but seeing benefits to others as the justification, and also forces C into rehab motivated by benefits to others but seeing benefits to C as the justification. An action-focused definition of paternalism must be specified so as to determine, for each of these actions, whether or not it is paternalistic. The reason-focused view, in contrast, will imply that it is paternalistic to be motivated by B's good to force her into rehab, and paternalistic to see and/or cite C's good as a reason for forcing C into rehab, while neither action is paternalistic as such.

I have so far discussed what I have called "kinds" of rationales, divided into motives and justifications, each category arguably containing several subcategories, with normative reasons a possible third category. That one and the same action often has several different kinds of rationales is one complexity faced by anyone who aims to determine the status of an action based on its (overall) rationale. Another such complexity is that there are also often many rationales of any one kind, differing in content or substance. We may, for example, be motivated to force a person into rehab both for her own sake and for the sake of her family. On the reason-focused view, it is the motives themselves, in combination with the interference that they are motives for, that are paternalistic. Hence, one motive for an action can be paternalistic, while another is not. That there are many rationales of many kinds raises no special problems on this view. On the action-focused view, however, we must determine whether or not an action that has multiple rationales is paternalistic. Scholars have offered different proposals, including that an action is paternalistic if benevolence is its only rationale (Gray 1983: 90), its main one (Archard 1990), or even just *a* rationale, however marginal (Bullock 2015).

Actions with both paternalism-making and other rationales are sometimes called "mixed cases" and treated as a sort of exception. Authors who take this path typically go on to focus exclusively on unmixed cases. This leaves it an open question how mixed cases, i.e., most actual cases, should be treated. Kleinig has two proposals in this regard. One is that actions should be considered paternalistic *to the extent* that their rationales are paternalism-making (Kleinig 1983: 12; also endorsed by Clarke 2002: 82–83). It is not clear, however, how this should be spelled out. Most obviously, the extent to which the rationale for an action is benevolence can be measured either in absolute terms – how strong is this rationale taken in isolation? – or in relative terms – how strong is this rationale relative to other rationales for the same action? Neither of these specifications, however, takes into account how strong a rationale is required for the action to be all things considered motivated or justified. On either specification, actions can be quite paternalistic even if they are fully motivated and justified by non-benevolent rationales (as argued in Grill 2007: 446–448). Perhaps, therefore, "to the extent" should be understood in some kind of relation to how strong of a total rationale is required for action, in either motivational or justificational terms. Alternatively, as indicated by Kleinig's second proposal (Kleinig 1983: 12; also endorsed by de Marneffe 2006: 74), an action can be considered paternalistic if benevolence is required for the action to be *all things considered* motivated, or justified, or either motivated or justified, or both motivated and justified.

The complexities that arise because actions have multiple rationales (in terms of content) and multiple kinds of rationales are aggravated when paternalists are groups, because different individuals often have different rationales for their actions. Call this the *diversity problem*. Hundreds of lawmakers in parliament, for example, may all vote for the same intrusive law for slightly different reasons (e.g., benefits to those intruded upon, benefits to others, environmental benefits, loyalty to one's party, advancing one's career, etc.) Proposals for when mixed cases are paternalistic can be adapted from one-on-one cases to groups, though they may seem (even) less appealing in this context. On the strongest account, where benevolence must be the only rationale, this will supposedly apply to all members, classifying as nonpaternalistic group actions where a single member has some other rationale mixed into his set of rationales. On the weakest account, where benevolence need only be present as a rationale, it is supposedly sufficient that this rationale is present for a single member. On in-between accounts, the threshold for what counts as the "main" reason must in group cases be defined in relation to more than one person, raising new issues. For example, in determining what is the main rationale for the group, are we to go by how many members have benevolence as their main rationale, or are we to aggregate some other way, perhaps to take into consideration how strong are different members' total rationales for action (e.g., how strongly motivated they are)? On the "to the extent" account, similar issues arise concerning how to aggregate over members. I see no reason to believe that these issues are unresolvable. The point is that this is work that has not been done and that must be done before the various proposals can be applied to group cases.

The diversity problem is noted by Douglas Husak in the context of legal paternalism (Husak 2003: 389–391). Husak also notes that it is difficult to know what motivates lawmakers and that laws remain in place over time, and so the same law can be supported for different reasons as times change (2003: 391). Because of these problems, Husak considers something like the reason-focused view. However, despite many reservations, he insists in the end that laws can be paternalistic, and suggests that, because of the problems with motives, we should go by a law's "best rationale" (2003: 392).[2] However, as Husak admits, this view has the general problem, described above, that it will classify all actions and laws as nonpaternalistic if benevolence is in fact never a good normative reason for interference. This could possibly be avoided by referring not to actual reasons, but instead to facts that would be reasons in other contexts, such that the "best rationale" for an action is the rationale that would be best if, counterfactually, protection and promotion of a person's interests provided as strong normative reasons for interference as they in fact do for responding to requests for help, or some such construction.

More to the point, it is not obvious that invoking actual normative reasons will avoid the diversity problem, since each member of a group may contribute in a different way to some group action and may each have different reasons for their respective contributions. Consider the case where A, B, and C each contribute to building a wall that will protect D from foolishly balancing on the edge of a cliff. Perhaps A raises the funds, B makes the plans, and C does the actual construction. They may all correctly believe that the immediate outcome of their joint action – that there will be a wall – is a good thing because it reduces the risk of harm to D. However, their main and sufficient actual normative reasons for contributing may not be the effect on D, but rather, e.g., that they have promised to contribute, that they will themselves be morally better people for contributing, or that contributing will bring them resources – salaries, reputation – that will enable good deeds in the future (feeding their children, building greater walls). Depending on one's general normative views, it may or may not be the case that we always have some reason to contribute, when we can, to other people's welfare, e.g., by reducing risk to them. If we don't, A, B, and C may each have no benevolent reason whatsoever to build the wall. If we do, the benevolent reason they have may be relatively weak and redundant. Though having a redundant benevolent motive may possibly taint an action as paternalistic on motivational accounts, it seems extreme to hold that redundant normative reasons make actions paternalistic. Therefore, it may be true for each of A, B, or C that the rationale for their action, such as it counts in this context, is non-benevolent. It may of course also be true for each of them that their rationale is benevolent, and so the diversity problem remains unsolved.

We could set aside individual reasons and look more generally at the reasons for, e.g., there being a wall. Indeed, this approach is common when discussing laws; there are supposed to be reasons for and against laws as such, supposedly for and against their existence. Husak expresses himself this way, as does Joel Feinberg. For Feinberg, laws are paternalistic if their "implicit rationale" is benevolent. This is, Feinberg explains, a sort of general understanding regarding the function of the law, explaining why it remains in place (1986: 17). I think, however, that we do best to interpret talk of practical reasons for other things than actions as shorthand for reasons for actions, such as introducing a bill or voting for it, or financing, planning, or constructing a wall. If we allow ourselves to talk non-reductively of reasons for such things as the existence of laws and of infrastructure, it is unclear how this bears on agents and their reasons. With paternalism in particular, with its strong connection to interpersonal relationships, motivations, and attitudes, jettisoning this connection is quite radical.

On the reason-focused view, diversity does not pose as much of a problem, since it can be accommodated by counting as paternalism any combination of interfering actions and benevolent rationales for those actions, rationales of any kind (motives, justifications, etc.). When A and B together force C into rehab, for example, A's motive may be to help C, while his justification is to protect C's family (who are innocent, more vulnerable, etc.). B, on the other hand, is more motivated by concern for C's family, but his justification is to help C with his drug problem (since he suffers the most, etc.). Depending on the details of a reason-focused view, it may be that, in relation to the interfering action, A's motive but not his justification is paternalistic, while B's justification but not his motive is paternalistic. The example is a simple one, however, and for more complex cases, the reason-focused view will imply that paternalism is sprinkled over a vast net of actions and reasons, making for a very long and complicated answer to the question, "Is this paternalism?"

Before I go on to discuss paternalism *towards* groups in the next section, I will briefly note one complication that does not directly have to do with reasons for action. In the example with the building of the wall, three different agents have different reasons for action. However, they also contribute differently to the collective action of "building the wall." It is not obvious that all contributions amount to interference. Perhaps just raising the funds for the project is not interfering, or perhaps just doing the work that is ordered by someone else is not interfering. More generally, someone could perhaps contribute to an outcome that seems, on the face of it, to involve interference, without thereby interfering herself. Perhaps in such cases the interference can only be found on the level of collective agency, just as with collective actions that seem benevolent though none of the contributing agents act benevolently. If so, this is a complication for both the action-focused and the reason-focused views, since both presume an account of interfering action.

2 Paternalized groups I: preventing consensual interactions

The discussion so far has been focused on paternal*ist* groups. All issues that I have discussed can arise whether the paternal*ized* is one person or several. In this section and the next, I will focus on actions that target groups, whether or not the agent is an individual or a group. Such actions can be interfering for some and not for others, and can be benevolent towards some and not others. Therefore, not all cases I will discuss are cases of group paternalism. Instead, one of my aims is to clarify which cases are and in what sense. As part of my discussion I will refer to concrete examples, such as drug regulation, and assume that the prevented activity, such as buying and using drugs, is indeed harmful. I make this assumption only for the sake of argument, since interactions that are not harmful to anyone should rather obviously not be interfered with and so are not very interesting to discuss. I will throughout speak of "harm" and "benefit" as outcomes, with the understanding that how these outcomes are relevant to paternalism depends on one's view of the reason component – actual outcomes may provide normative reasons, believed outcomes may be motivating and invoked as justification, etc.

I emphasized above that there can be several rationales, with different content, for the same interfering action. This can be because there is more than one reason to interfere with the same person, e.g., both to promote her well-being and to respond to a request for help from her. More often, however, it is due to effects on more than one person. If we interfere with two persons who are doing something that will harm one of them, we may be interfering with one to prevent harm to self and with the other to prevent harm to others. Sometimes, as Gerald Dworkin explains, "in trying to protect the welfare of a class of persons we find that the only way to do so will involve restricting the freedom of other persons besides those who are benefitted" (1972: 68). Dworkin labels such cases "impure paternalism," where the impurity is the interference with some other party, in addition to the beneficiary.[3]

The standard case of impure paternalism is interference with *consensual interactions*, and in particular with such interactions as are harmful to one of the parties and not the other. Examples include such extraordinary interactions as consenting to being abused or killed, and selling oneself into slavery. More practically relevant examples include selling sexual services and buying unhealthy consumer goods, such as recreational drugs. Interference will benefit the party that is harmed in its absence. There is presumably no benefit to the other party, e.g., the seller of drugs or the buyer of sex. Therefore, assuming the beneficiary is a single individual, this is not a group case. Interference with other parties is just a means, perhaps a necessary means, to producing the benefit. These parties are not themselves thereby paternalized. It is an interesting question whether we have any reason to regret, for the sake of the harming party, interference with actions that harm consenting others – such as selling them drugs. This issue, however, is independent of issues to do with paternalism.

In many typical cases of impure paternalism, such as prohibition of the sale of unhealthy products, the interference is most obvious with the nonbeneficiary, i.e., the seller (who may be prosecuted and punished). However, what makes the case one of paternalism is that there is also interference with the buyer, who is prevented from acquiring the desired good or service. As Feinberg argues, if others are prevented from selling me what I want to buy, or aiding me in my pursuits, then *I* am interfered with and *my* freedom is limited (Feinberg 1986: 9). Or, in Mill's words:

> there are questions relating to interference with trade, which are essentially questions of liberty [. . .] where the object of the interference is to make it impossible or difficult to obtain a particular commodity. These interferences are objectionable, not as infringements on the liberty of the producer or seller, but on that of the buyer.
>
> (1859: V.4)

Some authors describe as impure or "indirect" paternalism cases where they claim there is no interference with the beneficiary/buyer, but only with the nonbeneficiary/seller (Pope 2003: 687–688; Le Grand and New 2015: 37). If this was a form of paternalism, it would perhaps be a special kind of group case, involving two different members in quite different roles – one is interfered with and another benefits. However, it is unclear what the paternalism would consist of in such cases. Interference with one person in order to benefit another person is the standard contrast class to paternalism (preventing assault has this structure). If there is indeed no interference with the buyer, then, I propose, there is no paternalism, but only interference to prevent harm to others (Bayles 1974 and Hansson 2005 argue that many prohibitions of consensual interactions should be understood in this way).

Here, I should make an exception to my loose use of "interference" and acknowledge that some authors offer very wide understandings of interference, or, in other words, reject the assumption that paternalism is interfering. For example, Cass Sunstein and Richard Thaler claim that it is sufficient that there is an attempt to "influence choices" (Sunstein and Thaler 2003: 1162), and Danny Scoccia claims that any benevolent influence on another via "nonrational means" is paternalistic (Scoccia 2013: 76, as well as his chapter in this volume). Such actions may not warrant the label "interference." In the present context, however, what is important is that however we specify the action component of paternalism, actions belonging to this type must be directed *at the beneficiary* for there to be paternalism. It is not sufficient that some other person is the target of such an action. If we require coercion, then there must be coercion towards the beneficiary. If we require only influence on choice, then there must be influence on the choice of the beneficiary.

Some harmful consensual interactions are symmetrical in the sense that both or all parties harm each other and also consent to being harmed. Boxing is the traditional example, mixed martial arts competitions a more recent one. A street fight with willing participants is a more clear-cut example, without the commercial and institutional context. Benevolently interfering with consensual fighting may seem a clear example of paternalism, since there is both interference with and benefit to all members of the group. However, the details are somewhat intricate. Suppose that in a group consisting of A, B, and C, A wants to punch B, B wants to punch C, and C wants to punch A. If D prevents any punching from occurring by separating A, B, and C, this seems clearly to interfere with each of them by stopping them from punching the person they want to punch while also benefitting each of them by protecting them from being punched. However, such prevention does not seem paternalistic because no person benefits from the interference *with her* (assuming they are not harmed by delivering a punch). They all benefit from the interference with other people, from which they are protected.

Now consider a street fight with two willing participants, perhaps supporters of opposing sports clubs who take pride in fighting "for their team." We could claim that keeping these brawlers apart prevents A from fighting B, for B's sake, and prevents B from fighting A, for A's sake, hence benefitting each party only by interference with the other. However, this seems less plausible than in the three-party case, since both A and B want the same thing to happen – for there to be a fight between the two of them. They want this, let us assume, because they consider the risk of harm a fair price to pay for the thrill and the social recognition they get from fighting. Like the willing buyer of drugs or seller of sex, each brawler *invites* another person to (potentially) harm him. Like in these other cases, therefore, the freedom of each brawler is limited by preventing him from engaging in a harmful consensual interaction.

I have argued that interference in the two-party brawling case is paternalistic, while interference in the three-party punching case is not. Let me expand on the relevant difference. It is not that the two brawlers would object to interference. The three punchers too may object, because they value the opportunity to punch the person they want to punch more than they value the protection from being punched. The relevant difference is not in the numbers either. Two people may both want the other to suffer a punch more than they want to be protected from being punched themselves, and so be opposed to interference. I believe interference in these variations on the punching case would not be paternalism. The important distinction is, I believe, between two sorts of cases: first, those where a person wants some harm or risk of harm for its own sake, or where this harm or risk is integral to what he wants, as in the brawling case – part of what the brawlers take pride in is exactly to risk harm "for their team" – and, second, those where a person is prepared to accept some harm or risk of harm in order to get something that he wants more than to avoid this harm or risk, such as to inflict harm on someone else. In the former cases, prevention is interference with the person who wants the harm or risk. In the latter case, prevention is interference only with the source of the harm or risk and not with the person who is prepared to accept it. This distinction is relevant only when the source of harm or risk is another agent (and probably only when that agent causes harm or risk intentionally). When the source of harm is the person herself, we can (normally) protect her only by interfering with *her*, which is paternalism.[4]

Special considerations apply when a person is opposed to being protected by others, even if she does not want the harm or risk that may result from being unprotected. Scoccia (2013: 81) notes that some people are "committed to an extreme ideal of self-reliance" and therefore oppose benevolent interference even with their own substantially involuntary action. People may for similar reasons oppose benevolent interference even with their own attackers – they want to "fight their own battles." The state of being unprotected, or independent, is one to which risk is an integral aspect – it is just this risk that one does not want others to remove. I therefore propose that interference with committedly self-reliant people is paternalism.

Similar considerations apply when interferences with consensual interactions are coordinated through systems, institutions, or laws. A law that prevents people from harming themselves, or from soliciting the aid of others in harming themselves, involves paternalism (i.e., implementing it, or doing so for certain reasons, may be paternalistic). As for a law that prevents people from harming others without their consent, but that is universally opposed, whether or not it involves paternalism depends on whether the opposition is based on a desire for self-reliance or on a desire for the opportunity to harm others. In other words, it depends on whether people want to be unprotected themselves or whether they want others to be unprotected. In many cases, of course, members of a group will have different aims, opinions, and ideals. This and other differences between individuals is the topic of the next section.

3 Paternalized groups II: different effects on different people

In this section, I will first note and illustrate how one action can affect different people differently in ways relevant to paternalism. I will then move on to discuss how this may influence our classification of various cases as paternalistic. The relevant differences concern, first, the two core components of paternalism already identified: who is interfered with and who benefits. A third relevant difference is also a third component of paternalism: the *will* component. Benevolent interferences are typically considered paternalistic only when they are against the will of the target, i.e., when unwelcome or not consented to. This component is arguably not essential since it can be integrated into the action component such that an action counts as interfering only if it is not consented to. However, it is typically treated as a separate component, and I started to treat it as such towards the end of the previous section, in considering whether interference with brawling and punching would be objected to.

For there to be paternalism, then, an action must be unwelcome, interfering, and benevolent. However, I argued in the previous section that this is not sufficient, because sometimes the benevolence towards one person is not connected to the interference with her, as when the three punchers are all prevented from harming each other. What is relevant is not whether the action is benevolent, but whether the interference is benevolent. It can also be questioned whether an unwelcome benevolent interference with a person is paternalistic towards her if her objection is not connected to the interference with her (e.g., A stops B and C from daring each other into jumping off a cliff; both B and C object to A's interference; B does not want to jump and would not object but for the fact that she really wants C to jump). There are, potentially, interconnections between the three components, and these connections can be different for different people affected by the same action. This indicates that we should allow, as I started to do in the previous section, that one and the same action can involve paternalism towards some and not towards others. This indicates important modifications of both the action-focused and the reason-focused view, as both of these otherwise deal only in actions and not their diverse effects on different people. On the reason-focused view, the modification can be integrated by holding that what is paternalistic is not combinations of actions and their rationales, but rather combinations of rationales and (unwelcome) interferences with particular people. This makes for an even more complex but arguably more accurate analysis of paternalism in group cases. On the action-focused view, similarly, the modification could be taken to imply that what is paternalistic is not in fact actions but rather interferences with particular people.

Feinberg (1986: 20) discusses this modification and dismisses it as "an unnecessary relativizing of the concept" of paternalism. However, his discussion of group cases is quite limited, as I will soon explain. Given the mentioned interconnections and also given the general individualism inherent in the liberal tradition, it makes much sense to base analysis of paternalism towards groups on the effects on individual members.

I will now survey some examples of actions towards groups that are paternalistic towards some members but where one of the three components of paternalism is missing in relation to other members. First, missing benefits: prevention of consensual interactions that are harmful to only one party is one sort of case; another is the subjection of a group to a measure that protects only those that are vulnerable to some harm. An example of the latter is removing the sleeping pills or cigarettes from a shared home, benefitting the cohabitant who is suicidal or a habitual smoker, but only harming the cohabitant who is neither of these things but sometimes has trouble sleeping or enjoys a single cigarette. Just as for prevention of consensual interactions, only the interference with the beneficiary is paternalistic. The interference with nonbeneficiaries is a sort of collateral damage.[5] Unlike some cases of preventing consensual interactions, those that suffer this damage are not harming anyone or doing anything morally problematic, and so the interference with them must be considered a negative.

Second, missing interference: speed limits supposedly benefit drivers by reducing the risk of accident for them, but also of course reduce risks for cyclists and pedestrians. Even if pedestrians (who never drive) are included among the intended beneficiaries, and even if they are for some reason opposed to the regulation, it is not paternalistic towards them because it does not interfere with them.

Third, missing objection: many people appreciate that product safety legislation prevents them from buying unsafe machinery or consumer goods. Others oppose such restriction of their freedom, on either pragmatic or principled grounds. Libertarian chainsaw buyers are paternalized by the state, while most other customers are not.

There is very little discussion in the literature of how, on the dominant action-focused view, we should classify the surveyed examples. However, a number of influential contributions in the 1980s converge on a view we may call the *willing majority view*: for actions that interfere with and benefit all members of a group, if a majority consents and the action is motivated by its benefits to them, then the action is not paternalistic, though it may be unfair to nonconsenters (Arneson 1980: 471–472; Dworkin 1983: 110; Feinberg 1986: 20). The view is motivated by discussion of such cases as the prohibition on dueling and the fluoridation of drinking water, where a large majority favors the policy and it is implemented for their sakes. The willing majority view can be specified to different motivational accounts (one motive, main motive, only motive, implicit rationale, etc.) and can be transformed into a justificational account by substituting, e.g., "taken to be justified" or "claimed to be justified" for "motivated." The view still draws scholarly support (e.g., Le Grand and New 2015: 21–22).

The willing majority view says that, in some cases where the rationale for an action is to benefit a majority of those affected and where this majority is not paternalized, the action is not paternalistic. However, the view is restricted to cases where all those affected are both interfered with and benefitted and where the reason the majority is not paternalized is that they consent. This is the third of the cases just surveyed – missing objection. The spirit of the view, however, indicates that its proponents would not mind a generalization that incudes also missing benefit and missing interference.

The willing majority view categorizes actions based on their different effects on different people. Given the idea that paternalism actually resides in interferences with particular people and not in actions, the view may seem superfluous. If we know that an action is an unwelcome benevolent interference with A and with B but not with C, it is not clear what additional information is conveyed by saying that the action itself is or is not paternalistic. Those nevertheless committed to pinning the predicate "paternalism" on actions have some work to do. Even the generalized willing majority view is applicable only to actions that are nonpaternalistic towards a majority. A simple addition would be to categorize as *paternalistic* those actions that are paternalistic towards a majority. However, the 50% cut-off point seems arbitrary. It might also seem that factors other than sheer numbers could be relevant, such as the size of the benefits involved. Suppose that the prohibition of some rare and dangerous drug will marginally reduce the already very low risk that the majority ever confront this drug. The prohibition is introduced and the majority welcomes it for this reason. However, the prohibition will also drastically reduce the high risk of drug abuse and ensuing harm to some minority, who are opposed to prohibition. This prohibition will count as nonpaternalistic on plausible specifications of the willing majority view, which may seem counter-intuitive (on the importance of relative benefits to group consent, see Grill 2009: 151–153).

A further problem with the willing majority view is that it disregards the reasons for *why* people consent to interference with a group to which they belong, as briefly indicated above. Proponents tend to assume that the consenters consent out of self-interest, but they may instead be altruistically motivated, consenting for the sake of the nonconsenters, whom they see as failing to act in their own best interest. This raises the question whether altruistic consent renders interference nonpaternalistic towards consenters. In addition, given that some people's altruistic consent makes blanket interference more likely, these people seem to be using their consent to indirectly paternalize nonconsenters. If the consenters are in the majority, the willing majority view would categorize the interfering action as nonpaternalistic. This seems very questionable. We could modify the view to require a majority of *self-interested* consenters, but there are many additional issues to consider, such as how to count members who consent partly for self-interested and partly for altruistic reasons, and how to count liberally minded altruists who do *not* consent, in order to protect consenters whom they see as failing to give proper priority to their own liberty and independence (on these and other issues around group consent, see Grill 2009).

Before I conclude, let me briefly mention a fourth potentially relevant difference in how people are affected by the same action: who, among those interfered with, is acting voluntarily or competently. Though it has been convincingly challenged (Hanna 2011), it is a quite dominant position that unwelcome benevolent interference is not morally problematic, or much less so, when and because it is directed at choices or actions that are below some threshold of voluntariness. Following Feinberg, such interference is often called "soft paternalism." If some of those interfered with are below the threshold and some not, this situation is analogous to that when only some group members are interfered with or benefitted or objecting, though now the difference is not between paternalism and nonpaternalism, but between hard paternalism and soft paternalism. We could then either be content to determine for each paternalized member of the group whether the paternalism towards her is hard or soft, or we could identify some rule for whether or not the interfering action is soft or hard paternalism, which should be suitably sensitive to differences among group members.

4 Conclusion

Though examples of paternalism in the conceptual and normative debate often include groups both as paternalists and as paternalized, the interesting and difficult issues that groups raise are seldom explicitly discussed or analyzed. What little has been said in the literature on paternalism by and towards groups is quite cursory. The topic deserves more thorough treatment. I have focused in this chapter on some issues that arise from the mere fact that more than one person is either paternalizing or being paternalized, setting to one side issues to do with collective agency and patiency in any stronger sense.

In Section 1, I presented the action-focused and reason-focused views on paternalism and explained how groups as paternalists complicate both views but provide greater challenges for the action-focused view, on which actions must somehow be categorized as either paternalistic or nonpaternalistic based on the often rich and diverse total rationale for group interference.

In Section 2, I argued that standard cases of impure paternalism are not paternalism towards groups because they only benefit (or are only believed to benefit) one party. I also argued that if they do not interfere with the beneficiary, they are not paternalism at all. I went on to consider interferences with people who mutually harm each other and argued that these are paternalistic only if the affected people benefit via interference with themselves, not others. I proposed that interference with two or more people bent on harming each other can be paternalism if they seek to be harmed or put themselves at risk of harm, or if they seek an activity where such harm or risk is an integral aspect. In other cases, interference is not paternalistic, even if objections to interference may have normative significance in other ways.

In Section 3, I argued that, because of interrelations between the three components of paternalism – interference, benevolence, and will – we should understand paternalism in terms of unwelcome benevolent interferences with particular people, rather than in terms of actions, which may paternalize some and not others (when an action only affects one person, we need not distinguish between interference and action). This shift from action to interference holds for both the action-focused and the reason-focused view. There may be a connection between this shift and the controversy between action-focus and reason-focus in that, if we give up trying to pin the predicate "paternalistic" on actions, we may as well accept the reason-focused view, modified to deal in reasons for interferences rather than reasons for action.[6]

Related topics

Hard and Soft Paternalism; Libertarian Paternalism, Nudging, and Public Policy; Paternalism and the Criminal Law; Perfectionism and Paternalism; The Concept of Paternalism.

Notes

1 Shiffrin (2000) is an exception to the near consensus that paternalism essentially involves the protection or promotion of the interests of the person interfered with. Shiffrin clearly states that the paternalism-making rationale can be the improvement of things under a person's control that are not, strictly speaking, her interests. However, she seems to presume that the rationale must be benevolent in some sense. Presumably, she would not categorize as paternalistic interference that is solely motivated by self-interest, or by sheer malice. I will throughout use standard examples that do not cohere with Shiffrin's view, but which can be reformulated so that they do.

2 The difficulty with group motives has been noted more often in the context of liberal neutrality. Neutrality can fruitfully be understood in terms of a constraint on what reasons may be invoked for political decisions (e.g., Larmore 1987: 44; Wall 1998: chap. 2; de Marneffe 2010: chap. 5, esp. 134). Antipaternalism can similarly be understood in terms of what reasons may be invoked for interference (Grill 2015). So understood, neither of these liberal "–isms" is dependent on classifying actions or laws as non-neutral or paternalistic.

3 Feinberg complains that "impure" sounds like "a watered down sort" of paternalism and proposes to use "indirect" instead (Feinberg 1986: 9). Both terms are used, sometimes with the same meaning and sometimes with slightly different meanings.

4 This analysis of consensual fighting partly contradicts that in my (2007: 453–455). A note on terminology: In my (2007) I speak of different effects of actions in the abstract, noting that effects can be individuated by what person is affected but leaving it open that other factors may also be relevant. This framework may sometimes be useful, but here, for ease of presentation, I bypass talk of effects in general and speak only of interference with and paternalism towards different people.

5 If interfering with nonbeneficiaries is "the only way" to attain the benefit to the beneficiaries, then these cases are impure paternalism on Dworkin's characterization. However, this concept is usually only associated with interference with consensual interactions.

6 Thanks to Jason Hanna and Lars Samuelsson for very helpful comments on more than one draft.

References

Archard, D. (1990) "Paternalism Defined," *Analysis* 50: 36–42.
Arneson, R. (1980) "Mill Versus Paternalism," *Ethics* 90: 470–489.
Bayles, M. D. (1974) "Criminal Paternalism," in J. R. Pennock and J. W. Chapman (eds.) *The Limits of Law – Nomos XV*, New York: Lieber-Atherton, pp. 174–188.
Bullock, E. C. (2015) "A Normatively Neutral Definition of Paternalism," *The Philosophical Quarterly* 65: 1–21.
Clarke, S. (2002) "A Definition of Paternalism," *Critical Review of International Social and Political Philosophy* 5: 81–91.
de Marneffe, P. (2006) "Avoiding Paternalism," *Philosophy & Public Affairs* 34: 68–94.
———. (2010) *Liberalism and Prostitution*, New York: Oxford University Press.
Dworkin, G. (1972) "Paternalism," *The Monist* 56: 64–84.

———. (1983) "Paternalism: Some Second Thoughts," in R. Sartorius (ed.) *Paternalism*, Minneapolis: University of Minnesota Press, pp. 105–111.
Feinberg, J. (1986) *The Moral Limits of the Criminal Law, Volume 3: Harm to Self*, New York: Oxford University Press.
Gert, B. and Culver, C. M. (1976) "Paternalistic Behavior," *Philosophy & Public Affairs* 6: 45–57.
Gray, J. (1983) *Mill on Liberty: A Defence*, London: Routledge.
Grill, K. (2007) "The Normative Core of Paternalism," *Res Publica* 13: 441–458.
———. (2009) "Liberalism, Altruism and Group Consent," *Public Health Ethics* 2: 146–157.
———. (2015) "Antipaternalism as a Filter on Reasons," in T. Schramme (ed.) *New Perspectives on Paternalism and Health Care*, Cham, Switzerland: Springer, pp. 47–66.
Hanna, J. (2011) "Paternalism and Impairment," *Social Theory and Practice* 37: 434–460.
Hansson, S. O. (2005) "Extended Antipaternalism," *Journal of Medical Ethics* 31: 97–100.
Husak, D. N. (2003) "Legal Paternalism," in H. LaFollette (ed.) *Oxford Handbook of Practical Ethics*, New York: Oxford University Press, pp. 387–412.
Kleinig, J. (1983) *Paternalism*, Manchester: Manchester University Press.
Le Grand, J. and New, B. (2015) *Government Paternalism: Nanny State or Helpful Friend?* Princeton, NJ: Princeton University Press.
Larmore, C. E. (1987) *Patterns of Moral Complexity*, Cambridge: Cambridge University Press.
Mill, J. S. (1859) *On Liberty*, London: J.W. Parker and Son.
Pope, T. M. (2003) "Counting the Dragon's Teeth and Claws: The Definition of Hard Paternalism," *Georgia State University Law Review* 20: 659–722.
Scoccia, D. (2013) "The Right to Autonomy and the Justification of Hard Paternalism," in C. Coons and M. Weber (eds.) *Paternalism: Theory and Practice*, New York: Cambridge University Press, pp. 74–92.
Shiffrin, S. (2000) "Paternalism, Unconscionability Doctrine, and Accommodation," *Philosophy & Public Affairs* 29: 205–250.
Sunstein, C. R. and Thaler, R. H. (2003) "Libertarian Paternalism Is Not an Oxymoron," *The University of Chicago Law Review* 70: 1159–1202.
VanDeVeer, D. (1986) *Paternalistic Intervention: The Moral Bounds of Benevolence*, Princeton, NJ: Princeton University Press.
Wall, S. (1998) *Liberalism, Perfectionism and Restraint*, Cambridge: Cambridge University Press.

5
SELF-PATERNALISM

Chrisoula Andreou

1 Introduction

Paradigmatic cases of paternalism involve an agent's being constrained from acting on her preferences where this is supposed to be for her own good. Though the term has negative connotations, some instances of paternalism seem permissible. As the term suggests, paternalism includes cases in which a mature guardian constrains the behavior of a child that he is responsible for in order to protect the child and promote its well-being. The justification for this constraint is supposed to be that the guardian is more rational and/or savvy about what is in the child's long-term interest. Such paternalism is widely accepted as appropriate. Certain other forms of paternalism are widely rejected as inappropriate, including, for example, cases of paternalism in which one adult constrains another based on the implausible assumption that he knows better than she does. Harder cases include those in which one adult has good reason to think she knows better than another, but in which constraining the other is presumptuous because the other is mature enough to take responsibility for himself and his decisions and there is something to be said for his acting autonomously.

Most discussions of paternalism focus on cases in which an agent is treated paternalistically by *another* agent or entity (such as a political or social institution). And, indeed, the idea of *self-paternalism* might seem cryptic.[1] Can there, for example, really be cases in which I know better than myself or am more rational than myself, and so I have to get in my way for my own good? The aim of this chapter is to discuss the nature and possibility of self-paternalism by reviewing several types of cases in which an agent has reason to take measures that will constrain or put pressure on him to follow a certain course of action because he has reason not to trust himself to make good choices.

2 Temptation

While it is paradoxical to claim that I am more rational than myself, there is nothing paradoxical about the claim that I anticipate being less rational at some point in the future than I am right now. Indeed, if I expect to face a temptation, and have a poor track record with respect to overcoming temptation, it might be foolish *not* to anticipate being less rational (or at least less well-positioned for rational action) at some point in the future than I am right now. And, in such cases, self-paternalism seems like it might be perfectly in order. I thus turn to a discussion of some familiar sorts of cases involving temptation.[2]

Akrasia

Cases of akrasia are typically defined as cases in which an agent voluntarily acts against her better judgment. There is some controversy about whether there are any genuine cases of akrasia. It is sometimes argued that it is impossible to voluntarily act against one's better judgment, and that cases that appear to be cases of akrasia are cases in which something else is going on. Perhaps the agent is compelled to act against her better judgment by some irresistible desire; or perhaps the agent's judgment is temporarily altered at the time of action, and so, although her action may conflict with her prior and future judgment, it does not conflict with her judgment when she is actually acting.

This is not the place to defend the possibility of akrasia. I will here simply note that, if there are any genuine cases of akrasia, and one can sometimes anticipate a potential case of akrasia, one might quite reasonably aim to avert it by taking steps to make the "offending" act unavailable. If, for example, I expect my later self to surf the net instead of writing, I might sign up for a service that disables my internet connection for a certain chunk of time.

This seems like a good candidate for a case of self-paternalism. But there are some interesting issues to think about, including the following: Is the self that is accountable for the constraint the same as the self that is being constrained? Does the answer to the preceding question matter in relation to the permissibility of the paternalistic steps? What does this case suggest in terms of whether there is any significant difference between self-paternalism and interpersonal paternalism?

Clearly, the human being that is accountable for the constraint in the case of interest is the same as the one that is being constrained. But is there really just one self here (or is thinking in terms of one self really the most helpful way of understanding this case)? The fact that the same human being is involved doesn't seem to settle this question. We often say things like "I was a different person then" or "I wasn't myself then" even when we believe the same human being has persisted throughout. Perhaps we should take talk about my earlier self constraining my later self as suggesting that, at least in relation to what matters normatively, there are two selves involved here, and that this case, and indeed any case of paternalism, involves two distinct selves, one constraining and one constrained. On this take, there is no single self to which the "self" in self-paternalism refers.[3] Alternatively, it might be suggested that we should understand an earlier and later self as the same self so long they are unified by certain core features. If, for example, we think of a self as defined by its evaluative judgments, then the case of interest is one in which the same self persists throughout.[4] On this take, the self that is accountable for the constraint is the same as the self that is being constrained. It is just that the initiation of the constraint and the experience of constraint occur at different times.

The question of whether it matters if the constraining self and the constrained self are the same is quite interesting. On the one hand, it might be supposed that it matters a great deal, since, if it is the same self, then consent on the part of the person being constrained has been obtained. On the other hand, it seems like the really interesting question is whether it is acceptable for the constraint to remain in place if the agent's consent is retracted. And here it is not clear that there is a disanalogy between the intrapersonal case and the interpersonal case, assuming the constraint is there for the agent's own good. For in both cases it might be suggested that constraining a mature agent for his own good requires (at least morally speaking) his *current* consent (or at least adequate evidence that his consent has not been retracted).

This brings us to the debate about the permissibility of "Ulysses contracts." In the classic Homeric tale, Ulysses, anticipating that, if he does not take precautions, he will be lured into danger by the sweet singing of the Sirens, commands his shipmates to tie him to the mast of the ship (and plugs their ears with wax), so that he may enjoy the Sirens' singing without being able to change his course. Accordingly, in a Ulysses contract, one seeks to be bound so that one's future options are constrained because one anticipates that, without the constraint, one will be lured into a self-destructive course. A crucial question, of course, is whether it is permissible for one to seek to constrain oneself in this way, expecting that one's consent will later be retracted. And given that, as in many other potential cases of self-paternalism, the assistance of others is enlisted and so the case is not purely intrapersonal, there is also the (perhaps inextricably) related question of whether it is permissible for one's "cooperators" to refuse to unbind one if, as anticipated, one's consent is retracted. Importantly, Ulysses contracts are useful precisely when the agent's consent will be retracted *and* the agent, at the time of retraction, seems mentally competent. Were the agent, at the time of retraction, clearly mentally incompetent, interference would be justified regardless of the contract. But then the circumstances that make Ulysses contracts useful also make them highly controversial. Interference with a mentally competent adult for his own sake seems like a violation of the agent's autonomy.

It might be suggested that, at least in cases of akrasia, an agent that constrains herself does not compromise her future autonomy because autonomy is compatible with constraint, and indeed requires that one is constrained by the core features that make one who one is, which include one's evaluative judgments. Still, insofar as the akratic agent is not compelled to act against her better judgment, but does so voluntarily, there seems to be some sense in which the akratic agent is governing herself and so some sense in which interfering with her behavior interferes with her self-governance. So there remains plenty of room for debate.

It is worth noting that, although Ulysses' case is fictional, interest in Ulysses contracts is far from "purely academic." Allowing for legally binding Ulysses contracts unleashes an extremely powerful tool – one that has the potential to help many who have trouble overcoming temptation to shape their lives in a way that fits with their evaluative judgments. Consider, for example, someone with bipolar disorder who dreads his tendency to stop taking his medication when he enters a manic phase. Feeling well and under control, his medications and their side effects may seem like a dreadful burden and he may be tempted to stay off them for a while. The result is often a spiraling into a self-destructive pattern of behavior. Having gone through this multiple times, he may wish to bind himself to being admitted for psychiatric care if his family notices he is failing to take his medication, even if he then resists and is still mentally competent enough for his resistance to be such that, were it not for a Ulysses contract, detaining him without his current consent would be out of the question. As indicated above, there is still room for the view that a Ulysses contract cannot make it permissible to detain him. He, however, might see that tool as his only hope of evading a vicious cycle, and, with some plausibility, suggest that it is the withholding of the tool that is paternalistic.[5]

Distortion

Cases of akrasia are supposed to be cases in which one acts contrary to one's better judgment. But not all cases of giving in to temptation take this form. In some cases, temptation prompts distortion, and giving in to temptation by, say, X-ing involves revising one's judgment about what would be best and acting accordingly.[6] In such cases, being forced not to give in to temptation occurs not just without one's current consent, but also without the concurrence of one's current evaluative judgments. Interestingly, where one's current evaluative judgments favor the tempting act (in this case X-ing), even voluntarily refraining from X-ing seems problematic, since it seems to qualify as akratic.[7]

Now consider the following case:[8] One is getting ready for dinner and pours oneself a drink intending to stop after one so that one can get some writing done after dinner. But one is worried that, after one drink, one will favor enjoying another, not just as an appealing alternative, but as best, all-things considered. To avoid this temptation, one locks the remaining liquor in a cabinet, puts the key in a self-addressed stamped envelope, walks to the mailbox down the street, and drops the envelope in. One then heads back home, resumes one's dinner prep and drinking, and, predictably, soon regrets having locked up the liquor cabinet.

In this case, the earlier self, who is doing the constraining, does not share the same evaluative judgments as the later self, who is being constrained, and so, again, it might be wondered whether this potential case of self-paternalism is really a case in which the self that is doing the constraining is the same as the self that is being constrained. The idea that it is might draw on the idea that a self can endure over a certain time span even if some of its evaluative judgments fluctuate during that time span, particularly if the fluctuation is due not to a deep change of values, but due to a change in view concerning the implications of the different options.

Suppose, for example, that in the preceding case the reason that one regrets having locked up the liquor cabinet after one finishes one's first drink is that, relaxed by the experience, one comes to think that one was being neurotic and that just one more drink would not interfere with one's getting some writing done tonight – to the contrary, it would, one now believes, really get one's creative juices flowing. Here the change in one's evaluation of the prospect of having one more drink does not reflect a change in one's deep values, and so there is, perhaps, no reason to think that one is no longer the same self as one was when one poured one's first drink.

Presumably, self-paternalism is not even potentially appropriate in this case if one's revised take on the situation is correct and one really was just being neurotic. If self-paternalism is warranted, it is presumably because one's later judgment is unrealistic and was predictably clouded by the temptation of having another drink. In this case, one's prior self really does know better than one's later self, and the prior self's paternalism seems less problematic than in cases where the later self has an accurate view of the implications of her choices. Here it can at least plausibly be maintained that, although the later self does not consent, she would if she had all the facts straight.

Consider next an alternative way of fleshing out the dinner/drink(s) case. Suppose the reason one regrets having locked up the liquor cabinet after one finishes one's first drink is that, relaxed by the experience, one comes to think that one is too focused on work and that one needs to make more time for the simple pleasures of life, including experiences such as having a few drinks with dinner without worrying about having a productive evening. Suppose this evaluation involves a significant shift in one's values. What's particularly tricky in this version of the case is that it is not at all clear that there is a correct stance. Perhaps both one's earlier stance and one's later stance are equally permissible. If so, then it is not clear that the earlier self's paternalism really is for the agent's own good. For paternalism to be even potentially acceptable in this version of the case, there must be some reason to believe that the later self's stance is distorted, corrupted, or in some other way less adequate than the earlier self's stance.[9]

It might be suggested that the real test as to the adequacy of the later self's stance is whether it endures – if it does not, and the agent reverts to her original stance after the experience of temptation, then the shift can be explained in terms of the experience of temptation and self-paternalism is warranted. But an evaluative shift's being temporary does not seem necessary or sufficient for its being misguided. It is not necessary since, insofar as an agent has an interest in rationalizing her behavior and avoiding cognitive dissonance, even a misguided shift can be sustained after the experience of temptation.[10] In some cases, regret is not forthcoming not because it is unwarranted but because the agent does not want to admit he has failed; it may be more appealing to revise one's values instead. An evaluative shift's being temporary is also not sufficient for its being misguided since it may be that the so-called "experience of temptation" is not distorting, but revealing. One might, for example, underestimate the badness of pain except insofar as one is experiencing it, and value painkillers appropriately only during the short times when one is in extreme pain, with one's enduring prospective and retrospective evaluations being the misguided ones.[11]

Procrastination

In cases where there is some prospect of akrasia or distortion, one may find that one cannot trust one's motivations to remain as they are. One's worry may be quite different in cases where there is some prospect of procrastination in relation to an important goal that can only be achieved via a series of individually trivial contributions each of which requires some sacrifice. In such cases, the concern is often that the incentive structure will continually motivate one to delay goal-directed action because any single instance of goal-directed action is trivial in relation to whether or not one will ultimately achieve one's goal.[12] When this is the case, constraint may seem necessary precisely because one has reason to believe that one cannot trust one's future self to *deviate* from one's current and seemingly rational motivational aversion to engaging in a burdensome action that is too trivial to make or break one's chances of achieving one's goal. Such deviation is, however, essential to achieving one's goal if the goal can only be achieved via a series of individually trivial contributions each of which requires some sacrifice.

Consider, for example, the case in which one has the goal of getting in shape. Other things equal, no particular exercise session will make or break one's chances of getting in shape, and yet one cannot get in shape without ever exercising. So long as one finds exercising aversive, one may understandably prefer delaying this particular exercising session. And such a delay need not be problematic, so long as the motivation to delay is not *consistently* present and effective. It is only if one anticipates the motivation being consistently present and effective that contemplating a system that will force one to exercise regularly may seem in order. (Perhaps one can leave oneself with no mode of transportation except one's legs.) If, however, constraint is justified, the justification will not be that one's earlier self is somehow more rational (or less tempted to deviate from rationality) than one's later self. To the contrary, the problem occurs because one expects one's state of mind to stay the same. If, therefore, the prior self's constraining action is a form of self-paternalism, it differs interestingly from classic cases. What it has in common with such cases is that the constraint is initiated in light of the agent's expectation that, in the absence of constraint, she will exercise her agency in a way that will fail to serve her own good. Perhaps this is enough to make the case one of self-paternalism and it should not be assumed that, in cases of temptation where self-paternalism might be called for, one's earlier self is somehow more rational (or less tempted to deviate from rationality) than one's later self.[13]

3 Incentives and implementation intentions

For reasons that are familiar from interpersonal cases, it is often inconvenient or infeasible to ensure that one will be forced to do or refrain from doing certain things. In some cases, it is more advisable to alter the incentive structure so that one's later self is willing to act in the way that is, at least prospectively, seen as desirable. Just as a parent can paternalistically change her child's incentive structure by arranging things so that the child can have a lollipop if she eats her vegetables, an agent may be able to self-paternalistically change her future self's incentive structure by setting up a reliable reward system. For instance, she might have a friend hold onto discretionary funds that are returned to her if she conforms to certain monitorable requirements.

In cases where the temptation is not too intense, the agent may need nothing more than a little "nudge" in the right direction. In such cases, defaults can make a big difference. Though a peckish child might whine for a cookie if he sees a box on the counter, he might not think to ask for one if he encounters a snack tray loaded with fruit, cheese, and crackers instead. In the intrapersonal case, powerful defaults can often be created via implementation intentions. In forming an implementation intention, one settles on the "when, where, and how" of goal-directed action (Brandstätter, Lengfelder, and Gollwitzer 2001). For example, if one has the goal of getting a head start on a paper, one might settle on drafting an outline of the paper in one's office right after one is done teaching and before one breaks for the day. This creates a default that cues one to appropriate action. This does not preclude the agent from reconsidering later; but, oftentimes, action on implementation intentions is automatic and reconsideration does not occur.

4 Conclusion

Though the idea of self-paternalism may seem paradoxical, there are cases in which an agent has reason to interfere with her own behavior because she has reason not to trust herself to make good choices. The cases in question include cases in which a temporally extended agent anticipates being tempted to act in a way that conflicts with her own good due to akrasia, distortion, or procrastination. Reflection on how these cases might be fleshed out and on how the self might be conceived suggests room for debate concerning whether, in paradigmatic cases of self-paternalism, the self that is doing the constraining is (best understood as) the same as the self that is being constrained. Relatedly, there is room for debate concerning whether there is any significant difference between self-paternalism and interpersonal paternalism, and whether the permissibility of paternalistic constraint varies depending on whether it is intrapersonal or interpersonal.[14]

Related topics

Deciding for the Incompetent; The Concept of Paternalism.

Notes

1 This worry is articulated and briefly addressed in, for example, Husak (1981).

2 For some interesting related discussion concerning temptation, personal identity, and paternalism, see Kleinig (2009), which focuses on "the argument for paternalism based on complexities in the idea of personal identity" (104).

3 Note that, while I will not delve into questions regarding personal identity, the reader might want to consider whether or which existing accounts of personal identity might best accommodate this stance, and whether accounts of personal identity are meant to track what is supposed to be normatively relevant here.

4 This notion of the self is suggested in Watson (1975). Watson revises his position in Watson (1987), where he distinguishes "valuing" and "judging good" and maintains that there are "perverse cases" in which one "fail[s] to 'identify' with one's evaluational judgements" (150). See Bratman (2007) for a sense of the preceding and follow-up developments in the debate on self-governance.

5 For some helpful discussion of Ulysses contracts, episodic mental disorders, and autonomy, see, for example, Dresser (1982) and Davis (2008).

6 See Holton (2009) for a discussion of this sort of temptation and for his related influential distinction between akrasia and weakness of will.
7 See Holton (2009) regarding "the problem of akratic resolution."
8 The case is adapted from a case in Bratman (1999 [1998]).
9 Another interesting case to consider in relation to this issue is Derek Parfit's (1973) case of a young Russian man who has socialist ideals but who anticipates eventually losing his ideals and changing his mind about how to best use the estates he will inherit.
10 See Holton (2009) for helpful discussion of this idea.
11 For a related example, see Andreou (2014: 290–291).
12 Andreou (2007a); Andreou (2007b).
13 Elsewhere (unpublished manuscript), I endorse abandoning the assumption that some deterioration in rational status must be in play in temptation-based cases that are genuine cases of self-paternalistic constraint, and then argue, more broadly, against the idea that paternalistic constraint is predicated on the assumption that the constrainer is in some way "normatively superior" to the constrained. Douglas Husak has an interesting and influential argument according to which "the characterization of paternalism as necessarily involving a relation between unequals is deficient" (1981: 41); but there are aspects of his reasoning that fall short relative to the position I see as most defensible.
14 I am grateful to the editors for helpful comments on an earlier draft of this chapter.

References

Andreou, C. (2007a) "Understanding Procrastination," *Journal for the Theory of Social Behaviour* 37: 183–193.
———. (2007b) "Environmental Preservation and Second-Order Procrastination," *Philosophy & Public Affairs* 35: 233–248.
———. (2012) "Self-Defeating Self-Governance," *Philosophical Issues* 22: 20–34.
———. (2014) "Temptation, Resolutions, and Regret," *Inquiry* 57: 275–292.
Brandstätter, V., Lengfelder, A. and Gollwitzer, P. M. (2001) "Implementation Intentions and Efficient Action Initiation," *Journal of Personality and Social Psychology* 81: 946–960.
Bratman, M. (1999 [1998]) "Toxin, Temptation, and the Stability of Intention," in *Faces of Intention*, Cambridge: Cambridge University Press.
———. (2007) *Structures of Agency*, New York: Oxford University Press.
Davis, J. K. (2008) "How to Justify Enforcing a Ulysses Contract When Ulysses Is Competent to Refuse," *Kennedy Institute of Ethics Journal*, 18: 87–106.
Dresser, R. S. (1982) "Ulysses and the Psychiatrists," *Harvard Civil Rights – Civil Liberties Law Review*, 16: 777–854.
Holton, R. (2009) *Willing, Wanting, Waiting*, Oxford: Clarendon Press.
Husak, D. (1981) "Paternalism and Autonomy," *Philosophy & Public Affairs*, 10: 27–46.
Kleinig, J. (2009) "Paternalism and Personal Identity," *Jahrbuch für Wissenschaft und Ethik* 14: 93–106.
Parfit, D. (1973) "Later Selves and Moral Principles," in A. Montefiore (ed.) *Philosophy and Personal Relations*, London: Routledge & Keegan Paul, pp. 137–169.
Thaler, R. H. and Sunstein, C. R. (2009) *Nudge: Improving Decisions About Health, Wealth, and Happiness*, New York: Penguin Books.
Watson, G. (1975) "Free Agency," *Journal of Philosophy* 72: 205–220.
———. (1987) "Free Action and Free Will," *Mind* 96: 145–172.

Further reading

Ainslie, G. (2001) *Breakdown of Will*, New York: Cambridge University Press.
Elster, J. (1984) *Ulysses and the Sirens: Studies in Rationality and Irrationality*, Cambridge and Paris: Cambridge University Press and Editions De La Maison des Sciences De L'Homme.
Frankfurt, H. (1971) "Freedom of the Will and the Concept of a Person," *Journal of Philosophy* 68: 5–20.
Schelling, T. C. (1984) "Ethics, Law, and the Exercise of Self-Command," in T. C. Schelling (ed.) *Choice and Consequence: Perspectives of an Errant Economist*, London: Harvard University Press.

PART II

Paternalism and ethical theory

6
PATERNALISM AND WELL-BEING

Jason Raibley

1 Introduction

A man removes a cigarette from a woman's hand to prevent her from lighting and smoking it. "It's for your own good," he says. A physician administers a life-saving blood transfusion to a physically incapacitated but mentally competent patient who opposes blood transfusions on religious grounds. "You'll die without it," he tells the patient. A police officer, acting as an authorized representative of the state, tickets and fines a motorcyclist for riding without a helmet. "You're a danger to yourself, riding that way," he says as he tears off the ticket. A physician, in spite of a patient's explicitly communicated preferences, withholds disturbing information about that patient's condition. "It will only make things worse for her," this physician tells a trusted associate.

These are all examples of the sort of paternalism that has interested moral and political philosophers.[1] Other chapters in this volume pursue the analysis of paternalism in fine detail. But for the purposes of this chapter, *paternalistic action* is conceived as action that (a) interferes with an agent's *autonomy*, i.e., her rational and informed self-determination or self-governance, and (b) is done for her own benefit (cf. Feinberg 1986; G. Dworkin 1988; Beauchamp 1977; Shiffrin 2000; Scoccia 2013). Relevant forms of interference, here, include physical force, coercion (i.e., the threat of harm), the attachment of fines or penalties to certain options, and deception or the withholding of information with the goal of changing a person's evaluation of her options.[2] It is normally held that some initiations of physical force, such as seizing a person to prevent her from crossing an unsafe bridge when there is no time to warn her, do *not* count as paternalistic (Mill 1991 [1859]). One plausible view as to why is that "interference does not violate one's autonomy if one would consent to it, given one's current preferences and values, if one were instrumentally rational and well-informed about relevant, empirically ascertainable causal or means-ends matters" (Scoccia 2013: 80). However, there are various complications.[3] While all instances of paternalism involve paternalistic *action*, policies, laws, and rules can be called "paternalistic" by extension.

Paternalism and well-being are importantly connected topics of philosophical inquiry. As stated above, paternalistic actions are (b) done for the target's own benefit. To be done for the target's benefit is to be done for the purpose of improving or increasing her personal *well-being* (or *welfare* – these terms are here used interchangeably). Well-being is a gradable property or condition: a person might instantiate a high, or a middling, or a low level of it. But to say that a person is doing or faring well – or that her life is going well for her – is to say that she instantiates well-being to some sufficiently high degree (the exact level of welfare expressed by saying that a person is "doing well" may vary with the conversational context). Or, on a connected but broader use that is explained below, it is to say that her life is in most important respects choiceworthy or desirable (similar points about contextual semantics apply here).

There are traditions of thinking about well-being on which an agent's attitude (or lack thereof) towards an action or its likely outcome has a direct bearing on its welfare-value. According to these traditions, the sense in which paternalistic actions interfere with rational self-governance – i.e., the sense in which the agent opposes them or finds them alien – may either rule out their being overall beneficial or else entail that they are simultaneously harmful to a significant degree. Some accounts of this kind imply that paternalistic action can *never* fully succeed or even occur, because the autonomous choice of an end is in some sense a condition of its status as a welfare-good. Other accounts say the frustration of desire, interference with rational self-direction, or the removal of options is intrinsically harmful, so that while a paternalistic action might be beneficial in some respects, it is simultaneously harmful in others. These ways of thinking about well-being – and the broader approaches to moral and political philosophy in which they are embedded – provide support for the idea that paternalism is morally objectionable.

Importantly, these traditions of thinking about personal well-being do *not* oppose paternalism by arguing that considerations of well-being are trumped, outweighed, or silenced by considerations of individual rights or autonomy. Nor do they argue that agents of paternalism lack the knowledge they need to succeed.[4] Nor do they argue that *aggregate* or *societal* well-being would be diminished by broad-scale implementation of paternalism (because, e.g., such implementation requires knowledge that would be costly or harmful to obtain, or dangerous expansions of legislative and executive power). Nor do they argue that paternalistic action is wrong in virtue of the fact that it is disrespectful, insulting, demeaning, or a form of domination. Instead, these traditions imply that, *given the very nature of personal well-being and its main determinants, paternalistic actions have hidden welfare-costs*. Therefore, there are reasons based on the welfare of the *target* of paternalistic action against paternalistic intervention.

This chapter first distinguishes between *narrow* and *broad* concepts of well-being, which are respectively labeled "well-being" and "the Good Life." It then presents, describes, and evaluates several proposals concerning these concepts and their relation to individual attitudes and choices. It explains the implications of these proposals for welfare score-keeping for particular interventions (i.e., determining how these interventions benefit and harm their targets, and how much they do so). This should provide a sense of the options when it comes to the precise definition of well-being as this bears on the understanding of paternalism itself, the application of antipaternalistic principles,[5] and the evaluation of welfare-based arguments against paternalism.

It will emerge that there are plausible accounts of personal well-being on which rational self-direction, autonomous living, and the endorsement of ends are either conditions on personal benefit, or else non-instrumental welfare-goods, or else key instrumental goods. However, there is no plausible account of personal well-being, the Good Life, or any of the goods just mentioned that supports the claim that paternalistic action cannot succeed. On any plausible account, there are cases where such action maximizes *overall* well-being (or maximally promotes the Good Life), even if it is simultaneously intrinsically or instrumentally harmful to some degree. Consequently, if there are decisive reasons against paternalism, they must derive from some other source – e.g., from the bad effects of implementing paternalistic policies, or from freedom or autonomy considered as final goods in their own right, apart from their relation to personal well-being or the most choiceworthy life. In this case, the value of freedom and autonomy is not fully explained by the degree to which they are *good for* individual agents.

2 Narrow and broad concepts of well-being

The elements of personal well-being are those things that, considered simply in themselves, make lives go well for those who live them. While there is broad agreement on this much, this agreement masks a controversy about which concept is picked out by the word "well-being." This term is used to express both narrow and broad concepts of well-being, which are here labeled "well-being" and "the Good Life," respectively. Since autonomy might be importantly connected to either (or both) of these concepts, and since the plausibility of the idea that autonomy is good for a person might depend importantly on which of these we have in mind, a few words will be said about each of them.

The narrow concept of well-being concerns *one* good-making feature of human lives, the only feature that a supremely self-interested person would care about. But it does not exhaust the features that might serve to make a life more choiceworthy, which might also include: how good it is from the moral point of view, how good it is for the world, how many valuable achievements it contains, and its value as an object of aesthetic contemplation. These features might have no bearing on a life's welfare-value, though it makes perfectly good sense to care about them and even to accept certain trade-offs between them and well-being.

By contrast, the broad concept of well-being is the concept of *the Good Life*, i.e., the maximally choiceworthy human life. This concept is used to make an "all-in" evaluation of a human life. Any feature that might serve to make a life more rationally choiceworthy increases its value in this sense. Consequently, if "well-being" and its cognates express the concept of the Good Life, then welfare-evaluations will require us to appraise *all* the good-making features of a life and to commensurate these features, i.e., to know the welfare exchange rates between moral excellence, worthwhile achievements, and other goods.

Many ethicists working in the analytic tradition interpret the well-being concept narrowly and associate it with the following conceptual role. Well-being is the thing that is advanced or increased when a person is *benefitted*, and it is the thing that is adversely affected when a person is *harmed*. What promotes well-being for a person is not necessarily what is good *according to* her, i.e., what she *believes* is good or beneficial (von Wright 1963). Though well-being likely involves happiness, "well-being" is not analytically equivalent to "happiness," which picks out a transient psychological state consisting of positive affective, emotional, or attitudinal states (Haybron 2008; Raibley 2012). Well-being is, however, intimately connected with what matters to a person, or what she cares about (Sumner 1996). The life high in personal well-being must be a life that she would find compelling or attractive, at least if she were rational and aware (Railton 1986; Rosati 1996). Virtually everyone cares about their own well-being, at least to some degree. It is conceptually connected with prudential reasons for action, i.e., reasons of self-interest, and so it features in first-personal deliberation. It is also conceptually connected with reasons of beneficence, and so it features in deliberation about how best to help friends and loved ones (Feldman 1988; Scanlon 1998; Darwall 2002; Brink 2013; Tiberius forthcoming). Well-being has at least some connection – weak, imperfect, and contingent though it might be – with measurable psychological properties such as a preponderance of positive over negative affect, the emotional state (or condition) of happiness, and judgments of life-satisfaction: these things are proxies for well-being in real-life contexts (Haybron 2008). Well-being also presumably has some connection with constructs from welfare economics, clinical psychology, and theoretical medicine (Angner 2016; Nordenfelt 1995). Well-being is not the exclusive province of adult human persons: adolescents, children, newborn infants, and animals can be benefitted and harmed, their interests advanced or thwarted (Kraut 2007; Raghavan and Alexandrova 2015).

The concept that answers to this sort of conceptual role arguably features in the thought of Hobbes, Hume, Bentham, Mill, Kant, and Sidgwick. It seems related to the concept of the human good deployed by Aristotle and Aquinas.[6] It is in any case the target of many contemporary theories of well-being, including hedonism (Feldman 2004, 2010), desire-satisfactionism (Brandt 1979; Griffin 1986), life-satisfactionism (Sumner 1996), aim-achievementism (Scanlon 1998; Keller 2009), values-fulfillment views (Tiberius 2008; Tiberius and Plakias 2010; Raibley 2010; Dorsey 2012), bifurcated views involving *both* happiness or pleasure *and* meaning or narrative-role fulfillment (Kauppinen 2013; Haybron MS), and some hybridized and objective theories that include a subjective element (Arneson 1999; Adams 1999).

It is important to note that this *concept* of well-being is not inherently subjective: it does not require that a person is the final authority on her own well-being, and it does not require that a person's welfare-level depends exclusively on her mental states or pro-attitudes. A person's well-being in this sense might depend on how the world actually is, on her physical and psychological health, and on whether the things she wants are worthwhile or valuable. That said, this concept of well-being is not plausibly analyzed in a *purely* objective way, where the things that are good and bad for one have nothing at all to do with one's evaluative perspective.

By contrast, in some parts of ethics and political philosophy there is a tendency to think of well-being in a more expansive way. Some writers in these areas can be understood as *abandoning* ordinary patterns of usage concerning "well-being," "good for," "happiness," "benefit," and "harm," and repurposing these words to pick out concepts connected with the Good Life, i.e., *the maximally choiceworthy life*. The Good Life might involve well-being in the normal sense alongside other goods (Nussbaum 2000; Kraut 2007; Badhwar 2014). But some theorists who insist on the importance of this concept hold that the Good Life does not involve well-being in the previous sense at all (Hurka 1993).

The idea that the Good Life is the most important concept in the neighborhood of personal well-being is encouraged by certain Aristotelian doctrines concerning *eudaimonia* ("*eudaimonia*" is a Greek word that is translated variously as "happiness," "well-being," and "human flourishing"). One of these states that a life high in *eudaimonia* is beyond improvement and lacking in nothing whatsoever. Another states that *eudaimonia* is the sole ultimate end, i.e., the unique ultimate source of practical reasons. Another states that *eudaimonia* is unlike honor or fame in that it is not easily taken from a person by others or by fortune. Many writers in later antiquity accept these views and so are driven to the view that *eudaimonia* is constituted exclusively by moral virtue. Contemporary eudaimonists embrace similar doctrines about well-being, e.g., that it is a conceptual truth that one always has most reason to do what advances one's well-being (Bloomfield 2014). They then take these doctrines to support not a narrow sort of egoism, but a highly moralized conception of the Good Life on which it is exhausted or fundamentally conditioned by moral virtue (Annas 2011; Russell 2012).

A related tradition, focused on the Good Life, is inspired by some remarks made by G.E. Moore in *Principia Ethica* (1993 [1903]). In one place, Moore seems to propose analyzing the value of a life in terms of intrinsic value *simpliciter*, i.e., the contribution made by its contents to the value of the universe (Heathwood 2003; McDaniel 2014).[7] So, what is *really* "good for" an individual is to instantiate intrinsically good properties, or to be a part of intrinsically good states-of-affairs – e.g., to instantiate pleasure and knowledge and to be part of states-of-affairs involving friendship, love, and the appreciation of beauty (Hurka 2011). This Moorean approach leads naturally to a strongly objective and pluralistic account of the Good Life on which this concept is linked with our deepest and most fundamental reasons (Hurka 1993, 2011; Rice

2013). Here, what is "good for" a person might have little to do with what she wants, cares about, or finds compelling and attractive.

The remainder of this article will focus on well-being in the narrow sense except where otherwise indicated. However, there will be several places where it will be important to consider key claims as concerning the broader concept of the Good Life.

3 Locating the welfare-costs of paternalistic action: general remarks

Is it possible for actions that interfere with an agent's autonomy to advance her well-being? Different substantive accounts of autonomy and well-being have different implications here. Many support the idea that there is an inevitable cost to paternalistic interventions. But while such interventions may be intrinsically or instrumentally harmful, this does not mean they are overall harmful.

This last claim involves the distinction between *intrinsic* and *overall* harms and benefits, which is crucial for thinking about these issues (Bradley 2009). To say that an event or condition is an *intrinsic* (or *non-instrumental* – these terms are here used interchangeably) *benefit* for a person is to say that, considered simply in and of itself, it raises her level of well-being. The pleasures caused by the ingestion of powerful opioids are usually held to be intrinsically beneficial, even though these pleasures also cause, produce, or lead to weighty intrinsic harms (e.g., those associated with emotional problems, dependency, and withdrawal).[8] By contrast, to say that an event or condition is *overall* beneficial for a person is to say that it is *net*-beneficial: if we take into account *both* the event or condition *and* everything that it causes, produces, or leads to, the person's life is overall better for her on account of including it – i.e., better than it otherwise would have been. Consequently, the ingestion of powerful opioids for recreational purposes is typically *overall* harmful.

We normally speak as though complex occurrences like actions, events, and situations are intrinsically beneficial and harmful. But speaking strictly, *intrinsic* benefits concern the *fundamental* bearers of welfare-value, i.e., the *basic* building blocks of personal well-being. These must be specified in such a way as to isolate the most fundamental welfare difference-makers, and therefore they cannot contain "superfluous information" (Feldman 2000). The welfare-value basics ought to be characterized so that they include all the features relevant to their welfare-value and nothing else besides. For example, if (as hedonists claim) pleasure is the only thing that is intrinsically good, and its value is a function of its intensity and duration, then the welfare-basics will be attributions (to subjects) of pleasure of some intensity, lasting for some period of time. Neither actions ("taking the opium") nor events ("the most recent ballgame") nor complex situations ("our being in school together") will count as welfare-value basics. Of course, each of these items is intimately bound up with various basics, and so, speaking loosely, they can be called intrinsically beneficial on account of the welfare-value basics they comprise, include, or involve (as opposed to those that they cause, promote, or lead to).

This way of thinking about welfare-basics lets us avoid double-counting benefits and harms. It also helps us avoid the paradoxical thought that the exact same thing is both intrinsically beneficial and intrinsically harmful. Some complex events can contain both intrinsically beneficial and intrinsically harmful parts. It is also possible for a condition that would *normally* be intrinsically beneficial to be part of some "organic unity"-type compound that is intrinsically harmful. Depending on one's theory, this condition might retain its ordinary intrinsic value, or this value might be canceled or transformed by its presence in the compound. It is also possible for a condition that has *no* intrinsic welfare-value to be part of *two* distinct compounds, one of which is intrinsically beneficial, and the other of which is intrinsically harmful. But in none of these cases is the same exact thing both intrinsically beneficial and intrinsically harmful.

4 Paternalism and simple theories of well-being: hedonism and desire-satisfactionism

With these distinctions and clarifications in mind, let us consider some simple theories of welfare to see if it is possible for autonomy-compromising actions to be beneficial overall. As already noted, hedonism holds that the only thing that is intrinsically beneficial is pleasure, and the only thing that is intrinsically harmful is pain. Different forms of hedonism understand the pleasures and pains relevant to well-being in different ways. Pain can be conceived as nociceptive pain, as negative affect, as "painful consciousness," or as intrinsic attitudinal pain (i.e., displeasure taken directly in some state-of-affairs that is believed to obtain) (Sidgwick 1981 [1907]; Feldman 2004; Crisp 2006). Pleasures are more beneficial – and pains more harmful – the more intense they are, and the longer they last. Hedonism is usually developed with an atomistic structure, so that the value of a time period (up to a whole life) for a person is equal to the total amount of pleasure that it contains *minus* the total amount of pain that it contains.

Violations of autonomy can take different forms, but these violations need not *always* be accompanied by pain. For one thing, autonomy can be violated without a person realizing it, as when the doctor hides important information from the patient in the fourth example, above. That said, if a given action violates a person's autonomy, she will likely be displeased by this fact whenever she becomes aware of it. Consequently, some forms of hedonism are *compatible* with the idea that paternalism typically comes with an intrinsic cost – the degree of which depends on the intensity and duration of displeasure associated with paternalistic action. Still, hedonism does not support a very strong form of antipaternalism. For even if violations of autonomy are sometimes quite painful, there is no reason to think they are always *overall* harmful. For example, while a person might be displeased by having to don a motorcycle helmet, if this same action prevents her untimely death – or later episodes of attitudinal pain – it will be beneficial overall.

Simple forms of desire-satisfactionism have similar implications here.[9] On views in this family, it is intrinsically beneficial for a person if she desires that some situation, *p*, obtains and *p* does obtain (this is an episode of desire-satisfaction). It is intrinsically harmful if she desires that *p* when *p* fails to obtain (this is an episode of desire-frustration). The degree of benefit or harm associated with an episode of desire-satisfaction or desire-frustration is proportional to the strength of the relevant desire. The welfare-value of a life depends exclusively – and usually, additively – on the values of the episodes of desire-satisfaction and desire-frustration that it contains.

Desire-satisfactionists differ among themselves on many points of detail: on whether a person must also *believe* that *p* obtains; on how to measure strength of desire; on whether to focus on occurrent or dispositional desires, and on global or local desires, and on motivational or affective desires; on whether non-self-involving desires count; and on whether addictive, compulsive, uninformed, or irrational desires ought to be excluded from consideration (Griffin 1986; Heathwood 2015). Those who argue that certain uninformed and irrational desires ought to be excluded are known as *rational* or *idealized* desire-satisfactionists; this is an important variation on the main desire-satisfactionist idea (Brandt 1979).

Desire-based theories can be developed to underwrite a strong connection between violations of autonomy and intrinsic harm. Since paternalistic action is perpetrated *against a person's will*, it is naturally something that she desires not to occur. Suppose, for example, that a person strongly desires to ride a motorcycle without a helmet: it feels good, and she is confident that she will be able to avoid an accident. If simple desire-satisfactionism is true, then paternalistic action that frustrates this desire will be intrinsically harmful to at least some degree. By contrast, if rational desire-satisfactionism is true, it may *not* be intrinsically harmful: if this desire would be extinguished by basic information and clear thinking, then preventing her from satisfying it would *not* count as harmful.[10]

But neither simple nor rational desire-satisfactionism implies that paternalism cannot succeed. A paternalistic action that frustrates a person's rational desires might *also* lead to a life that is much higher in desire-satisfaction overall. Rationality and more information will not necessarily render a person an infallible judge of her own long-term interests. Suppose that a healthy young person decides, after correctly evaluating all the risks under conditions of full information, that the expected utility of purchasing health insurance does not justify the expense. She has a *rational* desire to go without health insurance for the time being. But suppose her desire is frustrated: she is induced to purchase insurance by the state's threat of taxes and other penalties. Suppose further that, against all odds, she is then exposed to a rare airborne virus, becomes very sick, and needs health insurance to avoid financial ruin. Here, the paternalistic action (the state's threat of fines) would be *overall* beneficial – i.e., it would lead to a greater sum-total of rational desire-satisfaction over time – even though it is bound up with an intrinsic harm, viz., an episode of rational desire-frustration at the time that it occurs. For this reason, rational desire-satisfactionism is compatible with successful paternalism, and does not afford strong protection against it.[11]

5 Paternalism and sophisticated theories of well-being: self-direction, autonomy, and endorsement

Certain sophisticated theories of well-being provide stronger support for the idea that it is difficult to advance personal well-being through acts that limit or interfere with autonomy. Typically, such theories include among the building blocks of personal well-being the realization of values (desires the subject *identifies with* and *treats as action-guiding*) and/or hierarchically structured goals. Some theories introduce additional welfare-value basics alongside these (e.g., health states, pleasure, knowledge), or further qualifications (e.g., one's aims and values must stand up to *rational scrutiny*, one must *enjoy* their realization, they must also be objectively *worthwhile*). Some theories go further, alleging that unimpeded rational decision-making where one has a robust array of options is itself intrinsically beneficial to a high degree. Theories built around ideas like these can say much more about *why* paternalistic action is often bound up with intrinsic and instrumental harms.

One influential proposal is that *rational self-direction* and *self-realization* – i.e., rational deliberation and planning aimed at the cultivation of one's individual talents – is intrinsically beneficial (von Humboldt 1993 [1792]; Mill 1991 [1859]; Hamilton 1859; Green 2003 [1883]). In *On Liberty*, Mill's harm principle seems to have as part of its basis some claim about the intrinsic benefit of living by one's own lights. Mill (1991 [1859]: §3.3) states that "the distinctive endowment of a human being" includes "human faculties of perception, judgment, discriminative feeling, mental activity, and even moral preferences [that] are exercised only in making a choice." He appears to take the enjoyed exercise of these faculties to be particularly beneficial, especially when it is directed at the realization of individual aptitudes.[12] Similarly, Green holds that "the proper aim of deliberation is a life of activities that embody rational or deliberative control of thought and action," and that a life of such activities constitutes the Good Life (Brink 2003: xl). These sorts of views could be supported by W.D. Ross-style thought experiments: if two lives are equal in their total amounts of pleasure and desire-satisfaction, but the second life also features rational self-direction and self-development, this second life appears to be *better for* the person who lives it (Wall 1998: 146–150).

Paternalistic action seems to impede rational self-direction insofar as it substitutes the judgment of another for the judgment of its target. On Mill's view, paternalistic action therefore deprives its target of an intrinsic benefit: it prevents her from directing her own action through her own thought and choice in an unimpeded way. Other things being equal, the target of paternalism is worse off than she would have been if she had exercised deliberative control. To this extent, paternalistic action is a *comparative* harm (Bradley 2012). Additionally, paternalism may be instrumentally bad, insofar as it undermines the target's *general disposition* to direct her own life.[13]

But consider the case mentioned at the outset, where a physician administers a life-saving blood transfusion to a physically incapacitated but mentally competent patient who opposes blood transfusions on religious grounds. This act preempts deliberative control on the part of the patient. However, it is not true that the patient would have been *better off* without it. Maybe the patient would have been better off if she had concluded that she needed a blood transfusion and then requested one. But without the blood transfusion, she will die, and death will prevent her from achieving any further welfare-goods. On views like Mill's, paternalistic action is intrinsically valueless, but it can be overall beneficial, provided that the target's future life contains more intrinsic benefits than intrinsic harms. Our patient's life will likely contain additional episodes of the exercise of rational self-determination, as well as other goods. For even if self-direction in the pursuit of self-realization is the most significant source of intrinsic benefit, it is not the *sole* source. Mill himself appears to allow that other high-quality pleasures (e.g., enjoyment of scientific learning or great art) are quite beneficial. It is also plausible that self-respect, positive health, and positive relations with others count as intrinsic benefits. If the patient's future life contains such things, the physician may have benefitted her overall.

A stronger form of antipaternalism might be supported by claiming that the exercise of deliberative control is "lexically prior" to other welfare-goods, so that *no* amount of more mundane welfare-goods can compare with the welfare-goodness of a single episode of deliberative control. But this view seems dubious, and even if it were true, it would only secure the conclusion that the blood transfusion is overall harmful on two further assumptions: (i) that the agent's opposing it counts as an exercise of rational self-direction, and (ii) that there would be no additional episodes of rational self-direction in her life if she were to live, perhaps because she would sink into a catatonic state out of worry over her own salvation.

None of this is to deny that views emphasizing the prudential value of rational self-direction have important implications concerning paternalism. These views imply that, if coercive interference is a constant presence in a person's life, this will prevent her from living a life that is exceptionally good for her (even if it is not positively *bad*). Furthermore, there are contexts, like the selection of one's career, where a person is likely a more reliable judge of her own interests than anyone else; in these contexts, paternalism is likely to backfire.[14] Finally, some paternalistic interventions might damage one's self-respect, undercut one's confidence, or otherwise detract from one's ability to engage in deliberative control in the pursuit of self-realization, which will deprive one of substantial welfare-goods. However, these points do not add up to an absolute prohibition on paternalism based on the welfare of its targets. Note also that much of Mill's own case against paternalism appears to rest not on its impact on its target, but on its probable negative impact on aggregate well-being stemming from costs surrounding its implementation, its potential for abuse, and its potential to deprive society of valuable knowledge. He does not appear to think that his theory of personal welfare *alone* can support (e.g.) his view that it is wrong to prevent people from purchasing opium or poisons.

A related view locates intrinsic benefits in a slightly different place. On this view, *autonomous living* is intrinsically beneficial. This is understood to involve charting one's own course through life, controlling one's destiny by fashioning it through successive decisions from an adequate range of options. Joseph Raz and Steven Wall, who defend this idea, believe that autonomous living requires having and using many of the capacities that Mill takes to subserve rational self-direction (e.g., self-awareness, understanding of the world, valuing, planning, and choice). However, they emphasize that it also requires having *plentiful options* (Raz 1986: 372; Wall 1998: 128, 132). Their key idea is that it is intrinsically beneficial (and/or constitutive of the Good Life) to determine the course of one's life oneself by realizing some options and blocking others. The more options one blocks, the more one becomes responsible for one's life, fashions one's character, and gives one's life meaning, all of which may be beneficial. Additionally, given the prevailing "social forms" in highly industrialized societies, many pursuits and ends require autonomous choice by their very nature: for example, to participate in the Western institution of marriage, you must freely choose your spouse (Raz 1986: 308).

This sort of view, which emphasizes having plentiful worthwhile options, might be developed to support a strong form of antipaternalism. It must be emphasized, though, that Raz and Wall allow that narrowly targeted forms of paternalism are compatible with it.[15] Such efficacious paternalism will typically address itself to biologically determined welfare-needs like the maintenance of life and health (Raz 1986: 308), be embraced by its target at some later point (Raz 1986: 369), and promote autonomous living in the long run by preventing harm (Raz 1986: 420). Raz writes that it is

> senseless to formulate either a general pro- or a general anti-paternalistic conclusion . . . paternalism affecting matters which are regarded by all as of merely instrumental value does not interfere with autonomy if its effect is to improve safety, thus making activities affected more likely to realize their aim.
>
> (1986: 422)

He consequently voices support for seatbelt laws, safety and quality controls for manufactured goods, and licensure requirements in medicine and law, even though they deprive individuals of options. Still, he opposes prohibition of risky sports, because such sports are not merely instrumental goods but are valued as ends by those who pursue them.

But let us consider the proposal that it is intrinsically beneficial to have more worthwhile options, and intrinsically harmful to have fewer. Suppose that our motorcyclist intrinsically values riding without a helmet, and that this is a worthwhile activity. This proposal then implies that she is intrinsically harmed when the state takes away this option for her. The removal of such a worthwhile option does not merely preempt the exercise of deliberative control, preventing the occurrence of an intrinsic benefit. Rather, it introduces an intrinsic harm. So while this proposal does not establish that paternalism is overall harmful (and so ineffective), it provides additional support for the view that paternalism has hidden costs which may weigh against whatever benefits it provides.

But is the presence of worthwhile options really intrinsically beneficial, and is their removal intrinsically harmful? Simon Clarke argues that, if autonomy would be impeded in the motorcycle case, then it would also be impeded by societal changes involving the disappearance of options, e.g., the disappearance of silent films, socialist communes, and steam engine trains. Reasons of well-being relating to autonomy would then favor *preventing these options from disappearing*, which Clarke takes to be absurd: "if the values of autonomy do not provide justification for protecting options from disappearing, . . . then they cannot provide justification against paternalism that forces people away from the same options" (2012: 54–55).

Sarah Conly presses a related objection. Only part of what you do in your life

> is ever chosen, and when it is chosen, it is always chosen from a relatively narrow set of options: options constrained by knowledge, by talent, by physical capacities, by

> geographic location, by the years in which you live. And, by state actions. Constraint on choice is ubiquitous.
>
> (Conly 2016: 447)

Nonetheless, well-being is possible. While paternalistic action may constrain choice in some ways, this rarely rules out the formation of personal values, or integrity, or some degree of rational self-direction or self-governance. It takes a few options "off the table," but many options are always already "off the table" due to factors beyond one's control (Conly 2016: 445). The important thing is to have *some* valuable options – i.e., realizable aims and values that one finds genuinely attractive or appealing, and that are suited to one's personal nature – and to be able to exercise deliberative control to realize them. So long as paternalistic intervention does not conflict with *this*, it does not impede autonomous living in an intrinsically harmful way.

Even if this dispenses with the idea that the removal of worthwhile options is always directly harmful, some sophisticated theories of well-being incorporate a different idea, viz., that the *choice* or *endorsement* of a goal is necessary for its achievement to be beneficial. Many who advocate this idea draw inspiration from John Locke. In "A Letter Concerning Toleration" (2003 [1689]), Locke argues that it could not benefit a person to force her into religious beliefs or observances that she does not endorse – even if the relevant beliefs are *true* or the relevant practices necessary for her eternal salvation. Locke's conclusions, here, appear to be supported by several different claims, including the claims that (i) coercion cannot produce sincere belief, and (ii) since salvation requires sincere belief, insincere religious statements and practices are prudentially worthless. These ideas are not directly relevant here, but we ought to consider the broader notion that some putative welfare-goods require choice or endorsement to be genuinely beneficial. This idea features in the work of Will Kymlicka (1990) as well as L.W. Sumner, who holds that a condition of a subject's life is intrinsically beneficial only if "he authentically endorses it [and] experiences it as satisfying, for its own sake" (1996: 172). It has been developed and defended in somewhat greater detail by Ronald Dworkin (1991, 2000).

On R. Dworkin's account of well-being, or "critical interests," a person's well-being cannot be advanced by conditions or aims unless she *believes them to be valuable.*[16] This sort of endorsement does not merely *enhance* the value of aims when they are achieved; it partly *constitutes* their prudential goodness:

> the connection between conviction and value is *constitutive*: my life cannot be better for me in virtue of some feature or component I think has no value . . . a person's life [cannot] be improved by forcing him into some act or abstinence he thinks valueless.
>
> (R. Dworkin 2000: 268–269)

This view might seem to support a welfare-based prohibition on all forms of paternalism. But R. Dworkin denies this. He distinguishes between *critical* paternalism and *volitional* paternalism (R. Dworkin 2000: 216–217). Volitional paternalism occurs when a paternalistic act is done against the target's *will*, but not against the target's *conviction*. Seatbelt laws are a good example of volitional paternalism:

> The state makes people wear seat belts in order to keep them from harm that it assumes they already think bad enough to justify such constraints, even if they would not actually fasten their seat belts if not forced to do so.
>
> (R. Dworkin 2000: 268)

By contrast, examples of *critical* paternalism include compulsory voting laws as well as restrictions on religious liberty or sexual acts. Here, paternalistic action does not merely contravene the target's immediate desires, forcing her to take more effective means towards her own ends. Instead, it contravenes her considered judgments about the value of those ends. This is what R. Dworkin opposes; he

> rejects the root assumption of critical paternalism: that a person's life can be improved by forcing him into some act or abstinence he thinks valueless. . . . [I]t is performance that counts, not mere external result, and the right motive or sense is necessary to the right performance.
>
> (R. Dworkin 2000: 269)

However, even this is misleading. Critical paternalism can be both successful and permissible if the target's endorsement of the end promoted by the paternalistic action is *likely to be forthcoming*:

> It overstates the point to say that [my theory] rules out any form of paternalism, because the defect it finds in paternalism can be cured by endorsement, provided that the paternalism is sufficiently short-term and limited so that it does not significantly constrict choice if the endorsement never comes. . . . [H]owever, the endorsement must be genuine, and it is not genuine when someone is hypnotized or brainwashed or frightened into conversion. Endorsement is genuine only when it is itself the agent's performance, not the result of another person's thoughts being piped into his brain.
>
> (R. Dworkin 2000: 269)

Consequently, R. Dworkin's view does not "rule out compulsory education and other forms of regulation which experience shows are likely to be endorsed in a genuine rather than manipulated way, when these are sufficiently short-term and noninvasive and not subject to other, independent objection" (R. Dworkin 2000: 274). But it *does* rule out some paternalistic actions. Suppose the patient who receives the blood transfusion does not believe that *earthly life made possible by the ingestion of human blood* is valuable at the time the doctor administers the transfusion, nor does she ever come to believe it is valuable. Here, R. Dworkin's theory seems to imply that the transfusion does not benefit her overall. It saves her biological life, but her remaining life will not be a good one because a welfare-ideal life (and a Good Life) must be accompanied by conviction and ethical integrity, or at least a sort of coherence between one's convictions and activities (Wilkinson 1996). Consequently, her non-endorsement *cancels* the putative welfare-value of further life.

This view is difficult to evaluate. Some may protest that the patient simply *cannot* be instrumentally rational and well informed about the relevant causal facts and still believe that there are strong reasons against receiving the transfusion. If this were true, the case may not even count as a case of paternalism (using the definitions provided in Section 1, above). However, let us suppose that the patient counts as rational and informed. On R. Dworkin's view, she would be harmed by the inner conflict involved in living by means of the transfusion: the incoherence between her life-activities and her deepest values would be intrinsically harmful, and it would likely also prevent the occurrence of later welfare-goods. The question, though, is whether this harm, taken in itself, would be significant enough to outweigh all the welfare-goods that would be represented by continued life, in which case death would be *overall beneficial*. And here, R.

Dworkin's idea seems to be that the conflict would mar her life in such a way that it would *cancel* the value of later welfare-goods, even if they were to materialize.

This seems somewhat far-fetched. Perhaps the doctor ought to respect the woman's judgment and not provide her with the blood transfusion. But if so, this more likely stems from the fact that respect for her rational judgment is good in its own right, independent of its effect on her (or anyone's) welfare, and that the law ought to reflect this. R. Dworkin may be correct that coherence between one's actions and convictions is intrinsically beneficial – and its absence intrinsically harmful. However, coherence should only be counted as one intrinsic benefit among others, a component of one's well-being but not its sole determinant. Paternalistic actions that force ends on individuals (who will never come to endorse them) therefore involve intrinsic harm. But whether this harm outweighs the benefits conferred by paternalism will depend on the case.

6 Paternalism and perfectionism

The Lockean idea that some putative welfare-goods require choice or endorsement to be genuinely beneficial may yet prove correct, however. To see this, let us consider a theory of the Good Life that is a natural "friend" of paternalism. For while classical liberals, libertarians, and contemporary liberals adopt theories of well-being that emphasize the costs inherent in paternalistic action, there are other theories of the Good Life (and well-being) that defuse worries about its compatibility with paternalism.

One such theory is Hurka's perfectionism, which is influenced by ideas from Aristotle, Aquinas, Marx, and the British Hegelians. On this view, the value of a human life depends on the degree to which it develops human nature. Human nature consists in those properties that are essential to human beings and conditioned on their being living things (Hurka 1993: 15–16). According to Hurka, there are two main properties that answer to this description: *embodiment* and *rationality*. Rationality involves both the ability to "form and act on sophisticated beliefs and intentions, whose contents stretch across persons and times and that are arranged in complex hierarchies," and the capacity that "grounds beliefs in evidence" (Hurka 1993: 39–40). Consequently, human essence-development consists in physical perfection (health and athletic excellence), practical perfection (achievement), and theoretical perfection (knowledge). The intrinsic value of a particular life depends entirely on the number, quality, and balance of the forms of essence-development that feature in it. Since there is no such thing as negative essence-development – "an essential property cannot be realized to negative degrees" – no life has negative intrinsic value, and "every human life is worth living" (Hurka 1993: 100).

In the present context, the most important form of essence-development is achievement. For Hurka, achievements involve forming intentions to accomplish worthwhile ends with the justified belief that one will succeed. The quality of an achievement is a function of its extent and dominance, i.e., the degree to which "its content stretches across times and objects, including persons," and the degree to which it has other intentions subordinate to it as means (Hurka 1993: 115–116). Importantly, achievements do not require that one *choose* or *endorse* or even *favor* one's ends. All that is required is for one to intend ends with the rational expectation of success.

Suppose the state takes a person with special aptitude in a given field and develops her into a great and successful talent, bringing it about that her life is replete with athletic excellence, knowledge, and achievement. But suppose it does this by constantly cajoling her with threats of imprisonment and harm to her family. Suppose she intends and accomplishes a wide variety of ends without choosing, endorsing, or favoring them. Suppose she never takes more than minimal enjoyment in their accomplishment and often dislikes their pursuit. If Hurka's perfectionism were true, the state's actions would result in her leading a superlatively Good Life.

Of course, given the motivational make-up of actual human beings, it is unlikely that this could be effectively managed. As Hurka points out, "coercion is a clumsy device. Instead of encouraging rational evaluation, it implicitly tells citizens to obey unthinkingly, without considering alternatives. Far from encouraging informed decision-making, it discourages it" (1993: 157). He emphasizes that, on his view, paternalistic action is often instrumentally bad for its targets. Still, in cases like the one just described, it might also produce athletic excellence, knowledge, and achievement that is tremendously good, promoting the Good Life for its target. Consequently, Hurka's perfectionism licenses paternalism up to the point where it is incompatible with a person deliberating about and implementing rational plans of action.

This account of the Good Life does not seem very plausible. Whatever value is represented by this person's essence-development, it does not result in her leading the Good Life to an exceptional degree. Certainly, her essence-development does not augment her well-being in the narrow sense. In this context, we might extrapolate from Locke's discussion to another important insight, viz., that *certain* welfare-goods, including especially achievements, require choice or endorsement to be genuinely beneficial. Perhaps the absence of choice or endorsement significantly diminishes the welfare-value of a person's achievements, and may even cancel their value for her altogether. This is another respect in which paternalistic action may have hidden costs.

7 Conclusion

Paternalistic action aims to promote a person's own good in a way that interferes with her rational and informed self-governance. We have seen that, on a range of plausible views about the nature of personal well-being and the Good Life, the sense in which paternalistic action interferes with rational self-governance may entail that it is simultaneously harmful to some degree, even if beneficial overall. To the degree that agents are displeased or frustrated by paternalistic interference, it involves intrinsic harms, though these may not be especially weighty. To the degree that such interference damages a person's self-respect, undercuts her confidence, or subverts her ability to engage in practical reasoning (thinking and evaluating for herself), it is at least instrumentally bad, and perhaps even overall harmful, depending on the details of the case. If coercive interference is a constant presence in a person's life, this may prevent her from engaging in valuable forms of activity that are intrinsically beneficial. Furthermore, coherence between one's actions and one's convictions about what is worthwhile is itself intrinsically beneficial, and its absence is intrinsically harmful. Paternalistic interference may threaten this coherence in subtle ways, preventing intrinsic benefits and causing intrinsic harms. Finally, the intrinsic benefits represented by perfectionistic goods such as athletic excellence, knowledge, and achievement are diminished and may be even be canceled if a person does not *choose* to engage in them or *favor* them. For all these reasons, paternalistic action has hidden welfare-costs that must be weighed against its benefits when assessing reasons of well-being in its favor.[17]

Related topics

Consequentialism, Paternalism, and the Value of Liberty; Mill's Absolute Ban on Paternalism; Moralism and Moral Paternalism; Perfectionism and Paternalism.

Notes

1 However, some argue that political philosophy should focus on the more expansive, ordinary concept of paternalism (Quong 2011). This chapter follows Feinberg (1986) in setting this more expansive concept aside.

2 There are borderline cases where interference with autonomy seems very slight, such as state subsidies for public radio or television that are financed through taxation. It is unclear whether paying for these subsidies out of general tax revenues is coercive in the same way – or to the same degree – as more typical examples of paternalism. Interference with autonomy also seems rather slight in cases of "nudging," i.e., the adoption of laws that provide for the control of one's "choice architecture" where this is motivated by concern for one's own benefit (Thaler and Sunstein 2009). These manipulations of choice architecture do not involve the removal of options or the restriction of liberty. Though sometimes labeled as instances of "libertarian paternalism," it is unclear that they are genuine instances of paternalism at all.

3 It is unclear how to apply this principle to certain cases (Hanna 2012). Suppose the woman described above wishes that she did not desire to smoke. Suppose she has communicated this wish to all her friends and associates. Moreover, this wish is quite stable, and it guides her action most of the time. But very occasionally she is weak and desires to smoke. It is not clear whether it would interfere with her autonomy (and count as paternalistic) for someone who knows this to rip her cigarette out of her hand. For reasons such as these, the concepts of autonomy and paternalism may have blurry boundaries (cf. G. Dworkin 2013).

4 J. S. Mill (1991 [1859]: §4.4) writes that "with respect to his own feelings and circumstances, the most ordinary man or woman has means of knowledge immeasurably surpassing those that can be possess by anyone else" (cf. Arneson 1980).

5 For example, J.S. Mill's famous harm principle states: "The only purpose for which power can be rightfully exercised over any member of a civilized community, against his will, is to prevent harm to others" (Mill 1991 [1859]: §1.9). To apply this antipaternalistic principle, we must know under what conditions others would be harmed, and this will require an understanding of personal well-being itself.

6 This claim is denied by McDowell (1980) and Annas (1993), but it is accepted by others.

7 This is what R. Dworkin (2000) labels the "impact model" of well-being.

8 Some more holistic theories provide grounds for challenging the idea that opioid use confers intrinsic benefits; see, e.g., Raibley (2013); Bishop (2015).

9 Forms of subjective desire-satisfactionism are arguably extensionally equivalent to some forms of hedonism (Heathwood 2006).

10 If it is possible for a person to know all about the dangers of riding a motorcycle without a helmet, to be reasoning well, and still to want very badly to ride without a helmet, then rational desire-satisfactionism implies that paternalistic action that frustrates this desire involves an intrinsic harm.

11 This argument assumes that "full information" comprises the sort of knowledge a subject might actually acquire with a bit more time, instruction, or luck. If rational desire-satisfactionism is built around a more robust concept of full information that includes, e.g., humanly unknowable facts about the future, then the theory produces objectionable results in a variety of cases (Velleman 1988).

12 For further discussion of Mill's view of well-being, see Feldman (1995), Brink (2013), and Riley (2015).

13 A great deal of weight is put on this idea by libertarian writers; see Rand (1965), Rothbard (2006 [1973]), and Wright (forthcoming) for discussion. Though it may achieve small benefits, these writers hold that paternalistic action is always overall harmful because it disrupts a process that is of vital importance for personal well-being, viz., practical reasoning about what to do. Because it is coercive, paternalistic action makes one's practical reasoning ineffective, or (in extreme cases) prevents it from occurring. The plausibility of this idea depends on the severity and frequency of paternalistic interference. These writers seem to respond to this difficulty by arguing that there is no principled way for the government to limit coercive interference. Consequently, their opposition to paternalism also depends crucially on this latter idea, which goes beyond paternalism's impact on well-being.

14 However, there are some contexts where most people are unreliable judges of their best interests (Thaler and Sunstein 2009).

15 Wall also explicitly rejects the idea that having more options is always autonomy- and welfare-enhancing; he merely says it is important to have access to a sufficiently wide array of worthwhile options (1998: 186–189).

16 It is unclear whether R. Dworkin's theory of critical interests is a theory of well-being or a theory about an entirely distinct concept. On his "challenge model" of critical interests, the goodness of a life consists in its "inherent value as a performance" (2000: 267). Some of his formulations raise the worry that he is working with an aestheticized concept of a valuable human life distinct from both well-being and the Good Life.

17 I thank Jason Hanna for his perceptive and detailed comments on an earlier draft. I also thank Teresa Chandler, Dale Dorsey, and Kalle Grill for helpful comments and discussion.

References

Adams, R. (1999) *Finite and Infinite Goods*, New York: Oxford University Press.
Angner, E. (2016) "Well-Being and Economics," in G. Fletcher (ed.) *The Routledge Handbook of the Philosophy of Well-Being*, Abingdon: Routledge.
Annas, J. (1993) *The Morality of Happiness*, New York: Oxford University Press.
———. (2011) *Intelligent Virtue*, New York: Oxford University Press.
Arneson, R. (1980) "Mill Versus Paternalism," *Ethics* 90: 470–489.
———. (1999) "Human Flourishing vs. Desire-Satisfactionism," *Social Philosophy and Policy* 16: 113–142.
Badhwar, N. (2014) *Well-Being: Happiness in a Worthwhile Life*, New York: Oxford University Press.
Beauchamp, T. L. (1977) "Paternalism and Biobehavioral Control," *The Monist* 60: 62–80.
Bishop, M. (2015) *The Good Life: Unifying the Philosophy and Psychology of Well-Being*, New York: Oxford University Press.
Bloomfield, P. (2014) *The Virtues of Happiness*, New York: Oxford University Press.
Bradley, B. (2009) *Well-Being and Death*, New York: Oxford University Press.
———. (2012) "Doing Away With Harm," *Philosophy and Phenomenological Research* 85: 390–412.
Brandt, R. (1979) *A Theory of the Right and the Good*, Oxford: Clarendon Press.
Brink, D. (2003) "Introduction," in T. H. Green's (ed.) *Prolegomena to Ethics*, Oxford: Clarendon Press.
———. (2013) *Mill's Progressive Principles*, New York: Oxford University Press.
Clarke, S. R. (2012) *Foundations of Freedom: Welfare-Based Arguments Against Paternalism*, London: Routledge.
Conly, S. (2016) "Autonomy and Well-Being," in G. Fletcher (ed.) *The Routledge Handbook of the Philosophy of Well-Being*, Abingdon: Routledge.
Crisp, R. (2006) *Reasons and the Good*, New York: Oxford University Press.
Darwall, S. (2002) *Welfare and Rational Care*, Princeton, NJ: Princeton University Press.
Dorsey, D. (2012) "Subjectivism Without Desire," *Philosophical Review* 121: 407–442.
Dworkin, G. (1988) *The Theory and Practice of Autonomy*, New York: Cambridge University Press.
———. (2005) "Moral Paternalism," *Law and Philosophy* 24: 305–319.
———. (2013) "Defining Paternalism," in C. Coons and M. Weber (eds.) *Paternalism: Theory and Practice*, New York: Cambridge University Press, pp. 25–38.
Dworkin, R. (1991) *Life's Dominion*, New York: Vintage Books.
———. (2000) *Sovereign Virtue: The Theory and Practice of Equality*, Cambridge, MA: Harvard University Press.
Feinberg, J. (1986) *The Moral Limits of Criminal Law, Volume 3: Harm to Self*, New York: Oxford University Press.
Feldman, F. (1988) "On the Advantages of Cooperativeness," *Midwest Studies in Philosophy* 13: 308–323.
———. (1995) "Mill, Moore, and the Consistency of Qualified Hedonism," *Midwest Studies in Philosophy* 20: 318–331.
———. (2000) "Basic Intrinsic Value," *Philosophical Studies* 99: 319–346.
———. (2004) *Pleasure and the Good Life*, New York: Oxford University Press.
———. (2010) *What Is This Thing Called Happiness?* New York: Oxford University Press.
Green, T. H. (2003 [1883]) *Prolegomena to Ethics*, New York: Oxford University Press.
Griffin, J. (1986) *Well-Being: Its Meaning, Measurement, and Moral Importance*, Oxford: Clarendon Press.
Hamilton, W. (1859) *Lecture on Metaphysics and Logic, Vol. 1*, Boston: Gould and Lincoln.
Hanna, J. (2012) "Paternalism and the Ill-Informed Agent," *Journal of Ethics* 16: 421–439.
Haybron, D. (2008) *The Pursuit of Unhappiness*, Oxford: Clarendon Press.
———. MS. "Well-Being: A Millian Hybrid View."
Heathwood, C. (2003) "Review of Stephen Darwall's Welfare and Rational Care," *Australasian Journal of Philosophy* 81: 615–617.
———. (2006) "Desire Satisfactionism and Hedonism," *Philosophical Studies* 128: 539–563.
———. (2015) "Desire-Fulfillment Theories," in G. Fletcher (ed.) *The Routledge Handbook of the Philosophy of Well-Being*, Abingdon: Routledge.
Hurka, T. (1993) *Perfectionism*, Oxford: Clarendon Press.
———. (2011) *The Best Things in Life*, New York: Oxford University Press.
Kauppinen, A. (2013) "Meaning and Happiness," *Philosophical Topics* 41: 161–185.
Keller, S. (2009) "Welfare as Success," *Noûs* 43: 656–683.
Kraut, R. (2007) *What Is Good and Why*, Cambridge, MA: Harvard University Press.
Kymlicka, W. (1990) *Contemporary Political Philosophy*, Oxford: Clarendon Press.

Locke, J. (2003 [1689]) "A Letter Concerning Toleration," in I. Shapiro (ed.) *Two Treatises of Government and a Letter Concerning Toleration*, New Haven, CT: Yale University Press.
McDaniel, K. (2014) "A Moorean View of the Value of Lives," *Philosophical Quarterly* 95: 23–46.
McDowell, J. (1980) "The Role of Eudaimonia in Aristotle's Ethics," in A. Oksenberg Rorty (ed.) *Essays on Aristotle's Ethics*, Berkeley: University of California Press, pp. 359–376.
Mill, J. S. (1991 [1859]) "On Liberty," in J. Gray (ed.) *On Liberty and Other Essays: Oxford World's Classics*, Oxford: Oxford University Press, pp. 5–128.
Moore, G. E. (1993 [1903]) *Principia Ethica: Revised Edition*, Cambridge: Cambridge University Press.
Nordenfelt, L. (1995) *On the Nature of Health: An Action Theoretic Approach*, Dordrecht: Kluwer.
Nussbaum, M. (2000) *Women and Human Development: The Capabilities Approach*, Cambridge: Cambridge University Press.
Quong, J. (2011) *Liberalism Without Perfection*, New York: Oxford University Press.
Raghavan, R. and Alexandrova, A. (2015) "Toward a Theory of Child Well-being," *Social Indicators Research* 121: 887–902.
Raibley, J. (2010) "Welfare and the Priority of Values," *Social Theory and Practice* 36: 593–620.
———. (2012) "Happiness Is Not Well-being," *Journal of Happiness Studies* 13: 1105–1129.
———. (2013) "Values, Agency, and Welfare," *Philosophical Topics* 41: 187–214.
Railton, P. (1986) "Facts and Values," *Philosophical Topics* 14: 5–31.
Rand, A. (1965) "What Is Capitalism?" in *Capitalism: The Unknown Ideal*, New York: Signet Press, pp. 1–29.
Raz, J. (1986) *The Morality of Freedom*, Oxford: Oxford Clarendon Press.
Rice, C. (2013) "Defending the Objective List Theory of Well-Being," *Ratio* 26: 196–211.
Riley, J. (2015) *The Routledge Guidebook to Mill's On Liberty*, New York: Routledge.
Rosati, C. (1996) "Internalism and the Good for a Person," *Ethics* 106: 297–326.
Rothbard, M. (2006 [1973]) *For a New Liberty*, Auburn: Ludwig von Mises Institute Press.
Russell, D. C. (2012) *Happiness for Humans*, New York: Oxford University Press.
Scanlon, T. M. (1998) *What We Owe to Each Other*, Cambridge: Belknap Harvard Press.
Scoccia, D. (2013) "The Right to Autonomy and the Justification of Hard Paternalism," in C. Coons and M. Weber (eds.) *Paternalism: Theory and Practice*, New York: Cambridge University Press, pp. 74–92.
Sidgwick, H. (1981 [1907]) *The Methods of Ethics*, 7th edn., Indianapolis: Hackett.
Shiffrin, S. (2000) "Paternalism, Unconscionability Doctrine, and Accommodation," *Philosophy & Public Affairs* 29: 205–250.
Sumner, L. W. (1996) *Welfare, Happiness, and Ethics*, Oxford: Clarendon Press.
Thaler, R. and Sunstein, C. (2009) *Nudge: Improving Decisions About Health, Wealth, and Happiness*, New York: Penguin Books.
Tiberius, V. (2008) *The Reflective Life*, New York: Oxford University Press.
———. (forthcoming) *Well-Being as Value Fulfillment*, New York: Oxford University Press.
Tiberius, V. and Plakias, A. (2010) "Well-Being," in J. Doris et al. (eds.) *The Moral Psychology Handbook*, New York: Oxford University Press.
Velleman, D. (1988) "Brandt's Definition of 'Good,'" *The Philosophical Review* 97: 353–371.
von Humboldt, W. (1993 [1792]) *The Limits of State Action*, Indianapolis: Liberty Fund, Inc.
von Wright, G. H. (1963) *The Varieties of Goodness*, London: Routledge & Kegan Paul.
Wall, S. (1998) *Liberalism, Perfectionism, and Restraint*, New York: Cambridge University Press.
Wilkinson, T. M. (1996) "Dworkin on Paternalism and Well-Being," *Oxford Journal of Legal Studies* 16: 433–444.
Wright, D. (forthcoming) "Reason, Force, and the Foundation of Politics," in G. Salmieri and R. Mayhew (eds.) *The Philosophy of Capitalism*, Pittsburgh: University of Pittsburgh Press.

7

CONSEQUENTIALISM, PATERNALISM, AND THE VALUE OF LIBERTY

Sarah Conly

Consequentialists judge the moral rightness of an act by its consequences. While there are different consequentialist accounts of the exact relationship between the best consequences and the best act, a classic consequentialist theory, such as utilitarianism, would hold that if an act brings about the greatest amount of welfare possible under the circumstances, it is the right act. Given this, it might seem that the relationship between paternalism and consequentialism would be straightforward: since paternalist policies aim to promote the welfare of the individual, and consequentialists promote the welfare of people as a whole, the two would appear to be consistent. Indeed, paternalism might seem to be an application of consequentialism in a particular sphere, applying pressure on an individual to do what best promotes his welfare where he would not otherwise do that. Except in those relatively rare cases where the welfare of an individual is unavoidably at odds with the overall welfare of society, the two should be able to act in concert.

A view that interprets rightness as the promotion of a single value, well-being, seems to be most compatible with paternalism, since the loss of liberty often involved in paternalist policies is not, per se, a loss of anything intrinsically valuable. The value of liberty is entirely a function of its contribution to well-being. An intrusive paternalist regulation may itself have costs, but if, when those costs are counted, the act still maximizes well-being, it's the right act.

At least, so it would seem. We know, though, that the foremost proponent of consequentialism has also been the foremost opponent of paternalism, cited by antipaternalists across the philosophical and political spectrum. John Stuart Mill famously wrote in *On Liberty* that there is one principle that should

> govern absolutely the dealings of society with the individual in the way of compulsion and control, whether the means used be physical force in the form of legal penalties, or the moral coercion of public opinion. That principle is, that the sole end for which mankind are warranted, individually or collectively, in interfering with the liberty of action of any of their number, is self-protection. That the only purpose for which power can be rightfully exercised over any member of a civilized community, against

> his will, is to prevent harm to others. His own good, either physical or moral, is not a sufficient warrant. He cannot rightfully be compelled to do or forbear because it will be better for him to do so, because it will make him happier, because, in the opinion of others, to do so would be wise, or even right. . . . The only part of the conduct of anyone, for which he is amenable to society, is that which concerns others. In the part which merely concerns himself, his independence is, of right, absolute. Over himself, over his own body and mind, the individual is sovereign.
>
> (Mill 2003a [1859]: 94)

This articulation of what has come to be called the Harm Principle has been endorsed by many who see it as a celebration of individual rights against the intrusions of government and society at large. The question, however, is whether this opposition to paternalism actually makes sense for a consequentialist who takes well-being as the good to be maximized.

There are different accounts of paternalism, but for many people paternalism is conceived of as a policy where people are discouraged from doing what they would otherwise have done by means that make their initial course of action less attractive.[1] Paternalistic policies often make the action the individual initially wanted to take much more costly, or, in some cases, impossible. Thus, in many cases they are coercive. Just when increasing the costs of an action results in the individual being coerced is something of a question: Is a tax that makes smoking more expensive coercive if it only increases the costs of cigarettes 1%? If the increase is 200%? It may be hard to specify when exactly a smoker will find that a disincentive has become positively coercive. Wherever the cut-off lies, though, some paternalistic policies will qualify as coercive, as they will make the action in question something that a reasonable person will feel she must refrain from, precisely because of the costs added by the paternalistic regulation. On most accounts, this means that the individual's liberty of action is being interfered with. Even if it is physically possible for the person to do what she would otherwise have done, if she is sufficiently discouraged from doing that action by the interference of another which makes that action more costly, some of her liberty has been taken from her.

What is the problem, though, with restricting liberty? On a welfare-maximizing account, it looks as if in any case where we are better off without a given liberty, there is no cost to losing that liberty: it is the consequences in terms of welfare, after all, that count. Indeed, we might permissibly lose a lot of liberty, if the result is still an increase of welfare. If we construe liberty as an instrumental good, something that is good only because of its consequences, then it ceases to be a good when it fails to produce the best consequences. While non-utilitarians may argue that liberty is an intrinsic good of great value, indeed, value so great that it might be impossible for any gain in welfare to offset its loss, a utilitarian may argue that since nothing outside of happiness is intrinsically valuable, when liberty is useful as a means we should cherish it, but when it ceases to be useful it may be dismissed without concern.

This makes sense. That is, if we say, first, that the right action is the act that produces the best consequences, and second, that in a given case, sacrificing liberty produces the best consequences, it follows that sacrificing liberty is the right thing to do. The caveat, the reason people like Mill aren't willing to sacrifice liberty when the liberty in question concerns actions that directly affect only one's self, is that they deny the second premise is likely to be true. If denying liberty of action towards oneself is extremely unlikely ever to promote the individual's welfare, it's a type of action from which governments and societies should refrain. While policies that deny my liberty to rob a bank do promote welfare, policies that deny my liberty to treat myself in the way that I want just don't. Sacrificing the liberty to choose what one wants to do with one's own life, where that doesn't affect others, is not an effective means to well-being. Liberty in such cases is required instrumentally for maximizing an individual's well-being.

1 Hedonistic consequentialism

There is more than one argument that might be made here, because there is more than one conception of what it is for a person to be better off, and the arguments for liberty will depend on what precisely we want to promote. Since this is a paper and not a treatise, I will not try to consider every form of consequentialism, but will concentrate on three that illustrate why some might think that liberty is indispensable if we are to promote the greatest well-being. The simplest and most straightforward is that the right act is the act that creates the greatest happiness for the greatest number, where happiness is construed as a subjective state, a state where one feels pleasure.[2] This is the consequentialism we see in Jeremy Bentham, who famously argued that the value of all arts and sciences, be the activity in question pushpin or poetry, was "exactly in proportion to the pleasure they yield" (Bentham 2003 [1825]: 94). Such hedonistic utilitarians naturally will be inclined to consider a number of factors in evaluating an act – including its side effects, such as precedent effect on society, and the effect on the actor's character – but in the long-term the goal is just to make the greatest number of people feel as happy as they can.

Here, the argument for liberty is twofold. First, we may argue, as Mill did, that we will be happier when we choose for ourselves how to live because each individual will be best able to make the choices that will give him pleasure (Mill 2003a [1859]: chapter 3). Having even the most benevolent paternalist choose for someone else may not only fail to maximize that person's happiness but may result in misery because, as Mill emphasized, people vary in their tastes and talents. Some people, for example, like to cook. If, on that basis, they benevolently require of me that I take French cooking lessons and produce Julia Child-like dishes for my own consumption, they will have made a grave mistake. While I might like to eat at Julia Child's table, I would very much not enjoy cooking meals that require hundreds of steps, employ techniques, tools, and ingredients I have never heard of, and take a great deal of time. Few things are as irritating as well-meaning people who try to make you engage in an activity on the grounds that it's something they themselves enjoy. Thus, the classic antipaternalist argument is that the most efficient way to maximize happiness is to let individuals make their own choices about how they themselves want to live, since they are best informed about what will make them happy.

However, there is a great deal of evidence that in fact individuals don't choose all that well if their goal is to be happy. Left to my own devices, I may choose wrongly not only in small things to which I don't give much consideration but in the big decisions, the decisions to which I would presumably give the greatest consideration. We frequently choose, for example, the wrong job or the wrong spouse. While most people show evidence of wanting to live healthy and solvent lives, we choose things that make us unhealthy and leave us hopelessly in debt. Left with freedom, we very often do things that make us very unhappy. It may be that in fact we don't really know what we want (and Mill, of course, lived in a pre-Freudian time when people may have believed that the content of their minds was more transparent than we now believe). So, I think that I want a huge house because I will enjoy the vast space, when what I really want is a status symbol that I unconsciously (and vainly!) hope will finally assuage my feelings of inferiority. Or, this may be because knowing what we *want* isn't actually a good indicator that we know what will make us *happy*. I truly want to be famous, but it has burdens I hadn't foreseen. Or, it may be that even if our ends themselves would conduce to happiness, we are poor at choosing means – we are right that being rich would make us happy, but wrong that buying lottery tickets is a good means to that end. Or, of course, it may mean feeling happy isn't what we most want, as will be discussed below. While there may be areas where people reliably pick what makes them happiest, this is not something we can rely on. If society, with its collective store of knowledge and more objective perspective on what generally seems to produce happiness, can be more helpful in some areas, why not let it?

However, a hedonistic defender of free choice can reply that this is no argument for paternalism. The antipaternalist can rightly say that even if the paternalist produces a state of affairs as an end goal that makes the individual as happy as circumstances allow, this isn't sufficient justification for intervening. When we consider the happiness of the individual, we consider not only the end state to be achieved but the happiness or unhappiness involved in the process of getting there. The salient point may not be whether we choose well, but how much we enjoy choosing. Consider romance: even if some match-making agency in the sky were able to find us more compatible spouses than our own libido-influenced process may do, we enjoy looking for our own partners, even given the many mistakes we make in attempting to negotiate a successful romance. For many aspects of our life, we like reviewing the options, we like imagining the various outcomes of our various choices, and we like feeling that we do have options. Both the process of choice and the knowledge that we have the freedom to engage in that process can make us happy. When it comes to failed marriages, for example, few enjoy divorce, but many people enjoy the ups and downs of the various romances that lead there, and enjoy the knowledge that whether or not they get together and stay together is up to the people in the relationship.

There is some truth in this rebuttal of the paternalist. However, it is also true that we often adjust to a loss of liberty without lasting hardship, or, indeed, without any hardship at all. Some processes may be ones we never enjoyed to begin with. If we are pushed to choose from a more limited range of options, that may come as a relief, since surveying the vast number of possibilities in some cases is simply tedious, even when we know our choice is important. Very few people, for example, enjoy decision-making processes that involve statistics. Paternalistic policies that constrain choice in such areas may increase happiness, rather than reducing it, when we see that we don't have to consider all the possible safety variables involved in automobile construction, since the government makes sure they all pass a certain safety threshold. Other choices would involve knowledge that it would be a hardship to survey, even where the facts are available: we don't have to review the history of our food products to ascertain their source or arrange to test them for *E. coli* because the government does that for us. In each case, the government removes options, and thus liberty, but it wasn't a liberty we would have enjoyed exercising.

What about our resentment at losing a liberty, even if it was a liberty we didn't want to make use of? While we often complain when we lose a freedom (when, say, we are no longer allowed to smoke in a restaurant), it's also true that we often adjust fairly quickly. There is a transition cost to the newer, less free situation, but in many cases that cost is not all that high, since our dissatisfaction is temporary. Many Americans objected to the creation and subsequent expansion of the Food and Drug Administration, designed to prevent them from buying fraudulent medicines; they objected to Social Security, designed to prevent them from ending their lives in poverty, and to Medicare, designed to provide basic health care for the elderly (Light 2013). Now all three of these are extremely popular with American voters, so much so that they are regarded as almost untouchable institutions. Change is irritating, and our first reaction seems to be to cling to the status quo where we can enjoy tranquil inertia. Then, if the change is seen as actually bringing about a good effect and is not too difficult to comply with, we become accustomed to it. We no longer plan as if the option were available, so we are not disappointed at not being able to pursue it – we know we can't whip out a cigarette after our restaurant meal, and we get in the habit of wearing a seatbelt so that it feels strange to be without one, and generally no longer chafe at restrictions that we barely notice.

This is not to say that people don't ever mind losses of freedom in a way that continues to be painful, or which perhaps becomes more and more painful, over time. Rather, it means that some losses of freedom bother us, and some don't. The good paternalist needs to consider many things in devising a paternalistic policy. For one thing, she needs to have a sound belief about what end state will actually make people happy. For another, she needs to consider the costs of whatever paternalistic interference would get us to that state. Both of these can be difficult, to be sure. Many have pointed out that paternalists themselves are subject to the same errors they try to correct for, and we need to be sure proposals are well vetted and then open to amendment if they prove unsuccessful.[3] The conclusion to draw is not that paternalistic policies can't maximize happiness, though, but that paternalists need to be careful in the consideration of policies. This, of course is something consequentialists are used to, since they need to make just the same sorts of calculations in order to maximize utility. There is no reason to think that this will be more difficult when we consider paternalistic policies than when we consider regulations as to how people may treat one another.

2 Preference-satisfaction consequentialism

Some consequentialists may say that mere pleasure is not the good. We want to promote welfare, and we can't say that a person is well-off if he is not living in the way that he chooses. While Mill and Bentham believe pleasure to be the ultimate goal in life, others think this is simply wrong. Some people think that a person's well-being must reflect what that person himself wants out of life. That might be pleasure, but it very well might not: a person might prefer achievement, or any number of states of affairs, to his own feelings of pleasure. Welfare can only be achieved through the satisfaction of preferences: perhaps not all preferences, but at least some strong preferences must be fulfilled for a person to have achieved a satisfactory degree of well-being. The problem with paternalism, it is said, is that by its very nature it interferes with the satisfaction of preferences. After all, we have said that a paternalistic policy gets you to choose what you would not otherwise choose. You want a cigarette, and paternalists interfere with the satisfaction of that desire by taxing cigarettes prohibitively. You want lots of French fries, and paternalists make that more difficult by hiding them behind the counter and then limiting your portion. The paternalist doesn't leave you alone to do what you want to do, but intervenes. We may know that it is a fact of nature that there is many a slip between the cup and the lip, but we are particularly indignant if it is the paternalist who is standing between us and our soda. Imposing a foreign set of values on us cannot possibly make us better off.

Many paternalists tend to agree that welfare consists of preference-satisfaction. Indeed, I would say that this is the most dominant form of paternalism around now (Thaler and Sunstein 2008; Conly 2013). Their argument is that, in fact, paternalism increases the likelihood of our achieving whatever it is that we prefer to achieve. It is certainly true that paternalist policies are apt to change our decision about what to do in light of the costs and benefits the paternalist has imposed, but in the long run, this actually increases the satisfaction of our preferences. The argument is that we often choose poor means to satisfying our desires, so that while we do "want" to do something in the sense that it is the object of immediate choice, that same thing is antithetical to what we want overall. This is something that has been exhaustively discussed. A whole host of authors now analyze the ways in which we make poor instrumental choices, defeating our own ends through cognitive biases or other flaws in reasoning (for example, Kahneman and Tversky 2000; Tversky and Kahneman 2000a, 2000b, 2000c). We want to be healthy, and we choose the French fries, over and over. We want to be solvent, and we charge things again and again, to the point where we will never be able to pay off our debt. We are swayed by the way information is framed, we are irrationally optimistic that we will avoid the harm that is statistically likely, we are unduly attached to the status quo even when that is contrary to our interests, and so forth. In the face of bounded rationality, interference is the way to get us where we want to be.

Some will argue that this can't be right, because it misconstrues what it actually is that people want. Such a critic may say that the only way to see what people actually want is to see what they do. Classical economists have typically held that the best indicator of what someone wants is what action he voluntarily chooses to perform, and while there may be occasional anomalies, especially in unfamiliar or stressful situations, most decision-making conforms to this (Rebonato 2012). When the paternalist says the obese person would prefer good health to eating lots of high-fat food she errs – she attributes her own preferences to the obese person, or she is misled by the fact that the French fry eater, pressured by social standards, claims to want health more than fries, even when he doesn't. Thus, when the paternalist intervenes, she really does impose a behavior that is contrary to the preference of the person involved. If he wanted a healthy diet he would eat a healthy diet! If the individual seems swayed by irrelevancies in making a decision (choosing the surgery described as having an 80% success rate rather than the one described as having a 10% failure rate just because the word "failure" is so daunting), that is simply that person's desire. Just because we think a feature is irrelevant doesn't mean it is irrelevant to that particular person. These are functions of individual preference: some people care about what other people consider irrelevant. I've known people who argue that the color of a car should be immaterial to whether or not one buys it, whereas others seem to think it is the most significant feature.

This is possible. The paternalist must grant that it can be hard to know what someone wants, in the face of that person's conflicting behaviors. Still, to argue that we always make the choice that reflects our preference of the moment is peculiar. For one thing, it suggests that no one makes a mistake – you chose what you wanted, even if at the time you were saying "I shouldn't be doing this," even if it is inconsistent with your behavior in general, even if it fills you with regret. This, to me, is implausible, as many of us do seem to think we have sometimes chosen wrongly. We don't just think that our preferences later changed, but that we made a poor decision at the time.

Second, it ignores the full import of cognitive bias. Even the person who thinks our behavior manifests our actual desires, whatever we may say to the contrary, is unlikely to say that this is true if a person is not fully informed when she chooses what to do. That is, no one is likely to say "Yes, she wanted to die" of the person who ate the death cap under the impression it was a chanterelle. The argument that her behavior is an expression of her true preference depends on her having known what it was that she was doing. This is where the paternalist may object. If we are making a cognitive error when we choose to perform an action, it is arguable that we don't know what we are doing. That is, we know what we are doing in one sense (I am now eating French fries) but not in another sense (I am contributing to a lifetime of obesity and associated ill health). It is precisely these connections between means and ends that are in error.

It is not, of course, that we literally have no idea that French fries, or whatever our weakness may be, is a relatively unhealthy food. We may well know how many calories it contains, we may know that it is not part of the diet full of antioxidants we have, in theory, decided to embark on, etc. This is what makes the case complicated. Even Mill admitted that if we are literally ignorant of a fact, and in light of that ignorance are about to act in a way inconsistent with our ends, we may be forcibly interfered with, if there is no time to educate us as to the matter at hand. In Chapter V of *On Liberty* he admitted that if a person is about to cross a bridge in ignorance of the fact that the bridge is broken, interfering with his progress is permissible, because we may assume he doesn't want to fall. The case with many of our poor decisions may seem to be quite different, in that we have been taught that smoking is unhealthy, that it is better to eat the salad than the deep-fried food, etc.: that person eating the French fries may even say, "I shouldn't be eating these, given what they will do to me," as she eats. So, there is a sense in which we know exactly what we are doing.

This is where cognitive bias comes in. Mill seems to have assumed that knowing the relevant facts – that the bridge is broken, that falling from it will harm you, that you don't want to be harmed, etc. – would naturally lead you to avoid crossing the bridge, so the person who stops you is doing just what you would have chosen to do if you had the relevant information. In fact, our decision-making is more complicated, and more flawed, than Mill realized. What our knowledge of psychology tells us is that there is a difference between having knowledge of a fact and that knowledge actually being operative in our decision-making. We know, in some sense, that an operation with a 90% success rate must have a 10% failure rate, but at the same time we are more likely to choose it when it is described to us as having a 90% success rate. We can know all the facts about smoking, and about obesity and heart disease, in one sense and yet optimistically act on the belief that none of those bad things will actually happen to us personally. The relevant knowledge doesn't find its way into our reasoning process when we are choosing. Mill seems to have thought that a person would either be knowledgeable, and thus would use relevant information, or would be ignorant, and thus unable to use it. Now we see that these are not the only options. Our informed choice may be in a very relevant deep sense uninformed.

Granted, the critic of paternalism may deny the whole existence of cognitive error. It is not an error, he can say, that our estimate of outcomes is swayed by what others consider irrelevancies. This is a hard row to hoe, however, given the evidence that we have as to the existence of these biases. It's not something that comes up only when we are considering preferences and choices. We know there is bias – that is to say, a selective tendency to error – even when we are simply estimating statistics, regardless of our own preferences. You are, for example, much more likely to think a tsunami is very likely in the near future if there has been one recently, even if you know that for the previous hundred years there haven't been any. You think you are less likely than the average person to be hit by lightning, for no reason other than that you are you. Such conclusions are just wrong, and arise from mistakes in reasoning. Someone who insists that our actions always reflect our preferences has to argue that these biases, so well-documented in other areas of calculation, somehow disappear or are ineffective when we are choosing actions. This is simply implausible.

Paternalists do often interfere with liberty. The argument that when they do they don't help us satisfy our actual preferences, though, is weak. If consequentialists want to promote preference-satisfaction, paternalism is a strategy they should embrace.

3 Eudaimonistic consequentialism

Another line of thought is that having a certain sort of liberty is a component of welfare, and a significant enough component of welfare that its loss can't be compensated for by gains in other parts of a person's welfare. Having (certain) liberties just is part of what it is to be well-off. This would be a kind of eudaimonistic consequentialism, where the welfare we want to promote isn't simply feelings of happiness, or the satisfaction of subjective desires, but a state in which the individual engages in certain sorts of valuable activities and has certain sorts of abilities. Mill, while identified as a hedonistic utilitarian according to his own description, would probably have belonged in this camp as well, since while he promoted pleasure and the absence of pain as the good, he also famously said that not just any pleasure would do. In *Utilitarianism* he argued for the value of "higher quality," more intellectual pleasures over even a greater quantity of "lower" pleasures (Mill 2003b [1861]: 186–190), and in *On Liberty* he wrote that while the only criterion of right moral action is utility, this must be "utility in the largest sense, grounded on the permanent interests of man as a progressive being" (Mill 2003a [1859]: 95). This is not the place for an extended discussion of what Mill meant precisely by his description of man as a "progressive" being, but clearly mere feelings of contentment were, for Mill, not enough to make a successful human life. Many people seem to agree with this, to believe that what we want is to promote particular activities and abilities, including our evaluation of choices, in terms of a general conception of how we want to live. Liberty of choice is important because to hone such abilities, it is thought, we need to be able to act on our choices, at least when that action merely affects oneself.

That independent thinking requires independence of action is invoked against paternalism in particular. Paternalistic actions will sometimes prevent certain choices, and not just because those choices conflict with someone else's freedom of choice. Without freedom of action, our ability to think about what we most want will atrophy, it may be said. We will regress to the status of children, incapable of seeing what is best for us as unique individuals. We will find ourselves in a state of "learned helplessness," where we are unable to figure out what to do and must have our decisions made for us. If we are free, on the contrary, we will hone our intellectual abilities and moral sensitivity to the point where they enable us to make the best decision overall, and we will then make those decisions.

That freedom of choice begets better choices seems to be a relatively common opinion, and it is in part for this reason that paternalistic regulations are so often resisted. The problem, though, is that there is no empirical foundation for these claims. It is presumably true that if we had no choice whatsoever, if every action was prescribed, we would think less about what it is we want to do, since such thought would be frustrating and pointless. Paternalists, though, are not advocating intervention in every choice – far from it. They advocate interference in some areas, making some options impossible, some less attractive, and some more attractive, according to what is in the agent's own interests. The argument against paternalism, then, must not simply be that we need some liberty of action to become reflective choosers, but that the more personal liberty we have the better choosers we will become, and this is a claim for which there is no evidence. If liberty resulted in our refining our abilities to make decisions, we would find ourselves in a much better situation than we now are.[4] Often when we have the opportunity to exercise our ability to reflect and choose, we seem in fact to favor inertia and the closest short-term pleasures, even when we think we should be doing otherwise. Given an unprecedented wealth of fresh and healthy food all year around, we choose such amounts of sugar, salt, fat, and cigarettes that lifestyle diseases are fast becoming a major source of world-wide morbidity. Americans often express the view that self-government is one of their most important political values, but given the liberty to vote, about 60% of Americans stay at home in elections that don't involve the president, and 40% stay home even for that – despite the fact that political policies have such an impact, and often an impact they lament (DeSilver 2016). Even if we agree with the eudaimonistic consequentialist that we are better off as persons when we engage in the activity of reflective choice, there is no evidence that the unfettered ability to choose results in our choosing well, or even exercising choice at all in any robust sense. The default option doesn't seem to be the independence of thought that Mill celebrated but too often a collapse into inertia, which, given the social forces around us, is likely to result in adherence to convention and the status quo. Paternalism may well restrict our options for choice, but experience suggests that many of us avoid exercising these options even when we have liberty.

Some may say that this simply shows that unfettered liberty is not sufficient for reflective choices. It might be that it is still necessary. We just need something else in addition. But what might this something else be? One reasonable answer would be education. But even education will require paternalistic interventions. After all, people have the ability to educate themselves now about many significant public policies, and apparently fail to do that: facts about climate change, international events, and many things that greatly impact us are simply not interesting to many people. If we are to educate people effectively, we will apparently need to force education upon them. We defend the requirement of education for young people, after all, on the grounds that it will help them make better choices for themselves and for society, and in the implicit assumption that if education is not mandated young people won't pursue it. If we want adults to make good use of their liberty, and education is required for that, this simply requires a different form of paternalism: education that is paternalistically motivated and that adults will be constrained to pursue.

I don't know if we are capable of the kind of objective reflection that some people celebrate as our uniquely human endowment. If we are, though, it is not a capacity that simply appears when we are born, like the capacity to feel, or even one in later life that emerges of its own accord, like the hormonal changes of adolescence. It takes training, and training is a species of interference in liberty, even when that is done specifically to allow the individual to better use and enjoy his capacities for decision-making. If we want to improve people's ability to make choices, the laissez-faire approach of the antipaternalist is not the way to do that.

4 Consequentialists and non-consequentialists

The conclusion for consequentialists is clear. Mill was wrong. Paternalism in some form should be celebrated by any consequentialist with a plausible view of welfare. Mill was wrong about our ability to make ourselves better off by simply consulting our desires and acting upon them, and wrong about our ability to think independently if given the liberty to do so. Paternalism does interfere with liberty, but when the paternalistic policy is an effective one, that liberty is well lost. Such interferences, properly done, can promote happiness, or preference-satisfaction, independence of thought, or reflective choice; whatever it is that a plausible consequentialism would aim to promote. If we were perfect agents we would not need such paternalistic policies to thrive, but it should come as no surprise that we are not perfect agents.

This is not to say that liberty isn't ever important. It depends on which particular liberty we are talking about. In some cases liberty is an indispensable good. The liberty to vote, and to run for office, and to criticize the government are important: they bring about goods that can't effectively be brought about any other way. Not all liberties are like this, however. Some liberties, including some liberties to make personal choices, bring us no significant goods, and may do us great harm.

What does this mean for non-consequentialists? This has been a discussion of the compatibility of consequentialism and paternalism. However, much of the discussion here should be persuasive to a non-consequentialist; that is, it should persuade the non-consequentialist that paternalism can be acceptable. Non-consequentialists often argue that paternalism is disrespectful of the individual, but the question is what it is that is disrespected. The implication of the argument here is that in many cases, free actions are not expressive of reflection. They do not convey our sense of values, and do not augment our ability to live according to those values. They may be straightforwardly irrational, either in the general sense that they are poor means to ends or in the specific sense that they are products of unreason – actions that we take only because we have erred in our thinking, erred in ways even we ourselves would acknowledge as error if we were aware of them. To interfere in such actions is not a manifestation of disrespect, because such actions are not indications of what the agent would do if acting in accordance with his own values. It is not disrespectful if someone tries to guard himself against actions at odds with his goals by, say, deciding not to subscribe to an internet service so that he won't spend so much time playing online games, time that he would in fact spend if the internet were available. He interferes with his freedom of action so that he can avoid short-sighted actions and live in accordance with his own values. Sometimes protective actions aren't ones the agent himself can take, though, and if the state steps in and does that, that is not disrespectful, either.

Of course, a non-consequentialist might argue that voluntary action is simply valuable for its own sake. He may say that even if it is harmful to the agent's welfare even according to his own conception of welfare, even if it is not indicative of the agent's values and not an outcome of rational reflection, even if it is based on deep irrationality, voluntary action is to be respected. Such a non-consequentialist faces two problems, though. First, this will simply seem implausible to many. If all we care about is that an action is voluntary, why do we fail to respect the voluntary actions of children and animals when those are harmful to themselves? Typically we feel we can override them precisely because they are irrational, or harmful, or not expressive of values or reflection, etc. We simply don't think all voluntary actions merit respect. Second, how do we deal with the fact that we are quite ready to override voluntary actions when they are harmful to others? Even if we "respect" them in some sense of respect, we are happy to prevent violent or destructive actions when they are injurious to others. If it is permissible to do this because such actions conflict with the voluntary actions of others, the same reasoning applies to restricting certain actions against the self: they conflict with other things that the agent wants to do, and which he has endorsed more reflectively than he has the irrational act of, e.g., smoking a cigarette.[5] The mere fact that an act is voluntary does not mean it shouldn't be overridden.

The justifiability of paternalism, properly conceived of and properly applied, is something that the consequentialist and the non-consequentialist can agree on. What we value in life and what we value in agency are not only compatible with but promoted by appropriate paternalist policies.

Related topics

Kantian Perspectives on Paternalism; Mill's Absolute Ban on Paternalism; Paternalism and Education; Paternalism and Sentimentalism; Paternalism and Well-Being.

Notes

1 Some take "paternalism" to pertain to those policies that limit liberty. Others construe it more broadly, so that incentivizing actions also counts as paternalistic. For example, if in order to encourage (supposedly beneficial) visits to museums a government subsidizes entry fees, some will see that as paternalistic, since the intent is to change behaviors. Others may argue that since such a policy does not restrict liberty, it is not paternalistic. See Le Grand and New (2015: chaps. 2 and 3) for a discussion of this difference.

2 The term "happiness" may be construed in many ways, but since both Bentham and Mill use it as interchangeable with "pleasure," that is what I will do in this discussion of hedonistic utilitarianism.

3 See Conly (2013: chap. 4) for discussion of errors in governmental policies and arguments that they are less of a danger than errors occurring when individuals choose for themselves.

4 Given liberty as to how to spend our time, we Americans watch television for 4.5 hours a day, on average, where fictional lives substitute for our own (Koblin 2016). And that is a statistic for what is called "live" television – excluding services like Netflix and Hulu.

5 I am assuming an agent here whose overall goals are health and long life, not one who thinks a short, unhealthy life is worth the joy of smoking.

References

Bentham, J. (2003 [1825]) "The Rationale of Reward," in J. Troyer (ed.) *The Classical Utilitarians: Bentham and Mill*, Indianapolis: Hackett.

Conly, S. (2013) *Against Autonomy: Justifying Coercive Paternalism*, Cambridge: Cambridge University Press.

DeSilver, D. (2016) "U.S. Voter Turnout Trails Most Developed Countries," *Pew Research Center*, www.pewresearch.org/fact-tank/2017/05/15/u-s-voter-turnout-trails-most-developed-countries/.

Kahneman, D. and Tversky, A. (2000) "Prospect Theory: An Analysis of Decision under Risk," in D. Kahneman and A. Tversky (eds.) *Choices, Values, and Frames*, Cambridge: Cambridge University Press.

Koblin, J. (2016) "How Much Do We Love TV? Let Us Count the Ways," *New York Times*, 30 June, www.nytimes.com/2016/07/01/business/media/nielsen-survey-media-viewing.html.

Le Grand, J. and New, B. (2015) *Government Paternalism: Nanny State or Helpful Friend*, Princeton, NJ: Princeton University Press.

Light, J. (2013) "Déjà vu: A Look Back at some of the Tirades Against Social Security and Medicare," http://billmoyers.com/content/deja-vu-all-over-a-look-back-at-some-of-the-tirades-against-social-security-and-medicare/.

Mill, J. S. (2003a [1859]) "On Liberty," in M. Warnock (ed.) *Utilitarianism and On Liberty*, Malden, MA: Wiley-Blackwell.

———. (2003b [1861]) "Utilitarianism," in M. Warnock (ed.) *Utilitarianism and On Liberty*, Malden, MA: Wiley-Blackwell.

Rebonato, R. (2012) *Taking Liberties*, New York: Palgrave Macmillan.

Thaler, R. and Sunstein, C. (2008) *Nudge: Improving Decisions About Health, Wealth, and Happiness*, New Haven, CT: Yale University Press.

Tversky A. and Kahneman, D. (2000a) "Advances in Prospect Theory: Cumulative Representation of Uncertainty," in D. Kahneman and A. Tversky (eds.) *Choices, Values, and Frames*, Cambridge: Cambridge University Press.

———. (2000b) "Loss Aversion in Riskless Choice: A Reference-Dependent Model," in D. Kahneman and A. Tversky (eds.) *Choices, Values, and Frames*, Cambridge: Cambridge University Press.

———. (2000c) "Rational Choice and the Framing of Decisions," in D. Kahneman and A. Tversky (eds.) *Choices, Values, and Frames*, Cambridge: Cambridge University Press.

8

KANTIAN PERSPECTIVES ON PATERNALISM

Melissa Seymour Fahmy

1 Paternalism

My aim in this chapter is to articulate the Kantian moral opposition to paternalism. Assessing the moral legitimacy of paternalism is complicated by the fact that there are competing accounts of just what paternalism is. Some accounts of paternalism understand restricting another's freedom to be an essential component of paternalistic action (Kleinig 1984). Others defend much broader accounts of paternalism, arguing that being "liberty restrictive" is not necessary (Shiffrin 2000). Matters are less contentious when it comes to the paternalist's motive for acting. Most agree that in order for A's action toward B to count as paternalistic, A's motive must be concern for B's good. Of course, one's "good" can be understood in a variety of ways, and A's conception of B's good might be quite different from B's understanding of her own good. Finally, some contend that B's attitude matters as well, arguing that A's action counts as paternalism only if it is unwanted by B (de Marneffe 2006).

Because I believe that Kantian opposition to paternalism is not limited to acts which are liberty restrictive, for the purposes of this chapter I shall adopt a modified version of Seana Shiffrin's broad account of paternalism. Accordingly, A's behavior toward B is paternalist provided that (i) A intends to have an effect on B or B's sphere of legitimate agency, (ii) A intends to promote B's interests and this is A's sole or primary reason for acting, and (iii) A substitutes her judgment or agency for B's "on the grounds that compared to B's judgment or agency regarding those interests, A regards her judgment or agency to be (or as likely to be), in some respect, superior to B's" (Shiffrin 2000: 218).[1] The following example is illustrative. B is a long-time smoker who desires to quit smoking, but is understandably finding it difficult to do so. A is a friend of B's and is aware of B's desire to quit smoking. If A elects to help her friend quit smoking by hiding, stealing, or destroying B's cigarettes without B's consent, then A acts paternalistically toward B (Shiffrin 2000: 215). Notably, A's behavior is paternalist despite the fact that there is no fundamental disagreement between A and B regarding B's interests.

Ethicists have provided several distinct reasons for thinking that paternalism is, if not necessarily objectionable, at least *pro tanto* morally problematic. Some make the argument that paternalism is objectionable in virtue of the epistemic obstacles and limitations that even the best-intentioned paternalist faces. According to this line of thought, these epistemic obstacles make it unlikely that the paternalist will achieve her goal of promoting the good of another. An argument of this sort can be found in John Stuart Mill's *On Liberty*. There he contends that the strongest argument against public interference with purely self-regarding conduct is that the odds are great that the public will be swayed by its own preferences and judge poorly the good of others (Mill 1978 [1859]: 81). Paternalism is objectionable in virtue of its likelihood to be self-defeating.

Others, like Joel Feinberg, argue that the objectionable feature of paternalism resides in the fact that liberty-limiting paternalistic interference constitutes an unjustifiable violation of the individual's sovereign right of self-determination. According to Feinberg, restricting individual liberty on strictly paternalist grounds is morally offensive "because it invades the realm of personal autonomy where each competent, responsible, adult human being should reign supreme" (Feinberg 1986: 25).

Perhaps the most common complaint made against paternalism is that such interference is *disrespectful*. As Shiffrin describes it,

> Even when paternalist behavior does not violate a distinct, independent autonomy right, it still manifests an attitude of disrespect toward highly salient qualities of the autonomous agent. The essential motive behind a paternalist act evinces a failure to respect either the capacity of the agent to judge, the capacity of the agent to act, or the propriety of the agent's exerting control over a sphere that is legitimately her domain.
>
> (Shiffrin 2000: 220)

Stephen Darwall similarly maintains that the objectionable character of paternalism is "primarily a failure of respect, a failure to recognize the authority that persons have to demand, within certain limits, that they be allowed to make their own choices for themselves" (Darwall 2006: 268). And more recently, Michael Cholbi has identified the wrong-making feature of paternalism as "substituting others' judgment concerning what a person's good is for that person's own judgment concerning her good, thereby failing to respect the individual as a locus of rational agency" (Cholbi 2013: 117). The respect objection has a broader reach than the previously considered objections insofar as it purports to give us a reason to reject paternalism even when it does not violate an autonomy right or restrict individual liberty and the paternalist in fact does know better how to secure another's good.

In her recent book *Against Autonomy*, Sarah Conly defends the use of coercive paternalism (e.g., legislation) to force people to act, or refrain from acting, according to their interests. Conly's argument is informed by research from behavioral economists and social psychologists, which concludes that human beings routinely exhibit certain cognitive biases and reason poorly in predictable ways. The paternalism that Conly defends does not rely on an objective account of well-being or valuable ends. Rather, the paternalism she defends is one that "helps people act according to their own values" (Conly 2013: 12). Conly argues for "intervention in cases where people's choices of instrumental means are confused." In such cases, action is constrained "only in order to get the person to do what he would want to do if he were fully informed and fully rational" (Conly 2013: 43).

What intrigues me most about Conly's defense of coercive paternalism is that it speaks directly to the respect objection. Conly suggests that paternalistic interference may be *more respectful* in virtue of the cognitive deficiencies that plague human rational agency. According to Conly, "Those who say we should respect autonomy by letting people hurt themselves irreparably do not, on my view, show as much respect for human value as they purport to" (Conly 2013: 1–2). I shall refer to this line of thought as *Conly's Challenge*. We can frame Conly's Challenge in the form of a question: Why think that interfering in the lives of others for the sake of promoting their welfare is *less respectful* than standing by and allowing individuals to make poor choices that undermine their well-being?

Recent defenses of paternalism, like Conly's, provide a welcomed opportunity for antipaternalists to reconsider their opposition. As mentioned above, the value most often invoked in opposition to paternalism is respect, and this is certainly true for the standard Kantian opposition to paternalism (Baron and Fahmy 2009). We ought not treat adults in a paternalistic manner because, well, they are adults, and such treatment would fail to provide them with the respect we are obliged to show them. By itself this claim is not particularly helpful or convincing, especially in light of Conly's Challenge. Arguments defending the legitimacy of paternalism beckon us to step back and ask why. Why is paternalism disrespectful? What sort of respect are competent adults entitled to, and how does paternalism fall short of this ideal?

My intention in this chapter is to explicate the Kantian grounds for the moral rejection of paternalism. I argue that a thorough account of the Kantian grounds for opposing paternalism involves three distinct yet related points. The first and most fundamental point concerns the Kantian moral grounds for acknowledging the decision-making authority of competent adults. In the following section I provide a Kantian answer to the question *why think that paternalistic interference entails a failure to properly respect others?* The second point conjoins social and psychological facts about human moral development with a distinctly Kantian conception of the relative worth of particular ends. While happiness and discretionary ends are by no means unimportant in Kant's theory, they are subordinate to the supreme good, virtue. In Section 3 I argue that, even when we choose poorly, making our own decisions instrumentally serves our moral development. Insofar as paternalistic interference aims to promote the well-being of another at the possible expense of undermining the development of her rational powers, the paternalist wrongly prioritizes well-being over self-development. The third and final point concerns the innate right to freedom, which is the foundation of Kant's political theory. Forms of paternalism that restrict an agent's freedom solely for the sake of promoting her welfare are doubly wrong: they are rights violations as well as failures of respect.

What I aim to provide is an account of why Kantian normative theory is opposed to paternalism that goes deeper than the simple declaration that paternalism is disrespectful. In the course of providing this account I aim to explain why a Kantian would not be persuaded by Conly's recent defense of paternalism. The Kantian grounds for rejecting paternalism reflect commitments to a particular conception of persons, as well as particular moral goods, such as self-perfection, and the secondary importance of well-being. In this way, Kantian opposition to paternalism is distinct from a Millian or Feinbergian rejection of paternalism.

2 Kantian opposition to paternalism: the value of humanity

My intention in this section is to explicate, from a Kantian perspective, the very common complaint that paternalism entails a failure of respect. This will necessitate answering a pair of fundamental questions about Kantian respect. First, what is the *source* of the Kantian duty to respect others? And second, *how* are we obligated to respect others? In other words, what does Kantian respect look like in practice? The answer to the first question informs the answer to the second, and so we must begin there.

What is the *source* of the Kantian duty to respect others? What grounds this moral obligation? This question can be answered in a deceptively simple way. The source of the Kantian duty of respect is *humanity*. As Kant describes it,

> Every human being has a legitimate claim to respect from his fellow human beings and is *in turn* bound to respect every other. Humanity itself is a dignity; for a human being

> cannot be used merely as a means by any human being (either by others or even by himself) but must always be used at the same time as an end.
>
> (Kant 1996: 579; 6:462)

Humanity is a deceptively simple answer to the question – *What is the source of the Kantian duty of respect?* – because it matters very much what *humanity* refers to, and yet Kant offers different descriptions of *humanity* throughout his work. For instance, in an early section of *The Doctrine of Virtue*, Kant declares, "The capacity to set oneself an end – any end whatsoever – is what characterizes humanity (as distinguished from animality)" (Kant 1996: 522; 6:392). On the basis of this passage (as well as others) we might be tempted to draw the conclusion that it is simply the capacity to set an end that entitles human beings to respect. But this would be a mistake. It is not simply our capacity to set an end that grounds our claim to respect, but our capacity for morality. Kant makes this explicit in a later passage in *The Doctrine of Virtue*:

> In the system of nature, a human being (*homo phaenomenon, animal rationale*) is a being of slight importance and shares with the rest of the animals, as offspring of the earth, an ordinary value (*pretium vulgare*). Although a human being has, in his understanding, something more than they and can set himself ends, even this gives him only an *extrinsic* value of his usefulness (*pretium usus*). . . . But as a human being regarded as a *person*, that is, as the subject of morally practical reason, is exalted above any price; for as a person (*homo noumenon*) he is not to be valued merely as a means to the ends of others or even his own ends, but as an end in itself, that is, he possesses a *dignity* (an absolute inner worth) by which he exacts *respect* for himself from all other rational beings in the world.
>
> (Kant 1996: 557; 6:434–435)

The view that human beings possess dignity in virtue of their capacity for morality and not merely their capacity to set an end is also expressed in the earlier *Groundwork*. There Kant contends that "morality, and humanity insofar as it is capable of morality, is that which alone has dignity" (Kant 1996: 84; 4:435).[2] Note that it is our *capacity* for morality that has dignity, something distinct from our moral goodness.

To summarize: human beings – insofar as their rational nature is the source of the moral law and thus allows them to give the law to themselves – have dignity. To say that someone is a *person*, or that she possesses *dignity*, is just to say that she ought always to be treated as an end.[3]

Our answer to the first question has taken us some way in answering the second question – *How* are we obligated to respect others? First and foremost, we are negatively obliged to avoid treating others *merely as means*. But we are also positively obligated to treat others as ends, that is, to treat them in ways that acknowledge their dignity (Kant 1996: 579; 6:462). Over the past four decades Kant scholars have devoted a great deal of thought and energy to the task of explicating what must be the case in order to draw the conclusion that someone has treated another *merely as means* or has failed to treat her as *an end in herself*.[4] I cannot hope to do justice to either the subject or the literature here, yet something more needs to be said about these concepts if we are to arrive at an adequate appreciation of the Kantian grounds for rejecting paternalism. To this end, I will employ an account of treating others merely as means recently articulated by Samuel Kerstein.

According to Kerstein,

> an agent uses another (or, equivalently, uses another as a means) if and only if she intentionally does something to or with (some aspect of) the other in order to realize

> her end, and she intends the presence or participation of (some aspect of) the other to contribute to the end's realization.
>
> (Kerstein 2013: 58)

Using another as a means is not necessarily morally problematic. The key moral distinction for the Kantian is the distinction between using someone *as a means* and using her *merely as means*. Kerstein proposes and defends several sufficient conditions for treating another merely as means. One of these sufficient conditions is the Actual Consent Account (ACA), according to which an agent uses another merely as means if it is reasonable for her to believe that

either

a) the other was informed, before it occurred, of the agent's intended use of him and at the time voluntarily dissented from it, or
b) the other was not or could not, before it occurred, be informed of her intended use of him. But if the other had been so informed, he would have voluntarily dissented from it,

and

c) the other has not, prior to the agent's use of him, given his voluntary, informed consent to it or to a set of rules governing his and the agent's interaction, according to which her use of him is legitimate (Kerstein 2013: 118).

It follows from the ACA that many paradigmatic cases of wrongful interference – deception, coercion, and force – are cases of using another merely as means. I find the ACA to be quite plausible and, of the various accounts of using another merely as means that Kerstein presents, the ACA is the most relevant to the evaluation of paternalism.

We are now in a position to answer the question articulated at the beginning of this section: Why think that paternalism entails a failure of respect? Paternalistic interference is a failure of respect insofar as it entails treating a competent adult merely as a means to her own welfare, interests, or even her own ends. In order to see this more clearly, let us return to the example introduced at the beginning of the chapter. B is a long-time smoker who desires to quit smoking, but is finding it difficult to do so. B's friend A elects to help B quit smoking by hiding, stealing, and destroying B's cigarettes whenever she can do so without B's knowledge. A uses B insofar as she intentionally interferes with B's property in order to realize her end of helping B quit smoking. A's behavior counts as paternalist insofar as (i) it has an effect on B and B's sphere of legitimate agency, (ii) A's primary reason for acting is the promotion of B's interests, and (iii) it involves A substituting her judgment and agency for B's on the grounds that A regards her judgment and agency to be superior to B's in virtue of the fact that A does not suffer from nicotine addiction.

Features (i) and (iii), which allow us to identify A's behavior as paternalist, also make it the case that A likely satisfies the conditions for treating another *merely as means* according to Kerstein's ACA. In the example, A interferes with B's property without B's knowledge. Why does A not inform B of her plan to help B quit smoking? The most plausible explanation seems to be that A believes that if B knew what she was doing – hiding, stealing, and destroying B's cigarettes – B would object to this form of "help" and would attempt to thwart A's unwelcomed interference by not leaving cigarettes in plain sight. Thus it looks like A satisfies condition (b) of the ACA. Conjunct (c) stipulates that in order for A to treat B merely as a means it must be the case that B "has not, prior to [A's] use of him, given his voluntary, informed consent to it or to a set of rules governing his and the agent's interaction, according to which her use of him is legitimate." A's behavior in the example also satisfies this condition. Things would be different had A explicitly offered to hide, steal, or destroy B's cigarettes and B had accepted A's offer. In this case, it would not be reasonable for A to believe either (a) or (b), and thus A's action would not count as treating B merely as means according to the ACA. But A's action also would not involve substituting her judgment or agency for B's and thus would not satisfy condition (iii) of our account of paternalism. Thus there appears to be a very close connection between the third feature of the modified version of Shiffrin's account of paternalism and Kerstein's ACA.

Clearly not every case of treating another merely as means is a case of paternalism. In most cases, we use others merely as means for selfish purposes. This might lead us to think that the selfishness is an essential wrong-making feature of our behavior. But this would be a mistake. A selfish motivation is not a necessary condition for using another in a morally problematic way. We can be beneficently motivated and still treat another merely as means. This explains why even modest paternalism (i.e., paternalism that does not entail substituting the paternalist's perception of the other's good for the agent's own) will still count as disrespectful from the Kantian perspective.

As noted earlier, the coercive paternalism that Conly defends is modest. As she explains,

> The paternalism I promote here is not a paternalism about ultimate ends; that is, I do not argue that there are objectively good ends, or objectively rational ends, or ends objectively valuable in any way, which everyone should be made to pursue. I am arguing for intervention in cases where people's choices of instrumental means are confused, in a way that means they will not achieve their ultimate ends.
>
> (Conly 2013: 43)

The paternalist regulations that Conly defends "are designed to help us reach our own goals" and "constrain action only in order to get the person to do what he would want to do if he were fully informed and fully rational" (Conly 2013: 12, 43).[5]

I am arguing that even this modest form of paternalism fails to address the core Kantian complaint. The core objection is not that paternalistic interference fails to respect the agent's conception of her good, or her choice of ends, or even her rational capacities. Rather, Kantian opposition to paternalism is grounded in the claim that the paternalist fails to respect the *status* of the other as an end in herself, and being an end in oneself is not dependent on being perfectly rational.[6] As Kant puts it, "as a person (*homo noumenon*) [a human being] is not to be valued merely as a means to the ends of others *or even his own ends*" (Kant 1996: 557; 6:434–435, emphasis mine). The authority to select not only one's ends but also the means to these ends (provided neither is immoral) comes with the status of being an end in oneself. This is why attempts to justify paternalism by appealing to our imperfect rationality will not be persuasive to the Kantian. Our claim to respect from others was never dependent on the supposition that we are perfectly rational beings. Thus, at least where competent adults are concerned, paternalistic interference remains objectionable from the Kantian perspective even when the paternalist knows what the agent wants and how best to secure it.[7]

Thus far I have been talking about paternalistic *interference*. Some authors define paternalism broadly enough to include omissions. Shiffrin, for instance, describes the following case:

> Suppose B has no valid claim to A's assistance but asks A, an acquaintance, for help building a set of shelves. A refuses, but not because A is too busy or disinclined in help. In fact, A is eager to deploy her carpentry skills. She declines on the grounds that B

> too often asks for assistance to his own detriment: he is failing to learn for himself the skills that he needs, or perhaps he displays unwarranted insecurity in his own skills.
> (Shiffrin 2000: 213)

According to Shiffrin, A's omission is paternalist insofar as "A substitutes her judgment for B's about what B should aim for and works around B's agency to get B to act as A believes would be better for B" (Shiffrin 2000: 213). One might reasonably wonder whether Kantian principles oppose this form of paternalism, as well as the more familiar types of paternalistic interference addressed above. It is not obviously the case that A *uses* B, thus it might be difficult to argue that A uses B merely as means. The moral assessment of A's omission appears to depend entirely on A's motives, her reasons for declining to help B. For instance, if A declines to help out of concern that she use her own agency only in responsible ways – that she not inadvertently create dependence – then I think there is little for the Kantian to find objectionable in the scenario, but also less reason to regard A's behavior as paternalist. This, however, is not how Shiffrin understands A's motivation. Shiffrin describes A as primarily concerned not with how she uses her own agency, but rather with getting B to act in a particular way. Insofar as A aims to circumvent and manipulate B's agency, this appears to constitute a failure to respect B's status as an end in herself and would thus be objectionable on Kantian grounds.

Conly challenges the antipaternalist by asking why it is more respectful to allow another to choose poorly rather than interfering when this will most likely yield a better result according to the agent's own understanding of her interests. The Kantian answer to this question is that it is more respectful to acknowledge another's authority simply because she is *entitled* to this recognition in virtue of being a fully mature – albeit imperfectly rational – moral agent. Giving others the respect to which they are entitled does not preclude concern for their welfare. It merely precludes promoting the welfare of others in ways that attempt to circumvent their agency. This takes a few options off the table (e.g., manipulation and deception), but it hardly ties our hands.

3 Kantian opposition to paternalism: self-perfection and the subordinate value of well-being (happiness)

The previous section located the principal wrong-making feature of paternalistic interference, at least from a Kantian perspective, in what the paternalist does. Paternalism is disrespectful insofar as it entails treating a competent adult as mere means to her own interests, welfare, or ends, or otherwise endeavors to circumvent or manipulate another's agency. Some authors believe that paternalism can be made acceptable provided that the paternalist does not substitute her judgment of another's interests for the agent's own, that she acts to secure what the agent wants or would want if she were not such a flawed thinker. In the previous section I explained why this qualification on paternalistic interference would not satisfy the fundamental Kantian complaint. In this section, I want to focus on the paternalist's *motivation* for acting, namely a beneficent interest in promoting another's good. My suspicion is that those, like Conly, who defend paternalism operate with a conception of another's interests that is not compatible with fundamental Kantian ideas about the subordinate value of individual happiness or well-being. This adds an additional layer to the Kantian opposition to paternalism.

Despite the fact that Kant understands the human desire for happiness to be "a powerful counterweight to all commands of duty" (Kant 1996: 59; 4:405), he does not recommend that we attempt to purge ourselves of this desire. This would be futile. The desire for happiness is an inextricable part of the human condition. Kant could not be less ambiguous about this. In the *Critique of Practical Reason* he proclaims that "To be happy is necessarily the demand of every rational but finite being and therefore an unavoidable determining ground of its faculty of desire" (Kant 1996: 159; 5:25), and in a later text that "it is unavoidable for human nature to wish for and seek happiness" (Kant 1996: 519; 6:387). For better or worse, human beings are creatures that need happiness. This fact about our nature has implications for Kant's moral theory. Kant declares that others' happiness is an end that is also a duty and that one's own happiness might also be a duty if it weren't for the fact that it is contradictory to say that we are constrained to adopt an end we necessarily and unavoidably have (Kant 1996: 517; 6:385–386).

Kant clearly understands happiness to be important; however, it is not the most important end for a human being. The subordinate value of individual happiness is perhaps best observed in Kant's doctrine of the highest good. According to Kant, the highest good is a composite good made up of virtue and happiness. Kant is critical of the Stoics and the Epicureans for what he perceives to be their failure to recognize that happiness and virtue are "extremely heterogeneous concepts" (Kant 1996: 229; 5:111). Virtue makes one worthy of happiness, but not necessarily happy. Kant explains:

> Now inasmuch as virtue and happiness together constitute possession of the highest good in a person, and happiness distributed in exact proportion to morality (as the worth of a person and his worthiness to be happy) constitute the *highest good* of a possible world, the latter means the whole, the complete good, in which, however, virtue as the condition is always the supreme good, since it has no further condition above it, whereas happiness is something that, though always pleasant to the possessor of it, is not of itself absolutely and in all respects good but always presupposes morally lawful conduct as its condition.
>
> (Kant 1996: 229; 5:110–111)

Virtue is unconditionally good, the *supreme* good, but it is not the *complete* good for human beings. Human beings need happiness, so happiness must be part of the highest good; its goodness, however, is always conditioned by virtue.[8]

Thus we see that the relationship between happiness and virtue is complicated. Human beings need happiness, and we unavoidably pursue our own happiness; however, the value of happiness (its goodness) depends on the presence of a good will (Kant 1996: 49; 4:393). Our desire for happiness will at times tempt us to transgress the moral law, and this provides us with moral reason to "attend to one's happiness" (Kant 1996: 214; 5:93).[9] But the relationship between happiness and virtue is more complicated still. We get the Kantian picture wrong if we think that the desire for happiness is simply some unfortunate human liability. Barbara Herman has thoughtfully observed that there is a more symbiotic relationship between the pursuit of happiness and self-development. She explains:

> Because effective agency is not like getting one's adult teeth, it will not just happen with time and food, a moral theory that prizes the value of rational agency has to be especially sensitive to its social and material conditions as it goes about the business of parceling out goods . . . the vehicle that drives the development of human rational agency is the natural interest we have in our own happiness.
>
> (Herman 2002: 241)

Our desire for happiness drives the development of human rational agency, in part, because happiness is such a nebulous end. We desire to achieve a state of lasting satisfaction, and yet we do not know what would bring this about. Kant laments that "it is a misfortune that the concept of happiness is such an indeterminate concept that, although every human being wishes to attain this, he can still never say determinately and consistently with himself what he really wishes and wills" (Kant 1996: 70; 4:418).

The desire for happiness drives the development of rational agency because pursuing happiness requires working out a conception of happiness, as well as coming to terms with the fact that not all of our desires can be satisfied. We must deliberate about which desires to take up as ends and which to reject. When we are left to make our own decisions – decisions about ends as well as means – we experience the consequences of our good and poor choices. We learn from our mistakes. We develop judgment. Moreover, we learn that we can delay gratification and that we can say no to our inclinations. Greater self-mastery is a desirable by-product of the pursuit of happiness. The self-command we gain in pursuing happiness is instrumental to our self-perfection as moral agents.

An additional problem with certain forms of paternalism is that the paternalist risks undermining the development of self-command and self-perfection by not allowing agents to reap the benefits of making their own choices. While paternalistic interference may succeed in promoting the individual's well-being, if it does so at the cost of imposing an obstacle to self-development, this is not a good trade off, at least from the Kantian perspective.

At this point we should acknowledge that paternalistic interference need not aim at promoting another's *happiness*.[10] A paternalistically motivated agent might aim to promote another's self-development, as is the case in the previously considered example from Shiffrin. A declines to help B not because she believes it will make B happier, but rather because she believes B will develop his skills and confidence if he completes the project on his own. And A may very well be correct in her judgment. Paternalism that aims to promote another's self-perfection could succeed in achieving its end. This observation reveals the limits of this particular argument against paternalism. Paternalist behavior does not always undermine the development of self-mastery, and thus concern for self-perfection will not be a consistent reason to oppose paternalism. Concern for agent self-development and self-perfection thus merely serves as a supplement to the more fundamental Kantian opposition to paternalism discussed in the previous section, namely, that most forms of paternalism treat agents as mere means or otherwise fail to respect their status as ends in themselves.

4 Kantian opposition to paternalism: the innate right to freedom

The final point regarding the Kantian grounds for opposing paternalism is relevant primarily to liberty-restricting paternalist legislation. The cornerstone of Kant's political philosophy is the view that human beings possess an innate right of freedom. The innate right to freedom entitles agents to "independence from being constrained by another's choice insofar as it can coexist with the freedom of every other in accordance with a universal law" (Kant 1996: 393; 6:237). In other words, actions (or laws) are juridically right insofar as they are consistent with universal freedom in accordance with universal laws and juridically wrong insofar as they are not. Presumably, the only warranted restriction on individual freedom would be those restrictions which are necessary to protect or ensure universal freedom. Thus the state would be justified in restricting my freedom provided that this was necessary to preserve or protect the freedom of others. As Kant puts it,

> if a certain use of freedom is itself a hindrance to freedom in accordance with universal laws (i.e. wrong), coercion that is opposed to this (as a *hindering of a hindrance of freedom*) is consistent with freedom in accordance with universal laws.
>
> (Kant 1996: 388; 6:231)

The state may use coercive force to prevent me from assaulting my fellow citizens or interfering with their property or to enforce a legitimate contractual agreement, but not to compel me to make more rationally prudent decisions regarding my own interests.

Kantian political philosophy provides very little space, if any, for justified paternalism. Is this a liability? I am not convinced that it is. On the one hand, it is not clear that certain sorts of legislation can *only* be justified on paternalist grounds. Take, for instance, mandatory seatbelt laws, which Conly regards as an example of paternalist legislation that enjoys widespread acceptance (Conly 2013: 175–176). The paternalist justification for this law appeals to the interests of the one required to wear a seatbelt. In the event of an accident, the paternalist argues, wearing a seatbelt can protect you from serious injury or even death. Given that staying alive is likely a necessary condition for achieving your ends, wearing a seatbelt is what you would choose to do if you were behaving rationally. But we could conceivably justify mandatory seatbelt laws on nonpaternalist grounds. Those who are seriously injured in automobile accidents require medical treatment the cost of which is almost never borne exclusively by the patient. There are also concerns about lost productivity at work, and those who are fatally injured in accidents may leave behind minors who must then be cared for by someone else. Thus we might justify this minor intrusion into individual liberty by appealing to interests that go well beyond the individual's interest in avoiding bodily harm.

Enforcing a mandatory seatbelt law is not terribly costly; however, other examples of coercive paternalism are. A ban on cigarettes, which Conly endorses, would necessitate law-enforcement resources that are arguably comparable to the current ban on marijuana. According to some estimates, the ban on marijuana costs the government billions of dollars a year (Sledge 2013). In addition to justifying a restriction on individual freedom, at least some paternalist legislation faces the additional challenge of justifying the use of public funds to force individuals to act in their own best interest. Conly recognizes this, and I believe her response is telling. She writes,

> we do need to consider the costs to third persons of these interventions. . . . In most of the cases discussed in this chapter, I think the costs to third parties are not so great as to make the paternalistic intervention too big a burden. On the contrary, they will make other people come out ahead. For example, while there will be the costs of enforcement – making sure there are no illegal cigarettes out there – which we will all be paying, there will not be the $96 billion in medical costs that we now pay for smoking-related diseases.
>
> (Conly 2013: 181)

Rather than directly defending the use of public funds to prevent individuals from making imprudent choices, Conly refers to the public good of health care savings. If we are going to appeal to public goods to justify the enforcement costs, why not appeal to the same public goods to justify the restriction on individual liberty? My point is simply that paternalistically justified coercion may be neither necessary nor the best means for achieving desirable ends like reduced rates of smoking and automobile fatalities.

5 Conclusion

In this chapter, I have endeavored to explicate, in distinctly Kantian terms, the common charge that paternalism is disrespectful. I have argued that most examples of paternalistic interference are objectionable on Kantian grounds insofar as they treat competent adults merely as means. In virtue of the respect we are entitled to from our fellow moral agents, being treated merely as means is morally problematic even if one is treated merely as means to one's own ends. The beneficent motive does not absolve this fundamental failure of respect. I have suggested that even when paternalism takes a subtler form and does not use another merely as means it may still be objectionably disrespectful from a Kantian perspective provided that the paternalist seeks to circumvent or manipulate another's agency. I have further argued that many examples of paternalism are additionally problematic when they endeavor to promote another's happiness or well-being at the expense of undermining her self-development. While this charge is not applicable to all cases of paternalism, I suspect it is relevant to a good many. In making these arguments, I hope to have shown Kantian opposition to paternalism to be highly plausible. The fundamental Kantian lesson is that our beneficence ceases to be morally admirable when it fails to be respectful (Kant 1996: 568–569; 6:448–449). The burden is on the defender of paternalism to explain either how paternalism is compatible with a plausible account of respect, or, alternatively, why principles of beneficence may sometimes overshadow principles of respect.

Related topics

Consequentialism, Paternalism, and the Value of Liberty; Libertarian Perspectives on Paternalism; Paternalism and Duties to Self; Perfectionism and Paternalism.

Notes

1 Shiffrin's own account is broader than the one I adopt here. Her account recognizes actions solely and directly concerned with a third party's welfare as paternalist, as well as some actions not motivated by welfare concerns at all. A full account of paternalism will depend on some account of an individual's "sphere of legitimate agency." However, I think we can proceed with merely a rough intuitive sense of what this sphere likely includes.
2 See also Kant (1996: 85; 4:436).
3 See Sensen (2011) for a more thorough exposition of Kant's use of the concept of dignity.
4 See Hill (1980; 2003), O'Neill (1985), Korsgaard (1986), Kerstein (2011, 2013), and Parfit (2011).
5 Conly defends coercive paternalism by appealing to cognitive biases which undermine our ability to select appropriate means to achieve our goals. However, the same research suggests that we are just as bad at selecting ends as we are at selecting means. Conly's defense of a modest paternalism thus appears vulnerable to the charge of arbitrariness.
6 I maintain that there is a fundamental difference between respecting someone's judgment and respecting her authority or propriety even when her authority is dependent on the possession of particular capacities.
7 In cases where we have reason to believe that another's agency is impaired in some significant sense (e.g., someone suffering from mental illness, serious addiction, or under the influence of drugs), beneficently motivated interference may be far less objectionable from the Kantian perspective. See Fahmy (2011) and Cholbi (2013). However, we might argue one's "sphere of legitimate agency" is more limited in virtue of her impaired agency, and thus the interference does not constitute paternalism according to the account articulated at the outset of the chapter.
8 Note that Kant regards a product of autonomy (virtue) as the supreme or unconditioned good, not autonomy itself.
9 See also Kant (1996: 519–520; 6:388).
10 I am grateful to Jason Hanna for bringing this objection to my attention and for the many other helpful suggestions he and Kalle Grill provided.

References

Baron, M. and Fahmy, M. S. (2009) "Beneficence and Other Duties of Love in the Metaphysics of Morals," in T. Hill (ed.) *The Blackwell Guide to Kant's Ethics*, West Sussex: Wiley-Blackwell.

Cholbi, M. (2013) "Kantian Paternalism and Suicide Intervention," in C. Coons and M. Weber (eds.) *Paternalism: Theory and Practice*, Cambridge: Cambridge University Press, pp. 115–133.
Conly, S. (2013) *Against Autonomy: Justifying Coercive Paternalism*, Cambridge: Cambridge University Press.
Darwall, S. (2006) "The Value of Autonomy and Autonomy of the Will," *Ethics* 116: 263–284.
de Marneffe, P. (2006) "Avoiding Paternalism," *Philosophy & Public Affairs* 34: 68–94.
Fahmy, M. S. (2011) "Love, Respect, and Interfering With Others," *Pacific Philosophical Quarterly* 92: 174–192.
Feinberg, J. (1986) *The Moral Limits of the Criminal Law, Volume 3: Harm to Self*, New York: Oxford University Press.
Herman, B. (2002) "The Scope of Moral Requirement," *Philosophy & Public Affairs* 30: 227–256.
Hill, T. E., Jr. (1980) "Humanity as an End in Itself," *Ethics* 91: 84–90.
———. (2003) "Treating Criminals as Ends in Themselves," *Jarbuch für Recht und Ethik* 11: 17–36.
Kant, I. (1996) *Practical Philosophy*, ed. and trans. M. Gregor, Cambridge: Cambridge University Press.
Kerstein, S. J. (2011) "Treating Consenting Adults as Mere Means," in M. Timmons (ed.) *Oxford Studies in Normative Ethics* vol. 1, Oxford: Oxford University Press.
———. (2013) *How to Treat Persons*, Oxford: Oxford University Press.
Kleinig, J. (1984) *Paternalism*, Totowa, NJ: Rowan & Allenheld.
Korsgaard, C. (1986) "The Right to Lie: Kant on Dealing With Evil," *Philosophy & Public Affairs* 15: 325–349.
Mill, J. S. (1978 [1859]) *On Liberty*, Indianapolis: Hackett.
O'Neill, O. (1985) "Between Consenting Adults," *Philosophy & Public Affairs* 14: 252–277.
Parfit, D. (2011) *On What Matters*, Oxford: Oxford University Press.
Sensen, O. (2011) *Kant and Human Dignity*, Berlin: Walter de Gruyter.
Shiffrin, S. V. (2000) "Paternalism, Unconscionability Doctrine, and Accommodation," *Philosophy & Public Affairs* 29: 205–250.
Sledge, M. (2013) "Marijuana Prohibition Now Costs the Government $20 Billion a Year: Economist," *The Huffington Post*, www.huffingtonpost.com/2013/04/20/marijuana-prohibition-costs_n_3123397.html [Accessed 4 Jan. 2017].

9

PATERNALISM AND DUTIES TO SELF

Michael Cholbi

Ordinary moral thought, and much of contemporary moral philosophy, tend to assume a social conception of morality. On this conception, our moral duties are exclusively other-regarding, that is, all our moral duties are ultimately justified by appeal to morally salient facts about other people. Proponents of this social conception of morality disagree about the source and extent of our moral duties. Nevertheless, that morality is a social tool, a set of principles or practices concerned with how we treat one another, is the guiding assumption of this conception of morality.

No doubt a significant portion of morality is social in this way. Historically, however, many moral thinkers have maintained that in addition to our other-regarding duties, we also have self-regarding duties or *duties to self*. We shall have occasion later to state more precisely what duties to self are, but at a rough level, duties to self are duties the performance or fulfillment of which by individual S is owed to S. When an individual fails to fulfill a duty to self, it is she, and not other people, who is thereby wronged.

The dominance of the social conception of morality has meant that contemporary moral philosophy pays scant attention to duties to self, and what attention it has paid has largely been skeptical of such duties. In my estimation, duties to self have been inadequately analyzed by philosophers, and as a result, prematurely dismissed. Suppose instead that there are duties to self and that the social conception of morality is mistaken. This supposition would have many implications, but the implications would be of particular interest to philosophers interested in the moral justification of paternalism. The debate about paternalism primarily concerns whether it can be morally justified to intercede, without their consent or authorization, in the choices or actions of rationally competent individuals so as to make those individuals better off in some way. Those sympathetic to paternalism may see duties to self as another avenue through which to intercede in individuals' choices or action for their benefit, namely, so as to enable them to fulfill such duties.

The purpose of this chapter is to sift through the ramifications of duties to self for the moral justification of paternalism. In some respects, duties to self do not appear to introduce any new considerations into these debates. However, I shall argue that the distinctive nature of the goods associated with duties to self, namely, that the values of these goods to an individual turn crucially on that individual's playing a key role in their realization, entails that we have reasons to resent paternalism directed at aiding our fulfillment of these duties, reasons that we do not have for resenting paternalism directed at realizing goods typically associated with our welfare.

But of course these conclusions are of merely theoretical interest if, as many philosophers have maintained, there simply are no duties to self. Hence our first task is to clarify and rehabilitate duties to self. The first two sections thus offer a model of duties to self, based on Kant's account of such duties, and answer common objections to the existence of these duties.

1 Clarifying duties to self: the Kantian model

The preeminence of the social conception of morality may make duties to self seem exotic or obscure. Let us then first clarify the nature of these duties, making reference to Kant's account of duties to self, arguably the most systematic and fully developed account of such duties.[1]

The "first principle" of duties to self, according to Kant, is to "'live in conformity with nature,' . . . that is, to *preserve* yourself in the perfection of your nature; the second in the saying, '*make yourself more perfect* than mere nature has made you'" (Kant 1996: 175/6:419). As Kant saw it, fundamental to our nature is the fact that we are practically rational agents, capable of choosing our goals and the means to achieving those goals. Our capacity for rational agency is thus the source of a distinct class of self-regarding moral duties. The various duties to self, for Kant, therefore correspond to various ways in which that capacity is to be respected and cultivated.

Kant divides duties to self into three categories. (His motivations for this division need not concern us here.) The first category are duties we owe ourselves due to our "animal," i.e., physical, nature. The most fundamental of these duties is the duty of bodily self-preservation. This precludes our killing ourselves as well as our intentionally disabling or incapacitating our bodies (1996: 176–177/6:422). The other duties in this category are not to imbibe food or drink in ways that impede our powers of "skill and deliberation" (1996: 180/6:427). What unites the duties we owe ourselves as animal beings is that in violating them, we inhibit our capacity to exercise, and hence fail to respect, our rational agency.

A second category of duties to self concern our "moral" nature. These include a duty to avoid "avarice," the condition in which one does not make use of one's resources to meet one's basic needs, as well as a duty not to make oneself servile to others. (Selling ourselves into slavery or acquiring crushing debts are examples of the servility Kant condemns.) Whether through stubbornly hoarding one's resources (avarice) or through subjugating one's will to another's (servility), violations of these duties inhibit our capacity to pursue our chosen ends and so show a lack of respect for our rational capacities.

The final Kantian category of duties to self concern the pursuit of our "natural perfection." We have, according to Kant, duties to cultivate our mental and physical talents so as to enhance our ability to pursue our chosen goals. In so doing, we respect our rational agency by rendering it more efficacious. Kant terms this an "imperfect" duty, meaning that we are not required to develop all our talents to the greatest degree. Rather, this duty allows our choice of which talents to develop to be guided by our tastes and interests.

Note that in speaking of duties to self, Kant (and other proponents of such duties) assume that at least some of our duties are "directed," that is, their performance is *owed* to specifiable individuals (May 2015). This assumption is not one shared by all moral theorists. Adherents of some versions of impersonal consequentialism, for example, maintain that, at root, all our moral duties are duties to bring about particular states of affairs, namely, to perform the act that results in the best overall outcome. According to such theories, to speak of duties being directed or owed to particular individuals is somewhat misleading, for even if (say) a particular individual stands to benefit from the fulfillment of our duty to bring about the best overall outcome, the duty is not fundamentally owed to her. According to these theories, if we have a moral duty to provide disaster relief, this is because doing so will result in the best overall outcome, not because it is owed to the particular beneficiaries of that relief; if we have a moral duty to be beneficent toward our siblings, this is because doing so will result in the best overall outcome, not because such beneficence is owed to our siblings; etc. That at least some of our duties are directed is far from a trivial assumption, but it raises issues in moral theory too large to be addressed here. There are also other controversial aspects of Kant's particular account of duties to self that we have not pursued in detail here,[2] and a fuller defense of duties to self would involve comparing the strengths and weaknesses of Kant's account to others. Nevertheless, Kant's account is instructive, highlighting three crucial features of duties to self.

First, his account highlights the properties of duties to self that distinguish them from duties to others. All duties are duties *of* selves, i.e., duties that call upon individuals to act in prescribed ways. What licenses classifying a duty as a duty *to* self? A useful method for capturing the essence of duties to self is to distinguish between the *subject* of a duty, that individual required to fulfill a given duty, and the *object* of a duty, that individual to whom fulfillment of a duty is owed (Hills 2003: 132; Timmermann 2006: 506). An other-regarding duty is such that the subject and the object of the duty necessarily diverge. If S is obligated not to harm T, then S is the duty's subject, T its object. If S wrongfully harms T, then T is wronged by S. In contrast, if S has duties to self (duties, as Kant believed, to preserve our bodily powers, develop our talents, forego servility, etc.), then when S fails in these duties, S herself is wronged by these failures. Note that a duty to self is therefore something more than a duty to treat oneself in a particular way. For instance, suppose you promise your fiancée that you will get a haircut before your wedding ceremony. The duty requires you to treat yourself in a particular way, but it is a duty whose object is the person to whom you made the promise, your fiancée. This duty is actually other-regarding, but because it demands a certain treatment of yourself, some call these duties *concerning* oneself or *indirect* duties to self.

Duties to self thus have a distinct object. But Kant's account also highlights that duties to self have a distinct subject too. If S has a particular duty to herself, then no one but S has that duty (though of course others are subject to the same duties with regard to themselves as S is with regard to herself). In Kant's terms, our duties to preserve our bodies, etc., are, by their very nature, duties that are *not* morally incumbent on others. A duty to self, therefore, is a duty in which the subject and object converge, that is, a duty to self is one whose fulfillment is owed to the very same person with the duty to fulfill it.

Third and finally, note that on Kant's account of duties to self, these are not *welfare*-based duties, i.e., they are not duties grounded in requirements to promote one's own welfare. The point of Kant's duties to self is not that by fulfilling them our lives will go better or be happier, in any ordinary sense. Rather, their point is to show respect for ourselves as rational agents. The person who fulfills her self-regarding duties maintains a kind of "moral health," a condition wherein her capacities to choose her goals with rational lucidity and to follow the most suitable plans for realizing those goals are unimpaired and not harnessed to the rational wills of others. But the maintenance of our moral health not only does not require us to promote our own welfare – in fact, it may demand that we forego opportunities to promote our own welfare. No doubt individuals can find themselves in situations wherein their own goals are promoted by ending their lives prematurely via suicide or by selling themselves into slavery. Kant's duties to self morally preclude such actions, *despite* the likelihood that they can contribute to individual welfare.

2 Answering skeptics about duties to self

Armed with Kant's account of duties to self, we are now in a position to address skeptics about duties to self, especially adherents to the social conceptions of morality.

Perhaps the boldest argument for rejecting duties to self is to assert that morality conceptually precludes self-regarding considerations altogether. Stephen Finlay (2008: 140–142) claims that "only considerations arising from the interests of others" belong to the "ordinary" understanding of morality, and given the authority of this ordinary understanding, we should reject duties to self. Morality is "purely and essentially other-regarding," Finlay concludes. No doubt Finlay is correct that self-regarding considerations have a marginal role in ordinary moral understanding. However, the fact that moral understandings are historically contingent should give us pause. As Peter Singer (2011) has observed, the last several centuries have witnessed a slow but steady growth in our "moral circle," as the scope of our moral concern has expanded to include animals, distant strangers, future generations, etc. The decline of philosophical and popular recognition of duties to self – a shrinkage in our moral circle – is therefore puzzling, even ironic: Why should our moral circle not also include our selves? At the very least, given that duties to self were part of ordinary moral understanding as recently as Kant's time, we ought not conclude that current moral understandings, with their apparent exclusion of duties to self, reflect the correct concept of morality as opposed to being one possibly mistaken *conception* of morality.

Finlay offers a second reason for skepticism about duties to self: if there were such duties, Finlay argues, then morality would condemn actions in which a person engages in extensive sacrifice of her own interests, particularly in order to protect or advance the interests of others. But because morality does not condemn such actions, and in fact finds them praiseworthy, there must not be duties to self (2008: 140). Here we see Finlay incorrectly interpreting duties to self as requiring that individuals promote their own welfare, so that large-scale sacrifices of one's own welfare are morally objectionable. However, we saw in Kant's account of duties to self that such duties need not be oriented around the promotion of one's own welfare. Hence, we need not suppose either that duties to self include a duty to promote one's own welfare or that moral theories acknowledging duties to self oppose acts of extensive self-sacrifice. Indeed, it is a strength of Kant's account of duties to self that it does not ground these duties in considerations of welfare, and as a result, can affirm the intuition that morality permits, even lauds, foregoing one's own welfare.

A third skeptical argument is that it is always possible and morally permissible for the object of a duty to release its subject from its performance; but no one can release herself from a duty; thus, there can be no duties in which the subject and object are the same individual, i.e., no duties to self. This argument is vulnerable at a number of points. For one, this argument is most often pressed in the context of promissory obligations (Singer 1959; Hills 2003: 132–134). What meaning can be given, skeptics about self-regarding duties ask, to a *duty* to keep a promise one made oneself if one has unfettered liberty to waive the duty by releasing oneself from it? Note though that Kant does not propose that there can be duties stemming from promises to oneself, and there does not seem to be any particular reason to suppose that promises can generate duties to self unless one makes the unlikely assumption that each of our other-regarding duties must have a parallel self-regarding duty.

Moreover, even if this "releaseability" argument spells doom for self-regarding promissory duties, it may not extrapolate to other self-regarding duties. In fact, releaseability is *not* a paradigmatic feature of duties to self. Kant certainly did not hold that we could simply opt out of our duties to preserve our lives, abjure servility, etc.[3] Indeed, our ability to release others from the duties they owe us seems to imply that we have duties to self from which we cannot release ourselves. We are morally permitted to waive the duties others owe us, as when, by giving informed consent to a medical treatment, we waive our right that others not interfere with our bodies. But this moral power to alienate our interpersonal rights seems to require that there be some inalienable right in us, a right resting on some value we have as rational agents (Velleman 1999: 611–612).

Advocates of duties to self should therefore not take the fact that we cannot release ourselves from them as an embarrassment or an anomaly in need of explanation. Rather, their non-releaseability is a consequence of the nature of the value on which such duties logically depend.

A final related worry about duties to self is that it appears difficult to make sense of *accountability* with respect to such duties. Ordinarily, moral duties carry the implication that those objects wronged by their violation can and should hold the subjects accountable for these violations, by adopting certain attitudes toward them (blame, resentment, etc.) or by treating them in particular ways (punishment, for instance). Marcus Singer (1959) hints that we cannot coherently hold ourselves accountable for violations of self-regarding duties. I do not see that this is so, however. We can and do feel guilt at entering into servile relations with others in order to promote our goals; we can and do feel resentment at the slothfulness that leads us to neglect our talents or to permit our bodies to deteriorate; and we can and do "punish" ourselves for violations of duties to self by predicating future rewards on their fulfillment (Cholbi 2015). Granted, the network of normative concepts we use to describe our accountability for self-regarding duties is likely to differ from the network used to describe our accountability for other-regarding duties. Our vocabulary for the latter is more likely to reference harms, rights, etc., whereas the vocabulary for the former is more likely to reference disappointment, regret, or self-respect. But these conceptual differences only suggest that self-regarding duties have a different normative basis from other-regarding duties, not that they have no basis whatsoever.

Thus, none of these considerations provide compelling reasons to reject duties to self. Indeed, these skeptical arguments can be justly accused of not taking seriously the prospect of duties to self as a distinct deontic category. All proceed on the assumption that duties to self, to earn their philosophical credibility, must be modeled on duties to others, and because (allegedly) they cannot be so modeled, they must be rejected. The replies provided to these arguments show that this assumption largely begs the question against duties to self.

3 Paternalism and the fulfillment of duties to self

Assuming then that there are duties to self as Kant envisioned them, let us now turn to our central aim, namely, exploring the implications that self-regarding duties have for the moral justification of paternalism. Two caveats are in order before embarking on this exploration.

No doubt paternalism could be motivated by the desire that others fulfill their duties to themselves, and efforts to lead others to choose or act so as to fulfill these duties would count as paternalistic inasmuch as they are motivated not by the interests, rights, etc., of paternalists or of the community at large but by the aim of leading others to do something thought good for themselves, namely, fulfilling their duties to self. However, we must not overestimate the efficacy of paternalistic measures in leading others to fulfill their self-regarding duties. For as Kant pointed out, duties to self are "duties of virtue" rather than "duties of right" (1996: 31/6:239). We can be moved to fulfill duties of right via "external" compulsion or coercion because their requirements are essentially behavioral. For instance, the threat of punishment may convince someone to refrain from stealing. Her behavior thus fulfills her duty to respect others' property, and she need not act from any specific motive in order for that duty to be fulfilled. Duties of virtue, in contrast, demand that we behave in particular ways on the basis of certain reasons, or in Kant's terms, that we adopt a particular "end." Our duties to self, according to Kant, belong in this second category. Our duty of self-preservation, for example, is not fulfilled simply by keeping ourselves alive. It is fulfilled when we keep ourselves alive because keeping ourselves alive is among our ends. Because duties to self require that we act on the basis of particular reasons or ends, their fulfillment cannot be coerced or compelled by "external" means. The fulfillment of our duties to self must instead be a matter of "free self-constraint" (Kant 1996: 147/6:382). Note that this does not mean that we cannot be compelled to behave *as if* we have such ends. A person planning to end her life, for instance, could be compelled to preserve herself if someone threatened to harm one of her loved ones if she proceeded with that plan. But in such a case, she will have conformed her behavior to the duty of self-preservation but will not have fulfilled that duty, since her end was not self-preservation (or respect for her own rational agency more generally) but protecting a loved one. This fact implies that paternalistic efforts aimed at leading individuals to honor their duties to self cannot succeed in a strict sense. We cannot compel, coerce, etc., others to keep themselves alive, develop their talents, and so on, *for the sake of* their self-regarding duties. But paternalists can hope to influence individuals so as to behave in accordance with these duties and, over the long run, to shape their character in ways conducive to their fulfillment. Take Kant's duty to refrain from "stupefying" ourselves with food or drink. Paternalistic measures such as alcohol taxes, etc., cannot compel people to refrain from excessive drinking out of respect for their own rational agency. But they could encourage moderation, reduce temptation, etc., so as to engender or habituate such respect.[4]

The second caveat concerns the kind of paternalism that duties to self make possible. Paternalism can be good for its target in one of two ways. *Welfare* paternalism aims to improve its targets' well-being, happiness, life-satisfaction, etc. *Moral* paternalism aims to improve its targets' moral conduct or character. Duties to self do not, as we have observed, acquire their rationale from welfare considerations; the Kantian duties to preserve our bodies, avoid servility, and the like are duties that bind us irrespective of how honoring these duties may contribute to our own welfare. Hence, there cannot be welfare-based reasons to justify paternalistic intercessions based on a concern for targets' ability to fulfill their self-regarding duties. Such intercessions could only be justified as species of moral paternalism. This does not preclude paternalism that uses the promotion of welfare as a means of facilitating its targets' fulfilling their self-regarding duties. Paternalistic anti-poverty measures, for instance, could increase their targets' welfare so that the targets will find income-producing activities that involve servility (jobs requiring a person to forego basic liberties, for instance) less appealing.

These two considerations bring into clearer view the sort of paternalism that duties to self might license: measures that facilitate, encourage, or reward individuals coming to have their own rational agency as one of their basic ends by facilitating, encouraging, or rewarding self-regarding choices that in turn foster the fulfillment of duties to self. The question at hand is whether such measures are morally justified.

4 A dialectical cul-de-sac?

The prospect that duties to self invite suspect forms of moral paternalism has sometimes been invoked as a reason to reject duties to self. Some have reasoned that if there were duties to self, communities might well feel compelled to enforce those duties, which would in effect treat individuals as children needing protection not only from others, but also from themselves (Baier 1958: 250; Denis 2001: 4–5). Suppose, for example, that among our duties to self is a duty of self-preservation. A community acknowledging such a duty could take measures that seek to prevent individuals from engaging in suicide. Some of these measures would be morally benign (Cholbi 2011: 116–117), but others would involve significant intrusions into individuals' personal spheres of action (e.g., monitoring social media for evidence of suicidal thinking) or liberty (compelling suicidal individuals to take psychotropic drugs, undergo mandatory psychotherapy, or be involuntarily institutionalized). If such measures are the price of acknowledging duties to self, some antipaternalists believe, then that is too great a price to bear.

However, this antipaternalist reasoning moves too quickly. For one, as we have noted, paternalistic measures cannot compel the *fulfillment* of duties to self. At most, paternalistic measures can compel the behaviors associated with duties to self (keeping oneself alive, say) in the hopes of inculcating self-regarding virtues. But here antipaternalists may argue that irrespective of whether paternalistic measures can compel the fulfillment of self-regarding duties, paternalistic efforts guided by the aim of helping individuals to fulfill such duties are nevertheless morally prohibited. For if, as Mill (1859) asserted, the only basis for interfering with a person's choices or actions without her consent is "self-protection" – that the individual's own good, "physical or moral," is "not a sufficient warrant" for such interference – then paternalism in order to inculcate self-regarding moral character is ruled out. Indeed, whatever the case for moral paternalism in general, the case for this form of moral paternalism appears especially weak. The inculcation of *other*-regarding virtues (generosity, sympathy, a sense of justice, etc.) will shape how the members of a community treat one another. But the only individual whose fate is at stake in the inculcation of self-regarding virtues, on the other hand, is the individual. If there is anything in which other people, and the community at large, do *not* have a stake, it is whether an individual develops the virtues that will enable her to preserve herself, forego avarice, develop her talents needed to advance her own ends, and so on.

Here those more sympathetic to the paternalistic cause may concede that communities do not have a direct interest in whether their members fulfill their self-regarding duties but nevertheless have morally justifiable grounds for paternalism in this regard. Community members ought to care for one another; to care for one another is to desire what is good for others for their own sake (Darwall 2002); the development of the self-regarding virtues needed to fulfill one's duties to self is good for a person for her own sake; so community members ought to take measures to inculcate these virtues in one another, via paternalistic measures if need be; hence, morally paternalistic measures aimed at the inculcation of self-regarding virtues ought to be undertaken. If sound, such an argument would lend support to including the inculcation of the self-regarding virtues as one of the aims of interpersonal conduct and of social policy.

But again, opponents of paternalism may concede that we have good reason to *consent* to acts and policies that inculcate the self-regarding virtues. It still does not follow, they may argue, that we have good reason to endorse paternalism in this guise. Paternalism, after all, is *non*-consensual intercession in another's choice or action for their own good. That we ought to consent to others' aid in inculcating self-regarding virtues in us does not obviate the force of this antipaternalist complaint.

5 Historical and ahistorical goods

To this point, it may seem that duties to self do not shift familiar debates about the moral justification of paternalism so much as rehash them: given that there are self-regarding duties, they offer a new avenue for moral paternalism. But the considerations offered for and against such paternalism are not fundamentally different from those that drive extant disagreements about paternalism. Advocates of using paternalistic measures cite their potential to inculcate self-regarding virtues that facilitate individuals' fulfilling their duties to self, to their presumed "moral" benefit. Conversely, antipaternalists invoke their cherished thesis that intercessions in others' choices and actions without their consent must be governed solely by considerations about how those choices and actions affect others – and duties to self, as we have seen, are not the concerns of others.

I shall now argue that considerations regarding how the value of goods can be sensitive to how they are realized offer a way out of this dialectical cul-de-sac, and as a result, we have reason to oppose paternalism aimed at inculcating self-regarding virtues and the fulfillment of duties to self.

With respect to some goods, it is often of little importance to us how we come to enjoy them. Such is the case, I propose, with many of the goods commonly assumed to contribute to welfare. Take, for instance, the value of relief from pain. Its value to us does not seem to hinge on exactly how pain relief comes about (whether simply from its abating, via the administration of drugs, removal of environmental conditions contributing to the pain, etc.). What matters is that we enjoy freedom from pain, regardless of precisely how that freedom comes about. Such claims should not be exaggerated. It may matter very much *morally* how our pain relief is attained. We would be justifiably worried to learn that our pain relief came at the expense of causing pain to someone else, for instance. But just insofar as pain relief is good for us, its goodness is *ahistorical*, independent of how it is realized. So too, I propose, for many other goods that contribute to our welfare. The value of experiencing natural or artistic beauty, for example, does not seem to hinge on how we come into a position to enjoy these goods.

Other goods, however, have a *historical* quality, such that their value turns in part on how those goods are realized. Some goods, particularly those predicated on one's relationships with others, derive their value from norms expressing and governing our attitudes. Hence, the value of a good whose history contains choices or events at odds with those norms is thereby called into question (Anderson 1993: 38–43; Scanlon 2008: 128–141). For instance, there are aspects of the good of friendship that do not seem to depend on how the friendship is established or maintained. One can find value in a friend's companionship regardless of the friendship's history, for example. Nevertheless, the value of a friendship can turn on its history. Friendships that result from sharing a workplace with someone are not thereby diminished in value, but it is hard to fathom that learning that one became friends with someone only after she was hired by one of your rivals to surveil you would not imperil the value one attributes to that friendship. It would be reasonable to question whether a friendship with such origins can embody the norms of mutual respect definitive of friendship and whether the friend has the attitudes characteristic of genuine friendship.

But just as the histories of relationships with others can influence the value those relationships have for us, so too can the history of how we relate to ourselves influence the value of the goods associated with *that* relationship. In other words, many of the goods associated with our relationships to ourselves have a noticeably historical character. Gwen Bradford has recently homed in on one important class of such goods, *achievements*. Under Bradford's characterization, achievements are those products of our efforts that are difficult to attain but are attained nevertheless via our competently causing them (2015: 20). An achievement is therefore a state that can be distinctly attributed to our efforts and more specifically, Bradford argues, to the exercise of our volitional powers. When (say) a person completes her first triathlon, this result was not easy to achieve, was achieved through her competent efforts (sustained training, etc.), and required steadfastness. Achievements give our lives meaning and worth, then, in part because of the history of their realizations. Suppose that, against all probability, a person is able to complete the triathlon easily, through sheer luck and modest effort. Its value *qua* achievement would thereby be reduced accordingly.

That achievements matter to us shows that the value of some goods hinges on the history of how they are realized, and in particular, how our own choices and efforts contribute to their realization. For some goods, it is not good (enough) that P – we must make it the case that P in order for P to matter to us wholeheartedly. The desire for such goods to be realized through our own wills reflects an implicit normative understanding of our relationship to these goods and to our selves. That we instantiate these goods does not exhaust their value to us. Their value to us stems in part from our standing in an authentic relationship to them – that they in some manner are a reflection of us. As I suffer from a toothache, it matters little to me whether relief comes from my efforts or not – it is of little normative import whether the relief reflects positively on my character, will, etc. In contrast, as I prepare to compete in my first triathlon, it matters to me that I (and others) can rightfully acknowledge that I am able (thanks to my preparation, dedication, mental resilience, etc.) to complete all three of its component events.

We noted in Section 1 that duties to self are both owed to our selves but also owed by our selves. As we noted, duties to self are, in Kant's scheme in particular, duties of virtue, and so cannot be fulfilled by others, strictly speaking. This is the metaphysical explanation of why duties to self cannot be fulfilled by others. But even if this were not the case and others *could* fulfill these duties, there would be something ethically amiss about their doing so. For duties to self, I propose, rest on what I have been calling historical values. What matters to us about the fulfillment of duties to self – why it is apt for us to feel proud of how we care for our bodies, manage our resources so as to further our ends, develop our talents, retain relations with others that are independent rather than servile, etc. – is that we bring about their fulfillment. Indeed, the fulfillment of duties to self only seems to reflect positively upon us, and upon our respect for ourselves as rational agents owed authority over our choices, if their fulfillment is attributable to us.

Again, others cannot fulfill our self-regarding duties. However, insofar as we are the object of these duties – those to whom the duty is owed – we are not indifferent to who their subject is. That duties to self rest on historical goods thus implies that the division of labor between ourselves and others, as far as whose volitions are responsible for their fulfillment is concerned, should tilt heavily in the direction of the former. The more their fulfillment can be attributed to the choices and efforts of others, the less significant their fulfillment is to us. Resentment directed at paternalism aimed at the inculcation of the self-regarding virtues is thus understandable. For others to make the inculcation of one's self-regarding virtues their business is to undermine the conditions under which their fulfillment is normatively significant to the person to whom they are owed. Paternalists motivated by the aim of inculcating the self-regarding virtues in others are interceding in the most intimate relationship we can have: the relationship between our selves as agents and as patients, the relationship most central to our self-governance (Cholbi 2014).

Achievements, we noted above, involve historical goods. Often, the fulfillment of duties to self will be an achievement. After all, the laziness that precludes our developing our talents, the indiscipline that leads us to neglect our bodily health, etc., can be difficult to overcome. Some agents will be advantaged in this regard, however, endowed with the self-regarding virtues to an unusually high degree. But even they, I propose, have reason to desire that their self-regarding duties be fulfilled through their exercise of these virtues rather than from well-meaning paternalists. Of course, at the other end of the spectrum, sometimes others' assistance can help us develop our self-regarding virtues. But consenting to such assistance is evidence of the presence of the very powers associated with these virtues. Indeed, in knowing when others can assist us in developing these virtues, we exhibit the virtues associated with respect for ourselves: genuine knowledge of our own traits and aptitudes is itself a developed talent, a way of acquiring resources needed to achieve our ends and a way of being benefitted by others without courting servility to them. Hence, we ought not resent the assistance we accept from others in helping us meet our self-regarding duties in the way that we ought to resent unsolicited, i.e., paternalistic, assistance. After all, when we fulfill our duties thanks to the welcome assistance of others, their fulfillment is nevertheless ultimately traceable to the exercise of *our* volitional capacities.

The metaphysical identity of deontic subject and object in the case of duties to self thus renders their fulfillment valuable in a way that speaks against paternalistic efforts to inculcate the self-regarding virtues that facilitate these duties' fulfillment. Ahistorical goods, and in particular many of the goods standardly thought to advance our welfare, do not lose their value when others are largely responsible for their realization. This is not to say that we lack reason to resent others' paternalistic efforts to help us realize ahistorical goods.[5] Rather, historical goods, of which the fulfillment of self-regarding duties is an instance, are such that we have a distinctive reason to resent paternalistic efforts to assist us in their fulfillment, namely, that the relation between the subject and object of these duties is internal or constitutive, such that the extent that others are responsible for their fulfillment and the goods realized in their fulfillment stand in an inverse relation.

Let us briefly address a worry about how I have argued for the claim we have reasons to oppose paternalism aimed at inculcating self-regarding virtues and the fulfillment of duties to self. My arguments have leaned heavily on Kant's specific understanding of duties to self: that they are duties of virtue, requiring one adopt particular ends, not just perform certain acts or pursue certain outcomes; that because they are duties of virtue, duties to self are historical, requiring for their fulfillment our recognition of reasons to treat ourselves with respect; and that to the degree others are volitionally responsible for duties to self being "met," the duties become unmoored from the values that lend the duties their rationale. Of course, other conceptions of duties to self, including ones that see them not as duties of virtue but simply as requiring one to bring about particular outcomes, are possible. (A quasi-Kantian conception could require that we preserve our lives, for example, but not that we do so from the recognition that our lives and the rational agency they embody are worthy of respect.) But note that such conceptions render obscure why duties to self are incumbent specifically on oneself. For if the nature of the self demands that a particular outcome concerning the self be realized, why does the obligation to realize it fall on oneself rather than on others? At the very least, non-Kantian conceptions of duties to self, wherein they rest on ahistorical goods and are not duties of virtue, will struggle to explain why the duties in question are duties to self.

6 Conclusion

Duties to self are an underexplored topic in contemporary philosophy, and the implications of such duties even less explored. Here I have argued that we have distinct reasons to resent paternalism aimed at enabling us to fulfill our self-regarding duties. I am circumspect enough to recognize that my argument has merely opened the door to further discussion of these matters, but it appears at least to shift the burden onto those who would defend paternalism motivated by a concern that others fulfill their duties to self.

Related topics

Kantian Perspectives on Paternalism; Moralism and Moral Paternalism; Paternalism and Well-Being.

Notes

1 For a more thorough discussion of Kant's account, see Cholbi (2016: 54–60), as well as Jeske (1996) and Denis (2001).

2 These controversial claims include that lying and the pursuit of sexual gratification are violations of duties to self (see Cholbi 2016: 56–58, 175–182 for discussion), as well as what Timmermann (2006: 508–510) calls Kant's "primacy thesis," that duties to self provide the "foundation" or "preconditions" of all moral duty.

3 I take this to be a different matter than whether duties to self might be overridden by duties to others, a prospect that Kant took seriously (1996:177/6:423, 184/6:432).
4 And of course this does not preclude such measures being justified on nonpaternalistic grounds (e.g., that excessive drinking poses risks to those besides the drinker).
5 A still more complicated sort of case is when a given good has both a historical and an ahistorical value. Physical health presumably has ahistorical value, but its cultivation is also a Kantian duty to self resting on historical value. This presents a conflict between welfare paternalism and moral antipaternalism which I will not attempt to resolve here.

References

Anderson, E. (1993) *Value in Ethics and Economics*, Cambridge, MA: Harvard University Press.
Baier, K. (1958) *The Moral Point of View*, Ithaca: Cornell University Press.
Bradford, G. (2015) *Achievement*, Oxford: Oxford University Press.
Cholbi, M. (2011) *Suicide: The Philosophical Dimensions*, Peterborough, ON: Broadview.
———. (2014) "Agents, Patients, and Obligatory Self-benefit," *Journal of Moral Philosophy* 11: 159–184.
———. (2015) "On Marcus Singer's 'On Duties to Oneself,'" *Ethics* 125: 851–853.
———. (2016) *Understanding Kant's Ethics*, Cambridge: Cambridge University Press.
Darwall, S. (2002) *Welfare and Rational Care*, Princeton, NJ: Princeton University Press.
Denis, L. (2001) *Moral Self-regard: Duties to Oneself in Kant's Moral Theory*, New York: Garland.
Finlay, S. (2008) "Too Much Morality?" in P. Bloomfield (ed.) *Morality and Self-Interest*, Oxford: Oxford University Press, pp. 136–154.
Hills, A. (2003) "Duties and Duties to Self," *American Philosophical Quarterly* 40: 131–142.
Jeske, D. (1996) "Perfection, Happiness, and Duties to Self," *American Philosophical Quarterly* 33: 263–276.
Kant, I. (1996) *Metaphysics of Morals*, trans. M. Gregor, Cambridge: Cambridge University Press (References include original Berlin Akademie, 1901 pagination).
May, S. C. (2015) "Directed Duties," *Philosophy Compass* 10: 523–532.
Mill, J. S. (1859) *On Liberty*, London: Parker & Son.
Scanlon, T. M. (2008) *Moral Dimensions: Meaning, Permissibility, and Blame*, Cambridge, MA: Harvard University Press.
Singer, M. G. (1959) "On Duties to Oneself," *Ethics* 69: 202–205.
Singer, P. (2011) *The Expanding Circle: Ethics, Evolution, and Moral Progress*, Princeton, NJ: Princeton University Press.
Timmermann, J. (2006) "Kantian Duties to the Self, Explained and Defended," *Philosophy* 81: 505–530.
Velleman, J. D. (1999) "A Right of Self-Termination?" *Ethics* 109: 606–628.

10
PATERNALISM AND RIGHTS

Daniel Groll

Are there any deep or systematic connections between paternalism and people's rights? Consider the following case:

> *Fatima's Keys*: Fatima loves listening to loud, live music. This has taken a toll on her hearing. Her doctors have told her that if she doesn't start wearing earplugs at shows, she may well lose her hearing altogether in the coming year. Fatima has ignored their warnings to this point. Tonight, pioneering grindcore band Napalm Death is in town and Fatima is keen to go. Her friend, Rob, is very concerned about Fatima's hearing. So he hides her car keys, thereby preventing her from making it to the concert.

First, Rob's action is a paradigmatic case of paternalism. Second, in hiding Fatima's keys from her and so preventing Fatima from using her own car, Rob has violated one of Fatima's property rights (barring special circumstances). Finally, to the extent that Rob's action strikes us as in some way morally problematic, it seems natural to explain that by appeal to the fact that Rob has violated one of Fatima's rights (for her own good).

So here we have an example of paternalism that violates the paternalized subject's rights. We might think reflection on cases like *Fatima's Keys* shows us that there is some interesting connection between paternalism and people's rights. Perhaps the connection is definitional: part of what makes an action or policy paternalistic is that it violates a right. Or perhaps the connection is normative: paternalism is (always? often? only sometimes?) morally problematic because it violates people's rights (even if we don't define "paternalism" in terms of a rights violation).

My main goal in this paper is to argue for the normative connection. Part of the task will be to explain exactly what the normative connection is. That, of course, will involve answering the questions embedded in the claim as well as offering an account of the right(s) that is (are) connected to paternalism's normative status.

Before going further, I want to say something about how I understand the terms "rights" and "rights violation." I take rights-talk to be part of a family of concepts that form, to use Darwall's term, an "interdefinable circle" (2006: 12). These concepts include: authority, legitimate claim or demand, obligation, and sovereignty. So, if I say that Fatima and only Fatima has a right to her car keys, I take this to imply that, under normal circumstances, only she has authority with respect to how the keys are used; that, normally, only she can make valid claims or demands on others with respect to the keys; that she has sovereignty with respect to the use of the keys, etc.

1 Specifying the normative connection

The claim we're considering is that there is a normative connection between paternalism and people's rights: paternalism is (always? often? only sometimes?) morally problematic because it violates people's rights. There is, obviously, a very big difference between the claims that paternalism is always morally problematic, often morally problematic, and only sometimes morally problematic. So, which version of the normative connection am I arguing for?

To help us answer this question, consider the following case:

> *Permission Denied*: Fatima is 10 years old and loves listening to really loud live music without earplugs. She really wants to see Napalm Death, but her father, Rob, won't let her go because he is concerned about the effect on her hearing.

Let's stipulate that Rob is entitled to decide whether Fatima can go to the concert and that he's making a good decision. If that's right, then it looks like there is nothing morally problematic with his decision. And that shows that some paternalistic actions are not morally problematic at all. So the "always" version of the normative connection claim can be ruled out . . . provided we think Rob is acting paternalistically.

Is he? Seana Shiffrin would say "no." According to her, we act paternalistically only when we treat competent adults like children (Shiffrin 2000: 219 n.24).[1] Treating children (or incompetent adults) like children is not, according to Shiffrin, paternalistic.[2] If that's right, then *Permission Denied* does not give us reason to reject the claim that paternalism is always morally problematic.

Suppose, though, that we think Rob *does* act paternalistically in *Permission Denied*. An advocate of the view that paternalism is always morally problematic needn't be bothered. That is because he might insist that what *Permission Denied* really shows is that we need to distinguish between *soft* and *hard* paternalism. Soft paternalism is directed at individuals whose actions are "substantially nonvoluntary" or people who do not rise to some threshold level of competence for making the decision at hand.[3]

Hard paternalism, on the other hand, targets "human beings in the maturity of their faculties."[4] If the normative connection is meant to apply to all paternalism – soft and hard – then certainly the "always" version will be false: soft paternalism is usually not morally problematic at all. But we might invoke the soft/hard distinction as a way of setting aside cases like *Permission Denied* altogether so that we can focus on the more normatively interesting cases involving hard paternalism. If we do that, then we can further specify the claim we're considering: *hard* paternalism is (always? often? only sometimes?) morally problematic because it violates people's rights.

Can we make a case for this claim? And which version: "always," "often," or "only sometimes"? To help us here, let's consider what it might mean to say that there is a systematic connection between hard paternalism's normative status and people's rights. Here are three possibilities:

> *Strong view*: Hard paternalism is *pro tanto* wrong inasmuch as it always involves a rights violation.
>
> *Moderate view*: Hard paternalism is always *presumptively* wrong inasmuch as there is always the justified presumption of a rights violation.

> *Weak view*: Hard paternalism is neither *pro tanto* nor *presumptively* wrong, but when it *is* wrong (either all things considered or *pro tanto*) that is because of a rights violation.[5]

If the Strong view is right, then hard paternalism is always morally problematic. If the Moderate view is right, then hard paternalism is often morally problematic, though in particular cases it will not be. If the Weak view is right, then paternalism is sometimes problematic, but there is no justified presumption of a rights violation in all instances of hard paternalism.

One might wonder whether the Weak view posits anything worthy of being called a "systematic connection" between paternalism and rights violations. I think it does: even if paternalism is often morally *un*problematic, it would be interesting to learn that *when* it is morally problematic it is because there is a rights violation. Even so, the Moderate and Strong views (and particularly the latter) posit a deeper, more interesting connection between paternalism and rights violations, so I want to explore their prospects. More specifically, I want to show that the Strong view has more going for it than it might seem.

2 Defending the Strong view part I: the definitional connection

One way to defend the Strong view would be to build the notion of a rights violation into our very understanding of what paternalism is. On this view, which I'll call *rv-Paternalism*, part of what makes an action paternalistic is that it violates the rights of the paternalized subject (for the sake of that subject's welfare or good).[6] If this is a compelling understanding of what paternalism is, then the connection between paternalism's moral taint and people's rights will be very strong indeed: it will be baked into our very understanding of the nature of paternalism.

Seana Shiffrin has provided an important contemporary version of something very close to *rv-Paternalism*. According to Shiffrin (2000: 218, emphasis added):

> Paternalism by A toward B may be characterized as behavior (whether through action or through omission):
>
> a) aimed to have (or to avoid) an effect on B or her *sphere of legitimate agency*
> b) that involves the substitution of A's judgment or agency for B's
> c) directed at B's own interests or *matters that legitimately lie within B's control*
> d) undertaken on the grounds that compared to B's judgment or agency with respect to those interests or other matters, A regards her judgment or agency to be (or as likely to be), in some respect, superior to B's.

There is a lot to say about Shiffrin's conception of paternalism.[7] What interests me here, however, is Shiffrin's idea that paternalism intrudes on the paternalized subject's "sphere of legitimate agency" and is directed at "matters that legitimately lie within B's control." She makes the connection between these ideas and people's rights clear when she notes that "a full account of paternalism will depend upon a fleshed-out account of autonomy rights – over what an agent (B) generally has proper domain, just in virtue of being an agent" (Shiffrin 2000: 218–219).[8] If Shiffrin is right about what paternalism *is*, then the Strong view will be vindicated since paternalism's morally problematic nature will have been built into our very understanding of the phenomenon.

We've already seen one hurdle *rv-Paternalism* needs to clear: cases like *Permission Denied* appear to present us with instances of non-rights-violating paternalism. But we've also seen how an advocate of *rv-Paternalism* can respond. First, like Shiffrin, they might deny that cases like *Permission Denied* evince paternalism. The other option is to reformulate *rv-Paternalism* as a claim about *hard* paternalism. This is the route preferred by Norbert Paolo (2015: 136), who says, "An act is a case of hard paternalism when the carer infringes a right of the cared for and tries to justify this interference with an interest or right of the cared-for."

But even if one thinks these responses are viable, *rv-Paternalism* faces another problem: there appear to be clear cases of (hard) paternalism that don't involve any rights violation. For example:

> *Rob's Keys*: Fatima loves listening to loud, live music. This has taken a toll on her hearing. Her doctors have told her that if she doesn't start wearing earplugs at shows, she may well lose her hearing altogether in the coming year. Fatima has ignored their warnings to this point. Tonight, Napalm Death is in town and Fatima is keen to go. Her friend, Rob, is very concerned about Fatima's hearing. The only way for Fatima to get to the concert is to borrow Rob's car. Rob knows she'll take the car whether he gives her permission to or not, so he hides his keys, thereby preventing her from making it to the concert.[9]

Fatima is not entitled (other things being equal) to use Rob's car. He does not (it seems) wrong her at all by hiding *his* keys. They're his! On the face of it, then, there is no rights violation here. Nonetheless, many people will count Rob's action as paternalistic and obviously so.[10] But according to *rv-Paternalism*, Rob's action is not paternalistic. And advocates of *rv-Paternalism* cannot appeal to the soft/hard distinction here since if Rob's action is paternalistic it is an example of *hard* paternalism.

rv-Paternalism is forced to say that cases like *Rob's Keys* offer merely purported instances of non-rights-violating paternalism: either the action is not paternalistic or there is some rights violation involved. The first route is unacceptably question-begging. The second requires that the advocate of *rv-Paternalism* give an account of the right (or rights) that is (are) violated. In the next section I argue that, contrary to initial appearances, we actually can plausibly posit a rights violation in cases like *Rob's Keys*. If that argument is successful, then *rv-Paternalism* may be safe, and it seems we'll have successfully defended the Strong view.

But *rv-Paternalism* faces another, in some ways more fundamental, challenge. This challenge is articulated by Dworkin (2016):

> As a matter of methodology it is preferable to see if some concept can be defined in non-normative terms and only if that fails to capture the relevant phenomena to accept a normative definition.

One might reject the methodological assumption as applied to concept definition in general, but think that it nonetheless applies to defining "paternalism." Given widespread disagreement about when, if ever, paternalism is justified, it seems preferable to not build the presence of a moral taint *into* our understanding of what paternalism is. But this is precisely what *rv-Paternalism* does.[11] The methodological challenge, then, gives us reason to look for another way of defending the Strong view.[12]

3 Defending the strong view part II: the normative connection

Let's take stock. We're trying to defend the Strong view that hard paternalism is *pro tanto* wrong in virtue of always involving a rights violation. If we can do that, we'll have shown that there is a systematic connection between paternalism and people's rights that makes hard paternalism always morally problematic.

However, we're now looking for a way to establish the connection without *defining* (hard) paternalism partially in terms of a rights violation. This leaves open the task of figuring out just what paternalism *is*, which I leave to others in this volume to tackle. The idea we're after here is just this: whenever (hard) paternalism is on the scene, so too is a rights violation. To paternalize a competent adult involves violating one, or some, of their rights. Crucially, this is not now intended as a definitional move. Rather, it is a substantive claim about some right(s) that people have as competent adults and how hard paternalism always abrogates it (or them).

We've already confronted the biggest obstacle to making good on this substantive claim: cases like *Rob's Keys* appear to involve paternalism with no rights violation. Really, there are two challenges to the Strong view one could extract from *Rob's Keys*. First, and most obviously, one might take *Rob's Keys* to provide a straightforward counterexample to the claim that hard paternalism always involves a rights violation (and so, to that extent, is always morally problematic).[13]

But you might take the case of *Rob's Keys* to show something else. Rather than seeing it as a case of unproblematic paternalism, you might see it as a case of *problematic, non-rights-violating* paternalism. That is, you might agree that cases like *Rob's Keys* exhibit morally problematic paternalism, but deny that the moral taint comes from a rights violation. More generally, you might think that identifying a rights violation in every instance of (hard) paternalism seems like a very heavy-handed way of accounting for the moral taint of paternalism. Does it seem plausible, in advance of examining particular cases, that the wrong of paternalism always consists in the violation of the paternalized subject's rights? Some want to say "no,"[14] that we can see the *pro tanto* wrongness of at least some instances of paternalism without having any thoughts about rights violations. So, according to this challenge, even if the Strong view is right in claiming that all instances of hard paternalism are morally problematic, it often (or at least sometimes) gets the *source* of the moral taint wrong.

Both challenges make clear what a defender of the Strong view needs to do. The Strong view is plausible only to the extent that it can identify some right (or rights) that is (are) violated in cases like *Rob's Keys*. Can we do this?

To start, we might ask: Is the claim that paternalism always involves the violation of a right the claim that it always violates *some right or other* or the claim that it always violates *some particular right*?

If we go the first route, the Strong view starts to look wildly implausible: if there is no one right that is violated by paternalism, why think that paternalism always involves a rights violation? It seems remarkable that it would![15] Suppose, then, we go the second route and attempt to identify one particular right that is (presumptively) violated in every instance of paternalism. There are two questions. First, what could this right possibly be? And second, why think that we have it?

One answer to the first question comes from Mill (1978 [1859]: 9):

> In the part [of his conduct] which merely concerns himself, [a person's] independence is, of right, absolute. Over himself, over his own body and mind, the individual is sovereign.

Joel Feinberg (1986) defends a version of this claim in his articulation and defense of a "right to autonomy." According to Feinberg, a person's life:

> Is after all *his* life; it *belongs* to him and to no one else. For that reason alone, he must be the one to decide – for better or worse – what is to be done with it in that private realm where the interests of others are not directly violated.
>
> (59)[16]

Feinberg's thought here is that my life is *properly* mine: I am the one rightly in charge of me, and so when matters involve only me, I am, by right, the one who gets to make decisions about those matters. If there is this kind of right, is it now plausible to think that Rob violates Fatima's rights in keeping his keys from her?

Answering this question depends on resolving some well-known issues with the Mill/Feinberg approach. The most notable is figuring out which actions or domains of action "merely concern" the acting individual or figuring out the boundaries of "the private realm where the interests of others are not directly violated." With respect to Mill's formulation, Fatima's desired course of action does not merely concern herself. It also concerns Rob. More generally, the worry is that the set of actions that only concern the acting individual is vanishingly small.[17] With respect to Feinberg's formulation, clearly Fatima would directly violate Rob's interests if she *stole* his keys. But that's not what we're wondering about here. The question is whether, in refusing to give Fatima his keys *for her own good*, Rob is interfering in Fatima's "private realm" of interests.[18]

I think we can sidestep these questions, and in doing so isolate a right that is plausibly violated in both *Fatima's Keys* and *Rob's Keys*, by switching our focus slightly. Let's call the potentially paternalized subject the "target agent" and the potential paternalist the "acting agent." Instead of asking whether the target agent's action is self-regarding or merely of concern to herself, we should ask, "Why is the *acting agent* doing whatever it is he is doing?" My thought is that if the answer to this second question is, "Only (or overridingly) for the sake of the target agent's good," then the acting agent is not only acting paternalistically but violating a right (assuming that the target agent is competent). What is that right? It is the right of competent individuals to be the only ones to act *only* (or overridingly) for the sake of their own good.

This way of putting things takes the focus away from whether the action in question is "self-regarding." The action might have all kinds of consequences for other people; it might be well outside the "private realm." What matters is *why* the acting agent is stepping in. If he is stepping in only, or overridingly, for the sake of the target agent's welfare, then the action violates a competent agent's right to be the sole agent that is authorized to act *only* for her own good. Everyone else must, in acting for the good of that person, be authorized to so act.

The claim then is that competent agents are sovereign with respect to acting for their own good. This does not mean that others cannot act for our good without wronging us! That would be an absurd view. The claim, rather, is that others wrong us when they act only or overridingly for the sake of our good. In so acting, they are unconstrained by consideration of whether their attempt to help is authorized. Crucially, if the target agent is not competent, then the acting agent may still well be doing something paternalistic[19] – whether authorization is given is not relevant for whether the action *counts* as paternalistic (in this way my view is not a version of *rv-Paternalism*). It only matters for the moral assessment of the paternalistic act. Lack of authorization only counts against the act if the target agent is competent.

If competent agents are sovereign over their own good in the way I've claimed, then, contrary to initial appearances, *Rob's Keys* does involve a rights violation. Inasmuch as Rob is acting for Fatima's good without due concern for whether his intervention is welcome, he is wronging Fatima. It is true that the car is his. So he does not violate any of Fatima's property rights in refusing to loan her the car. He *does* violate Fatima's property rights in *Fatima's Keys*. But it turns out that the absence or presence of property rights is not what is most salient for explaining what is problematic about Rob's paternalism in the above cases. There is a more fundamental right of sovereignty over our good at play that explains the moral taint in both *Rob's Keys* and *Fatima's Keys*. As noted above, this story explains why there is no moral taint in *Permission Denied*: the right in question only attaches to competent agents. We do nothing wrong – either *pro tanto* or presumptively – in acting for the good of our children or incompetent patients in our charge.

But won't this conception of the right that is violated in acts of hard paternalism make all kinds of ordinary actions paternalistic and rights-violating? Surprise parties, gift giving, and charitable donations come immediately to mind. The general worry is that the account of the right that is violated in hard paternalism threatens to morally taint all kinds of actions that we tend to think have no moral taint at all.

The general answer to this concern is that in the kinds of cases just mentioned, and others like them, there is normally a presumption that our benevolent actions have been implicitly authorized. And to the extent that there is *not*, to the extent we believe or don't care that our benevolence is not authorized by the benefactee, then these actions *do* acquire a moral taint.

By way of illustration, consider charitable giving. Typically, we think that this action has no moral taint. And typically charitable giving is motivated by a desire to help other people, to improve their welfare, to make their lives better. In giving to Oxfam I am no doubt motivated, probably even primarily motivated, by the good of the people I aim to help. But my impulse to help is plausibly construed as constrained by the belief that the help is wanted. Were it not, I would be far less inclined to give. If nothing else, the fact that the help was not wanted (perhaps it is not perceived as help) would give me serious pause (Groll 2014: 189). In general, when we aim to help other people – particularly competent adults – we are sensitive (however implicitly) to whether our help is welcome. If we act without concern for this question, or our concern for a person's good overrides concerns for whether the help is wanted, then I think it is plausible that there is something morally problematic with our action. The same point holds for gift giving and surprise parties: we typically take ourselves to have implicit authorization to do these things and if we really didn't think that we did (or didn't care), then the practices start to look morally problematic.

Even so, we might worry that this move only gets us so far. There seem to be all kinds of cases where I act for the good of others – particularly loved ones – without *any* thought of whether I have implicit authorization: think of the myriad actions, small or otherwise, that we do for romantic partners all the time without considering, first, whether some kind of authorization has been granted. Am I committed to saying that all such actions evince hard paternalism and involve a rights violation?

No. The reason is that being in this kind of relationship involves the parties granting each other a degree of leeway, a limited kind of authority, to look out for the good of the other, to care for the other. Indeed, I think this is part of what it means to enter into this kind of relationship. This doesn't mean that one can do just anything for the good of one's partner. But being in this kind of relationship involves a kind of limited "authority swap" wherein each partner is given a very open-textured permission to act for the good of the other. So, when we act unthinkingly for the good of those we are most close to, it is (usually) against the backdrop of knowing that we are part of a relationship that effectively authorizes such actions. Our actions are, typically, neither paternalistic nor rights-violating (though they can certainly be both).

A version of this point applies to some non-intimate relationships. To take one example: a professor might make students hand in drafts of papers for their own good even though the students would really rather not write a draft. Is the professor acting paternalistically and violating the right I identified above?[20] On my view, the answer to both questions is "no": professors and students stand in a distinctive relationship which is defined, in part, by the students authorizing the professor to structure the class at least in part with an eye to the students' good. Again, that doesn't mean the professor can do just anything for the sake of the students' good. But actions that would clearly count as paternalistic outside the relationship are not paternalistic in the context of the relationship since the relationship is partly defined in terms of one party granting a limited authority to the other.

Some may be concerned that this way of defusing the initial concern (that my view of the right violated in hard paternalism will result in all kinds of ordinary behavior being morally problematic) comes too close to a "just so" story: in cases where people seemingly act only or overridingly out of concern for another's good but where we are disinclined to say there is anything morally problematic, we simply invoke the idea that implicit authorization was (believed to be) given as part of being in a certain kind of relationship. But this isn't right. The proper lesson is that whether a particular action or policy is (hard) paternalistic, and so morally tainted, depends on the fine details of the case: on whether there was a prior relationship between the parties; on the nature of that relationship; on whether some form of consent or authorization was granted in the past. It will no doubt turn out that in looking at these fine details, some actions that take place within the context of the kinds of relationships that involve a transfer of authority will nonetheless turn out to be instances of hard paternalism and, on my view, rights-violating.

So: we have identified a right that plausibly might be thought to be violated in every instance of (hard) paternalism and in doing so we have lent some credence to the Strong view.[21] Some will greet this idea – or the idea that there is any right of this sort (including those identified by Mill or Feinberg) – with unbridled skepticism (for example: Conly 2013: 36). Certainly, a full defense of the view would need to offer an account of *why* or *how* we have such a right. Is it basic? Or is the right itself an outcropping of more general concerns about promoting well-being or something else of value?[22]

First, we might follow Mill (and some contemporary followers of Mill, like Dworkin 1993) by helping ourselves to an interest theory of rights and pointing to some important interest, or set of interests, that are protected or promoted by there being such a right. Broadly speaking, this would be to offer a *good-based* account of the origin and force of the right in question.

Second, we might try to show that the right in question is basic or fundamental, something we have in virtue of some feature possessed by mature adults. We might pursue this line of thought in a Kantian vein by arguing for the moral importance of our status as rational creatures with a capacity to set ends. Perhaps a story can be told according to which our rational end-setting capacities ground a kind of fundamental authority in each of us with respect to acting for our own good.

Far more needs to be said to flesh out either of these possibilities. What I want to emphasize now, however, is that adopting the Strong view needn't necessarily lend credence to an even *stronger* view, namely that (hard) paternalism is always and everywhere unjustified. This would only follow if we accept either the claim that all rights are absolute, i.e., "never permissibly infringed" (Shafer-Landau 2005: 189), or that the particular right I've identified is absolute.[23] The Strong view then does not, by itself, leave us with the Very Strong view that hard paternalism is always impermissible.

4 Conclusion: the challenge of avoiding rights-talk

I hope to have made somewhat compelling the idea that hard paternalism is always *pro tanto* wrong because it always involves a rights violation. But just how plausible it is will depend, in part, on the plausibility of competing accounts of when/why paternalism is morally problematic. By way of concluding, I want to very quickly examine some of those alternative accounts since, I will suggest, it is not clear that they are really alternatives at all.

How else might we account for the moral taint of paternalism? We need only consider some other common ways of conceiving of paternalism to see the possibilities. Perhaps paternalism (always or presumptively) interferes with the paternalized subject's liberty (Feinberg 1986) *and it is pro tanto or presumptively wrong to interfere with the liberty of another person.* Or perhaps hard paternalism (always or presumptively) diminishes the paternalized subject's autonomy (Enoch 2016; Dworkin 1988, 2016) and *it is pro tanto or presumptively wrong to diminish the autonomy of another person*. Or perhaps hard paternalism (always or presumptively) substitutes the paternalist's judgment for the paternalized subject's (Dworkin 1988; Shiffrin 2000)[24] *and it is pro tanto or presumptively wrong to do that.*[25]

Each of these suggestions comes with its own conception of paternalism which we could (and should) scrutinize independently. But I want to leave that aside and focus on the following question: Do the above suggestions actually avoid appealing to the idea of a rights violation in their account of paternalism's moral taint? The worry is that attempts to explain what it means to interfere with the liberty of another person, or to limit her autonomy, or to substitute one's judgment for another person's are implicitly appealing to the idea that the paternalist is acting in an *unauthorized* way.

Consider the idea that paternalism's moral taint stems from the fact that it "constitutes an attempt to substitute one person's judgment for another's, to promote the latter's benefit" (Dworkin 1988: 123). The problem is that it's not clear that we can spell out the idea of A substituting his judgment for B's in the way required to explain why (hard) paternalism is so often morally problematic without implicitly helping ourselves to the idea that A is not *authorized* to substitute his judgment for B's. Indeed, the idea of A substituting his judgment for B's suggests that the judgment in question was *properly B's to make in the first place.*

I grant we need not hear it this way. Talk of substituting one thing for another can indicate a mere temporal ordering: Bob Gainey begins the game at left wing for the 1986 Montreal Canadiens and then is substituted out for Mats Naslund 45 seconds later. There's no implication here that Naslund is taking something from Gainey that is properly Gainey's. Rather, talk of "substituting Naslund for Gainey" indicates, merely, "First Gainey, now Naslund."

But that's not what's going on in Dworkin's talk of "substituting judgment." First, in many cases, the paternalized subject will not have first made a judgment which is then supplanted by a *later* judgment by the paternalist. The timing of the competing judgments is not relevant for understanding how A substituted his judgment for B's. The substitution is not temporal but, metaphorically speaking, geographical: A is inserting a judgment where it does not belong. The territory in question belongs (at least presumptively) to B. That is why when A's judgment carries the day we say that A has substituted his judgment for B's. But if *B's* judgment were to carry the day, we would not say that B's judgment substituted for A's.

A similar point holds true for some of the other conceptions of paternalism that purport to explain paternalism's moral taint without appealing to the idea of a rights violation: Can we make sense of the idea of interfering with liberty or limiting autonomy without helping ourselves to some notion of what decisions, or domains of decision-making, *properly* belong the paternalized subject?[26] Maybe we can. The point I'm after here is just that alternative routes to explaining what so often makes (hard) paternalism morally problematic might not be as distinct from my version of the rights route as they initially seem.

Having said that, the nature of the right that is violated might be very different from how I have construed it: perhaps, for example, it is a liberty right and not the broader right I have identified. This could have implications for whether the Strong view is correct since it might turn out that not all instances of hard paternalism violate the liberty right (indeed, this seems to be the case in *Rob's Keys*). But the key thought now is just that avoiding rights-talk *altogether* when attempting to explain (hard) paternalism's moral taint (whether it always has it or not) is harder than it seems.

If all this is correct – if there is an independent account of the place of rights in paternalism that is somewhat compelling *and* if alternative accounts of paternalism's moral taint implicitly invoke the idea of a rights violation – then we have found a systematic connection between paternalism and rights without baking the idea of a rights violation into our understanding of paternalism.[27] And if my conception of the right in question is correct, we will have found that the moral taint of paternalism stems from the fact that paternalism toward competent adults involves violating the right of competent agents to be the only ones authorized to act solely for the sake of their own good.[28]

Related topics

Hard and Soft Paternalism; Libertarian Perspectives on Paternalism; The Concept of Paternalism.

Notes

1 Beauchamp (2009: 82) also goes this route.

2 Although she admits that this will "seem terribly counter-intuitive to those who regard our treatment of children and mentally disabled people as paradigm cases of justified paternalism" (Shiffrin 2000: 219 n.24).

3 More broadly, it is also directed at people where there is a question about whether their actions are voluntary or the result of minimally competent agency (Feinberg 1986: 12).

4 Although attempting to draw a sharp distinction between soft and hard paternalism in terms of different kinds of agents (competent v incompetent) is none too easy. For criticisms of leaning too heavily on the distinction see Conly (2013); Arneson (2005); Levy (2014); and Begon (2016).

5 There is also a fourth possibility, the Very Strong view, according to which hard paternalism is always all things considered unjustified. I say a little more about this below, but I don't consider it a plausible view.

6 The bit in parentheses is what I call the "welfare motive." It is accepted by almost everyone to be part of what makes an action paternalistic. Shiffrin, however, rejects it.

7 Including the fact that Shiffrin rejects the idea that paternalism must, by definition, be motivated by a concern for the paternalized subject's welfare.

8 Having said that, however, Shiffrin immediately adds:

> Although the characterization of paternalism that I am suggesting depends on an independent specification of an agent's rights, paternalist behavior cannot be reduced to some subset of behavior that violates these rights. For it is possible for A to act within her own rights, but be motivated purely by a disrespect for B's agency about matters that lie within B's purview and aim to maneuver around or manipulate B's agency. When this is A's sole motivation, she behaves paternalistically, I think, even if her behavior does not violate any distinct rights.
>
> (2000: 218)

This is what leads me to say that Shiffrin's view is very close to *rv-Paternalism*. I confess to being confused about why, given her account of paternalism, Shiffrin admits of the possibility of paternalism without a rights violation. It is true, as she notes, that there are examples of paternalism that do not violate "distinct" rights (like a particular property right for example). But it seems natural, given her view, for Shiffrin to posit that B has a claim against A attempting to substitute her judgment for B's with respect to spheres of interest that are legitimately within B's control. I discuss the possibility of there being such a right, and it playing a role in establishing a systematic connection between paternalism and rights violations, below.

9 Similar cases: Dworkin (1988: 122), Quong (2011: 79), Vallentyne in this volume.

10 Although a lot will depend on how we understand Rob's motive. If he is centrally, or equally, concerned with not implicating himself in Fatima harming herself, then his action may not be (straightforwardly) paternalistic. Thanks to Micah Lott for pointing this out. This is a point pursued in-depth in Shiffrin (2000).

11 More precisely, our fealty to *rv-Paternalism* falls afoul of the assumption if we do not attempt to formulate an adequate non-normative conception of paternalism.

12 If you are not at all sympathetic to the methodological challenge, then my defense of the Strong view in the next section could serve as a defense of *rv-Paternalism* as well.

13 Vallentyne, this volume. Notice that this is consistent with the Weak view, namely that when (hard) paternalism is morally problematic that is because it involves a rights violation.

14 Including an earlier time-slice of me. See Groll (2012).
15 The Moderate view is immune to this worry, but susceptible to a weaker version of it: Is it at all clear that there is even a presumption of a rights violation whenever paternalism occurs if there's no one right that is on the table as presumptively violated?
16 I'm indebted to Arneson's (2005) discussion of Feinberg.
17 For one account of how we should understand Mill on this point see Gray (1983).
18 I think Feinberg would probably say that insofar as Rob is not limiting Fatima's liberty he is not interfering with her at all, and so Rob's action is neither problematic nor paternalistic. As I go on to argue, I think there is a sense in which Rob could be said to interfere with Fatima, not necessarily in virtue of limiting her liberty but by violating one of her rights.
19 And indeed, I think they are simply in virtue of acting only or overridingly for the target agent's good. This is my current understanding of what paternalism is.
20 Thanks to Jason Hanna for raising this example.
21 Or the Moderate view if you think that the right I've identified is not in play in some circumstances.
22 Feinberg (1986: 59) briefly surveys the options.
23 As Shafer-Landau makes clear, there are parts in Feinberg that suggest that Feinberg believed that the rights constitutive of personal sovereignty are absolute and so that hard paternalism was always unjustified. But Shafer-Landau argues, convincingly in my view, that this view is neither consistent with some of Feinberg's other remarks on justified rights violations nor plausible on its own.
24 Dworkin (1988) seems to treat talk of "substituting judgment" and "violating autonomy" interchangeably.
25 Another option, suggested by Quong (2011), Groll (2012), and Shiffrin (2000), is that acting paternalistically involves acting with an insulting or disrespectful motive and that this is what accounts for paternalism's moral taint. This route avoids the problem I discuss below, but I am now (more or less) convinced by the criticisms of this view advanced by de Marneffe (2006) and Enoch (2016).
26 Beauchamp's (2009: 82) purportedly "value neutral" account is, I think, particularly susceptible to this challenge.
27 We certainly haven't baked it into a conception of paternalism if we take paternalism to include both soft and hard paternalism. Have we baked it into a conception of hard paternalism? I don't think so. The idea that competent individuals are sovereign, in some way, with respect to their own good does not figure into what makes an action paternalistic. Rather, it is a substantive, further commitment that explains why acting paternalistically toward competent adults is morally problematic.
28 This paper has been much improved by very helpful feedback from Jennifer Lockhart, Micah Lott, Norbert Paulo, Michael Fuerstein, and the two editors of this volume, Kalle Grill and Jason Hanna.

References

Arneson, R. J. (2005) "Joel Feinberg and the Justification of Hard Paternalism," *Legal Theory* 11: 259–284.
———. (2015) "Nudge and Shove," *Social Theory and Practice* 41: 668–691.
Beauchamp, T. L. (2009) "The Concept of Paternalism in Biomedical Ethics," *Jahrbuch für Wissenschaft und Ethik* 14: 77–92.
Begon, J. (2016) "Paternalism," *Analysis* 76: 355–373.
Conly, S. (2013) *Against Autonomy: Justifying Coercive Paternalism*, Cambridge: Cambridge University Press.
Darwall, S. L. (2006) *The Second-Person Standpoint: Morality, Respect, and Accountability*, Cambridge, MA: Harvard University Press.
de Marneffe, P. (2006) "Avoiding Paternalism," *Philosophy & Public Affairs* 34: 68–94.
Dworkin, G. (1988) *The Theory and Practice of Autonomy*, Cambridge: Cambridge University Press.
———. (2016) "Paternalism," in E. N. Zalta (ed.) *The Stanford Encyclopedia of Philosophy* (Summer 2016 Edition), http://plato.stanford.edu/archives/sum2016/entries/paternalism/.
Dworkin, R. (1993) *Life's Dominion: An Argument about Abortion, Euthanasia, and Individual Freedom*, New York: Vintage.
Enoch, D. (2016) "What's Wrong With Paternalism: Autonomy, Belief, and Action," *Proceedings of the Aristotelian Society* 116(1): 21–48.
Feinberg, J. (1986) *The Moral Limits of the Criminal Law, Volume 3: Harm to Self*, New York: Oxford University Press.
Gray, J. (1983) *Mill on Liberty: A Defence*, London: Routledge.
Groll, D. (2012) "Paternalism, Respect, and the Will," *Ethics* 122: 692–720.
———. (2014) "Medical Paternalism – Part 1," *Philosophy Compass* 9: 186–193.

Levy, N. (2014) "Forced to Be Free? Increasing Patient Autonomy by Constraining It," *Journal of Medical Ethics* 40: 293–300.
Mill, J. S. (1978 [1859]) *On Liberty*, ed. E. Rapaport, Indianapolis: Hackett.
Paulo, N. (2015) "The Bite of Rights in Paternalism," in T. Schramme (ed.) *New Perspectives on Paternalism and Health Care*, New York: Springer, pp. 127–141.
Quong, J. (2011) *Liberalism Without Perfection*, New York: Oxford University Press.
Shafer-Landau, R. (2005) "Liberalism and Paternalism," *Legal Theory* 11: 169–191.
Shiffrin, S. V. (2000) "Paternalism, Unconscionability Doctrine, and Accommodation," *Philosophy & Public Affairs* 29: 205–250.

Further reading

The locus classicus for discussions of paternalism (although the term never appears) is Mill's *On Liberty*. After that, J. Feinberg's *Moral Limits of the Criminal Law, Volume 3: Harm to Self* (New York: Oxford University Press, 1986) is the first place to go for discussions of paternalism, followed by R. Arneson's super helpful "Joel Feinberg and the Justification of Hard Paternalism," *Legal Theory* 11(2005): 259–284. S.V. Shiffrin's "Paternalism, Unconscionability Doctrine, and Accommodation," *Philosophy & Public Affairs* 29(2000): 205–250 is characteristically brilliant. For a recent, excellent overview of work on paternalism, see J. Begon, "Paternalism," *Analysis* 76(2016): 355–373.

11
PATERNALISM AND SENTIMENTALISM

Michael Slote

1

Every well-known theory of normative ethics has something to say about paternalism, but approaches that focus on human welfare seem on a priori grounds to be more favorable to paternalism than approaches, like the Kantian, which treat individual autonomy and dignity as fundamental to morality. Like classical utilitarianism, the sentimentalist tradition puts individual welfare front and center, and if we examine historical and present-day sentimentalism, we indeed do find that it favors paternalism in certain cases where Kantian rationalists would be hesitant or more than hesitant to recommend it. But paternalism is not a dirty word, and the fact that a view or set of views favors it in cases where another group of views does not is hardly an argument against the former and in favor of the latter. Everything depends on how plausible the arguments for or against paternalism are in a given case or a range of cases, and that is something I mean to explore here.

I am not going to examine sentimentalism from a historical perspective, though. I shall focus on present-day versions of sentimentalism, which presumably have had time to learn from the limitations and even errors of earlier forms of sentimentalism like Hutcheson's and Hume's. I am first going to show you why and how sentimentalism can argue against certain forms of paternalism. The rationalists' talk of respect, autonomy, and dignity can get us important anti-paternalist conclusions for many kinds of cases where, intuitively, we want to come out against paternalism. But rationalists tend to ignore the resources available to the sentimentalist for adjudicating many or most of those cases in a similar fashion. Sentimentalism has its own methods and bases for questioning the moral permissibility of various sorts of paternalistic intervention; and I shall be spelling that out at some length in what follows. Then later I shall explain why the same methods and bases plausibly allow the contemporary sentimentalist to advocate paternalism with regard to a whole range of other cases.

2

Present-day sentimentalism comes in two main forms, and those forms are related. The initial major impetus to present-day sentimentalism came from the work of the care ethicists Carol Gilligan and Nel Noddings. Gilligan's book *In a Different Voice: Psychological Theory and Women's*

Development (1982) was the first to raise the possibility of a care ethics opposed to Kantian ethics in anything like the terms in which such a possibility is now considered. Gilligan, however, never made it clear that care ethics opposes the idea of basing morality on Reason or reasons – she pretty much avoids those notions in her book. But in *Caring: A Feminine Approach to Ethics and Moral Education* (1984), a book that was published two years after Gilligan's, Nel Noddings made it very clear that she regarded caring and the ethics of caring as *not* based in reasons or rationality.

Noddings worked out some of the rudiments of care ethics in her book, and others have sought to develop things further (Noddings's own views have subsequently evolved). I am among the many who now call themselves care ethicists, and the present chapter will seek to apply care ethics, as I understand it, to issues concerning paternalism. However, I am also a virtue ethicist – of a sentimentalist variety. And virtue ethics is far from being the same thing as care ethics. Most virtue ethicists today follow Aristotelian models of ethics, and care ethics is not Aristotelian in its fundamental assumptions. Aristotle is, after all, an ethical rationalist.

But there are also forms of virtue ethics that avoid rationalism, that emphasize something other than Reason or rationality. One can find such types of virtue ethics in China (especially in the writings of Mencius and, much later, of Wang Yangming), and St. Augustine, with his emphasis on agapic love, is also, arguably, both a virtue ethicist and a sentimentalist. However, in recent times and in the English-speaking philosophical world, sentimentalist virtue ethics is mainly seen as historically embodied in the writings of David Hume (especially his *Treatise of Human Nature*). Thus, although I am a care ethicist, I also consider myself to be following in Hume's footsteps and sometimes call myself a Humean virtue ethicist (often also a *sentimentalist* virtue ethicist). Care ethics is, naturally, based on the importance of caring, but Hume, who never, to my knowledge, spoke (colloquially) of caring, based much of his ethical philosophy on sentiments like compassion and benevolence that are closely conceptually related to caring. And like the care ethicists after him, Hume considered the morality of actions to depend mainly or substantially on whether they express or manifest these sorts of sentiments. Virtue ethics ties the morality of actions to inner states of character and motivation, and for that reason it is not implausible to see Hume as a kind of virtue ethicist (despite the proto-utilitarian elements in his moral philosophy). So the question then remains why care ethicists don't typically regard themselves as also being virtue ethicists.

The reason isn't that virtue ethics has to be tied to rationalism and Aristotle. Care ethicists don't believe that for a moment, and neither, in the light of what was said just a moment ago, should they. Rather, most care ethicists don't think of themselves as virtue ethicists because they place more foundational importance on relationships than on virtues. Caring is the name of a praiseworthy individual sentiment/motive and thus of a virtue, but it is also the name of a kind of relationship, and care ethicists think the value of caring relationships is philosophically more basic than the value of individual caring sentiments, motives, or virtues. This issue concerning the foundations of care ethics is an important one, and I for one disagree with the majority opinion among care ethicists, holding as I do that care ethics can be plausibly viewed both as sentimentalist and as a form of virtue ethics. (Our mother's house has many mansions.) But for the purposes of the present chapter, this issue cuts no ice. I think the arguments about paternalism I shall be offering in what follows have force whether or not one holds that care ethics should be grounded in relationships. So I am not even going to refer the reader to the places where I defend care ethics as a form of virtue ethics based in caring as an individual virtue. What unites all prominent care ethicists is a belief that caring understood in non-rationalist terms is basic to morality, and that assumption will be central to my discussion of paternalism.

3

Almost everyone agrees that paternalism is most justified with respect to children – hence the name. But I want to show you now that if we look closely at what happens or can happen between parents and children, we can see the moral limits of paternalism as well. To do this I need to introduce a notion that will play a central and crucial role in what a sentimentalist wants or should want to say about paternalism. That notion is empathy.

Psychologists and others nowadays in fact speak of two different kinds of empathy. There is so-called projective empathy (sometimes referred to as simulation) that involves putting oneself in(to) the position (the "head" or the "shoes") of another person, and then there is what is variously called associative, emotional, or receptive empathy, which is a matter of receiving or taking in the attitudes, emotions, etc., of others. These two forms of empathy move, so to speak, in different directions, but (following Hume here) I am inclined to rely most on the second, more emotionally charged kind of empathy. In any event, I think we can use our ideas about empathy to gain a strong foothold on the notion or notions of respect that I think any realistic or plausible discussion of paternalism needs to rely on. Respect is not a foundational notion for care ethics, so unless care ethics can bring it into the picture, its picture of morality, in a non-tendentious and convincing way, it is going to have a very difficult time dealing plausibly with the moral issues surrounding or endemic to paternalism.

Well, I want to argue (and I believe I am the first care ethicist or sentimentalist to have thus argued) that empathy gives us all the purchase we need on the concept and reality of respect for others (Slote 2010: chap. 8). But in order to make this work, I think I need to bring in some examples having to do with parenting. There are times when parental paternalism seems eminently justified, as when a parent takes a very sick but protesting child to the doctor. And no one thinks such parental actions show a lack of respect for the child. But there are other times when parents impose their ideas about what is good for their child on that child in a disrespectful way, and one needs to be able to specify and explain the difference. This is something that both rationalists and sentimentalists need to do, and the rationalists haven't, I think, been all that successful in explaining these differences (you can judge for yourself by looking at the other relevant chapters in this volume). But I think the care-ethical sentimentalist can do this successfully using the notion of empathy, and let me begin with the clearest cases where paternalism on the part of parents shows a lack of respect for their child.

Empathy with another involves seeing and even feeling things from the standpoint of another rather than imposing one's own feelings and attitudes on that other. Overinvolvement with another person can mean that one does impose one's own views and that one fails to see at all clearly how differently the other feels or may feel on various issues, and this is something we (unfortunately) often find between parents and their children. Parents with a weak sense of self or self-worth often seek to live through (the successes of) their children and have a difficult time mentally separating their own needs from those of their children. Such parents ipso facto have difficulty empathizing with the individual point of view – the needs, wishes, ideas, and fears – of their children. For example, the parent may want the child to practice the piano for four hours every afternoon (in order to become the concert pianist they want him to become); and if the child ever protests and says he wants to play baseball with his friends instead, such a parent will often say, and believe, something like, "You can't really want to do that; I can't believe you are so ungrateful that you want to throw away what we are trying to do for you."

This sort of overinvolvement has recently been labeled "substitute success syndrome," and there is a large literature on that topic. But what should be obvious here is that a parent who responds in the way just mentioned fails to empathize with the point of view of their child. They don't see or don't allow themselves to see that the child really does want (sometimes) to play baseball, that they aren't dedicated to a future concert career the way the parent is on their behalf. (We see similar things with so-called stage-mothers.)

What is also obvious is that such a parent is failing to respect their child's independent wishes, desires, ideas, etc. So I want to propose that we unpack the latter failure in terms of the former, that we see the lack of respect manifested toward the child as a matter of lacking empathy with their child. A sentimentalist care ethicist and/or virtue ethicist can also use these terms to explain or clarify what is *morally wrong* with treating a child with a lack of empathic respect. If caring is incompatible with a lack of empathy, then they can say that a parent who treats a child in the above manner acts in an uncaring or somewhat uncaring way toward their child, and for the care ethicist this can be the basis for morally criticizing such a parent and/or some of their actions. The parent may demonstrate a caring attitude with regard to other issues (like feeding the child properly), but demonstrate a lack of caringness in what they do vis-à-vis the issue of a future career – e.g., if they browbeat the child into not, ever, playing baseball with his friends or if they physically prevent him from doing so.

4

However, we don't want to say that all paternalism toward children is morally criticizable or unjustified, and we don't want to say, for example, that a mother who takes their child to the doctor over the child's screaming protests is necessarily showing a morally criticizable lack of respect for that child. So how do such cases differ from those brought under the rubric of substitute success syndrome?

I want to say that such cases differ because they precisely don't manifest the kind of overinvolvement with one's child that is shown by the parent who lacks empathy with their child's independent ideas, desires, and aspirations. If the parent who brings the child to the doctor isn't showing an overinvolved lack of empathy for their child and what the child feels (and thinks), then, given the terms being emphasized here, the parent is being respectful to the child, and their paternalism in this instance can't be morally criticized.

But here an objection will occur to some readers. The substitute success syndrome parent shows a lack of respect for their child's desires and beliefs, and it isn't, at least on the face of it, clear why we shouldn't say that the parent, the mother, who takes their screaming, protesting child to the doctor is similarly unempathic with respect (or should I say with lack of respect?) to their child's desires and beliefs. If we do say this, then given the terms introduced above, we have to say that it is wrong and disrespectful for the mother to insist on taking the child to the doctor; and this, of course, goes against what we all believe about such cases. So what should the sentimentalist who emphasizes empathy say?

They can say, to begin with, that the mother who takes her child to the doctor need not lack empathy for what her child is feeling. She may take the child to the doctor while at the same time being very upset with what her child is feeling and with what she feels she has to do. And this may involve deeply empathizing with what is happening to or in her child. OK. But then the rationalist may ask how the mother who is thus assumed to be fully empathic with what her child is feeling can then, nonetheless, decide to take him to the doctor. If it is reason that tells the mother she must do so, then how is the sentimentalist going to be able to account for that in properly sentimentalist terms?

Well, let's specify the case a bit further. Assume that the child is really very ill, very much in danger: if they aren't taken to the doctor in time, they are likely to lose their sight or become lame for the rest of their life. How does a mother, a *mother*, engage with such dire facts? Won't her learning of the danger to the child's whole future engage her empathic juices and not just her reason (we shouldn't too quickly assume that these are entirely or even substantially separate)? Psychologists (and also Hume in the *Treatise*) tell us that our empathy is more strongly engaged by what is present than by what lies in the future, but they also tell us that the sheer size of a danger or form of suffering (e.g., how many people are involved) also tends to affect the degree or strength of our empathic reactions. In that case, the mother in our example may empathically feel her child's diminished or tragic future when she is told what is likely to happen if the child isn't immediately taken to the doctor. She may have a strong, vivid empathic sense of what it will be like for her child to have *their whole future life* limited or marred in the ways that now threaten to occur, and this can lead her to take the child to the doctor despite her simultaneous empathic sense of what the child is *presently* experiencing. There is nothing in the example for the sentimentalist to fear or for the rationalist to exult over, and we can now see how different it is from any case of substitute success syndrome parenting. It is at least partly because the mother here isn't overinvolved with her child, can see the child's adult future in a way that doesn't conflate it with her own needs and her own future, that we can say that her insistence on taking the child to the doctor isn't disrespectful to the child, or unempathic with regard to the *sum* or *aggregate* of the empathy-relevant factors involved in the situation. (One doesn't necessarily show a lack of empathy if one has to choose between helping one person and helping many and chooses to help the many.) So this kind of paternalism can be morally justified via the sentimentalist approach being advocated here.

However, we have so far focused solely on easy cases, cases where everyone or almost everyone agrees that we have an instance of justified paternalism or else agrees that the case is one of unjustified paternalism. Everyone thinks the mother should take her child to the doctor, and everyone or almost everyone agrees that the parent who imposes a possible career as a concert pianist on a largely unwilling child is acting wrongly, disrespectfully. But what about cases where opinion is divided, cases involving – finally! – adults interacting with adults? I want to limit myself to individual morality here and won't be speaking of political paternalism even though I think everything we have said and shall say can be applied to political or legal issues or examples. So let me mention a paradigmatic controversial case of possible or defensible paternalism, and let's see what the distinctions and concepts introduced so far allow us to say about such a case. The case is not unique, there are many like it, but the case I shall choose bears the relevant issues of paternalistic morality very much on its face and will be useful to us for that reason.

5

Philosophers and others who argue against many forms of paternalistic intervention don't usually object to such intervention when it doesn't interfere with the autonomous choice of the individual being interfered with. There is typically no objection to interfering with or against most suicide attempts because it can be held that such attempts don't manifest an individual's rational autonomy or autonomous choice, but predictably come from states like despair or depression in which an individual is less capable than most of us of exercising their autonomy. However, in other cases what an individual does to threaten their own overall future welfare can't so confidently be classed as showing a lack of autonomous decision-making.

Consider motorcyclists who say they like the feeling of the wind blowing through their hair and prefer to ride without wearing a helmet, even though they know how dangerous that can be. This kind of preference and intention cannot so easily be classified as showing a lack of autonomous decision-making, and the case therefore is one where philosophies grounded in notions of autonomy will typically insist that no individual has the right to interfere with a cyclist's desire to ride his motorcycle without a helmet. Of course, many states have passed laws making this sort of thing illegal (though not my own state of Florida), but such state intervention is far from a pure example of paternalism. The state intervenes regarding such cases in order to prevent individual harm, but also in an effort to keep down hospital costs for accident victims, and the latter motive involves considering the effect that allowing people to ride without helmets has on the public welfare generally, not just on the well-being of the cyclist. (The same point applies to the requirement to wear a seatbelt in one's car.) States or countries have (fiscal) concerns for general welfare that don't typically affect or engage private individuals, and the cases, therefore, that are viewed with most suspicion by those who emphasize individual autonomy are cases where a private individual interferes with the freedom of another adult for their own good. With the example of a motorcyclist this would involve someone, say, a parent, intervening in their adult son or daughter's life in order to make it impossible for them to ride their motorcycle without a helmet.

Imagine, for example, that a son has expressed a preference for riding without a helmet on the grounds that it feels good to have the wind rush through his hair; and, for the sake of easy use of pronouns, let us assume that his mother strenuously objects to this and makes her objections known to her son. The son, however, is not persuaded. What is the mother to do? Some will say that she has no right to paternalistically, or "maternalistically," interfere with her son's choice or decision; she has done all she can or ought to do to prevent his riding without a helmet. But what if she thinks otherwise? What if she knows a way of forcing her son to wear a helmet: she can pay someone to hook up the helmet to the motorcycle's engine in such a way that the engine won't run unless the helmet registers brainwaves from inside it. (The son can't afford to have this work undone.) So she pays someone to do this over the vehement and angry protests of her son. Has she acted wrongly and, in particular, has she shown disrespect for her son and his autonomy?

The liberal or Kantian who emphasizes freedom and autonomy in their philosophy will obviously object to what she does, but I don't think it is at all obvious that we should morally object to this sort of intervention (do you?); and what was said earlier about the relationship between respect and empathy can help make the case here. Like the mother who drags her fearful and protesting child to the doctor, the mother who undermines her adult son's ability to ride his motorcycle without a helmet may know full well how her son feels about her intervention. Her intervention or interference may not at all reflect a lack of empathy with her son's joyous experiences of riding with the wind blowing through his hair. But it presumably involves a more vivid (or weightier) sense than her son has of the danger and possible consequences of his riding in his preferred way.

The case is also different from that of the mother taking her child to the doctor because one can hope and assume that the child who is forced to go to the doctor will someday be able and willing to appreciate the rightness of what their parent had earlier forced them to do. By contrast, we can assume that the motorcycle-riding son is wedded, and known to be wedded, to his preference for freedom and certain pleasurable sensations while riding over his own safety and the peace of mind of his mother and/or father. Unlike the mother who takes her protesting child to the doctor, the mother who intervenes in the motorcycle example can't and doesn't assume that her son will eventually change his mind about her successful attempt to force him into riding with a helmet. She may still understand things very well from the standpoint of her son and, despite what she fears may be a permanent disagreement or even rupture between them, force him to wear the helmet.

This doesn't seem morally wrong to me, and I'll bet it doesn't seem wrong to many of you readers. However, the rationalist/liberal tradition that bases so much on considerations of autonomy and freedom (as opposed to sheer welfare) doesn't allow us to reach this result, and it is only the sentimentalist who offers a clear path to the right philosophical/moral conclusion(s) about the present sort of example. The sentimentalist can say that the mother doesn't act disrespectfully toward her son if she is fully empathic with his reasons for wanting to ride without a helmet, but decides to intervene nonetheless. (Again, this is similar to what we have said about the mother who takes her protesting child to the doctor.) Therefore, if we think the mother probably should intervene in the way and for the reasons I have described, we have reason to (re)conceive the morality of paternalism along sentimentalist lines and allow for this kind of individual interference with the wishes or preferences of another adult. I have described one particular case, but the reader will recognize that there are many other actual or potential cases that substantially resemble it and that, to a lesser or greater extent, depending on the degree of resemblance, call for a similar moral conclusion.

But our sentimentalist defense of the mother's intervention is in fact not yet complete. Care-ethical sentimentalism places great emphasis on relationships, good relationships, and for that reason it will want to say that the loving mother who forces her son to wear a helmet and who recognizes what her son feels about what she has done will not drop the matter or evade further discussion or argument with him after she has succeeded in her purpose. (I am indebted here to discussion with Nel Noddings.) Rather, the concern she has for their relationship and for her son's feelings (not to mention her own) will impel her or at least persuade her to try to continue some sort of dialogue with her son on the subject of what she has done and how he feels about it. No matter if the son is, in the wake of her successful intervention, too angry with her to want to speak further about the matter or even to want to speak with her at all. She can wait, and if she loves him, she *will* wait and will bring the matter up if and when he is later willing to discuss it (even if he is still angry with and resentful toward her). Their relationship has been at least temporarily damaged even if she has done the right thing in forcing him to wear a helmet, and an empathically caring mother will want to do something to make amends to her son, to try once more to make him understand where she is coming from and, more generally, to reestablish what I am assuming has been their previous good relationship. Once one sees that her obligations in this matter aren't just an issue of whether a single act at a single time was morally justified but also concern how, morally and intelligently, to continue on, or reestablish continuity, with her son, the original decision to make him wear a helmet may seem more justified than it would in the absence of these larger surrounding sentimentalism-friendly motivations and future actions (or dispositions to action) on the mother's part.

I hope it can be seen, therefore, that care-ethical and/or virtue-ethical sentimentalism has the resources for dealing with the morality of paternalism across a wide range of individual cases and in a morally plausible manner. This of course leaves political issues concerning paternalism largely unexplored from within the perspective of the present approach. Still, the concepts and arguments deployed here can be generalized and applied to political or legal examples, though it would take a good deal of work involving a variety of examples and political/legal circumstances to make good on this promise or prediction. So the case for a sentimentalist approach to paternalism has not been fully made here in this chapter, but I hope the reader can already at least see the promise of such a more general sentimentalist view of these matters. However, there may be more general worries about sentimentalism that could concern the reader at this point, and let me conclude this chapter by addressing some of those worries.

6

Even if sentimentalism can deal persuasively and intuitively with issues of paternalism, it may be felt, in the light of the whole recent history of ethics, that sentimentalism is implausible on other sorts of grounds. For one thing, sentimentalism, unlike Kantian ethics or rational intuitionism, may be thought to have nothing plausible to say about deontology. How, one might ask, can an approach to morality that emphasizes concern for the welfare of others tell us that there are times when we are not morally permitted to kill or harm an innocent person even if that will serve the general welfare? I have addressed this question at considerable length in previous work (most especially in chapter 1 of the book *Moral Sentimentalism* (2010)). For reasons there is no time to enter into here, it turns out that empathy is highly sensitive to the distinction between doing and allowing: we empathically flinch from a harm or pain we could ourselves inflict much more than from a harm or pain another person might bring about. A longer tale needs to be told and is told in the work I just referred to, but I should also mention something else pointed out in that book. If you look at rationalist attempts to justify the deontological distinction between killing and letting die or, more generally, between doing and allowing harm, you will see that they are very inadequate on their own terms. It would be ironic if sentimentalism could say more in favor of deontology than rationalism has proven itself able to do, yet that is something that at this point I feel very much inclined to conclude. But the reader should look at *Moral Sentimentalism*.

There is also the (related) issue of whether sentimentalism can do sufficient justice to our antecedent (to doing philosophy) strong sense of the objectivity of morals or morality: our sense, for example, that the wrongness of torturing babies doesn't just reflect how we humans are inclined to feel about the issue, but has a truth or validity that isn't relative to human nature. Previous sentimentalists, and most importantly David Hume, have treated moral questions as depending on our nature or as completely non-cognitive, but a sentimentalist doesn't have to accept such a take on the nature of morality. *Moral Sentimentalism* is largely devoted to showing that a sentimentalism that relies on empathy can help us justify the original strong sense (or feeling?) we have of the complete objectivity of moral distinctions. It argues that empathy helps us fix the reference of moral concepts and distinctions in a way that is as compatible with moral objectivity as Saul Kripke's use of reference fixing allows for and implies the objectivity of claims about the color of things (and about other "natural kinds"). Again, I refer the reader to the discussions of the just-mentioned book.

The argument of the present chapter is thus hostage, to some extent, to what one can plausibly say, in sentimentalist terms, about philosophical issues that go way beyond and are in some sense deeper than questions solely about paternalism. But I hope what has been said here about paternalism seems plausible in its own right, and I must simply send the reader concerned with other issues that affect the validity of sentimentalism to broader or deeper philosophical discussions that occur elsewhere.

Related topics

Paternalism and Education; Paternalism and Intimate Relationships.

References

Gilligan, C. (1982) *In a Different Voice: Psychological Theory and Women's Development*, Cambridge, MA: Harvard University Press.

Noddings, N. (1984) *Caring: A Feminine Approach to Ethics and Moral Education*, Berkeley: University of California Press.

Slote, M. (2010) *Moral Sentimentalism*, New York: Oxford University Press.

12
PATERNALISM AND AUTONOMY

Suzy Killmister

1 Introduction

It's often taken as given that autonomy and paternalism conflict; so much so that the conflict can seem analytic. This assumption is particularly noticeable if we consider how the two concepts are standardly defined. For instance, Gerald Dworkin (2014) defines paternalism as "[involving] some kind of limitation on the freedom or autonomy of some agent." John Christman (2015), meanwhile, observes that "[a]utonomy is the aspect of persons that undue paternalism offends against" (sec. 2.2).[1] So we are invited to understand paternalism as a limitation of autonomy; and we are invited to understand autonomy as that against which paternalism offends. Given such definitions, it is perhaps unsurprising that the burden taken on by defenders of paternalism has typically been understood in terms of showing how paternalism could be justified, *given that* it purportedly conflicts with autonomy (see, e.g., Feinberg 1986).

My goal in this chapter is to put some pressure on the assumption that paternalism straightforwardly conflicts with autonomy. Whether or not paternalism conflicts with autonomy, I suggest, depends to a significant extent on what we take autonomy to be – and that is far from a simple matter. Moreover, we will also need to be attentive to the wide variety of ways in which autonomy and paternalism can be said to conflict. That variety is already evident in the two quotations offered above: Dworkin speaks in terms of paternalism *limiting* autonomy; Christman speaks in terms of paternalism *offending against* autonomy. These are quite different forms of conflict, and it very much matters which we have in mind.

Here's how I'll proceed: In Section 2 I briefly sketch the distinction between hard paternalism and soft paternalism. In Section 3 I consider the nature of the purported conflict between autonomy and paternalism, teasing apart four distinct *prima facie* wrongs that paternalism might do to an autonomous agent. Further distinctions are introduced in Section 4, where I consider the range of theories of autonomy currently in circulation. Armed with all of these distinctions, Section 5 directly addresses the question of whether, and how, autonomy and paternalism could be seen to conflict. My goal in this section, I should stress, is not to provide a definitive answer to that question, but rather to explore how different theoretical commitments towards autonomy shape the range of possible answers to that question.

2 Paternalism: soft or hard[2]

Both hard and soft paternalism typically involve interventions against the agent's express will.[3] Where they differ is in the goals of that intervention. Hard paternalism has one core goal: the promotion or protection of the intervened upon agent's own good.[4] While soft paternalism shares this goal of protecting/promoting the agent's own good, it also aims to bring the intervened upon agent's action into line with what she "truly" wants. Soft paternalistic interventions thus presuppose a mismatch between the agent's express will and her authentic will. This mismatch may be due to factors such as misinformation (the agent expresses a desire to cross the bridge, but only because she doesn't realize that it is unsafe (Mill 1991 [1859]: 107)); weakness of will (the agent expresses a desire for another piece of cake, but deep down he would prefer to maintain his diet); manipulation (the agent expresses a desire to be a housewife, but only because of oppressive socialization); or cognitive bias (the agent expresses a desire to undergo surgery, but she would have refused it if the information had been presented in terms of probability of fatality rather than probability of survival (Thaler and Sunstein 2008: 36–37)).

For hard paternalistic interventions, there is no claim to be acting in accordance with the agent's autonomous will: the overriding motivating factor is the agent's well-being. Because of this feature, hard paternalism has often been thought to be straightforwardly autonomy-undermining. As we will see in Section 5, however, this assumption becomes complicated if we adopt certain theoretical frameworks for autonomy. With soft paternalistic interventions, by contrast, the motivation involves concern for the agent's autonomous will. Because of this feature soft paternalism is often thought to be compatible with autonomy. Indeed, Scoccia (2008) goes so far as to define soft paternalism as intervention that does not violate autonomy, and hard paternalism as intervention that does violate autonomy. Rather than build such commitments into the definition of hard and soft paternalism, my approach in this chapter is to treat the categories of hard and soft paternalism more loosely, and leave it open just what their relationship to autonomy might be. This brings us to the next important distinction: what, exactly, is the *prima facie* wrong of paternalism supposed to be?

3 The wrongs of paternalism

In order to make a start on this question, it will be helpful to take note of a point made by Joel Feinberg (1986: 27–51); namely, that the term "autonomy" can in fact refer to a number of different things. Three of these will be particularly relevant to our purposes here. First, autonomy can refer to what Feinberg calls a "condition." This sense of autonomy is often taken to attach to an agent's actions or desires: for any action an agent performs, or any desire she experiences, we may question whether that action or desire is autonomous. We may also ask more generally whether the agent herself is autonomous (in the condition sense), where that typically involves making an overall assessment of her patterns of desires and actions. As we will see in Section 4, it is this sense of autonomy that has received the most theoretical attention. Autonomy can also, however, refer to a capacity. In this sense, we are typically asking whether the agent has the necessary competencies to achieve autonomy as condition. An agent can have the capacity for autonomy without realizing that capacity in her condition. For instance, an agent may fail to act autonomously because she experiences weakness of will, or because her desire is the product of manipulation. But so long as those failures are not systemic, they are compatible with her being autonomous in the capacity sense. Finally, autonomy can refer to a right. Much like political sovereignty, autonomy as right appeals to the authority agents have to determine their own ends and pursue those ends unimpeded. Autonomy as right is typically taken to depend simply on having the capacity for autonomy, rather than being contingent on the autonomy of the relevant action.

The nature of the wrongs paternalism purportedly does to autonomy map onto these three different meanings of autonomy. With respect to autonomy as condition, one *prima facie* wrong of paternalism is that it *frustrates* the achievement of the condition. Take an agent whose lifelong goal is to climb Mount Everest. In the absence of intervention, we can suppose, she will successfully climb the mountain, and doing so will be an autonomous act. If she is paternalistically prevented from climbing the mountain, however, then she will be denied the opportunity to be autonomous (with respect to her mountain climbing).[5]

The relationship between paternalism and autonomy as capacity is more complicated. On the one hand, it is sometimes claimed that paternalism (particularly when it is systematic) *corrodes* the capacity for autonomy. The thought here is that the competencies necessary for autonomy must be developed through exercise. If the space in which agents are able to attempt autonomous action are curtailed, their autonomy capacities will wither. On the other hand, it is sometimes claimed that the wrong of paternalism lies in the *disrespect* it displays towards the agent's autonomy as capacity. The disrespect purported to be at play here corresponds to what Stephen Darwall (1977) calls "appraisal respect." Appraisal respect involves a positive evaluation of some feature of the agent; hence, appraisal *disrespect* involves a negative evaluation of that feature. In the context of paternalism, then, it might be said that paternalistic intervention displays an inappropriately negative evaluation of the agent's capacity for autonomy.

Finally, we have autonomy as right.[6] Here, the *prima facie* wrong of paternalism is clear: to prevent the agent from acting in accordance with her will, when her action would do no harm to others, is to fail to recognize her right to decide for herself. Such failures to recognize autonomy as right can also be understood as a form of disrespect, though this time of the recognitional variety – the paternalistic intervener is failing to give appropriate consideration to the target's status as an autonomous agent (cf. Darwall 1977: 38).[7] Importantly, giving appropriate consideration to the agent's authority involves more than just weighing her decision heavily; it involves treating it as decisive (Groll 2012). As we will see in Section 5, while the claim that paternalism disrespects autonomy as right is straightforward for hard paternalism, it becomes much more contested for soft paternalism.

To take stock: there are two kinds of paternalism, soft and hard. For each, we need to consider whether it a) frustrates autonomy as condition; b) corrodes autonomy as capacity; c) disrespects autonomy as capacity; and d) disrespects autonomy as right.

4 Varieties of autonomy

In order to even begin addressing these questions, we first need a theory of autonomy. Unfortunately, this is far from a straightforward matter. At the most basic level, there is general agreement that autonomy has something to do with self-determination: the autonomous agent is one who is "in charge of" her own activity. Once we try to unpack what it means to be "in charge," however, disagreement immediately arises. Rather than try to adjudicate those disagreements, my goal in this section is to provide a basic roadmap of the philosophical terrain.

At the risk of oversimplification, we can identify three broad theoretical approaches in the contemporary philosophical literature on autonomy. (As noted above, the primary focus of this work has been on autonomy as condition.) First, and arguably the most dominant, are what have come to be known as "proceduralist" theories of autonomy. One of the earliest, and most influential, such theories was that offered by Harry Frankfurt (1971).[8] According to Frankfurt, for an action to be autonomous it must be the product of a desire for which the agent has a second-order desire that it be effective in action. To put this more simply, my action of eating a chocolate chip cookie is only autonomous if I have a desire that my desire to eat the cookie be the one that moves me to action. Such an action is *not* autonomous if I desire to eat the cookie, but I also have a higher-order desire that this first-order desire not prevail (perhaps I have a higher-order desire that my first-order desire to maintain my diet be the one that "wins out" in this case). So an autonomous action is one in which the motivating desire "lines up" with my higher-order desire.

Frankfurt's theory is not without its problems.[9] Nonetheless, it has inspired a range of views that follow its central claim that to be autonomous is to in some way identify with the desire that moves the agent to act. For instance, John Christman's theory requires that the agent not be alienated from her motivating desire, and that she would remain unalienated from it were she to reflect on it in light of its history (Christman 1991, 2009). Michael Bratman's (2007) theory, meanwhile, requires that the agent's motivating desire accord with her self-governing policies (i.e., policies about what to treat as reason-giving in her practical deliberation). Moving away from such hierarchical construals of identification, Marilyn Friedman's (2003) theory requires that the agent reflectively endorse her motivating desire, where this means that her "whole self" takes a positive stance towards that desire. What's distinctive about all proceduralist theories is that they are content-neutral. In other words, according to proceduralist theories, any motivating desire can be autonomous provided it has arisen through an appropriate procedure and/or occupies an appropriate place in the agent's psychological profile.

The second approach is "substantive" theories of autonomy, which reject content-neutrality: for a substantive theory, whether or not an agent is autonomous depends very much on the content of her desires. For instance, Natalie Stoljar (2000) argues that a motivating desire is non-autonomous if it is the product of false and oppressive norms. Similarly, Paul Benson (1991) has argued for a "normative competence" view of autonomy, whereby an autonomous agent must be able to reflect on a potential course of action in light of the objective reasons that apply.

The final approach I will consider here are "dialogic" theories of autonomy.[10] Paul Benson (2005) and Andrea Westlund (2003) have each separately proposed theories of autonomy that require agents to take themselves to be answerable for their actions. In other words, an autonomous agent is one who is willing to "stand behind" and explain her actions in response to requests for justification.

5 Autonomy and paternalism

My goal in this section is to consider what resources the various theories of autonomy have to answer the four questions laid out in Section 1 with respect to both hard and soft paternalism. Doing so will not only cast light on the complexity of these questions, but also suggest some strengths and weaknesses of the different approaches to autonomy.

Does paternalism frustrate autonomy as condition?

It can seem obvious that hard paternalism frustrates autonomy as condition. After all, by definition hard paternalism prevents the agent from doing that which she expressly wills to do. With our three varieties of autonomy in view, however, that confidence should start to seem misplaced. Autonomy does not simply equate to doing what we want. As such, being prevented from doing what we want does not necessarily frustrate autonomy.

The clearest resources for arguing that hard paternalism frustrates autonomy come from proceduralist theories. For a proceduralist, there are no restrictions on what the agent can autonomously desire or pursue. This means that actions typically targeted by hard paternalistic interventions (smoking, not wearing a motorcycle helmet, selling oneself into slavery, etc.) could be autonomously performed; all that is required is that the agent have the appropriate second-order volition, or self-governing policy, or so on. Were such an agent to be prevented from acting as she willed, then her autonomy would indeed be frustrated. On a proceduralist model, paternalism thus functions much like weakness of will: the agent has authorized a particular action (through her second-order desire, etc.), and to be autonomous requires going through with that action.

Nonetheless, not all hard paternalistic interventions can be construed as frustrations of autonomy. Although proceduralist theories refuse to rule any particular class of actions to be necessarily non-autonomous, they still hold that some actions agents perform are non-autonomous. If an action is weak-willed, for instance, many proceduralist theories will consider it to be non-autonomous. From the perspective of such theories, hard paternalistic interventions that prevent weak-willed actions will not frustrate the agent's autonomy.[11] More broadly: for a Frankfurt-style procedural account, any action that is not motivated by a second-order volition will be non-autonomous, and hence preventing it will not frustrate the agent's autonomy; for a Bratman-style procedural account, any action that conflicts with a self-governing policy will not be autonomous, and hence preventing it will not frustrate the agent's autonomy; for a Christman-style procedural account, any action that is motivated by a desire from which the agent is alienated will not be autonomous, and hence preventing it will not frustrate the agent's autonomy; and so on.

If we turn to either of the other two kinds of autonomy theory, the situation starts to look even more problematic for the assumption that hard paternalism necessarily frustrates autonomy. Indeed, it could be plausibly argued that on a strong substantive theory of autonomy, virtually *no* hard paternalistic intervention frustrates autonomy. If we stipulate that the intervention is necessary to protect the agent's well-being, and that among the normative reasons the autonomous agent must recognize are those relevant to her well-being, then the actions hard paternalism would typically prevent are actions that would not be autonomously performed. As such, it would not frustrate the agent's autonomy to be prevented from performing them.

This is not to say that a substantive theory of autonomy has nothing to say against hard paternalism. In order to defend the claim that hard paternalism frustrates autonomy, a substantive theory could show how the intervened upon action is autonomous, despite its conflicting with the agent's well-being. For instance, Stoljar could point out that not all self-harming behavior is informed by oppressive norms, and hence not all such behavior is non-autonomous. Likewise, Benson could argue that sensitivity to the relevant normative reasons is compatible with choosing a self-harming course of action. Even with these responses in hand, though, it seems clear that a proponent of a substantive theory of autonomy would be led fairly inescapably to the conclusion that at least some hard paternalistic interventions do not frustrate the agent's autonomy.

Let's turn now to dialogic theories. As we saw above, such theories require that an autonomous agent be willing to "stand behind" her action. While a dialogic theory would potentially make quite different determinations about the autonomy of a particular action than a procedural theory, it does share the feature that *any* action could potentially be autonomous. This means that there is a straightforward connection between paternalism and frustration of autonomy: if the agent would have been willing to stand behind a particular action, and she is prevented from performing that action, then her autonomy has been frustrated. Moreover, unlike procedural theories, dialogic theories have the resources to say that even interventions in weak-willed action are potentially frustrations of autonomy; all that is required is that the agent would have been willing to stand behind her weak-willed action. That said, there will always be some actions that do not count as autonomous on such an account. Any action that the agent would not be willing to stand behind is not autonomous; hence intervention that prevents such action cannot frustrate the agent's autonomy.

We have seen that most theories of autonomy have the resources to support the claim that at least some instances of hard paternalism frustrate autonomy as condition. We have also seen, though, that for all theories of autonomy, there are at least some hard paternalistic interventions that do *not* frustrate autonomy. This should hardly be surprising. After all, the whole point of these theories is to differentiate between autonomous and non-autonomous action. Provided the latter is not an empty set, there will always be actions that can be prevented without the agent's autonomy thereby being frustrated.

What, though, of soft paternalism? Here, the claim that autonomy is frustrated is even harder to substantiate. What each kind of theory (procedural, substantive, and dialogic) share is the presupposition that there is a gap between what the agent wills in the moment, and what it would be autonomous for her to do. Importantly, soft paternalism shares this presupposition: by definition, soft paternalistic intervention is intervention in the service of getting the agent's action to line up with what it would be autonomous for her to do. If the soft paternalist shares a particular theory's understanding of what autonomous action requires, it cannot be accused by that theory of frustrating the agent's autonomy.

A more general diagnosis suggests itself here. Soft paternalism could most clearly be said to frustrate autonomy from the perspective of a theory that takes what the agent wills in the moment to necessarily be autonomous, even if it diverges from her long-term values or plans. From that perspective, any intervention pursued under the banner of soft paternalism will necessarily frustrate autonomy.[12] While this is a claim that is not infrequently leveled by soft paternalism's critics, it's important to stress that it relies upon a theoretical framework that is rejected by most contemporary philosophical approaches to autonomy – none of the canvassed theories take what the agent decides to do in the moment to necessarily be autonomous, and hence none can say that soft paternalism necessarily frustrates autonomy.[13]

That said, there is one potential avenue available to dialogic theories that is worth flagging. Recall, for dialogic theories an action is only autonomous if the agent can "stand behind it," where that is understood in terms of being prepared to answer for that action in a dialogic exchange. At first blush such theories seem to be in the same position as procedural theories, in that actions that wouldn't be autonomous could be prevented without frustrating the agent's autonomy. If we focus in on nudging as a form of soft paternalism, however, a potential response presents itself.[14]

What is distinctive about nudging is that it occurs below the agent's conscious awareness – nudging exploits an agent's cognitive biases in order to bring her actions into line with her longer-term desires and values. It could be contended that when an agent has been nudged into a particular action, she will not be able to answer for that action because the causal reason for that action lies beyond her awareness. Hence, nudging frustrates the agent's autonomy.

There are two possible responses available to the nudger. First, a pro-nudger could point out that nudging only frustrates autonomy if the action the agent would have performed *sans* nudge was itself one the agent could have answered for. Since nudges typically push agents away from actions that are themselves causally motivated by factors below the agent's conscious awareness, they do not relevantly change the situation with respect to the agent's autonomy. Second, who is to say that an agent *wouldn't* stand behind an action, even if it was caused by factors below their awareness? Agents are not necessarily good at identifying the causes of their actions, and are all too willing to construct a post-hoc rationalization for actions they perform (see, e.g., Fine 2008).[15]

Does paternalism corrode autonomy as capacity?

In order to answer the question of whether paternalism corrodes autonomy as capacity, we first need to have some idea of what this capacity consists in. As a starting point, we can say that the capacity for autonomy depends upon basic rationality. Beyond this, different theories of autonomy are going to foreground different requirements. For procedural theories, what matters is that the agent is capable of bringing her psychological states into the relevant structural arrangement. For some such theories, such as Bratman's or Friedman's, capacities for critical reflection on one's ends and values will be paramount; for others, such as Christman, critical reflection plays a much less prominent role. For substantive theories, normative competencies are paramount: the capacity for autonomy depends crucially on sensitivity to normative reasons. For dialogic theories, finally, discursive competencies come to the fore: the agent must be capable of participating in the give and take of reasons.

Interestingly, this is one aspect of autonomy for which soft paternalism may pose a greater threat than hard paternalism. Hard paternalistic regulations (at least within liberal democratic states) are almost always targeted at fairly narrow domains, such as health or retirement savings. Even if an agent were to lack the opportunity to exercise the competencies that constituted her autonomy capacity in these domains, there would be countless other domains in which she would be free to exercise those competencies. Assuming, as seems plausible, that these competencies are transferable between domains, there is no reason to think typical hard paternalistic interventions corrode agents' capacities for autonomy. (Admittedly, systematic, all-encompassing paternalistic intervention could corrode any of the above-mentioned competencies, but this is not what's typically being proposed by advocates of hard paternalism.)

The situation looks slightly different if we turn to soft paternalism, though. Precisely because soft paternalism presents itself as compatible with – and potentially even enhancing – agents' autonomy, it offers a strong rationale to intervene widely. As we saw above, most theories of autonomy allow that agents frequently perform non-autonomous actions. They may be weak-willed, or misinformed, or socialized in oppressive ways; intervening to prevent them from performing such non-autonomous actions could easily be seen as a boon to their autonomy as condition. The very success of soft paternalism in enhancing autonomy as condition, however, could be appealed to in arguing that more systematic forms of soft paternalism corrode autonomy as capacity.

While preventing you from skipping the gym this morning is unlikely to have any significant effects on your agential capacities, preventing you from *ever* acting in a weak-willed manner, or *ever* acting absent-mindedly, could potentially be more damaging – even if each one of those interventions enhances the agent's autonomy as condition. If autonomy requires willpower, or attention, or introspective skills (cf. Meyers 1989), then those skills may well need to be exercised – even if exercising them involves short-term failures of autonomy. Circumventing the various skills that constitute the agent's capacity for autonomy may render her ever more dependent on external scaffolding, and her own capacities might then wither away.

Now it might be responded that, if we can integrate soft paternalistic practices such as nudging into our lives, we won't need these capacities to be as highly developed. Even if such a response were correct, though, it would concede the point: to claim that it doesn't matter that a capacity is corroded because that capacity is redundant is to acknowledge that the capacity is indeed corroded.

Does paternalism disrespect autonomy as capacity?

Paternalism is frequently objected to on the grounds that it is disrespectful. For instance, Jules Holroyd (2009: 327) states that paternalistic intervention involves a "failure of respect for the agent and her competence to make decisions for herself." Somewhat more forcefully, Elizabeth Anderson (1999: 301) claims that paternalistic interventions "effectively [tell] citizens they are too stupid to run their own lives." Recall, the kind of respect at stake here is *appraisal* respect – it involves a positive estimation of some feature of the agent (in this case, her capacity for autonomy). If hard paternalism is to disrespect autonomy as capacity, then it must communicate insufficient regard for whatever range of competencies underpin that agent's capacity for autonomy.

The question of whether hard paternalism disrespects autonomy as capacity turns on how much respect the intervened upon agent can plausibly command. Sarah Conly (2013: 40–41) has argued that paternalism is not disrespectful precisely because we cannot, in fact, plausibly command very much respect on the basis of our autonomy capacities. She writes:

> When someone accurately assesses my abilities . . . and finds me lacking in some respects, it is very hard for me to argue that I have been degraded, and thus disrespected. . . . The suggestion here is simply that we should treat people in accordance with their real abilities and their real limitations.

Conly bases this claim on the prevalence of cognitive biases – since people are such poor reasoners, they cannot command the kind of respect that would block paternalism.

The difficulty with Conly's argument is that it appears to conflate autonomy as capacity with autonomy as condition. Even if agents only rarely achieve autonomy as condition, it does not follow that they have an impaired capacity for autonomy. As we have just seen, different theories of autonomy will identify different basic skills as necessary for the capacity for autonomy; but none of these sets of skills need be cast in doubt were the agent to occasionally – and perhaps even regularly – fail to achieve autonomy as condition in any given domain. Indeed, virtually all theories of autonomy take it as given that the vast majority of adults have the capacity for autonomy, even though they typically allow that much of human action falls short of being autonomous.

Let's take for granted, then, that the agents upon whom paternalism would be visited have autonomy as capacity. The question then becomes whether paternalistic intervention communicates insufficient respect for that capacity. Here, the distinction between autonomy as condition and autonomy as capacity has bite in the other direction, allowing for pushback on the claim that paternalistic intervention is disrespectful. Since failures of exercise need cast no doubt on the agent's capacity for autonomy, it could be argued, isolated interventions in the agent's actions need likewise cast no aspersions on the agent's capacity for autonomy. Faced with an accusation of disrespectful behavior, the intervener could sincerely respond as follows: "I have absolutely no doubt as to your capacity for autonomy, and I hold it in very high esteem; it's just that you were about to make a very poor decision."

The foregoing discussion may seem to miss the point. The accusations of disrespect leveled against hard paternalism are typically framed in terms of failing to treat the agent as competent to decide how best to live her own life (the quotes above from Holroyd and Anderson are representative in this regard). This is a competency much more tightly connected to successful exercise than the bare capacity for autonomy; as such, paternalistic intervention is much more plausibly seen as communicating a low appraisal of the capacity to make good decisions. As forceful as this kind of objection to paternalism is, however, it is important to stress that it is not an *autonomy*-based objection to paternalism. While the capacity to decide best how to live one's life is sometimes presupposed in defenses of autonomy as right (see, e.g., Mill 1991 [1859]), it is not standardly taken to be a necessary condition for autonomy as capacity. Autonomy is not orthonomy: one can be fully autonomous – according to most accounts – even while making choices that would, from an objective perspective, make one's life go worse.[16]

Let's turn, then, to soft paternalism. In what way might soft paternalism be said to disrespect the capacities of autonomous agents? Soft paternalism is largely motivated by the assumption that agents either lack relevant information (as in Mill's bridge-crossing case) or are likely to fail to process information rationally (as in cases of nudging). Assuming that one of the skills that constitutes the capacity for autonomy is information gathering and processing, this would be a natural place to look for the relevant disrespect.

George Tsai (2014) has persuasively argued that even the mere provision of information can sometimes be problematically disrespectful of the agent's capacities to gather or weigh information. His central case involves a woman making a decision about whether to apply to law school or grad school in philosophy, and whose father presumptuously presents her with an array of data he has collected comparing the two paths, thus demonstrating a lack of confidence in her ability to assess her options under her own steam. While the example as presented does not in fact fit with the definition of soft paternalism I have been working with, it can easily be tweaked so that it does: assume now that the daughter is on her way to the post office to send off her application. Now her father intercepts her to present her with the information he has gathered, requiring that she hold off on executing her decision until she has engaged with the reasons he sees as relevant. The daughter has a strong case here for accusing her father of disrespecting her agential capacities. Likewise, if the state requires all agents to undergo information sessions before engaging in purely self-regarding activities, a case could be made for the state evincing disrespect for citizens' agential capacities of gathering and weighing reasons.

As we saw for hard paternalism, though, the burden facing those wishing to make such a case is to show that the motivation for the intervention communicates a negative judgment of the agent's capacities, and in addition that this negative judgment is unwarranted. Returning to Mill's bridge case, we see that not all soft paternalistic interventions communicate a negative judgment of the agent's ability to gather information; they may simply communicate a judgment that the agent has lacked the *opportunity* to gather relevant information. One class of soft paternalistic intervention lacks this escape clause, though. It is built into the justification for nudging that agents are susceptible to cognitive biases. In other words, nudging necessarily presupposes a judgment that the intervened upon agents are unable, or at least unlikely, to independently weigh and act on the reasons they have. As such, a strong case could be made that nudging disrespects autonomy as capacity.

Interestingly, disrespect for autonomy as capacity is one of the few ways in which at least some forms of soft paternalism are plausibly *more* in conflict with autonomy than hard paternalism is. Recall, the case for hard paternalism disrespecting agential capacities ran aground because hard paternalistic interventions need not presuppose any autonomy-related incapacities in the intervened upon agents. By contrast, nudging necessarily presupposes such incapacities.

Does paternalism disrespect autonomy as right?

This is the sense in which paternalism most straightforwardly conflicts with autonomy. If agents have a right to decide self-regarding issues for themselves, grounded simply in their capacity for autonomy, then hard paternalism violates that right for all agents with the capacity for autonomy. Hard paternalistic intervention treats the agent's well-being as the only relevant factor in justifying the intervention; as such, it does not treat the agent's own decision as decisive, and hence fails to respect the agent's authority. In this way, it disrespects autonomy as right.

Importantly, this argument is open to all the various theories of autonomy. While each theory will draw different conclusions about the autonomy of particular acts, or the capacities most central to autonomy, all nonetheless accept that the vast majority of adults have the capacity for autonomy, and hence all have the necessary resources to posit autonomy as a right for the vast majority of adults. If the argument can be made that the capacity for autonomy grounds a right to autonomy – and I should stress that this is by no means a trivial argument – this would substantiate the claim that hard paternalism typically fails to respect autonomy.

This leaves us with one final question, namely whether soft paternalism disrespects autonomy as right. For this question, it will be useful to treat interventions that temporarily disrupt action to provide missing information, such as Mill's bridge-crossing case, separately from interventions that alter the presentation of information in order to affect which action the agent performs, as in nudging.

The first category of soft paternalistic intervention – i.e., the provision of information – cannot plausibly be said to disrespect agential authority. The central assumption of such interventions is that agents sometimes need to be interfered with in order to bring information to light that is pertinent to their decisions to act (Feinberg 1986). Such an assumption is not only compatible with respecting the authority of the agent to be the ultimate decision-maker; it actually *presupposes* such authority, insofar as it provides the information to the agent in her capacity as decision-maker.

The second category of soft paternalistic intervention is more fraught. On the one hand, nudging does technically leave the decision in the hands of the agent. It is for precisely this reason that its advocates refer to it as "libertarian paternalism" – the agent remains free to choose however she wishes to choose. In that regard, it seems to respect autonomy as right. On the other hand, nudging seeks to surreptitiously alter what the agent will choose (if it didn't have this effect, it would fail to serve its purpose). Such circumvention of agential decision-making processes sits uncomfortably with the injunction to respect autonomy as right.

What this discomfort suggests is that the kind of respect autonomy as right commands potentially goes beyond acknowledging authority to make a decision. Recognition respect involves responding appropriately to a particular feature – in this case, an agent's capacity for autonomy. Thus far, we have focused on taking the agent's decision to be decisive as the appropriate response. But respect for autonomy as right may go beyond this to include refraining from certain kinds of interference in the agent's decision-making process. At the very least, this looks to be a plausible avenue to explore for those drawn to the idea that nudging fails to respect autonomy.

6 Conclusion

I have sought to show that the relationship between autonomy and paternalism is far from straightforward, despite the ubiquity of claims to the contrary. Whether and how paternalism conflicts with autonomy depends on the nature of the conflict at question, whether the paternalism is hard or soft, and the theory of autonomy that we are presupposing. How we choose to combine these three variables will significantly affect the available analyses of paternalism's relationship to autonomy.

Related topics

Hard and Soft Paternalism; Libertarian Paternalism, Nudging, and Public Policy; Libertarian Perspectives on Paternalism; Paternalism and Rights.

Notes

1 I should note that Christman does go on to complicate the relationship between autonomy and paternalism in this text, and in his (2014).

2 I am using these terms in accordance with Feinberg (1986). Scholars differ, though, in precisely how they draw this boundary. For a distinct use of hard and soft paternalism, where the hardness tracks the intrusiveness of the intervention, see Conly (2013: 5–6).
3 Exceptions to this include when the paternalistic intervention occurs before the agent has formed a decision, in order to shape that decision. See Groll (2012).
4 But cf. Shiffrin (2000), who rejects the requirement that hard paternalism be motivated by the agent's own good.
5 This raises the question of how paternalism relates to autonomous desires, as opposed to actions. Since paternalistic intervention typically leaves autonomous desires intact, simply frustrating their realization, theories of autonomy that focus exclusively on autonomy as a condition of mental states, rather than being a feature of actions, will struggle to explain the wrong of paternalism. (I develop this line of criticism more fully in Killmister (2013a)).
6 This conception of autonomy has received significantly less attention in recent years (though for a notable exception see Anderson (2003)).
7 According to Dworkin (1988: 120): "Seeking consent is an expression of respect for autonomy in the way that apology is an expression of regret."
8 While Frankfurt doesn't actually use the term "autonomy" in this paper, many have read his description of "freedom of the will" as equating to that concept. Also, Frankfurt uses the term "second-order volition" to refer to the relevant kind of second-order desire.
9 Two central and interrelated lines of criticism are: 1) that the theory leads to an infinite regress, since whatever reason there is for granting authority to a second-order desire also gives us reason to grant authority to a third-order desire, and so on (see, e.g., Watson 1975); and 2) that the theory cannot cope with the problem of manipulation, since a skillful manipulator could simply control the agent's second-order desires to ensure they line up with whatever the manipulator wants the agent to do (see, e.g., Christman 1991; Benson 1991).
10 These approaches are typically understood as "relational" theories of autonomy, which is a broader category, and includes theories that require agents to stand in certain socio-political relations with others (see, e.g., Oshana (2006). I am restricting my attention to dialogic approaches since they are more conducive to the issues I am exploring in this chapter.
11 To be clear, just because an intervention prevents a non-autonomous action, it doesn't therefore count as soft paternalism. Recall, soft paternalism has to be motivated by promotion of the agent's autonomy. An intervention that accidentally prevents non-autonomous action while aiming exclusively at promoting the agent's good is still hard paternalism. It's these kinds of cases that proceduralist theories will have trouble with.
12 I say "pursued under the banner of soft paternalism" because, from this perspective, the agent's long-term values and plans do not speak for her, and hence ensuring they are satisfied cannot plausibly be done in the name of her autonomy.
13 I develop an alternative theory of autonomy, which I think has the resources to show that soft paternalism at least partially frustrates autonomy, in Killmister (2017).
14 A nudge is "any aspect of the choice architecture that alters people's behavior in a predictable way without forbidding any options or significantly changing their economic incentives" (Thaler and Sunstein 2008: 6).
15 This response exposes a potential weakness in dialogic theories: if people are willing to stand behind their actions even when those actions are the result of blatant manipulation, it's unclear that dialogic theories are tracking the relevant feature of autonomous action. I develop this objection further in Killmister (2013b).
16 The exception to this are some substantive theories. See especially Wolf (1993).

References

Anderson, E. (1999) "What Is the Point of Equality?" *Ethics* 109: 287–337.
Anderson, J. (2003) "Autonomy and the Authority of Personal Commitments: From Internal Coherence to Social Normativity," *Philosophical Explorations* 6: 90–108.
Benson, P. (1991) "Autonomy and Oppressive Socialization." *Social Theory and Practice* 17: 385–408.
———. (2005) "Taking Ownership: Authority and Voice in Autonomous Agency," in J. Christman and J. Anderson (eds.) *Autonomy and the Challenges to Liberalism: New Essays*, Cambridge: Cambridge University Press.

Bratman, M. (2007) *Structures of Agency: Essays*, Oxford: Oxford University Press.
Christman, J. (1991) "Autonomy and Personal History," *Canadian Journal of Philosophy* 21: 1–24.
———. (2009) *The Politics of Persons: Individual Autonomy and Socio-historical Selves*, Cambridge: Cambridge University Press.
———. (2014) "Relational Autonomy and the Social Dynamics of Paternalism," *Ethical Theory and Moral Practice* 17: 369–382.
———. (2015) "Autonomy in Moral and Political Philosophy," in E.N. Zalta (ed.) *The Stanford Encyclopedia of Philosophy*, https://plato.stanford.edu/archives/spr2015/entries/autonomy-moral/.
Conly, S. (2013) *Against Autonomy: Justifying Coercive Paternalism*, Cambridge: Cambridge University Press.
Darwall, S. (1977) "Two Kinds of Respect," *Ethics* 88: 36–49.
Dworkin, G. (1988) *The Theory and Practice of Autonomy*, Cambridge: Cambridge University Press.
———. (2014) "Paternalism," in E. N. Zalta (ed.) *Stanford Encyclopedia of Philosophy*, http://plato.stanford.edu/archives/sum2014/entries/paternalism/.
Feinberg, J. (1986) *The Moral Limits of the Criminal Law, Volume 3: Harm to Self*, New York: Oxford University Press.
Fine, C. (2008) *A Mind of Its Own: How Your Brain Distorts and Deceives*, New York: W.W. Norton & Co.
Frankfurt, H. (1971) "Freedom of the Will and the Concept of a Person," *The Journal of Philosophy* 68: 5–20.
Friedman, M. (2003) *Autonomy, Gender, Politics*, New York: Oxford University Press.
Groll, D. (2012) "Paternalism, Respect, and the Will," *Ethics* 122: 692–720.
Holroyd, J. (2009) "Relational Autonomy and Paternalistic Interventions," *Res Publica* 15: 321–336.
Killmister, S. (2013a) "Autonomy and False Beliefs," *Philosophical Studies* 164: 513–531.
———. (2013b) "Autonomy and the Problem of Socialization," *Social Theory and Practice* 39: 95–119.
———. (2017) *Taking the Measure of Autonomy: Self-Definition, Self-Realisation, and Self-Unification*, New York: Routledge.
Meyers, D. T. (1989) *Self, Society, and Personal Choice*, New York: Columbia University Press.
Mill, J. S. (1991 [1859]) *On Liberty and Other Essays*, Oxford: Oxford University Press.
Oshana, M. (2006) *Personal Autonomy in Society*, Burlington, VT: Ashgate Publishing.
Scoccia, D. (2008) "In Defense of Hard Paternalism," *Law and Philosophy* 27: 351–381.
Shiffrin, S. V. (2000) "Paternalism, Unconscionability Doctrine, and Accommodation," *Philosophy & Public Affairs* 29: 205–250.
Stoljar, N. (2000) "Autonomy and the Feminist Intuition," in C. Mackenzie and N. Stoljar (eds.) *Relational Autonomy: Feminist Perspectives on Autonomy, Agency, and the Social Self*, New York: Oxford University Press, pp. 94–111.
Thaler, R. H. and Sunstein, C. R. (2008) *Nudge: Improving Decisions About Health, Wealth, and Happiness*, New Haven, CT: Yale University Press.
Tsai, G. (2014) "Rational Persuasion as Paternalism," *Philosophy & Public Affairs* 42: 78–112.
Watson, G. (1975) "Free Agency," *The Journal of Philosophy* 72: 205–220.
Westlund, A. (2003) "Selflessness and Responsibility for Self: Is Deference Compatible With Autonomy?" *The Philosophical Review* 112: 483–523.
Wolf, S. (1993) *Freedom Within Reason*, New York: Oxford University Press.

PART III

Paternalism and political philosophy

13

MILL'S ABSOLUTE BAN ON PATERNALISM

Jonathan Riley

John Stuart Mill is commonly said to be absolutely opposed to paternalism. Richard Arneson speaks for many when he says that "Mill meant to assert . . . [an] absolute ban on paternalism" (1980: 470). This common opinion is only correct, however, if paternalism is understood as coercive interference with a competent individual's voluntary self-regarding actions that, by definition, do not harm others or do so only with their "free, voluntary and undeceived consent and participation" (Mill 1977 [1859]: 225; I.12). This understanding of what paternalism must mean for the received opinion to be valid is usually said to be implied by Mill's "one very simple principle," which holds that "the only purpose for which power can be rightfully exercised over any member of a civilized community, against his will, is to prevent harm to others" (1977 [1859]: 223; I.9). Strictly speaking, though, it is implied by his principle of self-regarding liberty, which slightly modifies the simple principle by stipulating that society properly does not have authority to even consider using coercion to prevent *consensual* harm to others.[1]

So, for Mill, obstruction of a competent individual's self-regarding actions, which don't cause any non-consensual harm to others, solely for his own good is never justified: "His own good, either physical or moral, is not a sufficient warrant" (1977 [1859]: 223; I.9). Such coercive interference, or credible threat of same, is always objectionable because, according to the principle of self-regarding liberty, the competent agent has a moral right to liberty in the positive sense of doing as he wishes and intends within his self-regarding domain. Any use of power, or significant threat of it, by government officials or others against his will to prevent him from intentionally harming himself or risking harm to himself violates his right of self-regarding liberty and is thus deserving of punishment.

Mill is also clear that incompetent individuals, who are incapable of rational persuasion, or cannot understand the consequences of what they do, or cannot form feasible intentions and carry them out, do not have a moral right of self-regarding liberty. Paternalistic interference with an incompetent person's self-regarding conduct is not wrongful in principle but instead may be justified solely to prevent him from unintentionally harming himself. There is no ban on paternalism with respect to children, deluded or insane or comatose adults, people in the grips of an addiction, wild-eyed barbarians who react with violence against any criticism of their behavior, and so on.

Notice that, for Mill, the motivation or reason for coercive interference with a competent person's self-regarding actions is irrelevant. The interference is wrongful in any case. Indeed, he suggests that paternalistic concern for the person's own good is often not the real motivation for the coercion. When a majority interferes with self-regarding conduct, he says, "it is seldom thinking of anything but the enormity of acting or feeling differently from itself" (1977 [1859]: 284; IV.12). So, while it is possible to distinguish between coercive paternalism and coercive moralism (motivated by the feeling that a self-regarding action is intrinsically immoral), the distinction is not of much practical importance for Mill. For convenience, I shall ignore it and define objectionable paternalism to include any coercive interference or credible threat of same, whatever its motivation, against a competent agent's self-regarding action.

Mill's absolute ban on paternalistic interference with a competent person's self-regarding actions strikes many critics as much too extreme. Arneson (2005, 2015), Danny Scoccia (1990, 2008), Sarah Conly (2013), Cass Sunstein (2014), and Jason Hanna (n.d.), among others, share the conviction that such absolutism is unreasonable and even fanatical. They reject what Sunstein calls Mill's "Epistemic Argument," which "insists that so long as they are not harming others, competent people should be allowed to act on the basis of their own judgments, because those judgments are the best guide to what will make their lives go well" (2014: 92). Even Joel Feinberg (1971, 1986), usually classified as a stalwart opponent of coercive paternalism, attributes this epistemic argument to Mill, and with a twist: he argues that Mill himself is forced to abandon it as untenable in some contexts. He agrees that Mill cannot possibly show that coercive paternalistic interventions cause greater harm than the harm they prevent "necessarily in *every* case" (1986: 59, original emphasis). In rare cases such as voluntary slavery, "Mill concedes [that] we are justified in interfering with [a competent person's] liberty in order to protect him from extreme harm." In such cases, he claims, Mill contradicts his robust assertion of antipaternalism: "At that point, Mill is finally ready to admit paternalistic reasons into his (otherwise) liberal scheme of justification" (1986: 59). In making this claim, Feinberg is reiterating a similar claim by Arneson (1980: 473, 485–489), who at the time regarded absolutist antipaternalism as a live philosophical option and thought that Mill was needlessly forced into self-contradiction by his neglect of nonpaternalistic reasons for coercive interference with voluntary slavery and other permanent abdications of liberty.

Against the critics, I argue that the crude epistemic argument attributed to Mill is a strawman that he doesn't embrace. Instead, he embraces what I call a provisional epistemic argument, according to which a competent individual, though not always prudent, is the best judge of her own good *as she conceives it*, and should be permitted to choose any self-regarding action which she judges is needed to attain it, *provided* she is in possession of any readily available public information (which others may need to supply through advice and warning) about the condition of external objects (such as a public bridge but also her own body and reputation) so that *she can do as she wishes in accord with her feasible intentions*.[2] As a result of the proviso, this argument is silent when the individual does not possess the relevant information, though she will evidently act as she wishes but may not intend if nobody is around to advise or warn her. My main claim is that the provisional epistemic argument is sound. A second important claim is that Mill never abandons *this* epistemic argument, not even in the special context of voluntary slavery contracts and other irrevocable long-term contracts. Contrary to the assertions of Arneson, Feinberg, and many others, he doesn't prescribe coercive interference with the relevant voluntary activities.

Mill's epistemic argument is a central concern not only because his critics often focus their attacks on it but also because he claims that it is "the strongest of all arguments against the interference of the public with purely personal conduct" (1977 [1859]: 283; V.12). He makes at least two other key arguments in support of his antipaternalism, however, and these may be called, respectively, the arguments from diversity and from individuality or self-development. While I cannot discuss them here in detail, an indication of them is needed to clarify his unusual notion of authentic personal happiness or "utility in the largest sense" (1977 [1859]: 224; I.11).

They help us to understand the justification for his preoccupation with a competent person's happiness *as she conceives it* rather than as others (including highly intelligent and informed others) conceive it.

According to the argument from diversity, competent people have diverse tastes such that one person's idea of personal happiness can be distinct from, and at odds with, another's. Mill stresses that a person learns what brings her happiness only by choosing her opinions and actions in accord with her own judgment and inclinations: "To give any fair play to the nature of each, it is essential that different persons should be allowed to lead different lives" (1977 [1859]: 266; III.9). It is not that each individual has a unique essence awaiting discovery but rather that different people inhabit different external circumstances, including different bodies with different susceptibilities to pleasure and pain, and their circumstances cause them to form distinct personal preferences over the sources of happiness. Their distinct preferences in turn lead to different ideas of what comprises personal happiness, as the preferred sources come through habit to be viewed as the concrete ingredients of happiness.

Given diversity, it is a dangerous delusion to think that only a single correct conception of personal good exists that justifies coercion of anyone who disagrees, or that only one rational ranking of (at least some of) the various concrete ingredients of personal happiness is permissible, or that only a few self-regarding lifestyles are worthy of a human being. On the contrary, it is reasonable for competent people in different circumstances to have conflicting ideas of their own happiness. Even suicide may be very desirable to someone near the end of a painful terminal illness, while it is abhorrent to a vibrant young person.

The argument from individuality holds that individuality, that is, an authentic character that reflects one's own understandings, desires, and impulses, is "one of the principal ingredients of human happiness" (1977 [1859]: 261; III.1). A process of self-improvement is possible, however, whose "only unfailing and permanent source . . . is liberty" (1977 [1859]: 272; III.17). By choosing as he wishes and intends, a competent person can improve his intellectual, moral, and imaginative capacities: "The mental and moral, like the muscular powers, are improved only by being used" (1977 [1859]: 262; III.3). Others can't accomplish his self-improvement for him by forcing him to behave this way or that. Rather, he can only accomplish it himself by choosing in accord with his own judgment and inclinations. In this way, his authentic character can be improved from a "miserable individuality" that is selfish and contemptible to become a "noble character" that is moral and admirable and, as such, facilitates general happiness (1969 [1861]: 213–214, 216; II.9, 14). Part of the argument is that competent people tend to learn not to repeat their mistakes. But there is much more to it than that. More specifically, the claim that liberty, protected by right with respect to self-regarding actions, promotes self-improvement in the direction of a noble individuality in which a love for equal justice takes priority over competing considerations, is tied to Mill's widely misunderstood doctrine of higher pleasures, as I discuss at length elsewhere (Riley 2015: 261–268; see also Riley 2013, 2014, 2016).

It emerges that an individual achieves his authentic happiness only if he develops an excellent character, in other words, a "Greek ideal of self-development" which Mill associates with Pericles (1977 [1859]: 266; III.8). Such a well-developed noble agent asserts himself by doing as he wishes and intends within his self-regarding sphere, but he also refuses to encroach on the equal rights of others, including their rights of self-regarding liberty. He voluntarily obeys reasonable social rules of justice that distribute universal equal rights and duties. General happiness "in the largest sense" can be maximized only if virtually all members of society develop this noble character. Consistent with this, any one noble person's idea of his own happiness has a content that is typically distinct from another's, even though all willingly comply with the same reasonable social rules of justice.

The equal right of the competent individual to choose among his self-regarding actions as he pleases is essential to his development of a noble character, including its self-regarding component that varies in content across different people. Along the path of self-development, he must struggle to improve and will inevitably feel frustration and regret over any foolish and contemptible self-regarding actions he takes as a result of his relatively undeveloped capacities. As Mill emphasizes, however, "happiness, and content[ment]" are "two very different ideas" (1969 [1861]: 212; II.6). Competent people learn to bear their discontent as the process of self-development unfolds: "Human beings have faculties more elevated than the animal appetites, and when once made conscious of them, do not regard anything as happiness which does not include their gratification" (1969 [1861]: 210–211; II.4).

True, competent people may need to be coerced to obey social rules of justice until they cultivate moral dispositions to do right which override their egotistical motivations: "As much compression as is necessary to prevent the stronger specimens of human nature from encroaching on the rights of others, cannot be dispensed with" (1977 [1859]: 266; III.9). Coercion is no doubt a source of dissatisfaction, but it is properly employed to prevent or punish other-regarding actions that cause wrongful non-consensual harm to others. Conduct that violates others' rights is not self-regarding conduct. In any case, as Mill goes on to say, "there is a full equivalent [to the individual] in the better development of the social part of his nature, rendered possible by the restraint put on the selfish part." Coercion is justified not only to prevent wrongful harm to others but also to help the individual to cultivate strong moral sentiments of justice: "To be held to rigid rules of justice for the sake of others, developes [*sic*] the feelings and capacities which have the good of others for their object" (1977 [1859]: 266; III.9).

Mill's arguments together make a compelling case against coercive interference with a competent person's self-regarding actions. The individuality argument insists that allowing her to choose among her self-regarding actions in accord with her own judgment and inclinations is necessary to develop a noble individuality associated with a human being's achievement of authentic happiness. The diversity argument holds that each person's idea of her own happiness can consist of distinctive concrete ingredients. And the provisional epistemic argument, which I aim to clarify, maintains that a competent individual is best placed to know her own self-regarding interests and the means of achieving them, provided she is suitably advised and warned about external objects relevant to her feasible intentions. In sum, these utility-based arguments all support the view that a competent individual has a moral right to freely pursue her own happiness as she conceives it.

Another theme woven into my discussion is that further work is needed to clarify what objectionable paternalism in Mill's sense entails and when it occurs. There is serious confusion about this because Mill's doctrine of self-regarding liberty is generally misunderstood. Many examples commonly viewed as illustrations of justified paternalism do not involve coercive interference with self-regarding conduct as he conceives it, and so do not constitute objectionable paternalism in his sense. If paternalism is understood in a non-Millian way as it commonly is, then it is misleading to paint him as a proponent of an absolute ban on paternalism even in the context of competent agents because he certainly endorses some of the interventions classified as paternalistic in non-Millian senses of paternalism.

1 Competence, intention, and choice

Mill is clear that his liberty doctrine applies only to competent individuals. Competence is conceived in terms of a threshold capacity "of being improved by free and equal discussion" (1977

[1859]: 224; I.10). It requires what Mill calls a "tolerable amount of common sense and experience" (1977 [1859]: 270; III.14) and "the ordinary amount of understanding" (1977 [1859]: 276; IV.3). As I read him, he is saying that a competent person has enough common sense to form feasible intentions or plans and can voluntarily take any movements (or direct others to take them) that she believes will implement her intentions, with the caveat that her beliefs are fallible. Strictly speaking, only competent people can choose to take *actions*. A feasible action just is a feasible intention combined with any voluntary movements expected to carry it out.[3] Incompetents can typically engage in voluntary *behavior* but they lack the ordinary understanding needed to form feasible intentions.[4]

Although it is not always easy to distinguish between competence and incompetence (especially because competent agents can become temporarily incompetent under the influence of drugs, disease, and so on), a competent person's voluntary actions are the feasible intentions she chooses to implement. To elaborate, her intention is a forecast of her movements (if any) to achieve a desired outcome X at some future time (which may be soon or distant), and her action (or course of action) Y is the intention plus the movements which she chooses because she believes they are needed to carry out the intention and achieve X. In short, her intention to achieve X by choosing Y is the conjunction of a wish to bring about X and a belief that X is likely to result from Y, where X may be defined comprehensively to include Y for convenience. For Mill, a wish or volition is determined by the strongest present desire, whereas a belief is a type of thought or idea which may be mistaken. So, a competent person's voluntary action is a feasible intention to achieve a feasible outcome that she most wants to achieve and thinks can be achieved by taking this action at present.[5]

In light of her fallibility, a competent individual welcomes advice and warning from others that help her to implement her feasible intentions and achieve her desired outcomes. She can appreciate the likely consequences of her actions so that any factual information of which she was previously unaware may persuade her to change her intentions. But she is rightfully the final judge of what to do within her self-regarding domain.

Competence doesn't imply full rationality: a competent person is not necessarily prudent or a skilled means-ends reasoner. Mill emphasizes that competent members of civil societies may make foolish or degrading self-regarding choices that arouse the displeasure and even contempt of other people: "Though doing no wrong to any one, a person may so act as to compel us to judge him, and feel to him, as a fool, or as a being of an inferior order" (1977 [1859]: 278; IV.5). But he insists that coercive interference with their foolish or degrading self-regarding choices remains illegitimate. According to him, an individual has no "socially obligatory" duties to be prudent or dignified in his self-regarding actions: "self-regarding faults . . . may be proofs of any amount of folly, or want of personal dignity and self-respect"; but by themselves they are not "a subject of moral reprobation" (1977 [1859]: 279; IV.6). If we speak of duties to oneself, such duties are nominal or metaphorical and cannot properly be enforced by others: "The term duty to oneself, when it means anything more than prudence, means self-respect or self-development; and for none of these is any one accountable to his fellow creatures." Thus, Arneson is on target when he says, "Mill is quite prepared to tolerate deviations from rationality that occur through a person's exercise of autonomous choice" (1980: 485).

Sunstein, Conly, Feinberg (before being corrected by Arneson), and others are mistaken to assert that, for Mill, competent people are fully rational agents who generally choose self-regarding actions that achieve prudent and admirable outcomes. On the contrary, he assumes that competent members of a civil society may often fail to exercise the higher mental capacities essential to making their own choices and instead rely only on "the ape-like one of imitation"

(1977 [1859]: 262; III.4). Many blindly imitate others (including their dead ancestors whose habits are embedded in customs) and do not carefully consider their personal goals or choose effective means of achieving them:

> even in what people do for pleasure, conformity is the first thing thought of; they like in crowds; they exercise choice only among things commonly done: peculiarity of taste, eccentricity of conduct, are shunned equally with crimes: until by dint of not following their own nature [i.e., their own desires and judgment], they have no nature to follow: their human capacities are withered and starved: they become incapable of any strong wishes or native pleasures, and are generally without either opinions or feelings of home growth, or properly their own.
>
> (1977 [1859]: 265; III.6)

Mill would hardly be surprised to learn that competent humans display cognitive biases and errors. He never confused the theoretical assumption of rational self-interest, made for scientific convenience in the context of abstract economic models, with the actual behavior of people in observed societies.[6] Perhaps behavioral psychology has made more precise the various sources of human fallibility and prejudice, but it isn't news that even competent people make foolish and degrading choices. Nevertheless, it is fair to ask what Mill means by his claim that "with respect to his own feelings and circumstances, the most ordinary man or woman has means of knowledge immeasurably surpassing those that can be possessed by any one else" (1977 [1859]: 277; IV.4). He cannot mean that a competent individual always chooses rational and prudent self-regarding actions that are in his best interests as a human being, whereas competent outsiders with a wish to intervene are invariably uninformed and foolish meddlers whose concern for the individual's good is always misplaced. Contrary to the critics, he is not saying that a competent person is always the best judge of which self-regarding actions are most likely to attain desirable human ends. Indeed, Mill is explicit about this: "If a person possesses any tolerable amount of common sense and experience, his own mode of laying out his existence is the best, *not because it is the best in itself*, but because it is his own mode" (1977 [1859]: 270; III.14; emphasis added). A competent person may choose foolish or degrading self-regarding actions, and yet these actions, unless he can be persuaded that they will lead to outcomes that he doesn't want or intend, should still be viewed as best for his own happiness *as he conceives it*.

Mill concedes that a competent individual may be foolish and impulsive, and that others can have superior knowledge which the individual should welcome in the form of advice and persuasion, even if she is left free to ignore it. Indeed, a competent person will take account of such information if she believes it saves her from taking an action that doesn't achieve the outcome she intends to achieve, or persuades her to change her intentions and achieve some other outcome. But she remains free to ignore it, and will ignore it, otherwise. So the individual may persist with her foolish plans, in which case her own judgments can lead her to take self-regarding actions that make her life go poorly and that she comes to regret. Still, Mill insists that she should have a right to do as she wishes and intends.

2 Knowledge of external conditions

Mill doesn't maintain that a competent individual always knows better than outside observers do about what may be called external conditions. External conditions include the state of external objects. For example, Mill implies that outsiders may know better the condition of a public bridge. They can advise the individual of its dangerous condition and thereby provide him with local common knowledge of which he was previously unaware. Mill also implies that outsiders know better their own feelings and circumstances. They can warn the individual of their distaste and even their contempt for his self-regarding conduct and thereby provide him with information about its consequences which he perhaps had not anticipated. Similarly, others may be better informed about the state of public opinion in his community regarding his self-regarding conduct. A competent person prefers to avoid unintentionally provoking the displeasure of others so "it is doing him a service to warn him of it beforehand, as of any other disagreeable consequence to which he exposes himself" (1977 [1859]: 278; IV.5). While identifying various "natural penalties," including loss of friendship, which the person may suffer "at the hands of others" for his self-regarding faults, Mill insists that the others have a right to freely avoid what displeases them and that the person has "no right to complain" about the natural penalties that are inseparable from their displeasure.

Mill is apparently saying that a competent person is the best judge of what may be termed his internal conditions. It is plausible to maintain that the individual has immeasurably superior means of knowing his own states of mind – his own thoughts, desires, motivations, attitudes, and the like – and of remembering the detailed history of his own self-regarding conduct, for example, while at the same time he is relatively unaware of others' dislike of his conduct. In that case, his self-regarding conduct may be imprudent even by his own lights if he values his reputation in the community. Similarly, a person has far superior means of knowing her own finances or career aspirations than outsiders do, and yet she can be relatively ignorant of how best to invest her money or to go about getting her preferred job. Unless she gets expert advice and acts on it, she may take financially injurious self-regarding actions that are at odds with her intentions and that even she believes were foolish and regrets.

As these examples illustrate, external conditions include not merely the state of objects such as public bridges but also the state of the objects which are the special personal concern of the individual such as his reputation and financial health. Even a person's own body is an external object about which others may know more than he does. A competent person has immeasurably superior means of knowing his own mental states as well as the states of his body insofar as they enter his consciousness as sensations. But he may need help from outside experts to learn about the physical conditions of his organs, brain, and nervous system to understand why he feels ill. Unless he receives advice and warning from outsiders, he may choose self-regarding actions relating to his own health which he doesn't intend to choose and which may cause him grave self-injury. So he has strong personal incentives to receive that advice and take it into account so as to critically assess his own knowledge of his internal conditions. This doesn't mean that he must accept the advice and act on it, although he may. Instead, according to Mill, the person's far superior knowledge of his own mental states, goals, and peculiar self-regarding circumstances entitles him to do as he wishes after receiving the advice and evaluating it in light of his intentions. His right of self-regarding liberty guarantees that he is free to act as he intends so he can ignore the advice if it conflicts with his intentions.

So Mill allows that outsiders may need to remedy competent people's comparative ignorance of external conditions before the people can do as they intend. He recognizes that a competent individual's relative ignorance of external conditions may lead him to take imprudent or degrading self-regarding actions that he would not take if he were made aware of knowledge possessed by others. The person's voluntary self-regarding actions in that case are unintentional and cannot confidently be viewed as what even the agent considers the best ways of achieving his own ends. If outsiders can and do provide the requisite information about external conditions through advice and warning, however, then the individual is at least saved from taking an unintended action. Yet he may reject their advice and intentionally carry on with the self-regarding conduct which they believe is foolish and which may in fact be foolish. Mill argues nevertheless that the individual's knowledge of his own purposes and of the self-regarding actions he intends to take to attain them "immeasurably surpass[es]" anyone else's so that a right to freely pursue his own happiness as he conceives it should be assigned.

3 Non-coercive interventions

Mill insists that he is not advocating "selfish indifference" and emphasizes that "there is need of a great increase of disinterested exertion to promote the good of others" (1977 [1859]: 277; IV.4). But "disinterested benevolence" cannot legitimately rely on coercion or threats of same to force competent adults to take prudent or virtuous self-regarding actions: "it is by . . . conviction and persuasion . . . only that, when the period of [compulsory] education [during childhood] is past, the self-regarding virtues should be inculcated" (ibid.). People "should be for ever stimulating each other to increased exercise of their higher faculties, and increased direction of their feelings and aims towards wise instead of foolish, elevating instead of degrading, objects and contemplations" (ibid.). But they have no moral right to coercively interfere with a competent individual's intentional self-regarding actions: "neither one person, nor any number of persons, is warranted in saying to another human creature of ripe years that he shall not do with his life for his own benefit what he chooses to do with it" (ibid.). Indeed, people have moral duties not to coercively interfere, which are correlative with the competent person's right to do as he wishes and intends within his self-regarding sphere.

Interventions that aim merely to ascertain whether a competent person is really doing as he intends and/or to persuade him to reconsider his intentions can be distinguished from coercive interference with his intended self-regarding actions. If an outside observer temporarily seizes an individual to advise him of a dangerous bridge when warning signs are not posted, for example, or if the outsider alerts the individual to a posted warning sign which he has apparently missed, then the outsider is legitimately aiming to supply the individual with information, of which the individual may be unaware, that is essential to his doing as he intends. If the individual doesn't intend to take on the risk of falling into the river to his injury or death, then he should be grateful for the intervention, which aids him in choosing a different action that matches his intention. Given the new information about the bridge, he will voluntarily find another way to cross the river or, if no other way is feasible, decide not to cross it at all. If he intends to take on the risk in any case, however, then he remains free to ignore the warnings as well as any attempts to persuade him not to cross the bridge because "he himself is the final judge" (1977 [1859]: 277; IV.4). There is no coercive interference with whatever he intends to do, even if some force has been employed to ascertain his intentions. Concerned outsiders have only checked that he possesses the information about external conditions which he needs for an ordinary amount of understanding of the likely consequences of his self-regarding actions. If he lacks such information, then his action may bring about an outcome that he doesn't intend or wish to achieve, and no competent agent wishes to do this. Only incompetent people can voluntarily behave in ways that will not implement their intentions or achieve an outcome they want.

Unlike these unobjectionable interventions whose purpose is to *facilitate* the competent person's voluntary self-regarding actions that realize his intentions when others doubt that he knows what he is doing, objectionable coercive interference *obstructs* his voluntary self-regarding actions even though there is no doubt that he has the information he needs to do as he intends. The unobjectionable interventions are not coercive because no competent person wishes to take actions that he doesn't intend to take, and so interference with such acts is not against his will. These non-coercive interventions are unobjectionable because they promote the competent person's self-regarding liberty: they don't prevent him from doing as he intends and thus achieving the outcome he wishes to achieve but merely check to make sure that he knows what he is doing and/or try to persuade him to alter his intentions.

Arneson apparently rejects this distinction between non-coercive and coercive interventions on the grounds that interceding to warn the competent individual of his dangerous self-regarding action is properly seen as coercive unless the individual consents to receive the warning. He claims that Mill agrees with him. According to his interpretation, Mill implies that a competent person who refuses to read a warning sign when he has the opportunity to do so should be free to proceed as he wishes: "if a person sees a warning sign on a highway, does not bother to read it, and subsequently crashes, his driving-so-as-to-crash is voluntary [and so should be free from interference]" (1980: 483). The person must have an opportunity to be warned, but his negative freedom doesn't require that he must attend to the warning before acting. By implication, a police officer who pulls the driver over to the side of the road and informs him of the danger engages in objectionable paternalism, even if the officer doesn't prevent him from driving on.

But there is no textual basis for attributing this unappealing view to Mill.[7] Contrary to Arneson, Mill does appear to countenance using an element of force to compel someone to receive a warning or listen to advice. He says that "considerations to aid his judgment, exhortations to strengthen his will, may be . . . *obtruded* on him, by others" (1977 [1859]: 277; IV.4, emphasis added). It is legitimate for one person to force herself upon another person to make sure that the other knows what he is doing, if she is uncertain about whether he is acting as he intends. A competent person cannot wish to act unintentionally: he chooses actions that he expects to fulfill his intentions, which is not to say that his intentions or plans are necessarily reasonable or prudent. So it is perfectly fine for others to offer information to him, and to make sure that he understands the information in light of his intentions, when they are unsure about what his intentions are and suspect that he may be acting unintentionally in ways that are likely to cause severe self-injury.[8]

Mill dismisses the idea that a competent person can wish not to have information essential to his doing as he intends:

> Such a precaution as, for example, that of labelling the [poisonous] drug with some word expressive of its dangerous character, may be enforced without violation of liberty: the buyer cannot wish not to know that the thing he possesses has poisonous qualities.
>
> (1977 [1859]: 294; V.5)

This claim is evidently of far wider application. Warnings to prevent people from unintentionally consuming arsenic which they mistakenly believe is sugar, or to prevent them from unintentionally venturing onto a dangerous bridge which they mistakenly think is safe, are not coercive interventions that prevent them from taking voluntary actions which they intend to take. The legitimacy of providing such advice is not contingent on the expressed consent of the recipient since no competent person can refuse to receive and consider the information: refusal is evidence of incompetence.

Nevertheless, a competent person may intend to do some very foolish things so that, even after being warned by others of the danger and being exhorted by them to avoid it, he wishes to carry on with his imprudent self-regarding actions. He may choose to incur the risk of grave self-injury or even death. Mill insists, however, that as long as he is competent, coercive interference with his chosen self-regarding actions is always wrongful, however foolish his actions may seem to others and may in fact be.

In his brief discussion of the dangerous bridge case, Mill assumes that there is a significant risk but "not a certainty" of injury or death if the person still wishes to cross the bridge despite being warned that it is unsafe. This leaves open the possibility that, in his view, the person should be "forcibly prevented" from crossing if there is a certainty that he will experience broken bones or lose his life (1977 [1859]: 294; V.5). It's reasonable to suspect that a person who wishes to severely injure or kill himself may be incompetent at the time, even if he is usually a competent agent. If he's found to be drunk or on drugs, or "in some state of excitement or absorption incompatible with the full use of the reflecting faculty," then forcible detainment is legitimate until it can be established that he has regained his competence, if he ever had any. In some cases, the assessment of competence even by experts is a difficult task and may require more than a brief period of detention.

Since the question of competence itself arises in this special context, apart from the issue of supplying information to a competent agent, Mill can concede that coercive interference is justified while the possibility of incompetence remains an open question. Not only in the bridge case but also in many similar cases, such as Arneson's example of the dangerous driver, government officials or other concerned people can legitimately use force to prevent the individual whose competence is reasonably suspect from engaging in self-regarding behavior that will cause severe self-injury or death *with certainty*. Paternalism is justified until the relevant authorities confirm that he has "the ordinary amount of understanding." If his competence is confirmed, however, then coercive interference with his intentional self-regarding actions solely to promote his own good is not legitimate. Against all advice and persuasion, a competent agent may decide to implement his feasible intention to commit grave self-injury or suicide. Outsiders may properly express disapproval of such pitiful self-regarding actions but cannot rightfully prevent or punish them, unless the actions at the same time violate moral duties to others, in which case they are taken out of the self-regarding sphere and placed into that of morality and law.

4 Why the provisional epistemic argument is sound

The crude epistemic argument attributed to Mill by his critics is merely a strawman, easy for defenders of paternalism to knock down. He doesn't make this crude sweeping claim that a competent individual is a fully rational judge who always knows better than others what a human being's personal happiness consists in, or which self-regarding actions will promote the personal good in some such objective sense.

The provisional epistemic argument really advanced by Mill is that a competent person, if in possession of readily available information about external conditions, is the best judge of her own happiness *as she conceives it* and so should be permitted to choose any self-regarding actions that she judges are needed to achieve it. Advice from others about external conditions is indispensable if readily available, and she may legitimately be compelled to receive it until she shows that she comprehends the information being conveyed, at which time she must be free to do whatever she wishes and intends. Once she has the available information so that she is not taking an unintended action, it seems hard to deny that she does know best which self-regarding actions are likely to achieve her subjective conception of her own good. She does have "immeasurably" superior means of knowing her own thoughts and feelings as well as her peculiar self-regarding goals and circumstances. What she intends to do may still be foolish and justifiably disliked and even condemned as offensive and disgusting by others. But it seems undeniable that her intentional self-regarding conduct is always the best guide to her own happiness by her own lights.

Mill also emphasizes that a competent person "is the person most interested in his own well-being: the interest which any other person, except in cases of strong personal attachment, can have in it, is trifling, compared with that which he himself has" (1977 [1859]: 277; IV.4). Society and government are not nearly so interested as he is in his own happiness as he conceives it. As a result, coercive interference can never be justified instead of advice and persuasion combined with leaving the person alone to do as he intends. Moreover, the competent person is far more likely to rectify his mistakes and reverse any self-regarding choices he regrets, than society and government are to admit their mistakes and change their laws and policies. Thus, the right to do as one wishes with respect to self-regarding conduct is a moral right: "In this department, therefore, of human affairs, Individuality has its proper field of action . . . in each person's own concerns, his individual spontaneity is *entitled* to free exercise" (1977 [1859]: 277; IV.4, emphasis added).

The soundness of the provisional epistemic argument depends on a showing that outsiders can only be worse judges of a competent person's happiness as he conceives it than he is himself, provided he receives advice and warning from others about external conditions when such information is readily available. Since it seems true that nobody knows *better* than a competent agent, who is duly informed about external conditions, which self-regarding actions reflect his own good as he conceives it, outsiders can never *improve* things for him by his own lights by coercively interfering with the self-regarding actions that he wishes and intends to take: "when it does interfere, the odds are that [the public] interferes wrongly, and in the wrong place" (Mill 1977 [1859]: 283; IV.12). Society and government may sometimes happen to get it right in the sense that the self-regarding action they demand is the same one that the individual would spontaneously take anyway. Even in those cases, however, the coercion or threat of it doesn't do any better than giving the individual a right to choose his own self-regarding actions. It could be replaced by non-coercive advice and persuasion that the individual would readily accept while pursuing his own good as he conceives it.

So, if there is any chance that coercive interference will *decrease* the competent and suitably advised individual's good as he conceives it, then general rules of self-regarding conduct enforced by society can only be objectionable. And Mill argues convincingly that there is at least a fifty-fifty chance that even a democratic society will reduce the individual's good as he conceives it by coercively interfering with his self-regarding conduct: "the opinion of the public, that is, of an overruling majority . . ., imposed as a law on the minority, on questions of self-regarding conduct, *is quite as likely to be wrong as right*" (1977 [1859]: 283; IV.12, emphasis added). Given that there is a considerable chance that coercive interference will get it wrong, whereas there is no chance that coercion will raise the individual's happiness as he conceives it, the provisional epistemic argument is sound, and the moral right of self-regarding liberty is justified.[9]

Unfortunately, the provisional epistemic argument cannot overcome some obvious problems, namely, that others may not always be present to supply information about external conditions, and that such information may not yet be possessed by anyone in the community. Under those conditions, which are frequent, even competent agents cannot avoid the risk of bringing about outcomes which they don't wish or intend to achieve. Again, government and concerned organizations can properly publish advice and warnings to carefully assess the risks before taking self-regarding actions, without coercively interfering with any competent person's choices.

5 Objectionable paternalism

By outlining the sphere of self-regarding action in which he says society has no direct interest and the competent individual is sovereign (1977 [1859]: 225–226; I.12), Mill provides a general picture of the kinds of coercive interferences that he regards as objectionable paternalism. These wrongful interferences include: legal or de facto bans against certain ideas or opinions such as atheism; similar bans of printed materials; bans on paintings, music, sculptures, or other artworks; bans of films or photographs, with the caveat that the materials may serve as evidence in a criminal trial if their content involves the exploitation of children or animals or the commission of real crimes such as murder; prohibitions of any goods or services, including alcohol, narcotics, prostitution, and gambling, that can be consumed or used in self-regarding ways; discrimination against polygamous, homosexual, or transgender lifestyles; bans on certain careers, or discrimination against individuals who wish to pursue them, as long as the jobs don't necessarily cause wrongful non-consensual harm to others; and bans against discussion, advice, warnings, persuasion, solicitation, and any other forms of expression that are "practically inseparable" from thinking, judging, and other self-regarding activities. This list is not meant to be exhaustive, but it serves to indicate what he has in mind, as he explicitly confirms at various points in *On Liberty* and other writings (Riley 2015). The liberty of doing as one wishes and intends with respect to these matters ought to be guaranteed by right not only to the competent individual but also to consensual groups of such individuals.

By contrast, many of the examples proposed by his critics don't involve coercive interference with self-regarding conduct. I cannot here illustrate the great frequency with which this occurs, so a very small sample of cases must suffice. Driving a vehicle on a public road is not self-regarding conduct. Rather, it is social or other-regarding conduct that poses a significant risk of non-consensual harm to others, and so society legitimately considers regulating it by enforcing general rules of the road. Prohibiting texting while driving, requiring fastened seatbelts in cars and trucks, and demanding that cyclists wear helmets may or may not be expedient social regulations, but they don't in principle infringe on self-regarding liberty. They are legitimately considered by society as measures to promote the driver's control of his vehicle and thereby promote a safe and effective flow of traffic. To be justified, any such measure must be shown to confer net social benefits, that is, cost less to implement than the amount of non-consensual harm it prevents that would otherwise be suffered by others, including other motorists, pedestrians, and taxpayers. Thus, for example, evidence might be gathered that when seatbelts are unfastened, drivers involved in mishaps tend to be thrown out of the driver's seat and lose control so that their vehicle careens into other vehicles and pedestrians, and that this happens sufficiently often that prevention of the non-consensual damage caused to others outweighs the costs of enforcing mandatory seatbelt legislation.

Or consider a case such as prohibited drugs and medicines which does involve self-regarding conduct but in which coercive paternalism, far from being justified as many critics insist, is objectionable. Instead of prohibition, why isn't it sufficient for government to compel competent people to get written evidence of advice and warnings for a reasonable fee from a qualified doctor or pharmacist, and then let the people be the final judge of whether or not to buy the drugs at a competitive price from any licensed seller upon presentation of the evidence of advice? It is immoral to prevent any competent adult, let alone a sick person near death, from consuming any narcotic, experimental drug, or herbal remedy he wants and is willing to pay for, as long as he has the ordinary amount of understanding, has been suitably warned, and intends to bear the risks.

6 Hard cases: irrevocable alienation of liberty

Arneson and Feinberg rightly emphasize that coercive paternalism looks to have moral appeal when it doesn't in cases where we fail to appreciate that nonpaternalistic reasoning provides the justification for coercive interference. But it is also important to acknowledge that antipaternalism faces troublesome cases in which there seems no appealing alternative but to accept paternalistic reasoning that justifies coercive interference with self-regarding conduct.

Arneson and Feinberg both attempt to deny that a competent person's choice to permanently alienate his liberty by becoming a slave is a problem for antipaternalism, for example, because, in their view, nonpaternalistic reasoning justifies coercive interference with such a choice. Arneson claims that Mill is needlessly driven to self-contradiction by "neglecting the possibility of nonpaternalistic rationales for prohibiting even voluntary slavery" (1980: 473), and Feinberg (1986: 58–59, 71–79) agrees. But they are both off-track on this point.

Arneson argues that prohibition of voluntary slavery can be justified by reasoning that those in "a very weak bargaining position" need to be protected from being exploited and harmed by others in stronger bargaining positions (1980: 472). This reasoning is allegedly nonpaternalistic because it would be welcomed by fully *rational* agents in the weak bargaining position. But this ignores the fact that competent yet foolish people may genuinely wish and intend to be slaves and thus will not welcome laws that ban voluntary slavery. Arneson then must decide whether to force these people to obey the laws for their own good or permit them to freely ignore the laws even though the laws are endorsed by rational individuals. Hard paternalistic reasoning is necessary to justify the first option, whereas the laws are purely nominal in the second option.

After expressing doubts about Arneson's argument on grounds that rational people may genuinely consent and participate in slavery in some contexts, Feinberg offers his own preferred form of nonpaternalistic reasoning to justify a ban against voluntary slavery. According to him, "an indirect application of the harm to others principle" justifies the ban because other people are otherwise forced to witness the misery of slavery or pay a lot of money to free the slaves: "There are certain risks then of an apparently self-regarding kind that persons cannot be permitted to run, if only for the sake of others who must either pay the bill or turn their backs on intolerable misery" (1986: 81). But voluntary slaves are not necessarily in misery or demanding their freedom, in which case taxpayers will not face the difficult choice posed by Feinberg. Like Arneson, he seems to be ignoring the fact that foolish yet competent people may genuinely wish and intend to be slaves. Prohibition of such voluntary slavery is wrongful paternalism.

Mill cannot deny that it is objectionable paternalism to prevent or punish a competent individual from choosing to be a slave. Imprudent yet competent people may well wish to participate in slavery arrangements, marriages (including polygamous marriages) without possibility of divorce, or other irrevocable contracts-in-perpetuity. Antipaternalists cannot avoid this fact by pretending that nonpaternalistic reasons exist to justify a ban against these self-regarding practices. And Mill doesn't try to avoid it. He does *not* prescribe coercion to eradicate any of the practices, including voluntary slavery. His ingenious solution to the difficulty of voluntary slavery is controversial, however, since some people will no doubt be offended and outraged by it.

7 Does Mill abandon the provisional epistemic argument?

Arneson asserts that Mill abandons his "normal presumption that individuals know their own interests better than outsiders" in the case of "contracts whose terms call for a long-term irrevocable forfeiture of liberty by one party," including "slavery contracts and marriage vows that disavow the possibility of divorce" (1980: 485). In his view, Mill believes that the state should refuse to enforce such contracts-in-perpetuity "because individuals are making judgments about what their future interests will be at some remote future time" (ibid.). And Mill certainly does believe, rightly, that people cannot be sure that their intentions will remain fixed into the indefinite future. He argues that it is very imprudent to enter into these contracts and so recommends that society should not endorse or enforce them. If competent but foolish people intend to enter into such agreements, however, then his commitment to absolute self-regarding liberty means that these people must be permitted to become slaves or married persons with no possibility of divorce if they wish.

Arneson, Feinberg, and many others think that Mill prescribes paternalistic interference with such self-regarding activities, with the implication that he admits in these cases that outsiders know better than the individual does her own good as she conceives it. As a result, these critics conclude that Mill contradicts himself since his robust commitment to absolute self-regarding liberty is blatantly at odds with a recommendation to prohibit these activities under law or custom. But Mill doesn't abandon his provisional epistemic argument in these situations. He never says that outsiders know better than the individual what is for her own good as she conceives it. He says only that the individual cannot be sure that her conception of her good, and of the self-regarding actions she intends to take to attain it, will remain fixed into the indefinite future. Others have no better understanding of what her future intentions may be: she may or may not continue to wish to be a slave, for example, or to be married to the same person. Not only she but everyone else is equally ignorant as to how she will conceive of her good in the future.

The critics are misled by their apparent assumption that contracts-in-perpetuity must be legal contracts, with the implication that voluntary slavery and no-divorce marriage must be understood as legal institutions or they lose their meaning. Prohibition of the legal contracts then suggests that these self-regarding activities are impermissible and properly stamped out as illegal. But this is another instance of the critics' misunderstanding of what Mill means by self-regarding conduct. It is not merely conduct permitted by law insofar as it doesn't violate others' justified legal rights. Rather, it is conduct that doesn't cause any non-consensual perceptible damage to others, even if the non-consensual harm is not of the sort which the law recognizes ought to be prevented as a matter of right. There is no reason to think that Mill rejects the idea of extra-legal self-regarding agreements or contracts-in-perpetuity, or that he dismisses the idea of de facto slaves and marriages as meaningless. Just as John Locke speaks of rights, property, slavery, and contracts in the state of nature, for example, we commonly use these terms outside the context of positive law.

Mill never advocates coercive interference with private agreements such as an agreement to become a de facto slave. He recommends that society should refuse to recognize or enforce the agreements as *moral promises* or *legal contracts*. In other words, he can be interpreted as saying that society should refuse to allow these self-regarding agreements to be moved into the sphere of morality and law, which is distinct from the self-regarding sphere in which no moral or legal duties exist. By keeping the agreement to be a slave within the self-regarding sphere, society ensures that the individual keeps her right of self-regarding liberty: she can always exit the slavery arrangement whenever she wishes. And society properly prevents anyone from holding her as a slave against her wishes: it retains its legitimate authority to prevent or punish wrongful non-consensual harm to others.

But what if the parties genuinely consent to move their private agreement into the sphere of morality and law? Is it not coercive to refuse to permit this move? It does seem plausible to argue that there is a coercive element in society's refusal to empower people to create enforceable slavery contracts or no-divorce marriage contracts. But society's refusal is a political decision about whether external sanctions should be employed to enforce the private agreements which people have a moral right of self-regarding liberty to negotiate with one another. That decision concerning social enforcement legitimately falls within the jurisdiction of society, whose political representatives can properly decide that it isn't generally expedient to use coercive legal penalties and/or public stigma to enforce private agreements in which one party permanently alienates his liberty. The refusal decision becomes self-regarding only in a very special case, to wit, the case where there is unanimity in society that such agreements ought not to be enforced by external sanctions. In that special case, however, there is no need for *coercion* anyway since everyone already agrees that these private contracts-in-perpetuity shouldn't be enforced by law or public opinion.

By refusing to use coercion to enforce the moral right to freely enter into private contracts-in-perpetuity, society guarantees that individuals and voluntary groups retain their moral rights of self-regarding liberty. There is no paradox in a coercive refusal to use coercion to enforce the right of self-regarding liberty in these cases, and thereby preserve that right for all. Consistent with this, society properly doesn't coercively interfere with those who consent to be de facto slaves. Individuals should not be punished for seeking and participating in de facto slavery if they can find it. True, by refusing to enforce slavery agreements, society and its representatives discourage the practice of voluntary slavery. After all, only foolish masters will offer any significant compensation for being a slave when the slave can and must be allowed to freely exit the arrangement. Some consensual slavery arrangements may nevertheless persist within the self-regarding sphere. But society is under no obligation to encourage voluntary slavery or to enforce private agreements of which it disapproves. Indeed, there is nothing objectionable about the government and other agencies offering advice and warning to any competent people who seek to be de facto slaves. But a competent person must be the final judge. Competent people who want and intend to be slaves must be free to find masters if they can, even if their self-regarding conduct seems foolish to others and is in fact foolish.

This interpretation accords with what Mill actually says. When he states his disapproval of voluntary slavery, he doesn't so much as mention coercive interference with it. He says: "The principle of [self-regarding] freedom cannot *require* that he should be free not to be free" (1977 [1859]: 300; V.11, emphasis added). In other words, the principle cannot require that society must endorse and enforce private contracts-in-perpetuity: "It is not freedom to be *allowed* to [permanently] alienate his freedom" (ibid.). By keeping slavery agreements within self-regarding spheres, society refuses to allow anyone to permanently alienate his moral right of self-regarding liberty, but it doesn't coercively interfere with his self-regarding actions. Instead, the person always has the moral and justified legal right to exit the slavery arrangement if he wishes.[10]

8 A live philosophical option

No attempt has been made in this chapter to give a full defense of Mill's absolute ban on paternalism understood as coercive interference with self-regarding conduct. He has powerful arguments besides the provisional epistemic argument to support his absolutism. Moreover, I have not addressed some of the challenging objections to his view put forth by his critics. Indeed, some commentators, including Arneson (2005: 264, 273–274, 278–284) and Scoccia (2008: 364–365), propose hypothetical examples which they claim demonstrate that absolutist antipaternalism is a species of fanaticism. While their claims are unpersuasive, in my view, responses to them must await another occasion, as must a more complete defense of Mill's position. In the meantime, it seems fair to reiterate that his absolutist antipaternalism is a live philosophical option, as Arneson at one time apparently believed: "Mill's absolute ban against paternalism may be right or wrong, but nothing in recent philosophical literature gives reason for rejecting it" (1980: 489). That remains true today and in all likelihood will always remain true with respect to competent members of civil societies.[11]

Related topics

Consequentialism, Paternalism, and the Value of Liberty; Libertarian Perspectives on Paternalism; Paternalism and Contract Law; Paternalism and Well-Being.

Notes

1 For my interpretation of Mill's doctrine of liberty, see Riley (2015), which also includes discussion of some influential competing interpretations. While I welcome criticisms, my interpretation has solid textual support.

2 This is not to say that the individual can do as she intends if her intentions cease to be feasible for one reason or another. For example, she cannot cross a bridge she intends to cross, despite being informed of its dangerous condition, if the bridge collapses before she starts onto it. The claim is that she is entitled to act as she wishes in accord with her feasible intentions without coercive interference from others.

3 Not all feasible actions require any movements, although those that don't are commonly called inactions. Inactions do implement feasible intentions.

4 Vain attempts to implement non-feasible intentions are a type of mere voluntary behavior rather than genuine actions. Incompetents are usually capable of voluntary behavior, including voluntary movements taken for no purpose as well as voluntary movements that, while intended to achieve an outcome, aren't feasible. Coercive interference with the voluntary self-regarding behavior of these incompetents remains possible, and is justified for their own good. Notice that if we speak of the behavior of all incompetents as nonvoluntary, as Feinberg and others do, then coercive interference with them is impossible: a person who has no will or cannot exercise the one he has cannot be forced to do anything against his wishes. In some cases, however, a person may be incompetent not because she lacks ordinary understanding but because she lacks any willpower to implement her feasible intentions or to direct others to implement them, as when someone is in a coma or in the grips of an addiction or disease so powerful that it destroys any capacity for choice. Coercive interference with these incompetents really is impossible.

5 The theory of action which I attribute to Mill is arguably a plausible version of what Michael Smith (2012) calls the "standard story" of action, with the caveat that a competent agent is not necessarily prudent or a skilled means-end reasoner. Mill is a hedonist so that desires are not ultimate motives; rather, they are motivated themselves by expectation of pleasure and relief from suffering. As Smith points out, however, "the [standard] story itself is *silent* about the origins of the desires and means-end beliefs that it says play a constitutive role in the explanation of action" (2012: 396, original emphasis). So Mill's hedonism "is neither here nor there, from the standpoint of the standard story."

6 Mill (1967 [1836]: esp. 321–323).

7 Arneson (2015: 679) also attributes a similar view to Feinberg, but he seems incorrect to do so. Feinberg argues that force is justified to ascertain whether a competent person's self-regarding choice is sufficiently voluntary: "the state has the right to prevent harmful self-regarding conduct when but only when it is substantially nonvoluntary, *or when temporary intervention is necessary to establish whether it is [sufficiently] voluntary or not*" (1986: 126, emphasis added). See, also, Feinberg (1986: 61, 127–134). It remains true that, as Arneson argues, Feinberg recommends coercive interference when Mill would not. For Feinberg, coercion can be justified on grounds that a competent person's self-regarding conduct is so harmful to self that it must be presumed insufficiently voluntary. By contrast, Mill prescribes liberty as long as the person's self-regarding action is intentional: the person must be permitted to gravely injure himself if he wishes and intends to do so.

8 It seems plausible to argue that people have a moral duty to provide advice and warning to the individual in such situations, if it is not too costly for them to do so – for example, when they are in the immediate vicinity of the individual and she isn't threatening them with violence. Such a duty is akin to the duty to prevent grievous non-consensual harm to others from third parties or from natural forces, when one can easily do so without danger to oneself.

9 The suggestion that coercive interference now is justified because the competent person will subsequently endorse it is vulnerable to the same fatal objection. The person is "quite as likely" not to endorse it. See, also, Arneson (1980: 473–474) and Feinberg (1986: 52–87).

10 Feinberg (1986: 71–79) points out that society's refusal to recognize and enforce a contract is not the same thing as a criminal law prohibition, which he regards as the epitome of coercive interference.

He apparently views the refusal as coercive to a lesser degree, which seems correct, but he goes on to confuse this with coercive interference with the self-regarding conduct per se, which is incorrect. He insists that Mill accepted coercive interference to prevent voluntary slavery (1986: 58–59), for example, and says that Mill's "solution" (i.e., society's refusal to endorse or enforce the contract-in-perpetuity) is "paternalistic in spirit" (1986: 72).

11 I am grateful to Paul Hurley and Molly Rothenberg for relevant discussions, and to the editors for their extensive comments on an earlier draft.

References

Arneson, R. (1980) "Mill Versus Paternalism," *Ethics* 90: 470–489.

———. (2005) "Joel Feinberg and the Justification of Hard Paternalism," *Legal Theory* 11: 259–284.

———. (2015) "Nudge and Shove," *Social Theory and Practice* 41: 668–691.

Conly, S. (2013) *Against Autonomy: Justifying Coercive Paternalism*, New York: Cambridge University Press.

Feinberg, J. (1971) "Legal Paternalism," *Canadian Journal of Philosophy* 1: 1–16.

———. (1986) *The Moral Limits of the Criminal Law, Volume 3: Harm to Self*, Oxford: Oxford University Press.

Hanna, J. (n.d.) *For Our Own Good*, Unpublished manuscript.

Mill, J. S. (1967 [1836]) "On the Definition of Political Economy," in J. Robson (ed.) *The Collected Works of J.S. Mill (CW), Vol. 4*, Toronto and London: University of Toronto Press and Routledge, pp. 309–339.

———. (1969 [1861]) Utilitarianism, in CW, 10: 203–259.

———. (1977 [1859]) On Liberty, in CW, 18: 213–310.

Riley, J. (2013) "Mill's Greek Ideal of Individuality," in K. N. Demetriou and A. Loizides (eds.) *John Stuart Mill: A British Socrates*, London: Palgrave Macmillan, pp. 97–125.

———. (2014) "Different Kinds of Pleasures," in A. Loizides (ed.) *Mill's System of Logic: A Critical Guide*, London: Routledge, pp. 170–191.

———. (2015) *Mill's On Liberty*, London: Routledge.

———. (2016) "Mill on Utilitarian Sanctions," in C. Macleod and D. E. Miller (eds.) *A Companion to Mill*, Oxford: Wiley-Blackwell, pp. 342–357.

Scoccia, D. (1990) "Paternalism and Respect for Autonomy," *Ethics* 100: 318–334.

———. (2008) "In Defense of Hard Paternalism," *Law and Philosophy* 27: 351–381.

Smith, M. (2012) "Four Objections to the Standard Story of Action (and Four Replies)," *Philosophical Issues* 22: 387–401.

Sunstein, C. R. (2014) *Why Nudge? The Politics of Libertarian Paternalism*, New Haven, CT and London: Yale University Press.

14

PERFECTIONISM AND PATERNALISM

Steven Wall

"Perfectionism" is a term of art, and it may not be the most apt term for the political view I discuss in this entry. As one writer observes,

> The term perfectionist suggests a conception of political justification according to which it is a legitimate aim of government policy to lead people to have the most perfect, most excellent, most admirable lives possible, and the government might therefore be justified in limiting a person's liberty or opportunities solely by the fact that this will lead them to have more excellent or admirable lives.
>
> (de Marneffe 2010: 15)

The conception of political justification expressed by these remarks is not precisely what I mean by perfectionism. It is instructive, however, to consider how it relates to the view I want to discuss. The conception of political justification I have quoted accents a maximal notion – most perfect, most excellent, most admirable. But the view to be discussed here, while it does not exclude political efforts to promote the maximal good, does not require them either. If one holds that it is a legitimate aim of government policy to help people lead sufficiently good lives, then one qualifies as a perfectionist on my understanding. Further, the quoted conception accents evaluative achievements – perfection, excellence, admirability – that may add to the goodness of a life but may not always contribute to making it good for the person who leads it. But the view to be discussed here includes a concern to advance the interests of people among the legitimate aims of government policy. Perfectionist policies, while they may promote excellence and admirable character traits, also aim to promote well-being.[1] Finally, the quoted conception accents the role that governments can play in restricting the liberties and opportunities of those subject to their authority. But the view I discuss in this entry is sensitive as well to the case for restricting government power to limit liberties and opportunities. Indeed, on the view I discuss, the case for restricting the authority of governments to limit liberties and opportunities is just as important as the case for expanding their authority to do so.

Bringing these contrasting remarks together, I can formulate an alternative conception of political justification to the one quoted above.

> The term perfectionist refers to a conception of political justification according to which it is a legitimate aim of government policy to help people lead sufficiently good

> lives, and the government might therefore be justified in expanding people's liberties and opportunities, as well as limiting them, on the grounds that doing so will help them to live such lives.

With this conception of perfectionist political justification in mind, we can turn to the central issues this entry addresses. These all concern the relationship between perfectionism and paternalism. I discuss four main claims.

1. Perfectionism in politics does not entail state paternalism.
2. A subset of state paternalistic policies has a perfectionist rationale.
3. Perfectionist justice can require perfectionist state paternalism.
4. Perfectionist state paternalism need not contravene a plausible principle of respect for persons.

I defend the first two claims by explaining why they are true. I defend the latter two claims by presenting some considerations that speak in their favor while acknowledging that these claims raise large issues that cannot be fully addressed in this entry. Before addressing the four claims, however, I say a few words about how I understand state paternalism, since the term "paternalism," like that of "perfectionism," admits of rival characterizations.

1 State paternalism

Not all paternalism is state paternalism. Private individuals can interact with one another paternalistically. Although obvious, this point is important. Paternalistic interference can be easier or harder to justify depending on the nature of the relationship between the interacting parties. In all likelihood, state paternalism is harder to justify than paternalism among friends or close acquaintances.

State paternalism, as I will understand it, consists of *governmental policies that restrict the options of those subject to them and are justified, in part, by appeal to the interests or good of those whose options are restricted.* Government policies restrict an option either by removing it from people's option sets, so that they no longer have access to the option at all, or less comprehensively by making engagement with the option costlier for them than it otherwise would be. This characterization excludes state policies that do not restrict options, but merely affect the order or manner in which they are presented to subjects (Thaler and Sunstein 2009). It also excludes state policies that augment options, or make certain options less costly to engage with, on the grounds that doing so will be good for those whose option sets are affected.[2] These exclusions are not meant to settle any conceptual issues. State paternalism, as I characterize it, involves policies that nearly everyone agrees are instances of paternalism, whereas the excluded policies are more controversially classified as paternalistic. The excluded policies are also, arguably, easier to justify, since they do not restrict options. If restrictive paternalistic policies can be justified, then these less restrictive policies presumably can be justified as well. Thus, by focusing on the focal cases of state paternalism, and avoiding the less central instances, we can present a less cluttered discussion of the four claims to be defended.

An example of state paternalism is a policy that criminalizes the sale and use of cigarettes. However, a restrictive policy of this kind would not be paternalistically imposed if the policy were justified solely on the grounds that it would prevent harm to others (out of concern for second-hand smoke, for example). Suppose now that the policy in question were justifiable, but could not be successfully justified by appeal to (non)paternalistic considerations alone.[3] To successfully justify the policy, policy-makers would need to appeal to the interests of those who would benefit from the restrictions imposed by the policy. Would the fact that the policy was not, in actuality, justified by appeal to paternalistic considerations render it a (non)paternalistic policy? My response is twofold. First, in this scenario, the policy would not be successfully justified, although it would be a justifiable policy. Second, while not successfully justified, the policy would not be paternalistically imposed, although it would be a paternalistically justifiable policy. Our concern in what follows will be with paternalistically imposed policies.[4] Government officials who put into practice paternalistically imposed policies characteristically believe that the policies are paternalistically justifiable, but if they are mistaken in this belief they still engage in state paternalism.

A final comment concerns the reference to "interests or good" in the italicized formulation above. Should interests/good be indexed narrowly, and fairly standardly, to well-being, or should interests/good be taken to include as well the (non)well-being aspects of good lives? An example brings out the point of the question. A government policy attempts to promote virtue in its citizens by requiring them to participate in jury trials and other civic functions, and by penalizing them for failing to participate. The policy restricts options, but the point of the restrictions is not to promote well-being, but rather to make people better people. For the perfectionist, at least as I have characterized him here, such policies are in principle justifiable, as they may help people lead sufficiently good lives. Whether or not they are aptly characterized as paternalistic is less important, but I propose to characterize them as so in this entry.[5]

2 Perfectionist antipaternalism and antiperfectionist paternalism

Some writers bring perfectionism and paternalism together. Perfectionist state policies, they suggest, ultimately rest on paternalistic judgments about what is good for people (Quong 2011: 73–74). Although not straightforwardly wrong, this suggestion is misleading. A perfectionist conception of political justification, depending on its substantive content, can support either state paternalism or state antipaternalism.

Mill's writings provide a compelling illustration of how perfectionist considerations can ground principled limits to state paternalism. The harm to others principle that Mill defends, which is a principle that explicitly rules out paternalism, is grounded on an appeal to utility, where utility is understood to include "the permanent interests of man as a progressive being" (Mill 1859: 10). This enlarged understanding of utility, Mill explains, encompasses the perfectionist good of individuality as "one of the leading essentials of well-being" (Mill 1859: 54). It is thus incorrect to characterize perfectionist policies as necessarily imposed on people against their will, for the perfectionist policies may be grounded on the perfectionist good of autonomy or individuality; and furthering this good may require respecting the will of people, even when they are disposed to make poor decisions. Similar antipaternalist conclusions are defended by contemporary writers who argue that state paternalism not only sets back autonomy, but also tends to impede rational evaluation and decision-making, which are key perfectionist goods (Hurka 1993: 157–158).

Just as perfectionism can support antipaternalism, antiperfectionism can support paternalism. In discussing justified state paternalism in *A Theory of Justice* Rawls observed that the parties in the original position would recognize that it is

> rational for them to protect themselves against their own irrational inclinations by consenting to a scheme of penalties that may give them a sufficient motive to avoid foolish actions and by accepting certain impositions designed to undo the unfortunate consequences of their imprudent behavior.
>
> (Rawls 1971: 218–220)

Antiperfectionist state paternalism, as understood by Rawls, has two features that are worth highlighting. First, it is not confined to soft paternalistic measures, where these are understood to include interferences and temporary detentions designed to protect people from the harmful consequences of decisions that are not truly their own.[6] Second, it is restricted to measures that respect peoples' "final ends and beliefs whatever they are" (Rawls 1971: 220). The first of these features is worth highlighting, since soft paternalism is sometimes considered not to be a form of paternalism at all (Feinberg 1986: 12). The Rawlsian view, however, encompasses both soft and hard paternalism (de Marneffe 2010: 99–101). The second feature is worth highlighting, as it reflects the antiperfectionist character of Rawlsian state paternalism. The state must refrain from judging peoples' final ends insofar as these ends are consistent with the principles of justice that Rawls defends.

Rawlsian state paternalism is thus a form of means-only paternalism. Each member of a well-ordered Rawlsian society is stipulated to have a higher-order interest in forming, pursuing, and revising a conception of the good life. This conception of the good life, in turn, defines each member's final ends. This last point is important. For notice that if we accent the interest each of us has in revising, or being open to the possibility of revising, our conception of the good, then we might not define our final ends in terms of the conception of the good that we currently accept. We might, instead, define our final ends in terms of the ends we would have if we were perfectly rational and perfectly well informed. This maneuver, in turn, could license a good measure of state paternalism that runs counter to the conceptions of the good that people actually accept. But Rawls insists that in addition to our higher-order interest in forming, pursuing, and revising a conception of the good life, we also have an interest in pursuing our actual conception of the good, even if it is not a fully rational conception of the good for us. Exactly how this interest in pursuing our actual conception of the good should be balanced against our interest in pursuing a fully, or more fully, rational conception of the good is not immediately apparent.[7] Those who follow Rawls in adopting means-only paternalism could depart from him in giving more weight to the final ends that we would have if we were more fully rational than to the final ends that we happen to have settled upon.

Be this as it may, the foregoing remarks bring into view a type of paternalism that has a distinctively perfectionist rationale. This type of paternalism rejects the means-only restriction. It allows that justified paternalism can target people's selection and pursuit of final ends. Rather than considering what people would or would not want to do if they were rational, this type of paternalism remains focused on what would or would not promote their interests, where their interests are not fixed by the final ends that they have settled upon. Consider, for example, the much-discussed case of voluntary slavery. While it is possible that no procedurally rational person would adopt as a final end the end of being a slave to another, this occurrence is not something that can be definitively ruled out. The most promising strategy for rejecting voluntary slavery contracts is not to hold that no rational person could enter into them, but rather "to appeal to some idea of what is a fitting life for a person." This will amount to "a direct attempt to impose a conception of what is 'good' on another person" (Dworkin 1988: 129).

Direct attempts by the state to impose conceptions of the good on others are often ill advised. They have a tendency to be counterproductive, since people tend to stick with their present values and concerns in the face of efforts by others to get them to give them up. And if people do not revise their conception of the good in response to state action designed to discourage it, then they will suffer the costs of the state action without reaping any of its benefits (Patten 2014: 129). There is, in addition, the setback to people's autonomy that can result from the state's efforts to directly promote the good. If autonomy is itself a perfectionist good, as many writers claim, then state efforts to directly promote the good by paternalistically imposing a conception of the good may be self-defeating.

These points explain why coercive state paternalism that targets final ends is often viewed as "a clumsy device" for furthering perfectionist ends. In the ensuing section, I will argue that these points, while generally sound, do not exclude all well-crafted coercive state paternalistic measures that have a perfectionist rationale. For now, it is instructive to return to the case of a restriction on voluntary slavery. Effective state paternalism here removes an option. No one is forced to avoid slavery. But the state effectively eliminates an option that would otherwise exist by refusing to enforce such contracts and preventing others from doing so. The justification for the state's action, moreover, can appeal directly to the perfectionist good of autonomy. The option to enter into such a contract is eliminated so as to ensure that people continue to have the freedom to lead their lives on their own terms.

Generalizing from the case of voluntary slavery, some may wish to restrict perfectionist state paternalism to measures that promote or safeguard autonomy. This position remains perfectionist insofar as autonomy is understood to be a perfectionist good, one that people have reason to realize irrespective of their actual concerns and values. Paternalistic restrictions on certain dangerous drugs, for example, may be justifiable on such grounds. Be this as it may, it is helpful to distinguish state paternalism that is perfectionist insofar as it is designed to promote or safeguard autonomy (Type 1 State Perfectionism) from state paternalism that is perfectionist insofar as it aims to steer people toward valuable options and away from disvaluable options in exercising their autonomous agency (Type 2 State Perfectionism) (Wall 1998: 197–198). Both types of perfectionism can ground paternalistic measures that have a perfectionist rationale.

3 Justice and paternalism

Critics of state perfectionism sometimes contend that perfectionist measures that favor some conceptions of the good over others are unjust. They start by assuming that a just distribution of resources has been achieved. From this benchmark subsidies or penalties designed to favor some activities over others, they contend, necessarily upset the just distribution of resources, unless these measures are unanimously favored. In this way, Rawls defended Wicksell's unanimity criterion on public expenditures to rule out spending public money on perfectionist goods (Rawls 1971: 249–250).

One can respond to this argument by claiming that sometimes perfectionist values should take precedence over justice. But most people who are sympathetic to state perfectionism will want to challenge the assumption that perfectionist considerations come into play only after a just distribution of resources has been determined. They will insist that justice itself is responsive to perfectionist judgment. Consider, for instance, the Type 1 State Perfectionism alluded to above. The state, it can be said, has justice-based duties to help its members lead autonomous lives. These duties direct the state to establish background conditions that facilitate the realization of autonomy. The duties may also justify a measure of state paternalism, such as policies that effectively eliminate options that are inimical to autonomy, such as the option to sell oneself into slavery.

The general structure of the justice-based argument for state perfectionism that we are now considering can be set out as follows:

i A just state must adopt measures that advance the fundamental interests of its members.
ii People have a fundamental interest in forming, pursuing, and realizing a valuable conception of the good life.

iii A valuable conception of the good life includes the realization of autonomy.
iv Therefore, a just state must adopt measures that promote or facilitate the realization of autonomy.

The argument can be challenged in multiple ways. Some will reject (ii) and insist that people have a fundamental interest in forming, pursuing, and realizing a conception of the good, whether it is valuable or not. Others, including some perfectionists, will reject (iii). Autonomy, they will claim, is not a necessary component of a valuable conception of a good life. Still others will grant all three premises, but deny the conclusion on the grounds that state efforts to promote autonomy violate valid constraints on state action. (In effect, these critics will hold that the "must" in (i) and (iv) is subject to the qualification that no valid constraint has been violated.)

My purpose here is not to present a full defense of the argument. It is presented merely to illustrate the justice-based case for state perfectionism, including the justice-based case for paternalistic measures that have a distinctively perfectionist rationale. Some writers who are sympathetic to state perfectionism may wish to limit perfectionist measures to the Type 1 variety. But the argument I have sketched can be extended naturally to include Type 2 measures as well by modifying iii.

iii* A valuable conception of the good life includes the realization of autonomy by way of engagement with valuable, as opposed to worthless, options.
iv* Therefore, a just state must adopt measures that promote or facilitate both the realization of autonomy and engagement with valuable options.

The third premise of the modified argument could be grounded on the claim that autonomy adds value to a life, but that fully good lives must realize autonomy by engaging in worthwhile pursuits. Alternatively, it could be grounded on the claim that autonomy's value is conditional, adding value to a life only if the life consists of engagement with valuable options (Raz 1986). If the latter, then the relevant perfectionist good is not autonomy per se, but rather valuable autonomous agency.

The modified argument might seem vulnerable to the two objections mentioned above. State perfectionism, including paternalistic measures of the Type 2 variety, may be self-defeating if they impede autonomy or impose high costs on the very people the measures are intended to benefit. Still, it is important to bear in mind that paternalistic measures can benefit some while setting back the interests of others. Restrictions on the sale and use of dangerous drugs, for example, may set back the autonomy of some, or they may fail to help some while imposing costs on them; but if the measures promote the autonomy of others by dissuading them from engaging in self-destructive drug use, then they may be justified on balance. Once the interests of all affected parties are taken into account, the measures may deserve support because they promote good lives and do so in a way that fairly balances the conflicting interests at stake.

The point can be generalized. State perfectionism, including perfectionist paternalism, aims to create or sustain a moral environment that helps those who live in it lead sufficiently good lives. A moral environment consists of a set of options to engage in activities that have ethical significance.[8] Plausibly, for any given moral environment, some who live in it will fare better than others relative to some alternative moral environment. So, in general, living in a given moral environment will have differential effects on the welfare of different people relative to the possible alternatives. Notice, moreover, that a society that undertakes no state perfectionism will still have a moral environment and that moral environment may disadvantage some of its members relative to alternative possible moral environments that have been shaped by state perfectionism.

If this is right, then there is no getting around the fact that interests between people will come into conflict (with respect to different possible and attainable moral environments), and there is a consequent need to think about how these interests can be fairly balanced or accommodated.

Both the proponent and critic of state perfectionism, then, must confront the issue of how to balance conflicting interests. Rejecting all state perfectionism is an option, but it too will result in a moral environment that disadvantages some members. The modified argument thus supports state perfectionism, but requires further modification. Premise (i) needs to be rephrased to read "a just state must adopt measures that advance the fundamental interests of its members, provided the measures in question treat its members fairly." This, in turn, yields a revised conclusion.

iv** Therefore, a just state must adopt measures that promote or facilitate both the realization of autonomy and engagement with valuable options, and it must do so in a manner that treats its members fairly.

Nothing so far has been said about how conflicting interests (with respect to different possible and available moral environments) can be balanced fairly. A full discussion of this issue cannot be ventured in this entry, but a few maxims can be briefly considered. Speaking broadly, there are two kinds of costs that need to be considered. There are, first, costs that the state engenders by the perfectionist measures it executes. If it prohibits a dangerous drug, and if legal access to this drug would benefit some of its members, then these members suffer a cost. Call this an *active cost*. But, second, there are also costs associated with the state's failure to undertake measures it could have undertaken. If it is foreseeable that if it does not prohibit a dangerous drug, then the lives of some of its members will be ruined, then these members suffer a cost. The cost here is not engendered, but countenanced. Call this a *passive cost*. Now one maxim for balancing the active and passive costs of perfectionist state action is a maxim of neutrality. Both kinds of costs should count equally. To apply the maxim of neutrality, one needs to gauge the magnitude of the respective active and passive costs in prospect, but, having done so, one should not then count one kind of cost more on the grounds that it is active or passive.

The maxim of neutrality seems quite favorable to the proponent of state perfectionism. For if state perfectionism engenders more harm than it prevents from occurring (if the active costs of state promotion exceed the passive costs of state inaction), then it plainly should not be undertaken. No clearheaded proponent of state perfectionism favors state action that is counterproductive. He favors only measures that can be expected to do more good than harm. The maxim of neutrality thus rules out measures that no clearheaded perfectionist would favor anyway, and it leaves in play the measures he is interested in defending.

The maxim of neutrality, however, can be challenged on the grounds that it ignores the responsibility of the different parties for the costs that they bear. Active costs are imposed on people by the state. Those who would benefit from having access to an option that is foreclosed by paternalistic restriction have not done anything, at least in the typical case, to bring these costs on themselves. By contrast, those who suffer passive costs due to the state's inaction may be, in the typical case, at least partly responsible for the costs they suffer. Someone who decides to take a dangerous drug that is not subject to paternalistic restriction, for example, is partly responsible for ruining his life, if that is what results. So, responsibility is a factor that could justify moving away from strict neutrality and toward a maxim that gives extra weight to active costs over passive costs. And the greater the extra weight that is assigned to active costs over passive costs the harder it will be to justify paternalistic restrictions. At the limit, one might propose an analogue to Blackstone's maxim about the criminal law: it is better to have ten people suffer a passive cost than to have one person suffer an active cost of the same magnitude.[9] Call this maxim *strong priority for active costs*. Acceptance of this maxim would not altogether rule out perfectionist state paternalism, but it would make it considerably harder to justify. Accordingly, one could accept the modified argument for state perfectionism, but insist that most perfectionist measures impose unfair costs on people.

There are good reasons to reject *strong priority for active costs*, however. For one thing, it is a mistake to assume that people are fully responsible for the passive costs they bear. Even if they are responsible for the bad decisions they make, they can still have a valid complaint against others for making available to them certain destructive options. Furthermore, it is very unclear just how responsible people are for the dispositions they have to make bad prudential decisions. As Arneson notes,

> Through no fault or choice of their own, some people by genetic endowment and favorable socialization have lots of [prudential decision-making ability] and a disposition to employ it fully on appropriate occasions of choice. Others have less ability and less disposition to deploy what ability they have.
>
> (Arneson 2005: 275)

With these points in mind, we might reject strong priority and accept a maxim that lies between neutrality and strong priority, one that we can call *weak priority for active costs*. This maxim directs us to reject a proposed paternalistic restriction if the active costs in prospect were greater than, equal to, or only marginally less than the expected passive costs of taking no state action. This maxim is plausible, but unlike the strong priority maxim, it does not establish a strong barrier to justified state paternalism.

The modified argument for state perfectionism, including paternalistic measures with a perfectionist rationale, can be challenged in multiple ways, as I have allowed. The preceding discussion sought to show only that state perfectionism should not be rejected on the grounds it would impose unfair costs on those who do not benefit from it. Some critics of state perfectionism, however, will insist that the modified argument goes wrong in its exclusive focus on promoting interests. On a familiar view, justice is not only a matter of showing proper regard for people's interests, but also a matter of respecting them as persons. Possibly, perfectionist state paternalism that would be justified by a due concern for people's interests is nonetheless unjustified because it shows undue respect for their capacities as rational agents.

4 Paternalism and respect

Just governments, let us grant, must respect those over whom they rule. The requisite respect can be derived from a deeper foundational idea. On some libertarian views of justice, people are self-owners, and self-ownership consists of a set of stringent rights that exclude perfectionist state paternalism. States that violate these rights thereby disrespect the persons whose rights are violated. Other views appeal to different foundational ideas, such as personal sovereignty or nondomination, to ground the rights in question. On all these views, respect for persons is a derivative notion. To establish that state action disrespects persons, one must first establish that it violates their rights; and it is the fact that the state action violates their rights that explains why it does not respect them.

A discussion of these rights-based, respect-derivative views cannot be undertaken here. They constitute an important challenge to state perfectionism, and state paternalism more generally. Some writers appeal to respect for persons in a different and more interesting way, however.

They suggest that respect for persons is a non-derivative moral demand. On this understanding, it is the fact that state action is disrespectful to persons that explains why it is unjust. I will be concerned here with this non-derivative understanding of respect for persons and the challenge to state perfectionism that it purportedly poses.

Some writers claim that it is part of the content of the respect for persons demand that no one should be subject to political coercion that cannot be justified to them, given their current conception of the good life, providing that conception of the good life is itself consistent with a sound antiperfectionist conception of justice (Larmore 1996: 137). By itself, this claim should not impress anyone. The content of respect for persons, understood as a non-derivative moral demand, is very much open to dispute. And it is not very compelling to argue for a conclusion [state perfectionism disrespects persons] by appeal to a moral demand [respect for persons] that already contains the antiperfectionist conclusion in its description.

The claim in question, however, can be defended by appeal to capacities that plausibly are essential to persons and that respect for which explains the antiperfectionist conclusion. Consider Nozick's statement that persons have the capacity "to form a picture of one's whole life (or at least significant chunks of it) and to act in terms of one's overall conception of the life one wishes to lead" (Nozick 1974: 50). To respect a person we may need to respect his possession of this capacity and his exercise of it. If so, then this demand of respect may ground a constraint against perfectionist state paternalism. Similarly, Rawls' claim that persons have a higher-order moral power to form, pursue, and revise a conception of the good likewise can be taken to refer to a capacity that is essential to persons and that explains why some forms of paternalism, such as perfectionist state paternalism, are disrespectful to persons.

The capacities that Nozick and Rawls single out are capacities of rationality. Persons are rational agents, and they exercise their rationality in forming a picture of their life as a whole or in forming, pursuing, and revising a conception of the good life. Rationality, however, can be understood either procedurally or substantively (Parfit 2011: 62). On the procedural understanding, persons are rational if they follow certain rules of thought, such as intending the means to their ends or avoiding inconsistences in their beliefs. By contrast, on the substantive understanding, persons are rational if they respond well to the reasons that apply to them. Naturally, a person can be procedurally rational while manifesting substantive irrationality. For example, a person can pursue a misguided conception of the good, one that runs counter to the reasons that apply to him, in a manner that is fully procedurally rational. Nozick and Rawls appear to have the procedural understanding of rationality in mind. However, it seems plain that, on the assumption that substantive reasons exist, persons have an interest in being substantively rational, as opposed to merely procedurally rational. That is why proponents of state perfectionism, as premise (ii) of the modified argument illustrates, maintain that people have a fundamental interest in pursuing and realizing a *valuable* conception of the good life.

The respect-based objection to state perfectionism, however, need not conflict with the claim that people have a fundamental interest in being substantively rational. For the objection holds that respect for a person is a matter of respecting the exercise of his rational capacities, not a matter of promoting his interests. And the relevant rational capacities, on the objection, are capacities of procedural rationality, specifically the procedurally rational capacities that are engaged when a person forms a picture of his whole life or forms, pursues, and revises a conception of the good.

Understood in these terms, the respect-based objection to state perfectionism cannot be definitively ruled out. But it can be said, in reply, that there is an alternative understanding of respect for persons, one that is formulated in terms of respect for people's possession and exercise of their rational capacities understood substantively. On this alternative understanding, state perfectionism need not be disrespectful to anyone, since it may help those who are subject to it to recognize and respond to the reasons that apply to them.

A respect-based objection to state perfectionism, accordingly, is not compelling unless a case can be made for rejecting the substantive formulation of the respect for rational capacities claim. Thus far, proponents of the objection have not carefully distinguished the two formulations of respect for rational capacities. Nor have they made, or even attempted to make, the case that the procedural formulation is superior to the substantive formulation, or that the substantive formulation is inadequate as an account of respect for persons.[10]

Before concluding this section, it will be helpful to consider briefly a final respect-based objection to perfectionist state paternalism. An often-voiced complaint against state paternalism is that it treats adult subjects as if they were children incapable of directing their own lives. By so doing, it disrespects them as autonomous agents. A state that tried to take over most of the important decisions that people need to make about how to lead their lives might effectively impede the ability of its subjects to lead their lives on their own terms. By treating them as if they were incapable of autonomous agency, such a state would plausibly disrespect its members. But state paternalism, of course, need not take this all-encompassing form. It can be discriminating and limited, aiming merely to close off, or raise the costs of pursuing, certain options while leaving many options in place for citizens to choose between. Such a state could not be justly charged with treating its adult members as if they were children. At most it could be charged with treating its adult subjects as if they, in general, were susceptible to certain irrational tendencies that could lead them on occasion to make poor decisions about their lives. But this view, far from being insulting or disrespectful, is evidently true and recognized to be true by virtually everyone.

Those subject to the limited and discriminating state paternalism here envisioned might agree that while people in general are subject to various irrational tendencies, they as individuals are not. They are capable of using dangerous drugs responsibly or engaging in reckless gambles without ruin, for example. The state treats *them* as children when it judges *them* incapable of making these decisions responsibly. But the proponent of state paternalism need not deny that some who are subject to paternalistic restrictions would be just fine in their absence. The case for state paternalism looks at the aggregate picture. It does not rest on the judgment that any particular individual is incapable of making responsible decisions with respect to any option it targets.

5 Paternalism and trust

Earlier I claimed that paternalism can be harder to justify depending on the nature of the relationship between the interacting parties. One reason for this is that trust between those who would impose paternalism and those who would be subject to it affects its advisability (Raz 1996; Clarke 2013). The justifying aim of paternalism appeals to the need to protect or promote the interests of those who would benefit from the interference, but just because paternalism has this aim does not mean that it fulfills it. What might be termed *false paternalism* is either paternalism that does not, in reality, protect or promote the interests of the intended beneficiaries, or paternalism that merely purports but does not actually aim to benefit them.[11] It is sometimes unclear whether paternalistic interference is false or genuine; and when it is unclear trust between the interacting parties becomes important to its justification. For if people distrust those who claim to interfere with them for their own benefit, and if this distrust is not unreasonable, then they may judge that purportedly genuine paternalism is false paternalism, and they may reasonably resent it. In this way, otherwise justified paternalistic interference can be called into question, as those who would impose paternalism have reason to avoid putting their intended beneficiaries in a position where they can with warrant resent what is being done to them.

These sketchy remarks, in need of development and refinement, highlight a feature of state paternalism that often distinguishes it from paternalism between friends and intimate associates. In large bureaucratic modern states, in which state officials are subject to all sorts of competing interests and pressures, citizens often come to distrust (with warrant) those who exercise authority over them. They judge state officials to be neither well placed nor well motivated to promote their good. In this context of distrust, even instances of genuine paternalism can be reasonably judged to be false, thus provoking warranted resentment.

The fact that state paternalism provokes warranted resentment does not establish that it is unjustified, but it raises the bar to its justification. Accordingly, the perfectionist state that aims to promote the good of its members through paternalistic measures must also be concerned with winning the trust of those over whom it rules. In circumstances of pluralism and significant disagreement over the good life, this is a formidable challenge.

6 Conclusion

This entry has considered the relationship between perfectionism and paternalism by discussing four main claims. These are that perfectionism in politics does not entail state paternalism, that a subset of state paternalistic policies has a perfectionist rationale, that perfectionist justice can require state paternalism, and that perfectionist state paternalism need not contravene a plausible principle of respect for persons. In discussing these four claims, the entry emphasized that not all state paternalism is perfectionist, but that there is a type of state paternalism that has a distinctively perfectionist rationale. It critically analyzed a range of considerations that speak in favor of, and against, this type of distinctively perfectionist state paternalism.[12]

Related topics

Kantian Perspectives on Paternalism; Paternalism and Intimate Relationships; Paternalism and the Criminal Law; Paternalism and Well-Being; Paternalism by and towards Groups.

Notes

1 Well-being has an objective dimension. It cannot be fully analyzed in terms of desire-satisfaction and/or pleasant mental states, or so I assume here.
2 These are perfectionist subsidies, and while they are funded by tax dollars, they do not directly restrict options, but rather create them or make engagement with them less costly.
3 To clarify: "justified" refers to the activity of offering reasons that are intended to provide sufficient support for a policy; "successfully justified" refers to the successful exercise of that activity; and "justifiable" refers to a policy that can be successfully justified, whether or not those who attempt to justify it succeed in doing so.
4 Paternalistically imposed policies are identified by the manner in which state officials justify them. Rather than referring to the justificatory practice of state officials, we could refer to their motives. But the motives of state officials are often unclear, and different state officials frequently have different motives for supporting the same policies. For this reason, I prefer the formulation in the text.
5 Paternalistically imposed policies, accordingly, are policies justified, at least in part, by appeal to the well-being and/or good of those subject to them. Of course, if people always have an interest in being good, then the two notions merge. But here I leave this issue open.
6 For discussion of the distinction between soft and hard paternalism see Feinberg (1986: 99–106). The significance Feinberg places on this distinction is open to challenge. See Hanna (2012) for one line of critique.

7 For interesting discussion of this issue see Grill (2015).
8 For more detailed discussion of this idea see Wall (2013).
9 For a weaker version of the maxim applied to paternalism see Dworkin (1972: 84): "better two men ruin themselves than one man be unjustly deprived of liberty."
10 Possibly, both the procedural and substantive formulations are valid, and an account of respect for persons must attend to both. The point in the text is that critics of state perfectionism have not shown that the substantive formulation on its own is inadequate.
11 To be a little more precise, false paternalism is false in virtue of the fact that the agent engaged in the paternalism makes false claims. Either the agent claims to benefit those subject to the paternalism (when in fact the paternalism does not benefit them), or the agent claims to be motivated by a concern to benefit those subject to the paternalism (when in fact no such motivation is present).
12 Thanks to Kalle Grill and Jason Hanna for excellent comments and suggestions.

References

Arneson, R. (2005) "Feinberg and the Justification of Hard Paternalism," *Legal Theory* 11: 259–284.
Clarke, S. (2013) "A Trust Based Argument Against Paternalism," in P. Makela and C. Townley (eds.) *Trust: Analytic and Applied Perspectives*, Amsterdam: Rodopi.
de Marneffe, P. (2010) *Liberalism and Prostitution*, Oxford: Oxford University Press.
Dworkin, G. (1972) "Paternalism," *The Monist* 56: 64–84.
———. (1988) *The Theory and Practice of Autonomy*, Cambridge: Cambridge University Press.
Feinberg, J. (1986) *Harm to Self*, Oxford: Oxford University Press.
Grill, K. (2015) "Respect for What? Choices, Actual Preferences, and True Preferences," *Social Theory and Practice* 41: 692–715.
Hanna, J. (2012) "Paternalism and the Ill-Informed Agent," *Journal of Ethics* 16: 421–439.
Hurka, T. (1993) *Perfectionism*, Oxford: Oxford University Press.
Larmore, C. (1996) *The Morals of Modernity*, Cambridge: Cambridge University Press.
Mill, J. S. (1859) *On Liberty*, Indianapolis: Hackett Press.
Nozick, R. (1974) *Anarchy, State and Utopia*, New York: Basic Books.
Parfit, D. (2011) *On What Matters, Volume 1*, Oxford: Oxford University Press.
Patten, A. (2014) *Equal Recognition*, Princeton, NJ: Princeton University Press.
Quong, J. (2011) *Liberalism Without Perfection*, Oxford: Oxford University Press.
Rawls, J. (1971) *A Theory of Justice*, Cambridge, MA: Harvard University Press.
Raz, J. (1986) *The Morality of Freedom*, Oxford: Oxford University Press.
———. (1996) "Liberty and Trust," in R. George (ed.) *Natural Law, Liberalism and Morality*, Oxford: Oxford University Press.
Thaler, R. and Sunstein, C. (2009) *Nudges: Improving Decisions About Health, Wealth and Happiness*, New York: Penguin.
Wall, S. (1998) *Liberalism, Perfectionism and Restraint*, Cambridge: Cambridge University Press.
———. (2013) "Moral Environmentalism," in C. Coons and M. Weber (eds.) *Paternalism: Theory and Practice*, Cambridge: Cambridge University Press.

15
LIBERTARIAN PERSPECTIVES ON PATERNALISM

Peter Vallentyne

Libertarianism is a natural-rights theory that holds that: (1) each agent initially fully owns herself, and (2) each agent has certain moral powers to acquire property rights in external things. The rights are typically understood as choice-protecting in the sense that rights are infringed when intruded upon *without the valid consent* of the rightholder. Thus, paternalistic actions that intrude upon the rights of an agent without her consent wrong her. This seems correct to me in cases where communication about consent with the rightholder is reasonably possible. Where, however, such communication is not reasonably possible, paternalistic intrusion without valid consent does not, I claim, always wrong the agent.

This chapter is nominally on paternalism from a libertarian perspective, but most of what I have to say applies (as explained below) to all choice-protecting rights-theories. My claim is that standard accounts are mistaken to hold that (e.g., paternalistic) actions that intrude upon a person's rights *without her consent* always wrong her. Libertarianism, and other rights-theories, should instead endorse *a choice-prioritizing* theory (sketched below) that requires actual valid consent when communication about consent/dissent is possible, but is sensitive to both hypothetical consent/dissent and the rightholder's interests when communication is not possible.

1 Libertarianism and wronging

An individual is *psychologically autonomous* (hereafter: autonomous) just in case (roughly) she has a sufficiently good capacity for rational reflection and revision of her beliefs, desires, and intentions. *Libertarianism* holds that each autonomous agent initially fully owns herself and has certain moral powers to acquire private property in external resources (see, e.g., Vallentyne 2000; Vallentyne and van der Vossen 2014). The details won't matter here. The crucial point is that each autonomous agent has some rights and thus can be wronged.

An individual is *wronged* when and only when her rights are infringed.[1] Conceptually, it can be permissible to wrong someone when there is an overriding justification (e.g., shooting someone in the leg when this is the only way to avoid a social catastrophe). If there is no overriding justification, however, it is wrong to wrong someone.

Libertarians tend to hold that rights are absolute (always conclusive). Thus, they tend to hold that it is always wrong to infringe a person's rights (since they deny that there are any overriding justifications). Other rights-based theories, however, can allow that rights are merely strong *pro tanto* considerations, which can be overridden by significant benefits of the right kind. In what follows, I shall leave open whether rights are absolute. My topic concerns the conditions under which paternalism *wrongs the beneficiary*, not the question of when such actions are wrong (impermissible).

I shall assume that we have some (e.g., libertarian) rights and investigate the conditions under which an *intrusion* (defined below) upon a person's rights *infringes* her rights in the sense she is *wronged* by the intrusion. If she is wronged, then the infringer owes her some kind of rectification (apology, compensation, etc.). I shall argue that the most plausible understanding of rights (in general, and for libertarianism) is such that, although many paternalistic acts do wrong the affected individual, many such acts that seem to wrong the rightholder do not in fact wrong her.

Paternalism is often defined as actions, motivated or justified by promotion of a person's good or well-being, that interfere with her, or limit her liberty or autonomy, against her will. From a libertarian perspective, however, whether a person is wronged does not depend on the agent's motives and intentions. It depends only on whether its causal impact infringes the person's rights. We shall therefore not address the notions of motivation, intention, interference, liberty, or autonomy. We shall focus solely on the question of when a rights intrusion (defined below) infringes rights (wrongs the rightholder). This will implicitly cover all cases of paternalist rights-intrusions however paternalism is understood. We won't address cases of paternalistic actions that do not infringe rights, since libertarianism does not judge such actions to be wrongful.

2 Rights: choice-protecting and interest-protecting

Libertarians tend to assume that rights are *choice-protecting* in the sense that one's right that someone not perform some action (e.g., poke a needle into one's body) is *not* infringed when done with one's *valid consent*. Thus, one is not wronged when the doctor gives one a vaccination with one's valid consent.[2]

There are many important questions about the nature of the relevant *consent*. Is it merely a mental act, which need not be publically expressed? If it is a public act, does there need to be uptake by the agent (or at least someone) in the sense that the consent is successfully communicated to the agent? I shall leave these issues open.

Whatever consent is, what is required for consent to be *valid*? Roughly speaking, the consent must be autonomous and suitably free and informed, but there are competing accounts of what this requires. Almost everyone agrees, however, that the validity of consent is undermined by lack of autonomy (e.g., as with young children and severely cognitively impaired people), by credible threats to wrongfully harm someone that one cares about if one does not consent (consent is not suitably free), and by intentional misrepresentation of significant facts that are relevant for consent (e.g., fraud). Many other factors may undermine the validity, but I shall not attempt to resolve this issue here. Instead, I shall simply appeal to the correct account of validity for consent (and dissent), whatever that is.[3]

Roughly speaking, choice-protecting rights make sense for agents that are psychologically autonomous (i.e., that can rationally reflect upon, and revise, their beliefs, desires, and intentions). Their consent, when suitably free and informed, reflects the exercise of their autonomy. For individuals who are not psychologically autonomous, however, it makes little sense to have constraints protecting their choices (in any robust way). Choice-protecting theorists often hold that such beings (e.g., young children) do not have rights, on the ground that rights (conceptually) *must* be choice-protecting. This seems quite mistaken. Rights can also protect individuals by *protecting their interests*.[4] Indeed, young children have such rights.

There are different ways that rights might protect interests, but the most direct, and most promising (I think), is for interests to play the same role that valid consent plays for standard choice-protecting rights. On this *directly interest-protecting* conception of rights, one's right that someone not perform some action-type (e.g., poke a needle into one's body) is infringed when and only when an action of that type is *against one's best interests*.[5] Thus, a child with the right not to have needles poked into her body is not wronged when the doctor gives her a vaccination, without her consent, and it is in her best interests. (Of course, parental consent may be required, but, if so, the parents, but not the child, are wronged when the vaccination is given without their consent.)

It's important to note here that directly interest-protecting rights do *not* give the rightholder a claim against *all actions* that are against her best interests. It is rather that individuals have claim rights against certain independently specified action-types (e.g., that others not use their bodies; e.g., by poking needles in them). The question concerns when such rights are infringed (e.g., not when valid consent is given). This requires both that the action be of a type protected by rights (a rights intrusion) and that the intrusion be against the rightholder's best interests (just as on a standard choice-protecting theory it requires both that the action be of the specified type and that it be without the valid consent of the rightholder).

In order to develop the interest-protecting account, let us say that an autonomous action *intrudes* upon someone's rights when it would wrong the person, if there were no *authorization* by the rightholder for the action. Intrusions can thus be wrongful (when there is no authorization) or rightful (i.e., no wronging or infringement, when there is authorization). Let us understand authorization to require valid consent for choice-protecting rights and to require not being against the rightholder's best interest (as clarified below) for interest-protecting rights. Thus, authorization is here understood broadly in the sense that it can be generated by a person's interests and not merely by expressions of her will.

There are many important issues concerning the relevant conception of interests for the most plausible version of a directly interest-protecting conception of rights. I find it plausible that the relevant interests typically include (1) more than a concern for one's own well-being (e.g., a parents' interests include their well-being and the well-being of their children), and (2) for autonomous agents, an interest in being in control of one's life (and not merely in what happens to one) and in being recognized and respected as an autonomous agent.[6] I do not, however, assume this in what follows.

Even if we fix upon a specific understanding of interests, there are important issues about what is required to be *in* someone's interest. I believe that the best understanding for directly interest-protecting rights is (1) based on the *objective* facts, and not on what the agent believes or reasonably should believe given her evidence, and (2) based on (objective) *probabilities* of the various possible consequences of the action (where, if determinism is true, all probabilities are 0 or 1). Thus, I assume that an individual can be mistaken about what is in her interests. A similar account can, however, be provided in terms of what the agent's evidence supports, or, less plausibly, in terms of the agent's beliefs.

Let us say that an agent's intrusion is *in the rightholder's best interests* (relative to the agent's feasible options that wrong no one else) when and only when the intrusion has an objective expected value (based on objective chances and the agent's interests) that is: (1) greater than the expected value, if the agent does not intrude upon the rights of the rightholder, and (2) such that no alternative *intrusive action* by the agent (a) is no worse for the agent (in terms of expected value for his interests), (b) wrongs no one else, and (c) has a higher expected value for the rightholder. The first condition requires that the intrusion be (objectively) in the rightholder's interest (better than non-intrusion). The second condition requires that it be in her "best" interests in the sense that no alternative intrusion by the agent is better for the rightholder, without being worse for the agent or wronging others. My intrusively saving your (positive value) life, in a particular way, at the cost of your losing a hand, is in your best interest if (1) you will die if I do not intrude upon your rights, and (2) I have no alternative way of intrusively saving your life that (a) wrongs no one else, (b) is at least as valuable for me (e.g., not more costly), and (c) is better for you. If, however, someone else will save you if I don't, with no associated harms, then my saving you (at the cost of your hand) is against your best interest. Or, if I have the alternative of saving your life at the cost of your losing only a finger (rather than your hand), and this would wrong no one, and cost me nothing, then saving you at the cost of your hand is against your best interest (in the stipulated sense).

As a first gloss, then, I claim that psychologically autonomous agents have choice-protecting rights, and non-autonomous individuals with interests have directly interest-protecting rights.

I shall now suggest that, even for autonomous agents, intruding upon their rights for their benefit without their consent does not always wrong them. This will involve a revisionary understanding of rights as (1) giving priority to the protection of autonomous choices, but (2) protecting their interests when they give neither (hypothetical or actual) valid consent nor valid dissent.

3 Choice-prioritizing rights

Below, I shall appeal both to consent and to *dissent*. The latter is the expression of the opposition of one's will to an action. Standard choice-protecting theories do not distinguish between the absence of valid consent and the presence of valid dissent, but I shall suggest that the latter has stronger moral force than the former.

There are several kinds of case where it is not reasonably possible to communicate with the rightholder about current valid consent to an intrusion. Standard (purely) choice-protecting theories hold that, in such case, intrusion without valid consent wrongs the rightholder. I shall argue, however, that such intrusions do not wrong her when (1) there is a certain kind of *hypothetical consent* from the rightholder, or (2) there is no *hypothetical dissent* and it is *not against her best interests*. I shall do this by claiming that rights are *choice-prioritizing*, where this allows, in certain cases, appeals to both the hypothetical consent/dissent of rightholders and to their interests.

To start, let us consider the case where it is *currently reasonably possible* to communicate with the rightholder about current valid consent to an intrusion. I won't attempt to give a full characterization of this notion, but the following should suffice to give its outlines. First, the issue only concerns the possibility of currently *communicating* a request for consent and of an answer being communicated back. The rightholder's disposition to *grant* consent is irrelevant. Second, the issue concerns communication about *current* (not past) consent/dissent. Third, it is *not* reasonably possible to communicate about valid consent when it is practically impossible. The rightholder may be *non-autonomous* (e.g., unconscious, demented, drunk, or having a panic attack). Or the rightholder may be autonomous, but it is practically impossible for the agent to communicate with her prior to the intrusion in question (e.g., where there is not enough time). Finally, when practically possible, it is *reasonably possible* to communicate with the rightholder about valid consent when the *costs* of doing so, without acting impermissibly, are not "unreasonable" for either the agent or for the rightholder. I here leave open the important account of what costs are reasonable, except to say that trivial costs are reasonable and enormous costs are not.

I assume, in agreement with the standard choice-protecting theory, that:

> *Choice-Prioritizing Rights, Version 1*: If it is currently reasonably possible to communicate with the rightholder about current valid consent to an intrusion, then the intrusion is authorized by the rightholder if and only if she has given unrevoked valid consent.

In what follows, I shall expand on this condition in ways that entail that an intrusion can be authorized (and thus not wrong the rightholder) both by a certain kind of hypothetical consent and by, in the absence of both hypothetical and actual valid dissent, not being against the rightholder's best interests.[7]

Let us now consider the case where it is *not reasonably possible* to communicate with the rightholder about current valid consent to an intrusion. It may be practically impossible (e.g., the rightholder is unconscious or there is not enough time to communicate) or possible but too costly (e.g., both the agent and rightholder will be seriously harmed if they communicate). To start, let us suppose that the rightholder has neither given *valid consent* to the intrusion, nor given *valid dissent* to the intrusion. A standard choice-protecting conception of rights holds that, if the rightholder is still autonomous, then there is no authorization in the absence of valid consent. This, however, seems mistaken. Suppose that you are about to be hit by a truck, and I have the time to push you out of the way, but not enough time to get your valid consent first. Given that there is no opportunity for consultation, surely my pushing you out of the way is authorized, and thus does not wrong you.

I (Vallentyne 2007) previously thought that, in such cases, intrusion is authorized if it is not against the rightholder's best interests, but I now think that this was a mistake. Consider a temporarily unconscious adult Jehovah's Witness who needs a blood transfusion (which intrudes upon her rights) to survive, where this is very significantly in her interests. Suppose further that she never gave valid consent or dissent to a blood transfusion. Nonetheless, suppose that, when she was last autonomous (an hour ago, say), she *would have dissented* (i.e., denied permission) to a blood transfusion, if asked for consent – even if she were given all the information available to the agent about the costs and benefits of a blood transfusion (including those of any possible afterlife). This is so, let us suppose, because she is mistaken about what is in her interests (she has mistaken beliefs about the consequences of a blood transfusion, and/or mistaken beliefs about her interests). Is a blood transfusion authorized for her (and thus does not wrong her)? I claim that it is not.

Although below I shall endorse an appeal to a person's best interests under certain conditions, I believe that a certain kind of hypothetical consent/dissent, if present, is *lexically prior* to interest considerations. If it is not reasonably possible to communicate about current valid consent, and there is no prior consent/dissent, then (1) such *hypothetical consent* is sufficient for authorizing an intrusion, even if against the rightholder's interests, and (2) such *hypothetical dissent* is sufficient for the lack of authorization. Let us unpack this idea.

First, the consent/dissent at issue is to *the specific intrusion* by the agent against the rightholder in the actual circumstances (where the rightholder may lack some relevant information). It is not to intrusions of that type in general. Second, although the consent/dissent is hypothetical, it is based on the rightholder's *actual consent/dissent dispositions*, relative to her stable beliefs, values, and disposition to revise beliefs and values in light of information, *when she was last autonomous* (which is the time of the intrusion, if she is currently autonomous).[8] If the individual was never autonomous, then there is neither hypothetical consent nor hypothetical dissent.[9] The disposition to consent/dissent is based on her beliefs and values (as revised in light of any hypothetically received new information). It is not based on the disposition to consent/dissent of people in general or of some highly idealized version of the rightholder. Moreover, her beliefs and values may be mistaken, and her values need not be limited to her well-being. Third, the only idealizations involved are something like the following: (1) The rightholder is informed of (but need not accept) the beliefs of the intruding agent (and those with whom he coordinates for the intrusion) about the extent to which the intrusion is in her interests, as well as his supporting beliefs, and (2) the manner in which new information is presented, and the conditions for reflection and revision, are adequate for rational assessment of the issue and for any consent/dissent to be valid. With respect to new information, the rightholder may deem it irrelevant, accord it minor weight, or take it very seriously. This is determined by her prior beliefs, values, and disposition for reflecting upon and revising belief, values, and intentions.[10]

I do not assume that the individual would either consent to a given intrusion or dissent to it. It may be that, even if the individual has perfectly coherent beliefs and values, she might consent and she might dissent (e.g., depending on different adequate ways that information is presented or different adequate conditions for assessment). Also possible is that she would withhold both. In these cases, hypothetical consent/dissent does no work.

A significant problem arises given that people's beliefs and basic (belief-independent) values can be (are!) incoherent. This is indeed a problem, and I shall simply assume that there are admissible ways of cleaning them up to be coherent. Hypothetical consent (or dissent) holds, I stipulate, when and only when, relative to *all* admissible ways of cleaning them up, the individual *would* consent (or: dissent) to the intrusion. Thus, if, on some admissible clean-up, the rightholder might not consent, and on some other admissible clean-up, the rightholder might not dissent, then hypothetical consent and dissent will be empty and play no role in that case. It may turn out, of course, that hypothetical consent/dissent will always be empty. I will proceed, however, on the optimistic assumption that at least sometimes they are not empty.[11]

Let us pause to be explicit about the relation between well-being, interests, values, and hypothetical consent/dissent. A person's *interests* (as I use the term) are determined by the facts relative to what she *has reason to value*. They almost always include her *well-being* (what makes her life go well for her) but almost always include more (e.g., the well-being of her children). A person's *values* need not be identical with what she has reason to value (due to ignorance or failure of rationality). She may value some things that she has no reason to value and may fail to value some things that she has reason to value. Hypothetical consent (as understood here), like actual consent, is based on the individual's hypothetically revised beliefs and values, not on her interests (about which she may be partly ignorant/mistaken) or merely her well-being.

The relevant notion of hypothetical consent/dissent will, no doubt, need to be fine-tuned, but the above should be enough to make the basic idea clear.[12]

Consider then:

> *Choice-Prioritizing Rights, Version 2*: (1) If it is *currently reasonably possible* to communicate with the rightholder about current valid consent to an intrusion, then the intrusion is authorized by the rightholder if and only if she has given unrevoked valid consent. (2) If it is *not currently reasonably possible* to communicate with the rightholder about current valid consent/dissent to an intrusion, *and the rightholder has given neither valid consent nor valid dissent* to the intrusion, then the intrusion is authorized if and only if (a) she hypothetically consents, or (b) she does not hypothetically dissent and the intrusion is not against her best interests.

This prioritizes consent over interests in two ways. First, where communication about current valid consent is reasonably possible, it is required for authorization. Second, where such communication is not reasonably possible, in the absence of actual valid dissent, hypothetical consent authorizes (even when against the rightholder's interests) and interests authorize only in the absence of hypothetical dissent.

There is one final case that we have not yet addressed. Suppose that the rightholder has given valid consent, or valid dissent, to the intrusion, but the agent has new information about the impact of intrusion, and it is not reasonably possible to elicit a reassessment by the rightholder.

A standard choice-protecting conception of rights views current valid consent/dissent as dispositive of authorization, even in such cases. This seems implausible. For example, suppose that you have given valid consent to a routine operation, but after you have been given general anesthesia, the doctor discovers that you have a special condition that makes it very likely that you will die. Suppose that you would dissent from the operation if you were given the new information. Surely the doctor would wrong you if he performs the operation. Respect for the rightholder's will, I claim, requires being sensitive to her consent/dissent in light of any new relevant information possessed by the agent. More exactly, hypothetical consent (or dissent) can sometimes *override* actual valid dissent (or consent), and not merely sometimes *fill in* in their absence.[13]

Consider then:

> *Choice-Prioritizing Rights, Final Version*: (1) If it is *currently reasonably possible* to communicate with the rightholder about current valid consent to an intrusion, then the intrusion is authorized by the rightholder if and only if she has given unrevoked valid consent. (2) If it is *not currently reasonably possible* to communicate with the rightholder about current valid consent/dissent to an intrusion, then the intrusion is authorized if and only if (a) she hypothetically consents (even if she validly dissented in the past), or (b) she does not hypothetically dissent, has not given unrevoked valid dissent, and the intrusion is not against her best interests.

This is the view that I wish to defend. It holds that actual consent is necessary and sufficient for authorization when it is reasonably possible to communicate with the rightholder about current valid consent. If it is not reasonably possible, then hypothetical consent or being in the rightholder's interest in the absence of both hypothetical dissent and actual valid dissent is necessary and sufficient for authorization. Interests, that is, determine authorization only where current actual consent is not reasonably possible, there is no hypothetical consent or dissent, and there is no actual valid consent or dissent.[14] For beings that were never autonomous prior to the intrusion, the appeals to actual and hypothetical consent/dissent are irrelevant, and their interests are the only factor for authorization.

It's important to keep in mind that the relevance of hypothetical consent/dissent, like that of actual consent/dissent, applies only where one still possesses the right that is being intruded upon. If one has contractually transferred it to someone else (e.g., irrevocably transferred decision-making powers to another person for this decision), or forfeited it due to wrongful behavior, then one's will is no longer relevant. Just as one's past and present actual consent/dissent (e.g., to a specific action/procedure) are irrelevant, so too are one's hypothetical consent/dissent.

4 A partial defense

The above account appeals to hypothetical consent/dissent, *when last autonomous*, and this temporal specification might be challenged in several ways: (1) If the person has been non-autonomous for a long time, why not just appeal to interests? (2) If the person has degenerated from highly autonomous to non-autonomous, why not appeal to the last time she was at *her highest attained level* of autonomy? (3) If the person will regain autonomy, why not appeal to when she is *next* autonomous? The basic answer to each of these questions is the same: authorization by hypothetical consent is sensitive to time in the same way that authorization by actual consent is. As long as the current person is the same person (in the morally relevant sense) as the past person, then: (1) The will of the past person takes precedence over the interests of the current person (e.g., your consent prior to the general anesthesia makes any appeal to your interests irrelevant). (2) The most recent expression of (revocable) valid autonomous consent/dissent takes precedence over earlier expressions, even if the rightholder was more autonomous at the earlier time. (3) Future consent (after the fact) does not authorize intrusion. The above account, I claim, rightly gives hypothetical consent the same priority to the most recent past.

The authorization conditions that I have articulated are in rough (see below) agreement with standard practice for informed consent and surrogate decision-making: one appeals to valid advance directives when they exist. When they do not exist, and it is not possible to consult with the rightholder, one appeals to the substituted judgment standard (roughly hypothetical consent/dissent). If that is indeterminate, then one appeals to the best interest standard.[15]

There is, however, one possible difference. Hypothetical consent/dissent can, I claim, *override* actual valid consent/dissent. Where communication is not reasonably possible, and the agent has crucial information not had by the rightholder, then, I claim, hypothetical consent/dissent (as characterized above) *better represents the autonomous will* of the rightholder than the less well-informed actual valid consent/dissent, or absence thereof. Above, I gave the example of valid consent to an operation being overridden by hypothetical dissent in light of new information. Here is a second example. Suppose that you have *validly dissented* from my interfering with your smoking cigarettes but that I now have information, not reasonably communicable to you, that the cigarette that you are about to smoke has been laced with poison. Suppose that, if you were made aware of the information of lacing, you would consent to this particular interference. Your will, I claim, is best protected by my physically stopping you from smoking that particular cigarette.

The choice-prioritizing account implies something even more radical: where there is no reasonable opportunity to communicate, prior valid consent/dissent can be overridden by hypothetical consent/dissent, *even when there is no new information*. This can arise when valid consent/dissent is given, and the rightholder later changes her beliefs, values, or disposition to consent/dissent, without revoking her consent/dissent. This is a controversial implication, but space limitations prevent me from discussing this further. The important point to keep in mind is that such overridings occur only when it is not reasonably possible to communicate with the rightholder.

As a practical matter, we should normally judge actual consent/dissent to be overridden by hypothetical consent/dissent only when we have strong evidence for the hypothetical consent/dissent. Thus, in practice, it may be relatively rare for the evidence to favor such overridings. Nonetheless, I believe that, when there is no reasonable opportunity for current valid consent/dissent, the most plausible conception of rights will hold that hypothetical consent/dissent overrides past valid consent/dissent.

Let us now consider an objection to combining choice-protection with interest-protection, and prioritizing the former. For simplicity, I here set aside hypothetical consent. For such theories, for autonomous agents, the will, when not silent, determines whether there is authorization for an intrusion, no matter how great the interest-benefit of the intrusion. For an individual who is *slightly below* the threshold for autonomy, however, her interests determine whether there is authorization for an intrusion, even when they are only trivially affected. This seems quite implausible (see, e.g., Scoccia 2013: 89 and Grill 2010: 14–18).

For conceptions of rights, such as the choice-prioritizing conception, that are sensitive to both choice-protection and interest-protection, this problem can be solved by holding that, for levels of autonomy below the threshold required for full choice-protection, the more autonomy that is present, the greater the role for the will. There are, of course, many ways that this might be so, and the following is merely an illustration of one relatively plausible way. For a given degree of autonomy, there may be a maximal setback to interests, relative to what is best for interests, over which the will has the authority. (The setback to interests will include any benefits of learning from the experience.) For individuals with no autonomy, the maximal setback may be zero. Only their interests matter. For individuals who are above the relevant threshold of autonomy, the maximal setback over which the will has authority may be infinite. (Alternatively, the threshold can be dropped and authority will be infinite only for "perfect" autonomy.) Their wills have priority. In between, the maximal setback increases as the individual's autonomy increases. The wills of young children may typically have authority over the color of their socks but not over whether to cross a busy street.[16] In short, an adequate theory of authorization will include some account of how the authority of the will grows as the rightholder's autonomy grows. Indeed, it will also include an account of the how the authority of the will shrinks as the rightholder's autonomy shrinks (e.g., in cases of dementia).

Obviously, a full defense needs to address many other issues. I hope that the above is enough to provide some motivation for the account.

Let us now consider whether this choice-prioritizing account is compatible with libertarianism.

5 Libertarianism and choice-prioritizing rights

Steve Wall (2009) has argued that libertarianism's endorsement of full self-ownership has very implausible implications for paternalism. Because the rights of *full* self-ownership are the strongest possible rights of self-ownership (e.g., rights to one's body), they are unconditional in application (apply under all conditions) and are absolute in force (always impermissible to infringe). Thus, the right not to have one's body touched holds no matter how great the benefits to the rightholder of being touched, and it is impermissible to infringe the right even when it greatly benefits the rightholder. I agree that these are problematic implications of libertarianism. On the choice-prioritizing view, however, the implications are much less problematic than on the standard choice-protecting view. Let me explain.

Rights of full self-ownership *apply unconditionally*, but this does not establish that it always wrongs the rightholder to intrude upon her rights, without her valid consent, in order to advance her interests. This is so on a purely choice-protecting conception, but not on the choice-prioritizing conception. On the latter view, protecting the interests of a rightholder can authorize an intrusion, when it is not reasonably possible to communicate about consent, and there is no hypothetical or actual valid dissent.

This leaves, however, one key objection from Wall (2009: 404). It may seem that *full* self-ownership leaves no room at all for interest-protection. Giving any role to interest-protection weakens the role of choice-protection, and that, it may seem, is incompatible with *full* self-ownership (the strongest set of rights over one's person). That, however, is a mistake. Full self-ownership is comprehensive in scope (includes all rights of self-ownership), unconditional in application, absolute in moral force, and with the strongest possible authorization conditions. The authorization conditions, however, have several relevant dimensions, and hence, as I now explain, there are several ways of being the strongest authorization condition.

Authorization conditions are stronger (1) when they provide a greater role (as sufficient conditions or as necessary conditions) to the rightholder's valid will (consent or dissent) and no lesser role for her interests, and (2) when they provide a greater role to the rightholder's interests and no lesser role for her valid will.[17] Having no authorization conditions is not maximally strong, since introducing either choice-protecting authorization conditions or interest-protecting authorization conditions would be stronger. A purely choice-protecting conception of self-ownership is indeed maximally strong, but so is a purely interest-protecting conception. Indeed, any consistent combination of choice-protecting and interest-protecting conditions is maximally strong, as long as it is not possible to strengthen one of them without weakening the other. The choice-prioritizing conception that I have articulated is maximally strong in this sense. It is thus compatible with full self-ownership.

Of course, traditional libertarians will insist that for autonomous agents, rights are purely choice-protecting. If that is so, then full self-ownership must be purely choice-protecting. I agree that this is how full self-ownership is normally understood. My claim is that understanding rights, and full self-ownership in particular, in terms of choice-prioritizing rights is both possible and more plausible than a purely choice-protecting account.

6 Conclusion

I have addressed the issue of when a rights intrusion is authorized and thus does not wrong the rightholder. Even if authorized, an intrusion may be impermissible because it wrongs someone else or is impersonally wrong. Even if not authorized, an intrusion may be permissible because there is an overriding justification. I have not addressed these two issues.

I have articulated and partly defended a choice-prioritizing authorization condition. It holds that, when it is reasonably possible to communicate about consent, then valid consent is necessary and sufficient for an authorization. When it is not reasonably possible to communicate about consent, it holds that hypothetical consent/dissent overrides actual valid consent/dissent and has priority over any appeal to interests. In the absence of hypothetical consent and dissent, and of valid consent and dissent, interests determine whether an intrusion is authorized.

Libertarians tend to endorse purely choice-protecting rights, but, I claim, a choice-prioritizing conception of libertarian rights is possible and indeed more plausible.[18]

Related topics

Deciding for the Incompetent; Paternalism and Rights; Paternalism and Autonomy.

Notes

1 Cornell (2015) argues that one can be wronged without having a right infringed, but I here equate wronging with right-infringing.

2 Of course, if past valid consent can make an intrusion rightful, this must be because the consenting person has rights over the later person whose rights are intruded upon (e.g., because she is the same person). The conditions under which this is so are, of course, controversial, and I shall simply assume that they hold in the cases that we discuss.

3 For superb discussions of what is required for valid consent, see VanDeVeer (1986: 45–58) and Feinberg (1986: 113–124 and chaps. 22–26). An important issue that I will not address is whether (as I believe) non-revocable consent (as in the case of Odysseus) can be valid. For excellent discussion, see VanDeVeer (1986: 294–301) and Feinberg (1986: 68–87).

4 For excellent discussion of choice-protecting versus interest-protecting rights, see Kramer, Simmonds, and Steiner (1998).

5 A slightly more demanding account holds that a right is not infringed when and only when the intrusion is in the rightholder's best interests (cf. not against them).

6 For further discussion of this issue, see Feinberg (1986: 57–62) and Dworkin (2014).

7 Throughout, we focus on consent and dissent in revocably authorizing an intrusion of a right currently possessed by the rightholder. Consent can also authorize transfers of rights to others (as when one sells or rents a car). When this is so, the transferor's consent/dissent ceases (at least temporarily) to be relevant to whether the right is intruded upon.

8 I appeal to the dispositions of the rightholder when last autonomous, rather than those she would have if she were now autonomous, because the latter might involve significant changes in the person's dispositions (e.g., the only way of restoring autonomy is an operation that radically alters her values and beliefs).

9 See the insightful Kuflik (2010), who also rejects the appeal to hypothetical consent for beings that have never been autonomous. He also appeals to interests.

10 For a contrasting view, see Enoch (2017), who allows all information about all facts compatible with the rightholder's deep commitments.

11 See Grill (2015) for insightful discussion of cleaning up incoherent preferences.

12 For related appeals to hypothetical consent, see Feinberg (1986: 173–188), VanDeVeer (1986: 75–81), Grill (2015), and Scoccia (2013). Unlike those accounts, I explicitly appeal to both hypothetical consent and hypothetical dissent. Like Scoccia (and perhaps Grill), I focus on when an intrusion is authorized (does not wrong the rightholder) as opposed to it being permissible (e.g., VanDeVeer). I follow Feinberg (1986: 324), but not the others, in (1) limiting the appeal to hypothetical consent/dissent to cases where it is not reasonably possible to communicate about current valid consent, and (2) appealing to the rightholder's disposition to consent/dissent when she was last autonomous, rather than to her hypothetical disposition if her "capacities for deliberation and choice were not substantially impaired." Scoccia (2013), and perhaps others, limits the relevant information to empirical information, whereas I would also allow evidence about anything, including the supernatural (e.g., God) and values.

13 Scoccia (2013: 83) and VanDeVeer (1986: 88) hold that hypothetical consent and actual valid consent are each sufficient. They thus leave no room for hypothetical dissent overriding valid consent, or hypothetical consent overriding valid dissent. Feinberg (1986: 181) seems to allow that hypothetical consent/dissent could override actual consent/dissent, but he holds that, as an epistemic matter, actual consent/dissent is non-overridable evidence of hypothetical consent/dissent. Grill (2015: 712) allows that actual consent can sometimes (e.g., if given under pressure) be overridden by hypothetical consent, but he rejects any priority rules.

14 VanDeVeer (1986), Scoccia (2013), and Grill (2015) make no appeal to interest-protecting conditions. I follow Feinberg (1986: 324) in holding that they are relevant when neither actual, nor hypothetical, consent/dissent holds.

15 I thank Daniel Groll for bringing to my attention the standard practice for informed consent and surrogate decision-making. For further discussion, see Jaworska (2009), Groll (2015), and Brock and Buchanan (1990: chap. 2).

16 I here ignore the important possibility that the level of autonomy may be choice-relative. Children may be highly autonomous with respect to choosing sock color but not with respect to choosing whether to cross the highway.

17 For simplicity, I do not here address hypothetical consent and dissent. This is yet another dimension on which authorizations can be weaker or stronger.

18 For helpful comments, I thank Kalle Grill, Daniel Groll, Jason Hanna, Joe Mazor, and the audiences of my talks at Hebrew University of Jerusalem, The Institute for Future Studies (Stockholm), and the meeting of the Italian Society for Analytic Philosophy (Pistoia, Italy).

References

Brock, D. and Buchanan, A. E. (1990) *Deciding for Others: The Ethics of Surrogate Decision-Making*, Cambridge: Cambridge University Press.

Cornell, N. (2015) "Wrongs, Rights, and Third Parties," *Philosophy & Public Affairs* 43: 109–143.

Dworkin, G. (2014) "Paternalism," in E. N. Zalta (ed.) *The Stanford Encyclopedia of Philosophy* (Summer 2014 Edition), http://plato.stanford.edu/archives/sum2014/entries/paternalism/

Enoch, D. (2017) "Hypothetical Consent and the Value(s) of Autonomy," *Ethics* 128: 6–36.

Feinberg, J. (1986) *The Moral Limits of the Criminal Law, Volume 3: Harm to Self*, Oxford: Oxford University Press.

Grill, K. (2010) "Anti-Paternalism and Invalidation of Reasons," *Public Reason* 2: 3–20.

———. (2015) "Respect for What? Choice, Actual Preferences, and True Preferences," *Social Theory and Practice* 41: 692–715.

Groll, D. (2015) "Medicine and Wellbeing," in G. Fletcher (ed.) *The Routledge Handbook of Philosophy of Well-Being*, New York: Routledge, pp. 504–516.

Jaworska, A. (2009) "Advance Directives and Substitute Decision-Making," in E. N. Zalta (ed.) *The Stanford Encyclopedia of Philosophy* (Summer 2009 Edition), http://plato.stanford.edu/archives/sum2009/entries/advance-directives/.

Kramer, M. H., Simmonds, N. E. and Steiner, H. (1998) *A Debate Over Rights*, Oxford: Oxford University Press.

Kuflik, A. (2010) "Hypothetical Consent," in F. Miller and A. Wertheimer (eds.) *The Ethics of Consent: Theory and Practice*, Oxford: Oxford University Press, pp. 131–162.
Scoccia, D. (2013) "The Right to Autonomy and the Justification of Hard Paternalism," in C. Coons and M. Weber (eds.) *Paternalism: Theory and Practice*, New York: Cambridge University Press, pp. 74–92.
Vallentyne, P. (2000) "Left-Libertarianism: A Primer," in P. Vallentyne and H. Steiner (eds.) *Left Libertarianism and Its Critics: The Contemporary Debate*, New York: Palgrave Publishers Ltd., pp. 1–20.
———. (2007) "Libertarianism and the State," *Social Philosophy and Policy* 24: 187–205.
Vallentyne, P. and van der Vossen, B. (2014) "Libertarianism," in E. N. Zalta (ed.) *The Stanford Encyclopedia of Philosophy*, http://plato.stanford.edu/archives/fall2014/entries/libertarianism/.
VanDeVeer, D. (1986) *Paternalistic Interference*, Princeton, NJ: Princeton University Press.
Wall, S. (2009) "Self-Ownership and Paternalism," *The Journal of Political Philosophy* 17: 399–417.
Zimmerman, M. J. (2008) *Living with Uncertainty: The Moral Significance of Ignorance*, Cambridge: Cambridge University Press.

Further reading (on paternalism generally)

Husak, D. N. (1981) "Paternalism and Autonomy," *Philosophy & Public Affairs* 10: 27–46. (Argues that paternalism is not incompatible with respect of autonomy.)
de Marneffe, P. (2006) "Avoiding Paternalism," *Philosophy & Public Affairs* 34: 68–94. (Argues that paternalism is not always disrespectful or wrong.)
Mill, J. S. (1991 [1859]) "On Liberty," in J. Gray (ed.) *On Liberty and Other Essays*, Oxford: Oxford University Press. (The classic anti-hard-paternalism statement.)
Shiffrin, S. (2000) "Paternalism, Unconscionability Doctrine, and Accommodation," *Philosophy & Public Affairs* 29: 205–250. (Analyzes the nature of paternalism and argues that the unconscionability doctrine in contract law need not involve paternalism.)

16

EGALITARIAN PERSPECTIVES ON PATERNALISM

Richard Arneson

Should egalitarian doctrines of social justice incline us to be more judiciously favorable toward paternalism or more relentlessly opposed to it? The answer is not obvious. It all depends, one might say. It depends on what version of egalitarian justice we are examining and what conception of paternalism we have in mind. This chapter examines the implications of two versions of egalitarianism regarding the moral status of paternalism. One version holds that justice requires bringing it about that all members of society have a fair opportunity to lead a life that is good for them, and holds also that fairness at least *pro tanto* requires either equality or priority (Temkin 1993; Roemer 1998; Cohen 1989; Arneson 1989; Parfit 1995; Dworkin 2000; Knight 2009; Segall 2013). Call this view *distributive egalitarianism*. The second version is *relational egalitarianism*, which holds that justice requires establishing and sustaining a society in which people relate as equals (Anderson 1999; Scheffler 2003). The examination aims to shed light on the question, under what conditions, if any, is paternalism morally permissible, or required. A secondary aim is to assess the plausibility of the two conceptions of egalitarianism – call them distributive and relational conceptions – by looking at the prescriptions they generate regarding paternalism. The two egalitarian doctrines might be viewed as rival or complementary, and the conclusions we should reach might be that we should accept one doctrine and reject the other, or accept both, or reject both.

Paternalism can occur in private interaction, as when one friend refuses to let another friend swim in a fast-moving stream that will likely drown this weak swimmer. For the most part this chapter concentrates on questions concerning whether and under what conditions governments are morally required or permitted to carry out paternalistic policies or prohibited from doing that.

As readers of this handbook are by now aware, the term "paternalism" is used to refer to different types of policies and behavior patterns. The broad theme is perhaps that paternalistic behavior treats competent adult individuals as though they were still children and incompetent to manage their own lives. Maybe "parentalism" would be a more appropriate label for this type of behavioral stance, but in philosophical discussions of the topic, the term "paternalism" is standardly deployed.

Although much printer's ink has been spilled in arguments about how best to define "paternalism," it should be clear that in philosophical discourse this is a nonissue (for a contrary view see Dworkin 2013; and see also Scoccia, this volume). Let each theorist define her terms as she likes. So long as the explanation of what is meant is clear and the theorist consistently sticks to her stipulations as to what notion is linked to each term, there should be no harm in letting a thousand flowers bloom definition-wise. To keep the conversation among theorists simple we can identify paternalism1, paternalism2, paternalism3, and so on. The disagreements that matter here are not definitional but rather normative: disagreements about which actions are morally forbidden, permissible, or required.

This chapter focuses on two conceptions of paternalism. One fits easily with Mill's (1978 [1859]) discussion of the legitimate limits of social control over the individual in *On Liberty*. On this view, paternalism is restriction of a person's liberty against her will for her own good. A second conception takes paternalism to be an act by one agent that (1) aims to improve the welfare, interests, values, etc., of another agent with respect to some particular decision or problem that agent faces and that (2) is motivated by a negative judgment about the ability of the person being helped (if well informed) to make the right decision and carry it out successfully (Quong 2011, following in part Shiffrin 2000).

Different conceptions of paternalism highlight different ideas of what is wrong with paternalism (see Stevenson 1938). The broad intent of the highlighting is that, at least in philosophical and perhaps in ordinary usage, the term has a negative connotation, and to chime in with usage, paternalism should be characterized so that identifying an act as paternalistic marks it as *prima facie* objectionable (even it if turns out that some or most instances of paternalism should turn out to be justified all things considered). Behind the restriction of liberty conception is the thought that people should be left free to act as they choose in the absence of serious countervailing reasons. Behind the negative judgment conception is the thought that one's status as an equal member of society requires that others treat one as competent to manage one's own life in the absence of serious evidence that one lacks this fundamental competence (or even regardless of evidence, as a fixed presumption). On the first view the issue is wrongful restriction of individual freedom; on the second, the issue is suffering treatment that is disrespectful of one's status as an equal among equals.

1 Mill's antipaternalism

One broadly egalitarian justice doctrine that tends toward favoring paternalism emerges from looking at John Stuart Mill's utilitarian argument to the conclusion that society should adopt a strict no-paternalism policy. Mill says society should adopt a strict rule to regard harm to non-consenting others as the only valid ground for restricting someone's liberty (Mill 1978 [1859]). Mill holds that adopting an exceptionless rule to this effect would be utility-maximizing (so better than making decisions on a case-by-case basis). Assigning to competent adult individuals the prerogative of deciding for themselves in self-regarding matters will be best on the whole because decisions by others or the state overriding the individuals' judgments would be more likely to be wrong than right. When it comes to deciding for oneself what one should do for one's own benefit, the individual usually has access to the information that is needed to make good decisions. The individual herself also has a stronger motive of self-interest to find out the necessary information than others who might presume to override her judgment for her benefit. The individual can also learn from her own mistakes. Moreover, others can learn from the experiments in living that are continuously taking place when people are free to venture on any path they choose when choosing how to live within constraints against causing harm of certain types to others of certain types. Mill must also maintain that state agencies will be inept at distinguishing the few cases in which paternalism would be utility-boosting from the many cases in which it would not.

Mill's claim that adherence to a strict policy of refraining from interference with individual liberty except to prevent harm to others would be utility-maximizing rests ultimately on empirical speculation. The speculation is likely false, but just suppose it is true. This supposition would not close the door on the possibility of justified paternalism. This door is open because Mill is assuming the correctness also of act utilitarianism, and this further assumption might be challenged. For purposes of public policy choice, utilitarianism says that for each candidate competing policy, determine the utility impact, positive or negative, of adopting that policy for each person who would be affected by it, sum the results, and select the policy with the highest positive total (or lowest negative total if all possible choices would have disutility).

2 Distributive egalitarianism

A range of broadly utilitarian principles would aggregate the utility or welfare impact of candidate policies on individuals by some more egalitarian formula. Rather than maximize the sum of individual welfare, one might opt for making the worst off as well off as possible, or equalizing welfare at the highest feasible level, or following a prioritarian rule, which would assign greater moral value to achieving a welfare gain for a person, the lower the person's lifetime welfare would be absent this gain, and select the policy that would maximize priority-weighted well-being. Another possible view would give priority to achieving some "good enough" or sufficient level of utility or well-being for all.

Consider a government that is contemplating laws that would restrict people's liberty against their will for their own good. The laws will vary in their likely impact on people, depending on how competent people are at choosing and acting in ways that effectively advance their good. There are many dimensions of competence that bear on the degree to which an individual acts with rational prudence. Some of us lack executive ability to carry through the decisions we have made; we suffer from weakness of will and allied defects. We fail to persist; we are feckless. Some of us lack the ability to gather information that is material to choices we will make. Some lack the ability to arrive at sensible evaluations of the value to us of the various outcomes to which actions we might choose might lead. Some of us lack the ability reasonably to integrate such empirical information and evaluative knowledge as we possess into the calculation and deliberation that issues in choice of action. And so on. These differences in different people's value-forming, belief-forming, decision-making, and decision-executing capacities all vary by degree. For simplicity we can divide individuals into good choosers and bad choosers according to their rational prudence capacities. Good choosers, with certain minor exceptions, will never stand to gain from paternalistic restriction of their liberty. They will never be tempted by bad options, so restricting their liberty to perform a bad option does not improve their prospects. Options that are fraught, sometimes useful but sometimes dangerous or disadvantageous, are taken up by good choosers only when that is in their interest. For example, if heroin use is left unrestricted, and heroin is available on the market, a good chooser foregoes heroin use except when that use would be, all things considered, advantageous to her. In contrast, bad choosers make mistakes, so their long-term welfare might be boosted by judicious restriction of their liberty in self-regarding matters.

The claim that good choosers cannot benefit from paternalistic restriction of their liberty is subject to at least two qualifications. One is that since we do not assume that even good choosers have limitless computing power, restriction of options available to a good chooser might help her economize on decision-making costs and in this way gain improved welfare. Consider a good chooser confronting a huge array of similar brands of toothpaste for sale, each one advertised in complicated but nonfraudulent ways that are time-consuming to take in and assess. I shall assume that this qualification turns out to be not very significant, given the tools that a market will likely make available to reduce these decision-making costs (government agencies might also provide information such as ratings of competing consumer products and services).

Another qualification is that individuals who are good choosers in the sense of being perfectly competent deliberators might not be always dedicated to seeking their own welfare. They might choose to sacrifice their own welfare for altruistic or other moral reasons. So one could boost the well-being of good choosers who are making decisions that are not aimed at maximizing their own welfare, by restricting their liberty in ways that induce them to be more narrowly prudent. However, if good choosers are stipulated to be those who are fully competent at practical reason, then when they act against their own interest, they have good moral reasons for doing so, and no correct moral doctrine will seek to override their own decisions in order coercively to bring about advancement of their own welfare.

In contrast, those who are bad choosers might be less than fully competent not only at determining where their own advantage lies but also at properly balancing the appropriate weight that their own advantage, competing with other moral concerns including altruistic concern for the welfare of others, ought to have in their personal decision-making. Bad choosers might wrongly underestimate or overestimate the weight that these moral concerns ought to have. When bad choosers make choices and pursue actions aimed at sacrificing themselves for the sake of others, they will sometimes be failing to exercise morally appropriate prudence, and here too, in principle, paternalistic restriction of their liberty could be devised that would reasonably work to their advantage.

The next premise linking pro-paternalism to egalitarianism states that those who are bad choosers are likely to be badly off over the course of their lives, compared to good choosers, insofar as all are seeking prudently to advance their interests. This is a tendency claim. Some good choosers will be unlucky, and become badly off despite their good choices; some bad choosers will be lucky, and become well off despite their bad choices. Some bad choosers will have high initial bank account wealth as they enter adult life, and other things equal, those with higher bank account wealth will tend to fare better. But on the whole and on average, we would expect that being a bad chooser – especially if one is very far from the ideal of a prudent agent – will bring it about that one leads a worse than average life.

There is then a potential conflict of interest between good choosers and bad choosers when prudent choice is at issue (Arneson 1997, 2005). Good choosers will be better off under a regime of no paternalism; bad choosers can be made better off in lifetime welfare terms if paternalistic policies are introduced, provided their administrative and other implementation costs do not outweigh the potential gains. If bad choosers are more likely to be below average in lifetime welfare prospects, then welfarist principles that give extra weight to aiding the worse off will be more pro-paternalist than principles that lack such an egalitarian weighting.

One might suppose that any potential conflict of interest between good choosers and bad choosers with respect to paternalistic public policies could be resolved by designing policies that apply separate rules to the two groups. (If we think of a range of competence levels ranging from good to bad, we could envisage a range of policies treating people with relevantly different characteristics in relevantly different ways.) Joel Feinberg (1971, 1986) provides an example of such policy differentiation. Suppose some people who decide to commit suicide are reasonably cutting short their lives when further life promises more harm than benefit, and some people who desire to commit suicide are making a mistake in calculating what is really in their long-term interest. Feinberg imagines a state responding to this situation by putting in place suicide control boards. One who is contemplating suicide applies to the board for a license. The board investigates and seeks to determine whether one's deliberative process leading to the decision for suicide is competent or not and on this basis gives or withholds a legal license for suicide. In many actual state policies, something similar occurs. An example: to be able legally to engage in hang-gliding in some jurisdictions, one must pass a test that establishes one has some competence to fly a hang-glider.

To clarify: devising policies that impose paternalistic restrictions only on bad choosers who are likely to benefit from them and that leave good choosers free to act as they like would not be eschewing paternalism, but rather embracing it. Such paternalism might include the hard as well as soft variety (hard paternalism prevents individuals, for their own good, from acting even on their substantially voluntary choices). A paternalistic regime that deployed only policies that separate good and bad choosers would be responsive to the welfarist egalitarian objection against strict antipaternalism currently under review.

Separating good from bad choosers and applying different rules to people depending on their level of prudent competence does not entirely promise to resolve the conflict of interest. There are at least two reasons for this. One is that laws and other public policies are and ought to be coarse-grained instruments: to function properly they must group fairly large categories of people together and treat the people classified together the same. Regarding paternalistic restriction of liberty, the administrative costs of operating procedures that sort people finely and specify subtle variations in treatment of people depending on subtle variations in their situations would be excessive by any reasonable moral accounting.

A second reason that limits the degree to which public policies should specify differential treatment of people in fine-grained ways is that such sorting can impose stigma costs that are morally excessive. Consider *stigma* to be a visible sign of low status. There are limits to the degree that we should distinguish more and less competent choosers and accord the more competent greater freedom. Sorting in this way brands the less competent as inferior in a public way, which makes the branded feel badly about themselves and also diminishes their sense of personal efficacy. Sometimes then egalitarian welfarism will dictate that there should be a blanket prohibition of a type of conduct such as recreational use of dangerous drugs like heroin for the sake of bad choosers at the expense of some good choosers who would be better off with no prohibition in place.

Putting the pieces of this argument together, we should acknowledge that egalitarian welfarism will give us extra reasons to favor paternalistic policies in some cases, beyond the reasons that a straight utilitarian welfarism would generate. Whether this amounts to a significant case for paternalism depends on the empirical facts in specific cases, but also more generally on whether (1) utilitarian welfarism is itself morally plausible and whether (2) the tweaking of this welfarism by an egalitarian distributive norm enhances its moral plausibility.

As to (1), I submit that welfarism is indeed plausible. *Welfarism* is the claim that the ultimate moral justification of any acceptable state policy consists entirely in the promotion of the welfare or well-being of individual persons and other sentient beings together with the fair distribution of welfare across individuals. The idea is that if the state is fair to everyone in these terms, it is fair to everyone, period. Perhaps the state sometimes is morally bound to assist people in their pursuit of goals other than their own welfare, but when this is so, the ultimate ground of the duty of the state to be of assistance is the welfare of those who will benefit from these non-self-interested endeavors. If the state is bound to help me in my quest to save the whales, this is not for my sake, but for the sake of the welfare of the whales whose life prospects are at stake.

Welfarism is strong medicine – a powerful and controversial claim. Distributive egalitarianism could incorporate a weaker version of the claim – *weak welfarism*, understood as the claim that an important component of the ultimate justification of the ensemble of acceptable state policies consists in the promotion of the welfare or well-being of individual persons and other sentient beings together with the fair distribution of welfare across individuals. Egalitarianism coupled with weak welfarism will tilt more favorably toward paternalism than would weak welfarism with no egalitarian commitment.

To give welfarism a fair hearing, one must conjoin the doctrine to the most plausible conception of welfare (or of what makes an individual's life go better for that very individual) that one can identify. If welfare is identified with the experience of pleasure, one might doubt whether it is really permissible to coerce someone in order to bring it about that the person experiences greater pleasure over the long run, when the person's considered judgment is that getting pleasure is less important to her than attaining significant achievement and maintaining loving relationships. The objection to coercing here might rest on the implausibility of a hedonistic view of welfare rather than on the in-principle wrongness of coercing someone for her own good.

Consider in this connection objections to paternalism that appeal to the value of individual autonomy. Some such objections, though not all, are dampened if autonomy is rightly regarded as a significant component of an individual's welfare. Suppose that a person lives better, *pro tanto*, if she pursues and achieves goals that she adopts for herself after reflective scrutiny and thereby attains autonomy. If this is so, then restricting her freedom for her own good at the expense of her autonomy will sometimes be wrong simply because the loss in autonomy she would suffer would outweigh the gains in other components of her welfare the contemplated restriction will bring about. If autonomy is deemed an important component of an individual's welfare, then promoting an individual's welfare and promoting her autonomy will often imply similar actions, though sometimes they will conflict. Moreover, sometimes promoting the person's autonomy and along with it her welfare will require paternalistic restriction. This can happen if the person left free to act as she chooses would make choices that will reduce her lifetime autonomy.

Objections to paternalism that appeal to autonomy might have a quite different character. They might be appeals to a deontological constraint on our pursuit of welfare for others. For example, one might hold that people have an interest in being autonomous, that is, conforming their conduct to norms they accept after critical scrutiny, and one might hold that everyone has a significant though not exceptionless moral right to act according to one's autonomous choices. Autonomy so construed could ground objections to paternalism even when it is uncontroversial so that the paternalism would advance the welfare of the individual subjected to it. Jane might have a moral right to act on her autonomous choice to sacrifice herself for the sake of her community or any other thing she thinks matters morally, even if this would lower her long-term welfare and this loss would not be offset by gains to other people's welfare. Jane's right to autonomy might underpin her moral right to act on her autonomous judgment, up to a point, even if her judgment is mistaken.

The autonomy objection to paternalism just mentioned depends for its force on the rejection of strong welfarism. The welfarist will have to insist that autonomy as reflective self-rule either is better regarded as a component of individual welfare or as a norm that is not in itself morally valuable but rather an important tool for bringing about long-term boosts in welfare fairly distributed across persons. When acting on one's reflective best judgment would worsen the long-term welfare of the individual without bringing about more than compensating gains for others, autonomy should not be thought to bar restriction of individual liberty for the individual's own good.

Return to (2), as stated seven paragraphs back. We first need to clarify the egalitarian idea in play here. Egalitarianism prefers more equal to less equal distributions of welfare across persons. (This is an incompletely determinate norm; not all distributions can be ranked on a scale from more to less equal.) I shall also count as egalitarian doctrines ones that attach greater moral value to achieving a gain in welfare of a given size for a person, the worse off the person would otherwise be in terms of lifetime welfare. These doctrines are called *prioritarian* (Parfit 1995). On both types of view, the guiding thought is that there is greater moral reason to bring about a small gain in well-being, or to avoid a small loss, for a homeless mentally ill person whose life is overall bleak than for a lucky wealthy person whose life is overall richly fulfilling. Egalitarianism so characterized might be qualified by a personal responsibility proviso, which stipulates that being worse off than others through no fault of one's own should trigger extra moral reasons to aid.

Some egalitarian distributive views that do not take welfare or individual well-being or good for a person as the equalisandum might also end up tilting more strongly in favor of paternalism than straight maximizing utilitarianism. But this will not generally hold. Suppose one believes that justice requires equalizing the resources or general-purpose means available to individuals as they begin adult life. Ronald Dworkin (2000) adopts a sophisticated variant of this justice ideal. Justice so conceived might be combined with rejection of paternalism: no restriction of someone's liberty against her will for her own good is ever morally permissible, one might hold. One might hold that beyond its responsibility to provide a fair distribution of resources, it is the job of each individual, and not the proper business of the state, to decide how to live so as to secure a good life.

3 Relational egalitarianism

Relational egalitarianism will lead to different, and often opposed, assessments of the moral status – permissible, mandatory, or prohibited – of paternalism. Relational egalitarianism identifies social justice, or at least a large component of social justice, with the creation and maintenance of egalitarian social relationships, relationships in which the participants relate as equals. This is a broad umbrella characterization, which covers many different ideals, depending on what counts as relating as equals and on what sorts of relationships are specified as ones that ought to involve relating as equals.

To illustrate how relational egalitarianism might issue in different verdicts about the moral status of paternalistic policies from those implied by distributive egalitarianism, consider the example of imposing blood transfusions on hospitalized adult members of the Jehovah's Witness religion, who believe that having one's blood tampered by transfusion will have disastrous consequences for one in the afterlife, and on this basis reject medical procedures that involve blood transfusions. A public policy of imposing blood transfusions whenever they are medically advisable and paying no heed to people's religious objections to the procedure might well result in avoidance of hospital care by Jehovah's Witnesses, to their medical detriment, so that respecting their will in this regard might be the best policy from a distributive egalitarian standpoint. But from a distributive egalitarian perspective, it is in principle acceptable to override the will of an individual in matters that affect only her, if overriding would boost her welfare without triggering other adverse consequences that would outweigh the welfare boost. A relational egalitarian might hold that this stance on the part of private individuals or public officials is incompatible with respecting the status of free and equal persons.

On the other hand, if relational egalitarianism is construed as requiring that the conditions in which all can relate as equals be continuously maintained, we can envisage possible scenarios in which individuals act in self-harming ways that undermine these conditions of social equality. Restricting people's liberty against their will for their own good might then work to sustain conditions of social equality. Policies that are paternalistic in the adverse-judgment sense might fulfill the same function. Relational egalitarianism can in this way justify paternalistic state policies. For example, Elizabeth Anderson (1999) holds that a just society sustains for each individual conditions that give her access to the capabilities needed for functioning as a full participating member of democratic society. Individuals left free might decline to save for old age, might engage in unhealthy lifestyles that greatly increase their chances of being afflicted by debilitating medical conditions, might let their job skills deteriorate to the point of threatening their ability to earn a livelihood, might gravitate toward idiosyncratic beliefs that lead them not to care about important components of their good such as friendship and family ties, and so on. Capability sufficiency for all might require policies that restrict or prohibit or mandate certain behaviors by them for their own good.

One might deny this last claim on the ground that whether a policy counts as paternalistic depends on its motivation, and a policy of restricting people's liberty so as to maintain capability sufficiency for all is not restricting people's liberty for their own good but rather precisely maintaining the social condition of sufficiency for all (Anderson 1999). However, this stance opens the door to lots of restrictions that usually get classified as paternalistic – barring heroin and other dangerous recreational drugs, barring greasy fried foods and sugary drinks that that are inimical to health, and barring lifestyles such as being a couch potato (too unhealthy) or playing extreme sports (too risky) might all be justified as necessary to maintain long-term capability sufficiency for all. If someone targeted for such restriction of liberty is not allowed to exempt herself on the ground that her considered judgment is that she prefers to live in ways that might result in her falling into a state of capability insufficiency, I submit the policy that takes this form is tantamount to paternalism.

If relating as equals is deemed to forbid paternalistic policies, the relational egalitarian facing the problems described in the previous paragraph would be constrained to address them by restoring conditions of social equality that self-harming behaviors erode without in any way restricting the self-harming behaviors. This takes us back to the issue whether or not respecting adult (normal, nonimpaired) citizens as free and equal, or as possessing equal basic dignity, rules out paternalistic interference or manipulation. Here is an argument to that conclusion. The argument is drawn from Quong (2011).

1 The state is morally bound to accord its normal adult citizens the status of free and equal citizens.
2 If the state accords its normal adult citizens the status of free and equal citizens, the state treats them as competent to manage their own lives.
3 Paternalistic action and policy carried out by the state treat some normal adult citizens as incompetent to manage their own lives.
4 Paternalistic action and policy carried out by the state fail to accord its normal adult citizens the status of free and equal citizens.

So,

5 The state is morally bound to refrain from imposing paternalistic action and policy on normal adult citizens.

This argument might be run either with the conception of paternalism as restriction of an individual's liberty against her will for her own good or with the conception of paternalism as action that aims to benefit an individual in some situation and that is motivated by an adverse judgment about the individual's ability to manage the situation competently for her own benefit. Either way, the relational egalitarian holds that in a society in which people relate as equals, an important component of this ideal is respectful treatment of everyone as free and equal by the state.

The relational egalitarian might hold that there is a stronger moral constraint against paternalism practiced by the state than against paternalism carried out among private individuals. The difference is that an individual, for example, will sometimes have reliable detailed information about what will serve his friend's interests in a situation, and will come by this reliable information in morally innocent ways. The individual can know enough to override the friend's judgment. In contrast, the state and its public officials will either not be in a good epistemic position to make a comparable judgment, or will be in the required good epistemic position only by taking actions that violate the person's right to privacy and hence should not be taken (Coons and Weber 2013: 14).

The idea of a free and equal person invoked in this argument is drawn from Rawls (1996). Ordinary citizens are free in having the moral powers to play fair with others and to adopt and follow and rethink a conception of the good, and equal in having these powers at a threshold or good enough level. So understood, premise 1 should be accepted.

However, premises 2 and 3 are problematic. The level of competence required to manage situations adequately varies from situation to situation. There are particular situations we may face, and types of recurring situations we may face, that may require a high level of competence, which some or many of us may lack. So the state's treating me as incompetent with respect to some area of my life need not involve denial of my status as free and equal as just characterized.

Moreover, according an individual the status of equal citizen, with an equal right to vote and stand for public office, along with other rights of basic citizenship, is compatible with judging the individual sufficiently lacking in competence for prudential agency to render some paternalism justifiable. So 4 should also be rejected.

To explore what the relational egalitarian might find to be problematic about paternalism, it will help to clarify the idea of equal status that might be thought to be threatened by paternalistic treatment. We should distinguish an offense against one's objective status as an equal citizen and member of society from one's subjective feelings of hurt or upset aroused by treatment accorded one by others that one perceives as insulting. The former involves an assurance, provided by how others regard one and treat one, that one is a member in good standing, entitled to basic justice and fundamental rights, someone whose interests should count for as much as anyone else's comparable interests in the determination of public policy. Jeremy Waldron affirms that one's dignity in this sense can be denied by hate speech such as racist epithets (Waldron 2012). Someone might hold that being the target of paternalistic treatment is in a similar way a denial of dignity and an offense against equal status.

But there is room to agree that dignitary harms can warrant legal protection to sustain the basic equality of citizens and yet to deny that paternalism, when based on reasonable adverse judgment about some aspect of some or all citizens' prudential abilities, and reasonably designed to prevent welfare shortfalls from arising from these deficits, qualifies as an offense against equal dignity (see de Marneffe 2006 for relevant discussion). The individual who is the target of paternalistic policy need not be regarded as lacking the traits that confer personhood status. Nor need her interests be discounted, as compared to the interests of other members of society – in fact counting her interests as just as worthwhile as the interests of others may require paternalism to promote her interests. The target of paternalism need not be denied equal citizenship status, or deemed unentitled to social justice. If she is being treated as less than an equal at all, this consists just in her being viewed as apt for paternalistic treatment – but it would beg the question flatly to assume that this violates the requirement to regard all persons as sharing an equal basic status.

The argument would have to proceed in the opposite direction. If on independent grounds one holds that each person has a right of self-sovereignty, a right to control her own life when her choices mainly affect herself, one might then infer that violating this right of self-sovereignty would be wrongfully disrespectful and should count as failing to treat the violated person as an equal (Feinberg 1986, and for criticism Arneson 2005). Or if paternalism is taken to be fundamentally treating someone as less than competent (in some particular ways), then one needs an argument that paternalism so construed is always wrong, before it would be plausible that paternalistic treatment of a person is in itself wrongfully disrespectful.

Regarding the latter construal, note that this negative judgment characterization of wrongful paternalism surely needs to be qualified, or else it is a nonstarter. Suppose I am climbing and break my leg. My partner quietly takes my pack from me and shoulders it himself as we limp down the mountain together. His treatment of me is motivated by a clearly and obviously justified negative judgment about my competence to make my way down the mountain without assistance. There's no insult here, much less wrongful insult, and no *prima facie* indicator of insult. We cannot easily fix the formulation by specifying that to qualify as paternalistic, the negative judgment motivating the treatment accorded to an individual must be unjustified (or unjustified given the evidence available to the agent according the treatment). The proponent of this idea of paternalism evidently has it in mind that within one's proper sphere of authority, others should not be questioning one's competence, whether their negative judgments would be warranted or unwarranted.

But why is questioning my competence wrongful to me if the judgment that I am incompetent in this way is well grounded? One might hold the contrary view, that treating someone on the basis of correct (or evidence-relative reasonable) empirical beliefs about his nature and circumstances is never in itself wrongfully disrespectful. If I am a bad parent, an ill-informed voter, a poor judge of what is good for me and how to attain what is good for me, and so on, treating me on the basis of these facts is never in itself inappropriate.

In passing, I note that the negative judgment characterization as stated here is too quick to dismiss the idea that provision of information that is material to choice, or more broadly, attempts at rational persuasion, cannot be problematically paternalistic. The negative judgment characterization manages the dismissal by requiring that the negative judgment about an individual that renders the treatment that follows paternalistic must be a negative judgment about the person's competence if fully informed. But treatment of a person might be thought to be objectionably paternalistic by virtue of proceeding from an unduly negative judgment about her competence to become appropriately informed about the situation (Tsai 2014). Trusting a person's competence to manage a situation usually includes trust in her competence to gather information and integrate it into her decision process.

Regarding the former construal, one might allow that forcing someone for her own good to do what she does not want to do can undermine her self-confidence and the confidence that others place in her, be experienced as unpleasant and demeaning, and often lead to other bads. Still, these acknowledgements would not plausibly ground a right of self-sovereignty, much less an absolute and exceptionless right. Taking into account all relevant negative and positive effects, restricting a person's liberty for her own good can make her life go better, in overall well-being, and sometimes radically improve her lifetime well-being. Since the right of self-sovereignty and the right to good well-being prospects are starkly opposed, supporting moral duties along the lines of the latter requires denying the claimed right of self-sovereignty.

Perhaps paternalism in itself (whether viewed as restricting someone's liberty against her will for her own good or as basing policies directed at people on adverse judgments about their competence) is not necessarily in conflict with the norm of treating all persons as social equals and as beings with dignity, but certain forms of paternalism might express or promote wrongful hierarchy according to relational egalitarian norms. For example, being denied wide freedom of choice in central areas of one's life such as romance and marriage, sexual activity, family relations, occupation and employment, friendship, and so on might be deemed incompatible with equal status. Those who are partial to the allowed narrow ranges of these types of activities are being treated as superior to those whose fundamental freedoms are tightly restricted. Another example might be narrowly targeted restrictions that give the plausible appearance of disrespectful animus toward particular groups.

It remains to consider the claim that even if paternalism in the relations among private individuals might be acceptable, state paternalism either will always be treating people on the basis of general presumptions, which might not apply to the individual case, or would have to involve state action to gather detailed particular information about individuals, which would violate their rights to privacy. Either way, the paternalism is disrespectful and wrong and has no place in a society of equals – or so it might be claimed.

This dilemma, I suggest, is not sharp. First, given that most laws must be coarse-grained and treat broad categories of people the same, it may be fair sometimes to prohibit a type of activity for the benefit of those whose liberty is restricted, even if some in the target group will be made worse off, not better off. The losses to these individuals may be fair in view of the gains to others. Second, it is an open question to what extent, in any given policy domain, it is possible or feasible for the government to gain relevant information about individuals who might be affected by policies it might choose that would sort these individuals into different categories receiving substantially different treatment. It also varies from case to case, to what degree the information that would be needed to sort people in a fine-grained way for policy purposes can be obtained, and if obtainable, can only be obtained by wrongful intrusion into people's private affairs.

Consider the possibility that by committing an antisocial criminal act that merits criminal punishment, an individual thereby incidentally reveals that he is probably making a mess of his life and failing also in prudential terms, and so may be punished, in part, with a view to improving his prudential prospects, even though it would have been wrong to impose coercively on him for his own good in this way had he not forfeited some of his rights by this criminal wrongdoing. Here the individual reveals by his conduct information about himself that arguably makes him more apt for paternalistic restriction. The state does not need to collect special information about him much less invade his privacy to get the information. Much the same would be true if the state established a standing policy whereby filing for personal bankruptcy triggers paternalistic state-imposed instruction in personal financial management.

4 Relational egalitarianism reaffirmed and restated

The discussion in the previous section has criticized claims that relational egalitarianism plausibly construed rules out all paternalism. Whether these criticisms succeed or fail, they do not gainsay the assertion that plausible versions of relational egalitarianism will oppose instances and types of paternalism that welfarist distributive egalitarianism embraces.

What is it to hold that people ought to relate as equals? This slogan requires interpretation. One might take an antipaternalist position to be component of what relating as equals involves. On this view, each person equally has her own life to live. She has duties, perhaps extensive and stringent duties, to refrain from harming others and indeed to assist them so they have freedom and resources to live in any of many significantly different ways. But each person has a non-delegable duty to form her own conception of the good life and to pursue it as she chooses by plans of her own devising (Dworkin 2000, chap. 6). Some paternalistic restriction of liberty might help an individual to live freely in her own considered way, so an exceptionless moral rule barring us from restricting a person's liberty for her own good or from acting on the basis of adverse judgments of her competence are implausible. But in a society of equals, people affirm and pursue different ideas of the good, and there is no single standard of objective well-being that can override the individual's choice of life plan pursuing goods as she weights them and goals that she herself embraces. Moreover, within a broad range, one's right to pursue one's own good in one's own way is not conditional on being prudentially competent. In a society of equals, we all extend the same basic rights of self-regarding freedom to all regardless of their choice-making and choice-executing talents.

From a welfarist standpoint, the world in which relational egalitarianism is perfectly fulfilled might well turn out to be one disastrously low in well-being. People leading squalid lives and people with poor native endowments of prudential competence will end up leading especially bad lives. We have strong duties to help save people in peril, and the relational egalitarian ideal interpreted in this antipaternalist fashion is just an ideology that distracts us from these duties or, even worse, wrongly tilts in favor of better-off individuals at the expense of the worse off.

The distributive egalitarian might insist that equality in power, rank, standing, and status are under modern conditions often instrumentally valuable means to distributive egalitarian goals, but not morally valuable in themselves. But this just restates the opposition between relational and distributive egalitarianism and does not resolve it. From the relational egalitarian standpoint, the welfarist egalitarian fails to register and respect the norm of equal liberty for all as the substance of justice, not merely as a possible tool for achieving it. As the philosopher Brian Barry once remarked, one man's *reductio ad absurdum* is another's *Q.E.D.*

Related topics

Paternalism and Education; Perfectionism and Paternalism; Paternalism and the Criminal Law; The Concept of Paternalism.

References

Anderson, E. (1999) "What Is the Point of Equality?" *Ethics* 109: 287–337.
Arneson, R. (1989) "Equality and Equal Opportunity for Welfare," *Philosophical Studies* 56: 77–93.
———. (1997), "Paternalism, Utility, and Fairness," reprinted in G. Dworkin (ed.) *Mill's 'On Liberty': Critical Essays*, Lanham, MD: Rowman and Littlefield.
———. (2005) "Joel Feinberg and the Justification of Hard Paternalism," *Legal Theory* 11: 259–284.
Cohen, G. (1989) "On the Currency of Egalitarian Justice," *Ethics* 99: 906–944.
Coons, C. and Weber, M. (2013), "Introduction: Paternalism – Issues and Trends," in P. C. Coons and P. M. Weber (eds.), *Paternalism: Theory and Practice*, Cambridge: Cambridge University Press.
de Marneffe, P. (2006) "Avoiding Paternalism," *Philosophy and Public Affairs* 34: 68–94.
Dworkin, G. (2013) "Defining Paternalism," in C. Coons and M. Weber (eds.) *Paternalism: Theory and Practice*, Cambridge: Cambridge University Press.
Dworkin, R. (2000) *Sovereign Virtue: The Theory and Practice of Equality*, Cambridge, MA: Harvard University Press.
Feinberg, J. (1971) "Legal Paternalism," *Canadian Journal of Philosophy* 1: 105–124.
———. (1986) *The Moral Limits of the Criminal Law, Vol. 3: Harm to Self*, Oxford: Oxford University Press.
Knight, C. (2009) *Luck Egalitarianism: Equality, Responsibility, and Justice*, Edinburgh: Edinburgh University Press.
Mill, J. (1978 [1859]) *On Liberty*, ed. E. Rapaport, Indianapolis: Hackett.
Parfit, D. (1995) *Equality or Priority?* Department of Philosophy: University of Kansas.
Quong, J. (2011) *Liberalism Without Perfection*, Oxford: Oxford University Press.
Rawls, J. (1996) *Political Liberalism*, New York: Columbia University Press.
Roemer, J. (1998) *Equality of Opportunity*, Cambridge, MA: Harvard University Press.
Scheffler, S. (2003) "What Is Egalitarianism?" *Philosophy and Public Affairs* 31: 5–39.
Segall, S. (2013) *Equality and Opportunity*, Oxford: Oxford University Press.
Shiffrin, S. (2000) "Paternalism, Unconscionability Doctrine, and Accommodation," *Philosophy and Public Affairs* 29: 205–250.
Stevenson, C. (1938) "Persuasive Definitions," *Mind* 47: 331–350.
Temkin, L. (1993) *Inequality*, Oxford: Oxford University Press.
Tsai, G. (2014) "Rational Persuasion as Paternalism," *Philosophy and Public Affairs* 42: 78–112.
Waldron, J. (2012) *The Harm in Hate Speech*, Cambridge, MA: Harvard University Press.

17
SHOULD THE CAPABILITY APPROACH BE PATERNALISTIC?

Serene J. Khader

Is the capability approach paternalistic? Should it be? The former question has attracted much attention, but the latter may be more apt. Capability theorists measure how well individual lives are going, and how just societies are, in terms of people's access to valuable abilities to be and do. Nearly all capability theorists claim that public policy should, in some cases, encourage people to value or exercise valuable functionings that they do not already exercise or value. Whether and how paternalistic the capability approach is depends on how often such policies are thought to be called for and whether such policies are justified as being in the interests of their targets.

Since the capability approach (hereafter CA) is still under construction, determining the extent of its paternalism is not a unidirectional exercise in which we discover the implications of an existing theory. The CA is a practical approach to measuring well-being, motivated by certain normative commitments about freedom and functioning. The CA aims to justify access to worthwhile functionings in a way consistent with respect for freedom, so the political philosophy behind it need not be worked out independently of paternalism concerns. The paternalism implications of a given view about freedom, well-being, or the principles of justice can count as a strike for or against incorporating that view into the CA.

How paternalistic the CA is depends largely on its explanations of why and when people deserve freedom to not engage in valuable functionings. In what follows, I lay out four potential views about the value of freedom to engage in disvaluable functionings and discuss how consistent each is with two normative commitments that motivate the CA: commitments to treating human beings as agents and nonideal theoretical commitments to opposing oppression and deprivation. Though I do not take a particular view of why capabilitarians should value freedom to engage in disvaluable functionings, I suggest that capabilitarians can value this freedom without taking it to be a first-order constituent of well-being. I also uncover a tension between thoroughgoing opposition to one form of paternalism (cost-soft paternalism) and the concerns about agency and nonideal theory motivating the CA. I call the freedom in question "content-neutral freedom," because the CA readily grants non-interference to those who want to exercise valuable functionings; we want to know what the CA has to say about whether, when, and why those who *reject* valuable functionings deserve non-interference.[1]

1 What is the capability approach?

The CA, developed by Martha Nussbaum and Amartya Sen, is a framework for assessing individual well-being and evaluating social arrangements. It is primarily a view about the currency, or metric, to be used in such evaluations. The preferred metric for understanding how well human lives are going is valuable abilities to be and do, rather than, for example, subjective well-being or income. Imagine a scenario where two people, Amina and Saba, are equally happy and possess equal finances. A road leads from the village Amina lives in to the school, but Saba lives in an isolated area with no access to education. Unsophisticated versions of subjective welfarism or resourcism say both are doing equally well, and perhaps that no injustice has been done to either. In contrast, the CA says (assuming education and mobility are abilities worthy of social promotion) that Saba is deprived. The CA also differs from a pure "functioning" approach, though there is some controversy about how robust the difference is (see Arneson 2000; Claassen 2014). Functioning approaches state that what matters is beings and doings, i.e., whether people actually engage in valuable functionings, rather than whether they *could*. On a functioning approach, Amina is deprived if she does not go to school, despite possessing the opportunity to go.

The CA is not a theory of distributive justice that gives rules about how capabilities should be distributed or institutions should be structured. Nussbaum's (2001) variant of the CA incorporates a sufficientarian distributive principle. She argues that societies do not meet minimal criteria for justice if they do not enable a threshold level of certain capabilities for their members, but she is open to additional distributive principles. Sen distances his variant of the CA from distributive justice altogether, seeing the focus on ideal distributive principles as a pragmatic impediment (Sen 2009).

Though the CA is not a theory of distributive justice, it assumes certain normative commitments, loosely describable as liberal. Nussbaum and Sen take protecting and enabling the exercise of freedom to be of very high political priority. Both refuse to specify capabilities in a way that tolerates only a narrow range of individual conceptions of the good life. Capabilities matter, for Nussbaum and Sen, partly because they help individuals live in ways they first-personally deem valuable. Both also argue that promoting capabilities rather than functionings is a way of respecting individual freedom.

Still, there is no consensus about how to articulate the foundational conceptual links between capabilities and liberalism. Sen usually speaks of the capabilities as (exhaustively) constitutive of freedom (Sen 1999, 2009), where Nussbaum describes them as supports for it (Nussbaum 1999). We might say that, in addition to endorsing a certain metric of justice, capabilitarians hold that people who have opportunities to exercise valuable functionings, and to deliberate about whether and how to do so, are, ceteris paribus, more free than those who do not. I discuss the specific capabilitarian articulation of these points in terms of agency in the next section. An important undertheorized question concerns whether the CA should be inserted into an existing liberal political theory (in the place, for example, that the primary goods occupy in Rawls' theory) or become a normative political theory of its own. On the former response, freedom constrains how the capabilities should be pursued and can be worth protecting even if it makes individual lives go worse; on the latter, freedom just is the possession of options to be and do that contribute to individual lives going well.

Capabilitarians do not agree about which capabilities are valuable, and, though Nussbaum proposes a universal list, Sen allows only context-specific lists (Robeyns 2003; Sen 2004; see Claassen 2011). Nussbaum and Sen agree that participatory and democratic methods should play a role in determining which capabilities ought to be pursued and how. The list disagreement should not be confused with a lack of logical commitment to the view that some capabilities are more valuable than others, since assessing any state of affairs in capabilitarian terms requires selecting some valuable capabilities.[2] To say that the society that eliminates malaria is better arranged than the one that makes malaria easily contractable requires taking a stance about value. Both Nussbaum and Sen extensively discuss abilities to be healthy, to move from place to place, to influence one's political community as an equal, and to exercise practical reason as examples of valuable capabilities. Nussbaum and Sen also both defend the CA on the grounds that social institutions should support more than negative freedom, because, as Nussbaum puts it, "choice is not pure spontaneity" (Nussbaum 1999: 45; see also Anderson 1999). Genuine choice requires having acceptable options, not merely not being interfered with. This suggests that the CA places a special value on functionings that are constitutive of basic well-being, rather than those constitutive of excellence.

2 Nonideal and agency motivations for the capability approach

The extent of paternalism capabilitarians can accept depends partly on why they have adopted the CA to begin with. The CA emerged in the 1980s and 1990s as an alternative to three other approaches to measuring well-being: the international development measure GDP, hedonist or desire-satisfactionist utilitarianism, and the Rawlsian notion of primary goods. Two cross-cutting concerns, agency concerns and nonideal theoretical concerns, motivated criticism of these views.

Capability theorists, especially Sen, allege that other metrics of well-being treat people as moral patients, or containers for happiness and functioning, rather than active participants in evaluating and changing their own lives. Utilitarian approaches treat people as vehicles for the satisfaction of their own desires (Sen 1973, 1980, 1999), and approaches that focus on commodities risk assuming the commodities are intrinsically valuable. Though Nussbaum does not make these particular arguments, she treats living in a way one chooses and evaluates positively, rather than one in which one engages grudgingly or unthinkingly in valuable activities, as necessary for a life that goes well (Nussbaum 2001). For both, an acceptable theory of well-being will highly value people's abilities to set, revise, and achieve ends. Within this emphasis on agency, we can identify four distinct points.

First, the power to set and revise ends possesses intrinsic value. Second, respecting persons requires accepting that people may have ends besides their own welfare, especially self-sacrificing and tuistic ends. To treat only welfare-enhancing ends as important is to fail to appreciate the importance of first-personal evaluation in human life (Sen 1973, 1982, 1999; see also Cudd 2014). Third, people should not be treated as incapable of revising their ends and perspectives. This, too, would risk treating people as passive bearers of goals (Sen 1999, 2002, 2009: 87–124). Fourth, agency is instrumentally valuable for securing valuable capabilities besides agency. For example, people who can influence their political environments are more likely to possess stable access to food, safety, etc. (Sen 1999). Much of Sen's critique of GDP as a development indicator consists of empirical demonstrations of the fact that approaches to poverty alleviation that undermine individual decision-making power and devalue democratic processes are ineffective. Undergirding all four of these views is also the notion that the ability to set and revise ends can be practiced and strengthened through collective deliberative processes. Individual agency can be enhanced through interpersonal practices of value-clarification. The CA incorporates these commitments to agency by a) treating abilities to be and do rather than actualized functionings as the currency of justice and b) envisioning a significant role for deliberative practices in securing the capabilities.

A second set of motivations behind the CA can be understood as nonideal theoretical. Nonideal theories, in my usage, focus on rectifying existing injustice rather than offering a picture of justice (see Mills 2005). Inadequacies in real-world poverty indicators motivated the CA initially. According to Sen, agnosticism about ideal justice is an advantage of the CA (Sen 2009). Nussbaum defends her argument for securing basic capabilities for all on the grounds that it is a feasible proposal for ending severe deprivation in our historical moment. Though it is possible to imagine a capabilitarian ideal theory, most of its advantages over resourcist views (views that assess well-being in terms of access to goods) lie in its usefulness under nonideal conditions.

Two of the most common arguments for the CA, the argument from adaptive preferences and the argument from differential conversion factors, gain much of their appeal from concern with real-world oppression and deprivation. The latter argument states that individuals convert goods into functionings at different rates, so equality of resources can result in unequal amounts of functioning. Differential conversion rates are supposed to motivate a worry that equality of resources is not the kind of equality we care about. Similarly, the argument from adaptive preferences states that the CA is capable of identifying a moral problem utilitarianism cannot: the fact that people diminish their desires in response to oppression and deprivation. Capability theorists say that justice should care about the well-being such people have lost, even if they themselves do not.[3] For example, Sen argues that a very poor person may make herself happy by contenting herself with "small mercies"; capability theorists will say she is deprived if she lacks key capabilities, even if she feels happy and does not desire more than she has.

But why see responding to adaptive preferences and differential conversion factors as a good thing? One answer is that responding to oppression and deprivation is morally urgent. The argument from differential conversion factors is particularly persuasive if we worry that a seemingly egalitarian resource distribution will perpetuate existing deprivation and inequality. The most popular example of differential conversion factors states that giving the same transportation budget to a person in a wheelchair and an ambulant person will result in different levels of mobility. The built environment favors the latter, but equal resource distribution will not reveal the injustice done to the former. The conversion factor argument is sometimes discussed as concerning human diversity rather than oppression, but a capability focus is especially likely to reveal ways existing background institutions are designed to make it easier for some members of society to meet their needs than others. The pull of the conversion factor argument on those who are not particularly concerned with oppression and deprivation is much weaker; it is merely a concern about making sure that differences caused by nature and luck are attended to, one that might be attended to relatively well by resourcism.[4]

We can make a similar case about the argument from adaptive preferences, and making it can help us reveal some particular features of capability theorists' notion of agency. Getting the "wrong" answer about the well-being of people who adjust their desires to bad conditions is not likely to count as a significant strike against an ideal theory, because many of the conditions that cause adaptive preferences will not exist in the world it imagines.[5] Adaptive preferences are especially a problem for assessing well-being in a world where oppression and deprivation exist. The tendency to see adaptive preference as a political problem, rather than to see the political prescription to respond to them as condescending and autonomy-denying, is likely to track a refusal to think of respecting existing beliefs and desires as the best way to respect agency. If one thinks that desires are not static or that people's existing desires and behaviors do not transparently reflect what they care about or could come to care about, one may wish to distinguish respecting agency from respecting existing desires. This idea about agency dovetails with the third agency motivation I described above; the idea that respecting people's existing desires may continue oppression and deprivation (and that this is a bad thing) can be consistent with respect for agency, as long as respecting agents is distinguished from respecting their existing desires.[6] This notion that adaptive preferences interfere with the type of agency that capability theorists seek to protect can be understood as undergirded by a refusal to idealize human agents. According to Mills (2005), nonideal theorists refuse to imagine away limitations on the capacities of actual human beings caused by oppression and deprivation.

Absent concern about oppression and deprivation, the case for simply revising other metrics of justice rather than adopting the CA is much stronger. For example, one might simply say that it is important to give extra resources, on a case-by-case basis, to people with unfavorable conversion rates. Thomas Pogge (2010) argues that this approach is preferable to the CA because it sidesteps the (in Pogge's view objectionable) claim that people who need more resources to achieve functionings are doing less well than others. Without a background context of oppression and socially caused deprivation that allows such judgments to concern social institutions rather than individual deficits in native endowments, Pogge's disrespect concern is more persuasive. Reasons to prefer the CA over utilitarianism also weaken absent concerns about nonideal conditions. Under ideal conditions, values and desires that do not promote individuals' well-being may appear as reasonable variations in conceptions of the good rather than potentially harmful adaptive preferences.

3 Perfectionism and paternalism

To know what it would mean for the CA to be paternalistic, we need to distinguish perfectionism from paternalism. Perfectionism concerns the endorsement of a conception of the good life; paternalism concerns the *means* public institutions use to promote what is taken to be in people's interests. Though perfectionists often specify that the good is the development of human nature or that societies should maximize excellence, the CA need not be perfectionist in these senses. Though they disagree about which capabilities are valuable, capability theorists cannot avoid the perfectionist view that some functionings are objectively good. Nor can they be neutral among views about what contributes to human well-being (see Deneulin 2002; Khader 2011; Claassen 2014; Khader and Kosko forthcoming).[7] To know whether one individual is doing better than another, or to know whether society is doing as well as it could, capabilitarians need to pick some set of functionings and see whether opportunities to engage in them are available.

It may seem that promoting capabilities rather than functionings eliminates the possibility of perfectionism. Yet consider the capability of bodily health. Note that even this framing is not evaluatively neutral; "health" designates a desirable functioning, and social institutions that promote the associated capability will make health readily available, rather than ensure that all bodily states (healthy and unhealthy) are equally available. Public neutrality about many capabilities may also be practically impossible. The capability of health requires access to medical care, nutrition, support networks, knowledge about how to care for oneself and others, etc. Making it harder to live unhealthily may just be an opportunity cost of making the health capability widely available. Conceiving of health as a capability does advise against forcing people to eat well or accept particular medical procedures. But it is one thing to permit opting out of a valuable functioning and another never to value any functionings over others.

Non-neutrality about what makes a life go well may seem at odds with the (not only) liberal value of living according to one's own lights. Yet protecting a multiplicity of ways of life may be enacted by selecting and formulating capabilities such that they can be justified and achieved in a variety of ways. Nussbaum adopts this strategy, arguing capabilities should be multiply realizable and justifiable (Nussbaum 2001: 75–77). She also limits perfectionism by arguing that the CA need only promote capabilities up to a threshold level,[8] such that the need for a stance about the relative value of getting vaccines or not getting vaccines does not require one about the relative value of being a dancer or a mechanic.

A society can be perfectionist without being paternalist (see Raz 1986; Khader 2011; Claassen 2014). Paternalism involves individuals' freedom being abridged for their own sakes, and consists in the choice of certain *means* to effect valued outcomes, rather than the mere valuing of those outcomes. To answer the question of what level of paternalism capability theorists should accept, we need a definition of paternalism. Let us say that paternalism occurs just in case a person experiences a nonvoluntary reduction in liberty in order to promote her interests.[9] Since some definitions of liberty eliminate the possibility of paternalism altogether by suggesting that a person cannot *really* will against their own good, let us also stipulate that a person can, in principle, will things that are in their interests and things that are not. Additionally, rather than drawing a hard line between what reduces liberty and what does not, we can understand the level of paternalism of a given policy as increasing as costs attached to an agent's desired course of action increase.

This definition counts some non-coercive acts as paternalistic. This means that capability theorists' common reply to paternalism allegations – that, by focusing on capabilities, they avoid *forcing* people to function – responds only to a narrow range of paternalism concerns. Incentives, taxes, fines, and social sanctions can all constitute paternalistic interference in the lives of those who do not already want to engage in valued functionings (see Deneulin 2002). For example, guaranteeing the capability of health may involve fining people who do not secure health insurance, taxing unhealthy products, making such products more difficult to access in the physical design of stores, or providing financial incentives for receiving certain treatments. Though none of these confronts the agent who does not desire health with imprisonment or the threat of physical force, and none (provided the agent is not financially desperate) makes it impossible to live an unhealthy life, each decreases the freedom of agents who wish to live unhealthily. A smoker's freedom to smoke – and to use their money in other ways – is lessened by taxes on cigarettes, even if it is not totally vitiated. Using this scalar understanding, we can understand paternalism as hard to the extent that it imposes high costs on behavior inconsistent with well-being and soft to the extent that it imposes low ones. Cass Sunstein understands paternalism on this type of continuum (Sunstein 2014). Sunstein's way of drawing the distinction between hard and soft paternalism does not map onto the traditional distinction. Soft paternalism is usually understood to consist only in acts that check for the voluntariness of actions or interfere with nonvoluntary conduct. In contrast, the cost-soft understanding I adopt counts low-cost interventions that aim at changing people's behavior and desires as soft.

4 The capability approach and content-neutral freedom

Concerns about the effectiveness and means of paternalistic intervention are not the only limitations on paternalism available to capability theorists. They may view freedoms not to engage in valuable functionings as worth protecting for independent moral reasons. Let us call the opportunity to have one's choices about one's own life be decisive regardless of whether they are choices to function well "content-neutral freedom." For capability theorists, difficulty respecting content-neutral freedom arises primarily in cases where an agent does not want to engage in a valuable functioning, or wants to engage in a disvaluable functioning. Agents who already want to engage in valuable functionings will face little, if any, paternalistic intervention in a capabilitarian society. Content-neutral freedom is nonidentical with autonomy understood as the ability to exercise the capacity for choice in one's life (see Raz 1986). An individual who lacks opportunities to engage in disvaluable functionings may retain opportunities to make decisions just by being able to choose how to combine and enact valuable functionings.

Capability theorists want to assign some value to content-neutral freedom; it is difficult to make sense of the focus on capability over functioning without it. Yet in order for this focus to be nonarbitrary, and in order to have a principled way of understanding when focus on functioning is appropriate, we need an understanding of *why* people deserve not to face high costs for not engaging in valuable functionings. An important question here concerns whether opportunities not to engage in valuable functionings are first-order constituents of well-being – that is, about whether an individual life can be said to go worse without a given opportunity to engage in a disvaluable functioning.

Before discussing four views capability theorists can take to valuing content-neutral freedom, we should put aside some common strategies that promise to eradicate paternalism from the CA almost entirely. According to one, functionings gain their value from being chosen. This is an unsatisfactory explanation of the value of content-neutral freedom for three reasons. First, only a small subset of valuable functionings, such as sexuality, must be unforced to retain their value. The value lost by force may also sometimes be outweighed by the value of the functioning itself (see Arneson 2000); being incentivized to eat right may be worse than eating healthy food completely freely, but the presence of strong incentives to eat healthily does not vitiate the value of health. If only some functionings lose their value when engaged in absent opportunities to reject them, and if the value lost may be outweighed by the value of functioning, then, without some other reason for valuing content-neutral freedom, the CA will sometimes endorse paternalistic freedom reduction to promote functioning.

A second strategy for avoiding paternalism is simply to claim that the only valuable functionings are ones an agent herself values. This strategy is often appealed to by capability theorists, facilitated by Sen's description of capabilities as freedoms, and a definition of freedom as the pursuit of what one values. But, because of the need to identify objectively valuable capabilities I described in the perfectionism section, a policy aligned with this strategy would be more likely to be resourcist, or even utilitarian, than capabilitarian. Additionally, Sen's more common locution is that capabilities are what people value and "have *reason* to value" (Sen 2002: 52). It seems impossible for this locution to avoid reintroducing the idea that some functionings are valuable irrespective of whether people actually value them (Khader and Kosko forthcoming).

5 Capability as such

There are four more promising routes through which capabilitarians might assign value to content-neutral freedom. I will discuss the scope of permissible paternalism each generates, as well as the extent to which each view responds to the agency-respecting and nonideal theoretical motivations of the CA. One, valuing "capability as such," is developed, though not endorsed, by Ian Carter in imagining an antipaternalist CA. On the capability as such view, the currency of justice is all abilities to be and do, rather than a putatively objectively valuable subset. Carter also stipulates that the view includes a sort of pluralism-maximizing principle stating that as many functionings as possible should be available (Carter 2014: 99). The view removes promoting individual well-being from the menu of legitimate state motives, and thus eliminates paternalism by eliminating much of the paternalist motive. If the defender of capability as such draws on autonomy as a justification for maximizing available capabilities, paternalism to protect people's psychological capacities for choice may sometimes be called for.

The capability as such view precludes much paternalism but may justify heavy restrictions on individual freedom for nonpaternalist reasons. The most practical way to ensure that a wide variety of ends are available may be to promote certain "all-purpose functionings" – like nutrition, education, and relationships.[10] Additionally, if people's desires happened to converge, the capability as such view would seem to support reducing people's freedom to reduce clustering and maximize pluralism. The extent of agency protected by the capability as such view thus depends largely on empirical factors. It is also unclear that the standards for interpersonal comparison offered by the capability as such view will track intuitions about what counts as deprivation, since only the quantity of capabilities matters, irrespective of whether these are capabilities to be nourished, go hungry, or paint masterpieces. The capability as such view may be a CA only in name, given that it is not particularly responsive to the nonideal and agency concerns of the CA – and given that it does not offer an intuitively plausible evaluative space for interpersonal comparisons.

6 Opportunities to engage in disvaluable functionings as components of well-being

A second way of theorizing the value of content-neutral freedom understands opportunities to engage in disvaluable functionings to directly contribute to a life that goes well. This view should be distinguished from the view that disvaluable functionings, such as malnutrition or illiteracy, themselves contribute to well-being. A view that disvaluable functionings were intrinsically valuable would both be highly counter-intuitive and, if the disvaluable functionings were thought to be as well-being-enhancing as their opposites, prevent the CA from offering political prescriptions at all. Rather, this second view states that, ceteris paribus, a person who has the ability to engage in a given functioning and the ability not to engage in it has a better life than a person who has only the former.

The extent of paternalism permitted by this view varies from functioning to functioning. It is plausible that the ability not to vote combined with the ability to vote enhances a person's life more than compulsory voting. It is less plausible that a person who has the opportunity be sick or healthy is better off than a person who lacks the opportunity for illness. To know just how much paternalism is permitted, we may need a relatively comprehensive view that ranks various capabilities, the choice not to engage in them, and their contributions to a life that goes well. However, intuition suggests that the view is implausible about functionings constitutive of basic well-being, such as nutrition. The view that opportunities to engage in individual disvaluable functionings makes a life go better will likely have to exclude opportunities to forego basic functionings to be plausible. A plausible version of the view will likely recommend strong paternalism where functionings constitutive of basic well-being are at stake.

The opportunities to engage in disvalued functionings view also seems inconsistent with the subset of agency concerns that have to do with allowing people to care about ends besides their own well-being – though it may be objected that people cannot care about anything at all without basic functionings. On the other hand, basic functioning paternalism may be consistent with some nonideal concerns. Nonengagement in basic functionings is especially likely to be a result of socially caused deprivation. Nussbaum argues for a presumption against the notion that people's nonengagement in functionings like nutrition and sanitation is voluntary (Nussbaum 2001: 44, 92–94), and I argue elsewhere that, even if nonengagement is voluntary, this is not a reason to assume that people's rejection of such functionings is durable in the face of information and good options (Khader 2011). A nonideal view of agency suggests that imposing some costs on nonengagement in basic functionings may even enhance agency by encouraging value reflection.

7 Content-neutral freedom as condition of an autonomous life

The third and fourth ways of valuing content-neutral freedom in a capability theory do not take individual opportunities to engage in disvaluable functionings to directly contribute to a life that goes well. Instead, content-neutral freedom is generally worth protecting because of the *enabling* role it plays in the existence of other goods, goods that are either constitutive of, or background conditions for, well-being. To motivate these other views, without yet describing either, we can think of cases where reducing an individual's content-neutral freedom to promote functioning seems objectionable. Consider a person who lacks the freedom to starve. The idea immediately conjures intuitively noxious images of a person being forced to eat.

There are ways to hold onto the intuitive noxiousness without claiming the opportunity to starve is valuable. For example, we might say that individuals deserve the prerogative not to eat, because individuals deserve to be able to live lives that are bad for them. This idea, unlike the above views about the value of content-neutral freedom, allows that a person's life is not directly improved by having the opportunity to starve. Lest the idea that opportunities to starve do not directly contribute to well-being seem paradoxical or condescending, we can remember that many valuable life-plans do not manifest well-being, even by their own authors' lights. Self-sacrifice (such as fasting) would not be self-sacrifice if it did not involve foregoing something self-beneficial. We may want to design social institutions in a way that allows the choice of morally valuable ends that are not well-being enhancing. Another path is to see the intuition that forced eating is noxious as triggered by something other than the intrinsic benefits to the agent of the opportunity foregone. So, for instance, the idea of force-feeding might conjure the image of a political prisoner, and protecting the opportunity to starve might be a way of protecting people from this type of control. Rather than saying that opportunities not to function well directly add to individuals' well-being, capability theorists might see content-neutral freedom as helpful in protecting against certain common threats to well-being.

My point so far is only to suggest that there are ways to think people deserve content-neutral freedom, and to value it alongside valuable functionings, without claiming that opportunities to engage in disvaluable functionings themselves contribute to lives going well. Now that we have seen this, I describe a third view about the value of content-neutral freedom. According to it, freedoms not to engage in valuable functionings are means to achieving autonomy as non-alienation. Serena Olsaretti, who defends such a view (2005), argues that endorsement is a necessary but insufficient condition of a life that goes well. Rather than claiming that any particular opportunity to engage in a disvaluable functioning contributes to well-being, Olsaretti claims that feelings of non-alienation improve lives that already exercise valuable functionings. Olsaretti sees certain functionings as necessary constituents of well-being, and an agent's endorsement of a life that exhibits such functionings as adding more well-being still. Endorsement, for Olsaretti, is an attitude toward one's life in general rather than specific functionings (this differentiates Olsaretti's view from the athletic, choice-focused view G.A. Cohen attributes to capability theorists),[11] and freedoms not to function well are valuable only insofar as they contribute to the good of endorsement. If people experience themselves as routinely prevented from doing what they want to do, Olsaretti claims, they will lose their sense that they are the authors of their own lives.

This non-alienation view takes content-neutral freedom to be a valuable means to endorsement, and so the extent of paternalism it permits will depend heavily on situation-specific empirical facts. Persistent cost-hard paternalism across a variety of domains of life seems likely to have cumulative endorsement-undermining effects, especially since people chafe most strongly at policies that are costly to them. On the other hand, domain-specific hard paternalism is likely to be permissible in some cases, especially when there is reason to expect that individuals will come to endorse its effects (and its means) later. Cost-soft paternalism seems especially permissible on the non-alienation view, since agents' feelings of violation are both likely to track costs and constitute costs themselves, and the non-alienation view locates the wrongness of paternalism in its evocation of subjective feelings of being infringed upon.

The non-alienation view leaves much latitude for paternalism of both kinds concerning basic functionings – especially if it exists alongside latitude to opt out of less basic ones. Since endorsement only adds value to a life that includes valuable functionings, it is difficult to argue that protecting opportunities to opt out of basic functionings, like health, is more important than promoting the functionings themselves (Olsaretti 2005: 105). Additionally, subjective endorsement is unlikely to arise in people who lack basic functionings, both because ill health, malnutrition, etc., are likely to produce life dissatisfaction and because the psychological capacities and states that enable endorsement require some actual functioning (Olsaretti 2005: 96–97).

The non-alienation view seems highly consistent with the capabilitarian emphasis on respecting and promoting agency. However, discouraging paternalism that people will notice and chafe at is potentially at odds with attending to the concerns about responding to oppression and deprivation I mentioned earlier. Because of adaptive preferences, ideology, and other status quo biases it may seem that those most likely to bristle at paternalism are those who are likely to benefit. A woman in a patriarchal society who believes she deserves less than men, for example, may feel arrogated by attempts to equalize her capabilities. Making the acceptability of paternalism responsive to people's subjective feelings also seems to give special weight to opposition to paternalism that is initially felt very strongly, and such a weighting may be morally arbitrary.

However, the non-alienation view itself suggests an answer to these worries: when a functioning is morally urgent, the means of paternalism should be tailored so as not to encourage alienation. The view may thus attempt to reconcile nonideal theoretical and agentic concerns by recommending increases in people's engagement in valuable functionings by means that will enable *them* to come to value those functionings. The pull here is toward agency-engaging cost-soft paternalism. The non-alienation view may also discourage paternalism in cases of voluntary self-sacrifice. The perception of social institutions as overreaching is likely to increase if valuable self-sacrifice is not permitted, and this is a threat to everyone's endorsement. Additionally, strong feelings of non-endorsement can count as reasons not to subject self-sacrificing people to paternalism – at least in cases where they do not sacrifice basic functionings.

8 Content-neutral freedom as institutionally valuable

A fourth view takes content-neutral freedom to be important for just institutions. Though promoting well-being is a goal of just institutions, not all values they promote must themselves be constituents of well-being. Justice involves coordination, and social institutions are patterns of relationship. There are therefore reasons to value opportunities that make the right types of institutions possible, irrespective of whether any individual life goes better with or without them. Such opportunities may even count as intrinsically valuable, despite contributing only instrumentally to a life that goes well. Opportunities to form the right types of relationships can be understood to be worth protecting because they express intrinsic values relevant to institutions or because they provide individuals with intrinsically valuable forms of standing – without being constitutive of individual well-being, and without contributing to individual well-being in every case. If content-neutral freedom protects the right types of social relationships, and these relationships have intrinsic value for political standing but instrumental value for well-being, capabilitarians may endorse the institutional view for one or both of two reasons: because the capabilities constitutive of well-being are threatened by the lack of content-neutral freedom in a wide enough range of cases to justify a general stance protecting it, or because of independent commitments to some egalitarian theory of justice.

The view of content-neutral freedom as institutionally valuable holds that non-interference in self-regarding engagement in disvaluable functionings is crucial for a society that rejects unjust patterns of relationship. Philip Pettit (2001) defends a focus on capability rather than functioning on the grounds that it protects individuals from domination. According to him, real-world cases where all but valuable functionings are removed from an individual's option set are inevitably ones where some power could turn against her at will. Being required to engage in valuable functionings also means being vulnerable to being forced *not* to function well. Sen himself makes different but compatible institutional arguments. He argues that freedom *from* governments is required to sustain background contexts in which valuable capabilities are available. According to Sen, poverty alleviation practices without (moderated) markets and democracy are likely to be ineffective because resources and political processes can be hijacked by elites. As Sen argues in his empirical work, famines only occur in non-democratic countries, because democracy ensures the sensitization of elites to the concerns of nonelites (Sen 1999). It is not clear, however, whether the freedoms he has in mind – freedoms of association, speech, press, etc., qualify as content-neutral freedoms. Surely these freedoms need to be expressible in content-neutral ways to exercise their protective function. Freedom of speech cannot protect against tyranny if it is protected content-dependently.

However – and raising this concern illuminates the scope of paternalism recommended by the institutional view – nondomination concerns may encourage paternalism relative to functionings that protect against domination and oppression. Some such functionings are necessary for democratic participation (see Anderson 1999), but given that domination can occur interpersonally, paternalism toward other functionings may be warranted. For instance, if income-generating activities protect women from domestic abuse, domination worries may recommend promoting them paternalistically.

The institutional view responds to agency concerns about treating people as means to their own well-being by preserving individual prerogatives to pursue non-well-being-oriented ends. Since the aim of the institutional view is to combat domination, it foregrounds the nonideal theoretical goal of ending oppression. Further, to the extent domination is a force that keeps people's worldviews partial and underdeveloped, as Sen clearly believes it is (see Sen 2009: 157–164, 242), the paternalistic interventions the institutional view recommends will recognize that unjust conditions pose limitations on human agency.

9 Conclusion

I have described four views about the value of content-neutral freedom capability theorists can adopt, the range of paternalistic intervention each justifies, and the extent to which each is compatible with concerns motivating the CA. Though I have taken no stance about which view capability theorists should adopt, and the last three views may be combined, my analysis here permits three observations about attempts to limit capabilitarian paternalism.

First, capability theorists who wish to limit paternalistic intervention aimed at promoting basic functionings will probably need to theorize beyond a list of first-order constituents of well-being. Making judgments about when to engage in paternalism on a theory that treats opportunities not to engage in valuable functionings as themselves valuable may require a comprehensive view of well-being in which weights are assigned to various valuable and disvaluable functionings. Such weightings are difficult, and potentially politically objectionable, to develop. Weightings would likely suggest that the well-being added by being paternalized into exercising a functioning will outweigh the well-being added by the opportunity to engage in its disvaluable opposite. Knowing why people should be able to pursue ends besides well-being also requires more than a list of first-order constituents of well-being. It may be necessary, in explaining why capabilitarians should value content-neutral freedom, to refer, as Olsaretti does, to value attitudes about one's life as a whole or to concerns about just institutional design.

Still, and second, limiting paternalism in the CA does not require generalized opposition to public intervention, or the view that functionings are worthless unless they are freely chosen. Concerns about oppression and domination, and concerns about the role of non-alienation, push against the presumption that paternalistic action is generally a bad thing. Instead, these nonideal concerns suggest reasons to choose the means of paternalism carefully. Forms of paternalism that threaten alienation do so precisely by imposing costs agents see as unacceptable. Forms that threaten domination do so by relying on means that are insensitive to the paternalized agent's wishes.

The fact that opposition to interference as such need not ground limitations on paternalism is good news for capabilitarians because of my next (third) observation: the nonideal theoretical motivations of the CA require some tolerance of paternalism. Of course the reductions in freedom required to reduce oppression and deprivation may be justified on other-regarding rather than paternalistic grounds; providing collective goods, such as a malaria-free environment (see Olsaretti 2005), or encouraging a more fair distribution of social advantages may be justified in terms of what citizens owe to one another.[12] Still, worries about replicating existing inequalities through facially egalitarian policies are at odds with a strong presumption against paternalistic intervention. Under real-world conditions, protecting people's freedom to do what they are already doing may just be a way of preserving existing oppression and deprivation. Accepting some paternalism to reduce oppression is consistent with the nonideal notion of agency to which capability theorists often appeal. Where it seems to libertarians and some liberals that respecting people's ability to form and revise conceptions of the good means non-intervention, capability theorists often suggest that non-intervention can also be autonomy-disrespecting. Treating people as vehicles for their existing preferences may itself be construed as a way of denying their agency. The idea that people may not truly embrace their existing plans and desires is associated with forms of positive freedom theory that define away paternalism by suggesting that only "real" desires are desires for well-being. But rather than suggesting that all rejection of valuable functionings is nonvoluntary, the nonideal notion of agency suggests only that people can come to revise their preferences and values when social institutions engage them as deliberators. The nonideal theoretical motivations of the CA seem to recommend cost-soft paternalism and even begin to provide a framework for justifying it.[13]

Related topics

Egalitarian Perspectives on Paternalism; Paternalism and Autonomy; Perfectionism and Paternalism.

Notes

1 See Carter (2014) for a discussion of how most capabilitarian notions of freedom are content-dependent.
2 See Khader and Kosko (forthcoming) on the implicit perfectionist commitments of Sen's view.
3 Khader (2011, 2012, 2013) and Begon (2015) offer paths to claiming adaptive preferences are unreliable that do not involve understanding people with adaptive preferences to be rationality deficient.
4 Rawls postpones concerns about differential conversion rates to the legislative and judicial stages rather than leaving them out altogether.
5 The capability literature does not focus on Elsterian adaptive preferences, wherein people unconsciously downgrade options that have ceased to be available (Nussbaum 2001; Khader 2011).
6 Views that emphasize people's right to live in ways they deem valuable, including liberal ones, utilitarian ones, and the CA, face challenges about how to handle adaptation cases of various kinds. Focusing

on what people might want under idealized conditions, as some utilitarians and liberals do, or giving weight to what people should or could come to want, as perfectionists do, all threaten to disrespect persons by suggesting that their existing values and reasoning processes are defective. A challenge for views that want to respect autonomy or agency while questioning existing desires and values is to offer some reason to weigh ideal or later desires and values over existing ones. Nussbaum discusses these challenges and the theories subject to them at length (2001: 111–161), and Sen offers one response (2009: 274–276).

7 It is worth noting that claiming that different lists will be useful in different contexts, as Sen does, does not eliminate the problem of having to prefer some functionings over others.

8 Nussbaum also attempts to limit perfectionism in more recent work by arguing that the capabilities are the potential topic of a political overlapping consensus (2001; 2011). This move successfully limits the reliance of capability theory on a comprehensive view of well-being; it does not do away with the need for claims about the value of certain functionings over others.

9 This definition of paternalism places paternalism toward cultures outside the scope of analysis.

10 The notion of all-purpose means draws from Rawls' notion of primary goods.

11 See Cohen (1993) for an argument that the CA assumes an athletic picture of human life in which all functionings must be actively chosen.

12 Nussbaum also argues that most functioning-promotion policies can be justified in terms of the shared burdens choices not to function well impose (2001: 94).

13 This essay has benefitted from the research assistance of Stephanie Sheintul as well as comments from the editors of this volume.

References

Anderson, E. (1999) "What Is the Point of Equality?" *Ethics* 109: 287–337.

Arneson, R. (2000) "Perfectionism and Politics," *Ethics* 111: 37–63.

Begon, J. (2015) "What Are Adaptive Preferences? Exclusion and Disability in the Capability Approach," *Journal of Applied Philosophy* 32: 241–257.

Carter, I. (2014) "Is the Capability Approach Paternalist?" *Economics and Philosophy* 30: 75–98.

Claassen, R. (2011) "Making Capability Lists: Philosophy Versus Democracy," *Political Studies* 59: 491–508.

———. (2014) "Capability Paternalism," *Economics and Philosophy* 30: 57–73.

Cohen, G. A. (1993) "Equality of What? On Welfare, Goods, and Capabilities," in M. C. Nussbaum and A. Sen (eds.) *The Quality of Life*, Oxford: Clarendon Press.

Cudd, A. (2014) "Commitment as Motivation," *Economics and Philosophy* 30: 35–56.

Deneulin, S. (2002) "Perfectionism, Paternalism, and Liberalism in Sen and Nussbaum's Capability Approach," *Review of Political Economy* 14: 498–518.

Khader, S.J. (2011) *Adaptive Preferences and Women's Empowerment*, Oxford: Oxford University Press.

———. (2012) "Must Theorising about Adaptive Preferences Deny Women's Agency?" *Journal of Applied Philosophy* 29: 302–317.

———. (2013) "Identifying Adaptive Preferences in Practice," *Journal of Global Ethics* 9: 311–327.

Khader, S. J. and Kosko, S. J. (forthcoming). "Reason to Value and Perfectionism in the Capability Approach," in L. Keleher and S. J. Kosko (eds.) *Ethics, Agency, and Democracy in Global Development*, Cambridge: Cambridge University Press.

Mills, C. (2005) "Ideal Theory as Ideology," *Hypatia* 20: 165–184.

Nussbaum, M. C. (1999) *Sex and Social Justice*, Oxford: Oxford University Press.

———. (2001) *Women and Human Development: The Capabilities Approach*, Cambridge: Cambridge University Press.

———. (2011) "Perfectionist Liberalism and Political Liberalism," *Philosophy and Public Affairs* 39: 3–45.

Olsaretti, S. (2005) "Endorsement and Freedom in Amartya Sen's Capability Approach," *Economics and Philosophy* 21: 89–108.

Pettit, P. (2001) "Symposium on Amartya Sen's Philosophy: 1 Capability and Freedom: A Defence of Sen," *Economics and Philosophy* 17: 1–20.

Pogge, T. (2010) "A Critique of the Capability Approach," in H. Brighouse and I. Robeyns (eds.) *Measuring Justice*, Cambridge: Cambridge University Press, pp. 17–60.

Raz, J. (1986) *The Morality of Freedom*, Oxford: Clarendon Press.

Robeyns, I. (2003) "Sen's Capability Approach and Gender Inequality," *Feminist Economics* 9: 61–92.

Sen, A. (1973) "Behavior and the Concept of Preference," *Economica* 40: 241–259.
———. (1980) "Equality of What?" in S. McMurrin (ed.) *The Tanner Lectures on Human Values, Volume 1*, Cambridge: Cambridge University Press.
———. (1982) "Rights and Agency," *Philosophy & Public Affairs* 11: 3–39.
———. (1999) *Development as Freedom*, New York: Knopf.
———. (2002) *Rationality and Freedom*, Cambridge, MA: Belknap Press.
———. (2004) "Capabilities Lists and Public Reason," *Feminist Economics* 10: 77–80.
———. (2009) *The Idea of Justice*, Cambridge, MA: Harvard University Press.
Sunstein, C. (2014) *Why Nudge?* New Haven, CT: Yale University Press.

PART IV

Paternalism without coercion

18

LIBERTARIAN PATERNALISM, NUDGING, AND PUBLIC POLICY

Muireann Quigley

1 Introduction

The behavioral psychology and economics literature demonstrates that we all display a range of cognitive biases, heuristics (rules of thumb), and other "tics" which affect our decision-making in a variety of ways (Kahneman 2011; Samson 2014). These "stumbles" or "errors" can arise as a result either of our automatic and reflexive cognitive processes or our more conscious and reflective ones (Amir & Lobel 2008). Take, for example, three effects and biases commonly found in the literature: (1) the way information is presented affects the decisions we make based on that information (framing effects); (2) defaults are sticky and we display inertia in changing them (status quo bias); and (3) we tend to underestimate the probability that certain bad things will happen to us and overestimate the probability that things will be alright (optimism bias) (Samson 2014). Because our choices and decisions are regularly and systematically affected by the presence of such biases and heuristics, we are (allegedly) what Dan Ariely (2009) calls predictably irrational.

Given these systematic and predictable cognitive quirks, we can influence people's actions or decisions by altering the choice architecture which surrounds them; that is, by altering the contexts in which they act or make decisions (Thaler and Sunstein 2009: 6). In their eponymous book, Richard Thaler and Cass Sunstein use the term nudge for "an aspect of choice architecture that alters people's behavior in a predictable way without forbidding any options or significantly changing their economic incentives" (Thaler and Sunstein 2009: 6). They also say that "a nudge is any factor that significantly alters the behavior of Humans, even though it would be ignored by Econs"; Econs being those who meet the idealized rationality assumptions of neoclassical economics (Thaler and Sunstein 2009: 8; Whitman and Rizzo 2015: 411–417; Hansen 2016: 6–8). The paradigm example of a nudge in the book (and followed in much of the literature) is the cafeteria queue. A variety of research has shown that altering the layout of food in cafeterias affects the consumption of particular items of food (Just and Wansink 2009; Rozin et al. 2011; Wansink and Just 2011). Thus the benevolent cafeteria choice architect could change how food is presented to nudge their customers to choose healthier items of food; for instance, by putting salad items in the more accessible locations and putting desserts out of view. Such actions would, according to Thaler and Sunstein, be forms of "libertarian paternalism" (Sunstein and Thaler 2003; Thaler and Sunstein 2009: 253–254). They are paternalistic in the sense that they represent legitimate ways "for private and public institutions to attempt to influence people's behavior . . . [and] to steer people's choices in directions that will improve the choosers' own welfare" (Sunstein and Thaler 2003: 1162). And they are libertarian insofar as the interventions and strategies at issue are "relatively weak and nonintrusive type[s] of paternalism, because choices are not blocked or fenced off" (Sunstein and Thaler 2003: 1160). As such, the authors claim that libertarian paternalism "should be acceptable to those who are firmly committed to freedom of choice on grounds of either autonomy or welfare" (Sunstein and Thaler 2003: 1160). Note that, although they are bedfellows, nudges and libertarian paternalism are not identical. Nudges refer either to the means or mechanism of achieving certain outcomes (i.e., policies and interventions), while libertarian paternalism refers to a particular underlying political philosophical justification for their use.

Reflecting on the recent enthusiasm for nudging and libertarian paternalism, this chapter draws on Jenny Steele's idea of "alluring concepts." An alluring concept refers to one which has "distinct appeal . . . [it] may very neatly capture a way of approaching a complex problem, or [it] may carry inherent normative appeal" (Steele 2016). Concepts can be alluring for a variety of reasons and can act as aids for debate, analysis, and decision-making. However, when we become *too* captivated by them – when they become *too* alluring – they may also be dangerous. They can potentially be distracting, diverting us down some (possibly overly narrow) routes and away from other more fruitful ones. They can thus obfuscate and muddy the analytical waters. My overall claim is that we have become too captivated by nudging and libertarian paternalism. The result of this is that we may well be obscuring both the wider picture regarding the legitimate scope of government action, as well as other more practical philosophical (and arguably more pressing) concerns. My contention is that a variety of academic literatures (ethics and philosophy, law, political science, policy studies, and so on), as well as politicians and policy-makers, would do well to largely abandon the dual nudge and libertarian paternalism project (at least as it is currently being approached). As we will see, this is for four interrelated reasons. First, simply labeling something as a nudge is not enough to do the justificatory work required to tell us if particular interventions are permissible or if other ones might be preferable, all things considered. Second, interventions which draw on the behavioral sciences are not limited to nudges or those which are libertarian paternalistic. We therefore risk missing important aspects of the use of the behavioral sciences in law- and policy-making if we continue to be captivated by these two alluring concepts. Third, there are as yet unanswered questions about the (types of) evidence required to implement behavioral law and policy and assess its successes, including nudges and other libertarian paternalistic policies. Finally, despite apparently challenging traditional philosophical, legal, and economic conceptions of the rational actor, the normative touchstone of both nudges and libertarian paternalism is a version of rationality which does not exist. This requires us to re-evaluate the normative significance we attach to claims (inherent in many nudge-libertarian paternalism arguments) that people are imperfect decision-makers. In making my arguments, this chapter takes a step back from some of the detailed ones regarding paternalism found in other parts of this volume (and elsewhere). I do this in order to take stock of the wider backdrop against which behaviorally inspired law- and policy-making takes place and to apply a critical lens to the way in which we have been examining this thus far.

2 Policy-making, politics, and the philosophical allure of nudge

Interest by governments and policy-makers in behavioral approaches to law, regulation, and policy is not new. Nevertheless, with the publication and popularization of *Nudge*, they have displayed renewed attention to behavioral research in an attempt to achieve a range of policy goals. This has led to the creation of units both within and outside government in order that insights from the behavioral sciences can be integrated into law and policy. These include the Behavioural Insights Team (BIT) in the United Kingdom, the Social and Behavioral Sciences Team in the United States, and Behavioral Economics Team of the Australian Government. In mainland Europe, several countries also have dedicated behavioral teams within government or are explicitly using behavioral research to inform policy-making (Joint Research Centre 2016). Moreover, at the European Commission level, the Directorate-General for Health and Consumers has set up the Framework Contract for the Provision of Behavioral Studies to EU policy-makers. Its purpose is to "facilitate the running of behavioral studies in support of EU policy-making" (van Bavel et al. 2013). To this end, in 2016 the Joint Research Centre (JRC) published an extensive report on the application of behavioral insights to policy in Europe (JRC 2016). The report contains numerous examples of law and public policy initiatives which have been influenced by or have utilized insights from behavioral science research. Initiatives include those in diverse areas such as consumer protection, employment, energy, health, finance, taxation, and transport. Let us take a look at just a few of these in order to better understand the types of initiatives being rolled out and the behavioral insights which lie behind them.

With increasing frequency numerous European countries are implementing strategies regarding the display and packaging of tobacco products. In Iceland and the UK, tobacco products can no longer be openly displayed. In Ireland and France there is now a plain packaging requirement, while similar legislation is in progress in Finland, Hungary, Norway, and the UK (JRC 2016: 26). These stratagems simultaneously decrease visual cues which trigger consumption and harness the power of social norms by sending a message regarding the (lack of) societal acceptability of smoking. A transport-related example can be found in Luxembourg. Here the government ran a road safety campaign which used messages incorporating framing and salience. One of their messages, "48% of deadly car accidents are due to excessive speed," aims to make the consequences of speeding more concrete for drivers (JRC 2016: 31). Finally consider employee pensions. It is now compulsory in the UK for employers to automatically enroll employees in a workplace pension (HM Government 2016). As noted earlier, people display inertia with respect to moving away from default options. Therefore, changing the default has the potential to increase workplace pension enrollments. There are numerous other initiatives which could be mentioned across the different policy areas, but I have chosen these as they are all examples of behavioral science-inspired measures which could be viewed as libertarian paternalistic. These policies are (at least partially) concerned with the welfare or interests of the individual being nudged. As Le Grand and New put it, "libertarian paternalists not only seek to intervene when individuals are subject to reasoning failure; they also actively manipulate reasoning failure to encourage individuals to make choices that support their own well-being" (2015: 133). Ostensibly, however, nudgees are still at liberty to choose. The committed smoker can still purchase cigarettes, drivers can decide not to reduce their speed, and while the automatic pension enrollment scheme is compulsory for employers, employees can individually opt out.

Part of the attraction to government and policy-makers of strategies which draw on behavioral insights is that they are seen (rightly or wrongly) as light-touch, low-cost ways of achieving particular policy goals. They are viewed as light-touch in the sense that they are not infrequently framed as alternatives to state regulation (HM Cabinet Office 2010: 4; HM Government 2010: 7; Le Grand and New 2015: 136). Such framing may appeal to those who see regulation as a symptom of an overly coercive and paternalistic nanny state. Presenting these approaches in non-regulatory terms may induce us to think about this type of government action in a certain way. It endeavors to send a message that these are measures that no one need worry about. The allure here is in part rhetorical. A "nudge" sounds like just a little thing; and when explicitly juxtaposed with "law and regulation" for political purposes, then it is surely unobjectionable. Moreover, in explicitly framing the nudge project as a form of libertarian paternalism, it is set as having the potential to appeal to both sides of the political spectrum, simultaneously capturing concerns regarding the preservation of liberty and those regarding the welfare and interests of the individual. Behavioral initiatives may additionally be perceived as being both light-touch and low-cost in economic terms from the point of view of government when compared to other potential stratagems; for example, requiring shops to hide tobacco products from view, thus reducing their visibility to individual consumers, may be less costly than extensive information and education campaigns about the dangers of tobacco consumption (of course whether or not behaviorally influenced law and policy is actually low-cost – all things considered – will depend on the details of the initiative or intervention at issue, as well as on how they are tested and implemented).

Leaving politics and policy-making aside, nudge as a concept, and libertarian paternalism as an underlying political philosophy, have also proved alluring to scholars in terms of philosophical analysis. Regarding "nudge," several have tried to refine Thaler and Sunstein's account (given in the introduction) (Hausman and Welch 2010; Saghai 2013; Mongin and Mikael 2014; Yeung 2016). Take for example Yashar Saghai's re-worked definition of a nudge. He says that "A nudges B when A makes it more likely that B will ϕ, primarily by triggering B's shallow cognitive processes, while A's influence preserves B's choice-set and is substantially non-controlling (i.e., preserves B's freedom of choice)" (2013: 491). This definition makes explicit elements which are less so in Thaler and Sunstein's original (although they are there by implication and in other things they say). Saghai's definition takes account of both the mechanism of action of nudges (shallow cognitive processes) and seeks to distinguish them from other types of influences (more controlling ones such as coercion or bans). While this latter aspect is explicit in the original, Saghai goes further and has developed a taxonomy of influences: choice elimination, compulsion, coercion, behavioral prod, disincentive, incentive, nudge, and rational persuasion. Behavioral prods, for instance, can be distinguished from nudges because they are substantially controlling: "A makes it more likely that B will ϕ, primarily by triggering B's shallow cognitive processes, while A's influence preserves B's choice-set but is substantially controlling" (Saghai 2013: 491).

We can see how challenging, dismantling, and refining the elements of nudge in such a manner might help us to achieve analytical clarity. First, we can ensure accuracy and precision regarding what is at issue; i.e., to make sure we are in fact talking about nudges and not other types of influence. This is especially important in debates where normative arguments are made with reference to empirical claims. In the current context, it is claims about a particular empirical reality regarding human rationality and cognitive processes which motivates some commentators' normative claims about how law and policy ought to be made. Secondly, refining definitions can give us a common language and jumping off point and thus aids in the development of broader frameworks for analyses. It can do this by shedding light on (ir)relevant aspects of debate.

Similarly the idea of libertarian paternalism has been attractive to scholars. By bringing together two types of political philosophical justifications which are frequently set up as opposing each other – libertarianism and paternalism – we are seemingly offered a new concept to approach the complex business of law- and policy-making. It is this promise of a "third way" which forms part of the allure of libertarian paternalism as a political philosophy (Thaler and Sunstein 2009: 253). It is, as Richard Arneson calls it, "an unstable normative position" (2015: 677). By this I take him to mean that the libertarian and paternalistic elements are not balanced; for him, the rationale behind the types of interventions proposed point more towards paternalism: "the background idea to which [Thaler and Sunstein] appeal suggests more paternalism, or anyway a wider range of paternalism, than they endorse" (2015: 678). Others, such as Le Grand and New, see it differently and highlight the libertarian aspects. They do not deny there is paternalism involved, but consider the impact on autonomy from nudges to be minimal, or at least much less than that from other types of interventions (2015: 145–146). Such normative instability is in itself alluring to philosophers. It encourages debates which focus on asking whether the interventions at issue are either libertarian or paternalistic, and whether the very idea is an oxymoron (Sunstein and Thaler 2003; Mitchell 2005). Examining these questions is valuable for a number of reasons, not least in deepening our understanding of the type of society we wish to live in and in shedding light on the relationship between government and the citizenry.

Despite these positive aspects, I am skeptical about whether the revision and specification of definitions helps us in debates regarding nudging and the use of the behavioral sciences in law and policy more generally. Likewise, while debates regarding the libertarian and/or paternalistic qualities of such approaches are alluring, ultimately they might be distracting us from broader political (philosophical) considerations, as well as practical ethical matters at the coalface of policy-making. The rest of this chapter is devoted to making this case.

3 A smokescreen of conceptual thinking?

In part, my skepticism regarding the dual nudge-libertarian paternalism project centers on the conceptual; in particular, whether definitions of "nudge" are intended to have descriptive utility and/or normative force. Note that Thaler and Sunstein's original definition is stipulative; it aims to "stipulate or impart a meaning to the defined term and thus involve no commitment that the assigned meaning agrees with prior uses (if any) of the term" (Hansen 2016: 6). Hence their definition of a "nudge" has descriptive content. It is descriptive insofar as they describe a *mode of action* of nudges. Nudges alter behavior through choice architecture and via mechanisms which would affect Humans but not Econs. Saghai's definition is also descriptive in this sense: nudges are those, and only those, influences that leave our choice set open and do so via a non-controlling method. Note, however, that depending on the interpretation these definitions may not simply be stipulative in a descriptive sense – specifying a new meaning, unrelated to previous usages of words – but may also contain particular normative prescriptions and commitments. Thaler and Sunstein's original definition certainly does. It is unselfconsciously designed to meet the requirements of their particular underlying political theory: libertarian paternalism. Nudges are those aspects of choice architecture which promote the person's own welfare or autonomy or serve their own interests (paternalistic), yet leave the choice set unchanged/do not forbid any options (libertarian). Their definition of a nudge is, therefore, normative in the sense that it entails the very normative moral criteria upon which nudges are purported to be justified. This interpretation seems reasonable since their arguments in favor of libertarian paternalism (Sunstein and Thaler 2003) came before they started employing the term "nudge," with the idea of a nudge being explicitly built upon a libertarian paternalistic justification (Thaler and Sunstein 2009). Correspondingly, Saghai's definition could be interpreted as having a normative spin. Nudges are influences that leave our *choice set open* and do so via a *non-controlling* method. These elements could be viewed as normative requirements regarding the permissibility of nudges (although Saghai himself leaves open questions of the permissibility of nudges and whether they are preferable to other types of influence). We can see that there is a circularity problem here, especially for Thaler and Sunstein's definition. When we ask whether nudges are justified, one response is "yes, because they are libertarian paternalistic" – they promote a person's interests and well-being while also respecting their autonomy. But this is a trifling response, because the very definition of nudge is itself based on this purported underlying justification.

A related difficulty can be illustrated by considering charges of coercion. Coercion is often characterized as involving the foreclosure of options to the coerced, and/or as infringing individual liberty in some way, and/or as being substantially controlling (Anderson 2014). As such, if nudges are defined as having an open choice set, not infringing liberty, not forbidding any options, and/or as being substantially non-controlling, there is no case to answer (Quigley 2014). There is no need for us to ask if nudges are coercive, since we already know by definition that they are not. When applied to actual laws and policies this is unhelpful. We cannot just label something a nudge and be done with it. While the label offers an attractive way to categorize interventions, we may end up obscuring important considerations. There is a danger that in calling something a "nudge" we might avoid doing the hard analytical work required to see if *particular* interventions are justified. This is because when we employ concepts, to some degree we end up generalizing, rather than focusing on specifics. Indeed, the ability of concepts to capture generalizations is part of what is alluring about them (Steele 2016). Nevertheless, as the coercion example demonstrates, a focus on the conceptual can distract us. When nudges are defined in this way, it does not aid us in addressing some possible objections to interventions defined as nudges. Depending on the definition, we might in effect define out certain normative concerns and, consequently, miss crucial analytical avenues. With regards to coercion, the correct question is not necessarily whether nudges writ large are coercive, but whether any particular law, or policy, or intervention is, and (more importantly) whether or not that impacts on its permissibility in different circumstances.

All of this notwithstanding, it is not in any case clear how having a nailed down definition of a nudge (whether normative or non-normative) is useful in law- and policy-making. Despite the rhetorical allure of nudge, practically speaking law- and policy-makers do not decide that they will implement "nudges" and proceed from there. The tools and methods they draw on are not reliant on labeling one way or another. Further, there is in fact no "nailed down" definition. Instead what we get is numerous variations and iterations of "nudge." We can understand why this is alluring for philosophers. We are at home with such iterative analytical processes in the search for clarity, coherence, and consistency. Nevertheless, rather than adding clarity, the proliferation of and focus on definitions might simply add to opacity by overly complicating matters and distracting us from other issues. As Andrés Moles notes, the

> political morality of nudges depends on substantive arguments: finding out what duties public officials owe to citizens and what their prerogatives are. These underlying principles are not available through conceptual analysis. Whether nudging is permissible or not depends on normative arguments, not on semantics.
>
> (2015: 648)

This is important because the terminology of nudge may be used to serve particular political ends. It may, for example, be used to describe policies or interventions which are not libertarian paternalistic and which may in fact simply be coercive. As we just saw, in such situations it is of little use to ask if nudges – or indeed other libertarian paternalistic policies generally – are coercive since by their very definition they are not. The danger is that when we become too captured in the thrall of new concepts (either nudge or libertarian paternalism), there is "the potential for conceptual thinking to amount to a smokescreen" (Steele 2016). It may distract us from asking relevant questions about the permissibility of the particular policies or interventions at issue.

4 Law, regulation, and libertarian paternalism

A further part of my skepticism centers on the possibility that we may simply be arguing about old wine in new casks (Grüne-Yanoff 2012). It is both the newness and oldness which makes nudge and libertarian paternalism so alluring. Turning our analytical gaze to these is appealing because we feel like we are tackling new concepts, problems, and arguments. But they are equally alluring because we can draw on the usual philosophical suspects for our analyses. Arguments regarding autonomy, paternalism, coercion, and so on feel familiar to us. With paternalism, for instance, we can point out that Thaler and Sunstein's characterization of paternalism does not fit with standard accounts found in the literature (see Scoccia's entry in this volume); for example, to the extent that the interventions they propose do not (if they do not) interfere with liberty or autonomy, or cannot be construed as being against the will of those nudged, then they cannot be counted as paternalism properly so-called. The analytic philosopher is at home with this type of analysis. But herein lies the danger. As "bread and butter" concepts in philosophy (and law to a certain extent), concepts like paternalism are comfortable philosophical stomping grounds. This is not necessarily a bad thing, especially if these concepts are appropriate to the moral evaluations of nudges and libertarian paternalism. Nevertheless, to some extent they represent argumentative low-hanging fruit. The danger is that this can lure us into narrowing our focus and conducting our analyses along particular well-trodden lines. In part this is a concern about argumentative laziness. Declaring that something is or is not paternalistic in nature may operate as a shorthand which distracts us from examining whether or not particular measures are justified in particularized contexts (regardless of the broader theoretical label we may give them). Relatedly, this may distract us away from broader considerations regarding the practical philosophical aspects of law- and policy-making. Let us consider why and how this might be the case.

Much of the concern regarding nudging and libertarian paternalism contains two strands. The first is a worry about the legitimacy of the state employing nudges as tools of government. The second is a concurrent concern about the effects on *individuals* of alterations to choice architecture. While both of these aspects are important, frequently arguments regarding the first are conducted through the lens of the second. Consequently, critiques of nudging and libertarian paternalism in the context of state action become reduced to individualistic concerns; for instance, claims that nudges are objectionable because they infringe autonomy, bypass rational decision-making, lack transparency (to the individual nudgee), may involve deception, are manipulative and potentially coercive, do not move us in the direction of our true interests, can lead to the cultivation of inauthentic values, and that their use may be infantilizing (i.e., that they may lead to a decreased ability for decision-making) (e.g., Bovens 2009; Hausman and Welch 2010; White 2013; Cratsley 2015; Glod 2015; Hanna 2015). At the core of each of these is a concern about individual liberty and autonomy. This is understandable. These are important aspects to consider in any critique of law, policy, and government action. Even so, it can result in partial analytical blindness.

Some of this comes from the fact that individualistic critiques can belie the complexity of the policy problems at issue. The aims of government are wide and varied; for example, enforcing compliance with legal duties, solving collective action problems, ensuring the availability or maintaining the stability of a variety of public and common goods, and providing resources and services in the public interest for the shared benefit of the citizenry. Moreover, the policy problems which the state and state actors must address are diverse, difficult, interwoven, and also highly politicized. This is the case across policy domains: finance, health, taxation, education, the environment, and so on. The societal challenges we face in each of these areas cannot be solved through an individualistic lens. If analyzed in this way, then we may miss other pertinent considerations. These include the welfare and well-being of other citizens, the potential time dependency of the preferences of those citizens, the collective responsibilities of citizens, and the broader responsibilities of government to the collective such as balancing competing needs and interests.

A related issue is that nudging seems to have become shorthand for a range of policy measures influenced by the findings of behavioral research. However, while nudge policies draw on insights gained from, for instance, behavioral economics, they are not one and the same (Hansen 2016). The application of behavioral insights in law- and policy-making involves a wider range of activities than strict nudges and/or libertarian paternalistic interventions. Take the example of tax compliance letters. A randomized controlled trial run by the BIT and UK tax office found that using messages which highlighted the already high rate of compliance with tax payments led to an increase in tax payments (thus utilizing research regarding the power of social norms). Such applications of behavioral insights are not (at least directly) for the benefit of the target of the nudge, nor aimed at improving their welfare or autonomy (something which is required on Thaler and Sunstein's account). They are not, therefore, libertarian paternalistic in nature. Although we might consider that helping individuals to pay their tax bill on time is a benefit from the viewpoint of keeping them on the right side of the law, this is not the overall aim of a system of taxation. The aim of the tax system is to ensure that the state can pay for public goods and services. Albeit this benefits citizens as a whole, it is a collective benefit to society, not one that flows to the individual as a direct result of being nudged.

Regulation and policy-making is not just about individual decision-making, and the aim of government is not always (or perhaps even often) about securing the welfare or liberty of the individual citizen. As such, the legitimacy of state action writ large cannot be determined solely by reference to the individual. We need to turn our attention to consider factors beyond the individualistic lens of nudging and libertarian paternalism to instead think about the *legitimacy* of the state employing different forms of choice architecture (Yeung 2016). This means going beyond autonomy-related concerns and taking account of the different aims and functions of government just noted, as well as the tools used to achieve these within different policy domains. To do this we need to address the *who-what-why-how* challenge; that is, to ask *who* is being regulated, by *what* means, *why* this is needed, and *how* it is being done. However, determining the interrelationship between these factors is not always straightforward, especially given the uncertain legal or regulatory status of some strategies. Take the plain cigarette packaging example given in Section 2. Although the tobacco industry is being directly regulated by traditional command and control legislation (and thus a coercive influence), consumer behavior is not subject to direct regulation (even though we can say they are being nudged). Yet it is the consumer, and not the tobacco industry, who is the actual end target. There is, therefore, more than nudging going on when we consider the intervention as a whole. Hence it is too simplistic to just label such measures as "regulatory" or "non-regulatory" (Brownsword 2008: 7; Quigley 2013: 595–599).

Having said all this, when commentators contrast nudges with regulation their intention may not be to contend that they can never fall within the ambit of state regulation. Indeed Thaler and Sunstein would not deny that nudges and other behavioral approaches could (and do) form part of the regulatory landscape. What is actually at issue is the permissibility or not of a variety of state sanctioned strategies and interventions. So sometimes the purpose of the regulatory/non-regulatory division is to draw a distinction between the lighter forms of paternalistic interventions which nudges represent and other harder forms of paternalism. Le Grand and New, for instance, explicitly contrast libertarian paternalistic interventions with other more coercive ones

(2015: 136). Specifically they juxtapose strategies such as changing defaults and using framing to affect information provision with ones which involve forced compliance or compulsion (2015: 136–138). For them, the advantage of nudge and libertarian paternalist approaches is that they can promote well-being by addressing individual reasoning failures, but compromise autonomy to a lesser degree than other types of paternalism (2015: 138). It is not enough, however, to favor such strategies because they are the least restrictive in terms of liberty and autonomy. We also need to take seriously questions of evidence and effectiveness regarding nudge-libertarian paternalism in comparison to other potential approaches. It is to these I now turn.

5 Biases, empirical evidence, and normative baselines

In general, there is a lack of understanding from scholars, as well as law- and policy-makers, regarding the empirical and methodological limits to translating research findings from the behavioral sciences into *effective* and *legitimate* law and policy. In essence the evidence base (or at least interpretations of this) underpinning behavioral law and policy is open to dispute. It is thus crucial that a variety of methodological and empirical concerns are considered, not in the least because any deficiencies in these impact on our normative analyses and conclusions.

The first thing to note is that behaviorally inspired policies and interventions display wide variation in terms of their empirical underpinnings. Some have been subject to rigorous real-world trials, some are the result of the wider application of small-scale studies (either in the real-world or laboratory-like experiments), and some have simply been extrapolated from or inspired by general behavioral science principles without ever being specifically tested (JRC 2016: 16). One consequence of this is that the evidence for certain interventions or initiatives may be empirically weak and misleading. They may be only tangentially and tenuously connected to actual evidence regarding causal mechanisms and/or the effectiveness of particular interventions. Relatedly, much of the relevant behavioral research is done in controlled laboratory-like settings. As such, there are uncertainties about the translation of research findings to the applied setting (Quigley and Stokes 2015: 72–73). The real-world environment is much more complex than the controlled experimental setting. Consequently, policies based on experimental studies which have not been tested in the real-world may not be effective, or there may be as yet unknown unintended consequences which flow from implementing behavior change policies (Quigley and Stokes 2015: 64–71). The potential gains or benefits of an initiative need, therefore, to be balanced with other types of considerations. A "light-touch, low-cost" nudge may not be the best route if all it will produce are marginal effects. In such cases, it could end up being a waste of money and other resources.

Questions have also been raised about the replicability (and thus robustness) of some research findings from behavioral psychology. It is undergoing a so-called "replicability crisis." An example of this are experiments on ego depletion. There are many variations, but the original involved putting a plate of chocolate chip cookies as well as a bowl of radishes beside student test subjects. They were then instructed to eat only the cookies or radishes. Afterwards they were asked to solve a puzzle. Those who ate the cookies persevered in trying to solve it for much longer than those who ate radishes. This led to theorizing that willpower is limited and can be depleted (by resisting eating cookies) (Baumeister et al. 1998). The effect was then found in many other studies. However, a recent large-scale replicability project involving 2000 subjects concluded that "if there is any effect, it is close to zero" (Hagger et al. 2016: 558). In this vein, the Open Science Collaboration tried to replicate 100 studies which had previously been published in psychology journals. They found that the replicability of these was around 47% (OSC 2015) and that a "large portion of replications produced weaker evidence for the original findings despite using materials provided by the original authors" (943). Remarkably, the results of the OSC study itself have been challenged, with claims that the OSC data do not support their own conclusions (Gilbert et al. 2016).

The difficulties and uncertainties just noted are not insignificant, yet they are not commonly acknowledged or discussed outside of the behavioral sciences literature. Despite the fact that the science itself does not yet have a "stable and critical core," it is frequently represented in other literatures, including law and philosophy, as if it does (Priaulx and Weinel 2014: 372). What is more, when utilizing empirical evidence for normative and policy ends proponents have a tendency to give only a snapshot of the relevant studies and choose those which confirm the position they are arguing for (Gigerenzer 2015: 366–371). Part of this may be that a gap exists between an "understanding" and "deep understanding" of the science needed to make behavioral law and policy: while behavioral science "may appear accessible and comprehensible, there is a 'glass wall' that prevents outsiders from being able to gain the kind of 'deep understanding' necessary to access or comprehend research undertaken within expert communities" (Priaulx and Weinel 2014: 383–384). This is something which may have profound implications for what Priaulx and Weinel term "the legal translation of 'insights' from the behavioural sciences" (2014: 368). If, for instance, individuals do not exhibit the kinds of biases or reasoning failures commonly discussed, or if these are of a different nature, or if the consequences of real-world interventions are not as anticipated, then the normative (legal and philosophical) arguments which have hitherto shaped both academic debate and public policy initiatives need serious reconsideration. To wit, the normative legitimacy of nudges and other behavioral science-inspired forms of libertarian paternalism is called into question.

These issues are compounded by a normative concern which has not yet been adequately addressed. While nudging and libertarian paternalism are being presented as new concepts (and as we saw earlier herein lies some of their allure), the normative baseline which they employ is not new. Notwithstanding the fact that the nudge-libertarian paternalism project is predicated on claims regarding humans' biases and reasoning failures as an everyday empirical reality, it nonetheless still utilizes the neoclassical rational actor as the normative touchstone (for critiques see Berg and Gigerenzer 2010; Gigerenzer 2015; Whitman and Rizzo 2015). By this I mean that the traditionally conceptualized rational actor continues to be the normative reference point; deviations (interpreted as irrationality) are measured in relation to this ideal. There is an obvious analytic and normative tension here. On the one hand proponents of the nudge-libertarian paternalism approach commonly promote empirical findings from the behavioral sciences as challenging standard normative models and assumptions within economics, law, and philosophy. Meanwhile, on the other, they measure people's decisions and actions against the very normative reference point they are challenging, something which by their own admission does not exist. Thus a particular normative status quo is being perpetuated, even as it is purportedly being challenged.

Why, if Econs do not actually exist, are we building entire definitions and theories around this baseline? Doing so is problematic because instead of disrupting old analytical ways, we simply reinforce them. And as a consequence we may fail to generate new insights regarding the problems we aim to address. One significant consequence of the continuing acceptance of the neoclassical normative baseline is that it perpetuates that idea that empirically observed deviations from that model mean that people act *irrationally*. One solution may be to view (at least some of) our cognitive biases as part of an "ecological rationality." This is the idea that human behavior demonstrates evolutionarily advantageous short-cuts. We have heuristics and biases that developed to fit with the kinds of quick decisions we need to make on an everyday basis (Gigerenzer and Brighton 2009; Gigerenzer 2015). We are, for instance, good at evaluating things like natural frequencies (numbers expressed in the form 1/10, 1/1000, and so on) and at sorting through information quickly for that which is most relevant to us. These abilities help to truncate decision-making processes. The case for "deviations from rationality" may, therefore, be overstated, with the result that we ought not to *generally* infer irrationality or ineffectiveness from the use of cognitive short-cuts. Thus far the law and philosophy literatures have largely neglected the ecological rationality view when discussing the implications of behavioral science research (although see Jones 2001 and Feldman and Lobel 2015). This is not altogether surprising since calls for an evolutionary behavioral economics (into which ecological rationality falls) are relatively new. Yet it warrants serious consideration. If (at least some of) the findings of behavioral science do not actually indicate *irrationality*, then we need to re-evaluate the normative significance we attach to claims (inherent in nudge-libertarian paternalism arguments) that people deviate from some sort of rationality baseline.

6 Concluding remarks

In this chapter I have argued that when nudging and libertarian paternalism frame and drive our analyses regarding the use of behavioral sciences in law and policy, we miss the wider picture with regards to law and governance, regulation, and public policy formation. Nudging and libertarian paternalism are alluring concepts because they seemingly capture a new way of addressing complex policy problems. They also, at least on the face of it, have normative appeal. However, alluring concepts, while attractive in some respects, can lead us down narrow analytical pathways. It has been my contention that, to some extent, this is what has happened. We have become captivated, and thus captured, by nudging and libertarian paternalism. We have become overly focused on analyses of definitions and on retracing well-trodden philosophical pathways with individualistic concerns at their core. To be clear, I do not claim that these are not important, but that by being too distracted by them, we may be exhibiting partial blindness and missing other analytical avenues which merit examination. I argued that mere labeling is not enough to do the justificatory work and, in any case, the use of the behavioral sciences in law and policy is wider than nudging and libertarian paternalism. As such, focus on these may distract us from analyses of state action writ large, including considerations of the complexity of the issues, actors, and choice architectures involved. In addition, I noted that there are as yet unanswered empirical and methodological difficulties in this area. If taken seriously, these require us to reconsider the normative bases, framings, and implications of behavioral law and policy, including nudge and libertarian paternalism.

Related topics

Hard and Soft Paternalism; Paternalism and Contract Law; Paternalistic Manipulation.

References

Amir, O. and Lobel, O. (2008) "Stumble, Predict, Nudge: How Behavioral Economics Informs Law and Policy," *Columbia Law Review* 108: 2098–2138.

Anderson, S. (2014) "Coercion," in E. N. Zalta (ed.) *Stanford Encyclopedia of Philosophy*, Spring 2014 edition, https://plato.stanford.edu/entries/coercion/.

Ariely, D. (2009) *Predictably Irrational: The Hidden Forces that Shape Our Decisions*, London: Harper Publishing.

Arneson, R. (2015) "Nudge and Shove," *Social Theory and Practice* 41: 668–691.

Baumeister, R. F., Bratslavsky, E., Muraven, M. and Tice, D. M. (1998) "Ego Depletion: Is the Active Self a Limited Resource?" *Journal of Personality & Social Psychology* 74: 1252–1265.

Berg, N. and Gigerenzer, G. (2010) "As-If Behavioural Economics: Neoclassical Economics in Disguise?" *History of Economic Ideas* 18: 133–165.
Bovens, L. (2009) "The Ethics of Nudge," in T. Grüne-Yanoff and S. O. Hansson (eds.) *Preference Change: Approaches From Philosophy, Economics and Psychology* London: Springer Dordrecht Publishing, pp. 207–220.
Brownsword, R. (2008) *Rights, Regulation, and the Technological Revolution*, Oxford: Oxford University Press.
Cratsley, K. (2015) "Nudges and Coercion: Conceptual, Empirical, and Normative Considerations," *Monash Bioethics Review* 33: 210–218.
Feldman, Y. and Lobel, O. (2015) "Behavioural Trade-Offs: Beyond the Land of Nudges Spans the World of Law and Psychology," in A. Alemanno and A-L. Sibony (eds.) *Nudging & the Law: A European Perspective?* Oxford: Hart Publishing, pp. 301–324.
Gigerenzer, G. (2015) "On the Supposed Evidence for Libertarian Paternalism," *Review of Philosophy & Psychology* 6: 361–383.
Gigerenzer, G. and Brighton, H. (2009) "Homo Heuristicus: Why Biased Minds Make Better Inferences," *Topics in Cognitive Science* 1: 107–143.
Gilbert, D. T., King, G., Pettigrew, S. and Wilson, T. D. (2016) "Comment on 'Estimating the Reproducibility of Psychological Science'" *Science* 351(6277): 1037.
Glod, W. (2015) "How Nudges Often Fail to Treat People According to Their Own Preferences," *Social Theory and Practice* 41: 599–617.
Grüne-Yanoff, T. (2012) "Old Wine in New Casks: Libertarian Paternalism Still Violates Liberal Principles," *Social Choice and Welfare* 38: 635–645.
Hagger, M. S. et al. (2016) "A Multilab Preregistered Replication of the Ego-Depletion Effect," *Perspectives on Psychological Science* 11: 546–573.
Hanna, J. (2015) "Libertarian Paternalism, Manipulation, and the Shaping of Preferences," *Social Theory and Practice* 41: 618–643.
Hansen, P. G. (2016) "The Definition of Nudge and Libertarian Paternalism: Does the Hand Fit the Glove?" *European Journal of Risk Regulation* 1: 1–20.
Hausman, D. M. and Welch, B. (2010) "Debate: To Nudge or Not to Nudge," *Journal of Political Philosophy* 18: 123–136.
Her Majesty's Cabinet Office Behavioural Insights Team. (2010) "Applying Behavioural Insight to Health," www.gov.uk/government/publications/applying-behavioural-insight-to-health-behavioural-insights-team-paper.
Her Majesty's Government of the United Kingdom. (2010) *The Coalition: Our Programme for Government*, London: HM Government.
———. (2016) "Workplace Pensions," www.gov.uk/workplace-pensions/about-workplace-pensions.
Joint Research Centre of the European Commission. (2016) "Behavioral Insights Applied to Policy: European Report," http://publications.jrc.ec.europa.eu/repository/bitstream/JRC100146/kjna27726enn_new.pdf.
Jones, O. (2001) "Time-Shifted Rationality and the Law of Law's Leverage: Behavioral Economics Meets Behavioral Biology," *Northwestern University Law Review* 95: 1141–1205.
Just, D. R. and Wansink, B. (2009) "Smarter Lunchrooms: Using Behavioral Economics to Improve Meal Selection," *Choices* 24(3): 1–7.
Kahneman, D. (2011) *Thinking Fast and Slow*, New York: Farrar, Straus & Giroux.
Kahneman, D., Knetsch, J. L. and Thaler, R. H. (1991) "Anomalies: The Endowment Effect, Loss Aversion, and Status Quo Bias," *Journal of Economic Perspectives* 5: 193–206.
Le Grand, J. and New, B. (2015) *Government Paternalism: Nanny State of Helpful Friend?* Princeton, NJ: Princeton University Press.
Mitchell, G. (2005) "Libertarian Paternalism Is an Oxymoron," *Northwestern University Law Review* 99: 1245–1277.
Moles, A. (2015) "Nudging for Liberals," *Social Theory and Practice* 41: 644–667.
Mongin, P. and Mikaël, C. (2014) "Rethinking Nudges," HEC Paris Research Paper No. ECO/SCD-2014–1067.
Open Science Collaboration. (2015) "Estimating the Reproducibility of Psychological Science," *Science* 349(6251): aac4716–4711–aac4716–4718.
Priaulx, N. and Weinel, M. (2014) "Behavior on a Beer Mat: Law, Interdisciplinarity, and Expertise," *Journal of Law, Technology, and Policy* 2: 361–391.

Quigley, M. (2013) "Nudging for Health: On Public Policy and Designing Choice Architecture," *Medical Law Review* 21: 588–621.

———. (2014) "Are (Health) Nudges Coercive?" *Monash Bioethics Review* 32: 141–158.

Quigley, M. and Stokes, E. (2015) "Nudging and Evidence-Based Policy in Europe: Problems of Normative Legitimacy and Effectiveness," in A. Alemanno and A-L. Sibony (eds.) *Nudging and the Law: A European Perspective?* Oxford: Hart Publishing, pp. 61–82.

Rozin, P. S., Scott, S., Dingley, M. S., Urbanek, J. K., Jiang, H. and Kaltenbach, M. (2011) "Nudge to Nobesity: Minor Changes in Accessibility Decrease Food Intake," *Judgment & Decision Making* 6: 323–332.

Saghai, Y. (2013) "Salvaging the Concept of Nudge," *Journal of Medical Ethics* 39: 487–493.

Samson, A. (ed.) (2014) *The Behavioral Economics Guide 2014* (1st edition), pp. 1–27, http://behavioraleconomics.com.

Steele, J. (2016) "Alluring Concepts and the Common Law," presented at *Fossilisation and Innovation in Law*, Newcastle University, Newcastle-Upon-Tyne, UK, 12 July.

Sunstein, C. and Thaler, R. H. (2003) "Libertarian Paternalism Is Not an Oxymoron," *University of Chicago Law Review* 70: 1159–1202.

Thaler, R. H. and Sunstein, C. (2009) *Nudge: Improving Decisions About Health, Wealth, and Happiness*, London: Penguin Books.

van Bavel, R., Herrmann, B., Esposito, G. and Proestakis, A. (2013) "Applying Behavioral Sciences to EU Policy-making," EUR 26033 EN, ftp://ftp.jrc.es/pub/EURdoc/JRC83284.pdf.

Wansink, B. and Just, D. (2011) "Healthy Foods First: Students Take the First Lunchroom Food 11% More Often Than the Third," *Journal of Nutrition Education and Behavior* 43(4): S1–S8.

White, M. D. (2013) *The Manipulation of Choice*, New York: Palgrave MacMillan.

Whitman, D. G. and Rizzo, M. J. (2015) "The Problematic Welfare Standards of Behavioral Paternalism," *Review of Philosophy and Psychology* 6: 409–425.

Yeung, K. (2016) "The Forms and Limits of Choice Architecture as a Tool of Government," *Law & Policy* 38: 186–210.

19

PATERNALISTIC MANIPULATION

Moti Gorin

1 Introduction

Liberals have long argued that paternalism is morally objectionable – much of the time, at least, if not always. While manipulation has not been featured as prominently in the writings of moral and political philosophers, it too is generally regarded as objectionable, sometimes for the same reasons cited in criticisms of paternalism. Both paternalism and manipulation are contested concepts with respect to their content as well as their normative significance. Philosophers disagree about what attitudes, actions, and policies should qualify as paternalistic or manipulative, and they disagree about whether paternalism and manipulation are best understood as morally neutral concepts or, rather, as concepts whose application entails a (perhaps defeasible) negative moral evaluation. They also disagree about what, if anything, can justify paternalistic or manipulative attitudes, actions, and policies.

In this chapter I will provide a brief overview of philosophical work on manipulation, explain how paternalistic manipulation differs from nonpaternalistic manipulation, and explain what makes paternalistic manipulation morally objectionable (when it is). On my favored account, manipulation is a form of interpersonal influence that deliberately fails to track reasons. Manipulation is paternalistic to the extent that it is motivated by the manipulator's desire to promote the manipulee's good, and paternalistic manipulation is wrongful when the target of the manipulation is a reasons-responsive agent whose authentic preferences – preferences that reflect the agent's deeply held values – would play a decisive role in determining the agent's action only in the absence of the manipulation.

2 What is manipulation?

We all are acquainted with manipulative characters. We may know them personally or we may encounter them in books, newspapers, films, or television. Often when we experience manipulation in our own lives we do not notice it immediately but become aware of its influence later, sometimes only after deciphering a pattern in a series of actions, none of which alone would have aroused our suspicions. Perhaps everyone reading this essay has at one time or another complained of having been the target of some manipulative action or policy. For example, you may have discovered that the bouquet of roses, or the extra day off work, or the ride to the airport was not given selflessly as a token of affection or appreciation or friendship, but rather as a unilaterally imposed debt to be collected under circumstances not of your choosing. Or, you may find yourself favoring one product brand over another for reasons having nothing to do with quality or cost but rather because of the amusing advertisements you recently viewed on the television during a commercial break. What are we noticing when we are inclined to apply the label "manipulative" (or its cognates)?

Intuitively, to be manipulated is to be treated in a way that involves being used. When we believe we have been manipulated we may feel that the manipulator failed to take us seriously as "our own" person, as an agent whose thoughts, desires, emotions, goals, or plans are in some important sense independent and authoritative. We may feel as if we were unwitting actors in a play written and directed by someone else or, worse, as if we were mere props in such a play. We may come to discover that whatever role we thought we were playing was not the role we actually were playing – others controlled events in a manner that rendered us ignorant of the full meaning of our own behavior. And if we could more or less see what was happening while we were manipulated, we may feel that the manipulation nevertheless circumvented or co-opted our will, making it difficult or impossible to do what we otherwise would have chosen to do.

3 Manipulation: philosophical analyses

Of course, such assessments are not always accurate. What seemed to us to have been manipulation may not have been. There is a difference between believing that one has been manipulated and being manipulated (Rudinow 1978: 338). What, then, makes it true that one has been manipulated?

The core commitment of most manipulation theorists is that manipulation involves the non-coercive bypassing or undermining of the manipulated agent's rationality. There is something quite compelling about this claim and thus it is not surprising that many writers have defended some version of it. Marcia Baron has argued that manipulativeness is a vice because it "reflects either a failure to view others as rational beings, or an impatience over the nuisance of having to treat them as rational – and as equals" (Baron 2003: 50). According to Patricia Greenspan, manipulation is "an interference with rational self-governance," while for Lawrence Stern "manipulation is subverting or by-passing another person's rational or moral capacities for the sake of some result" (Greenspan 2003: 164; Stern 1974: 74). Following Alfred Mele (2001), Eric Cave claims that manipulation "operates by bypassing an agent's capacities for control over her mental life" (Cave 2007: 133). According to T.M. Wilkinson, manipulators intentionally pervert the target agent's decision-making processes (Wilkinson 2013: 347). Thomas Hill has argued that manipulation is "intentionally causing or encouraging people to make the decisions one wants them to make by actively promoting their making the decisions in ways that rational persons would not want to make their decisions" (Hill 1984: 251). On Jason Hanna's favored view, "manipulation . . . involves the intentional use of nonrational means of influence, where these means affect a target's deliberation for the worse" (Hanna 2015: 630). For Robert Noggle, manipulators aim to get their targets to behave in ways that fall short of relevant ideals (epistemic, emotional, etc.) and that as a result "acting manipulatively violates rational agency and fails to respect the personhood of the victim" (Noggle 1996: 53).

I have argued elsewhere that manipulation need not bypass or subvert an agent's rational capacities (Gorin 2014). I will not rehearse that argument here, though I briefly will explain how my view departs from those that do take the bypassing or subversion of rationality to be essential to manipulation. On my view, manipulation is best understood as a process of influence that deliberately fails to track reasons, which failure need not entail the bypassing or subversion of the targeted agent's rationality.

On the view I propose, A manipulates B if and only if A deliberately and non-coercively influences B to x and one of the following conditions is met:

(1) A believes that B lacks sufficient reason to x.
(2) A believes B has sufficient reason(s) to x, but A is not motivated by this reason(s).
(3) A's influence of B is motivated by B's sufficient reason to x, but A deliberately leads B to x in light of some other reason.
(4) A exploits means of influence that do not reliably track reasons.[1]

The most common variety of manipulative acts fall under (1). These are cases in which the influencing agent cannot make an appeal to good reasons in order to motivate her target's behavior because the influencer does not believe there are such reasons. In these cases, a manipulator is often intending that the manipulee behave in ways that set the manipulee's interests back. Consider a kidnapper who lures a child into his car with candy or an aspiring, utterly self-interested tyrant who runs a fear-based political campaign to win office. Assuming these agents do not believe the targets of their influence have sufficient reason to do what the influencer intends they do, the influence takes the form stated in (1). However, some kidnappers and aspiring tyrants may believe that their targets have sufficient reason to act in the relevant ways. For example, a tyrant may believe that life under his rule will benefit everyone. In cases like this, the manipulation will take another form, such as that captured by (2) or (3).

Manipulators often will bypass or subvert the manipulee's rational capacities – this happens most clearly when manipulation takes the form stated in (1) – but, contrary to the established view, they will not always do so. Manipulation that does violate an agent's rationality has received fairly extensive discussion, so here I briefly will focus on manipulation that does not do so. Consider the following case, which is an example of manipulation taking the form specified by (3):

> *Playtime*: Time spent interacting with other people is very important for children's healthy psychosocial development. In speaking to the parents of a child with delayed social development, which delay counselors suspect is partly the result of insufficient "quality time" with her family, counselors have discovered that the parents are more informed about, and motivated by, the role of physical activity in promoting their daughter's health than they are about healthy psychosocial development. Knowing this, when advising parents about the importance of playing with their child the counselors emphasize the physical health benefits of physical exercise. The counselors do this even though they know that young children, who tend to move a lot, generally get sufficient exercise with or without the company of others.

The counselors do not lie to the parents. They do not emphasize their true motive – improving the child's psychosocial development – but there is no general obligation to inform others of all of one's intentions or beliefs, especially when one has good reason to believe that it will make no difference. We are in some contexts obligated to share relevant information, but typically relevance is cashed out in terms of considerations that may lead the agent receiving the information to pursue some alternative course of action (Faden and Beauchamp 1986; Nozick 1974: 31). For example, a physician who does not tell her patient about the risk of serious side effects has failed to discharge her moral and professional duty to provide relevant information. The counselors do not hide information of this sort. Rather, they deemphasize information they have good reason to believe the parents will judge to be irrelevant and that requires valuable time to elaborate, and whose upshot – more playtime – is identical to the upshot of the information they do share and which is most likely to motivate the parents. Nor do the counselors invoke powerful emotions or irresistible incentives that swamp the parents' deliberative capacities. The reason they cite – that exercise is important, that play is a form of exercise for children – really is a reason for parents to play with their children, and it is a reason these parents in particular recognize as such. After all, even if children tend to get sufficient exercise without their parents' involvement, they assuredly will get it with their parents' involvement. Moreover, parents generally have an interest in promoting their children's welfare, and the counselors are helping them promote this interest. It is safe to assume that if the parents were ideally rational agents, i.e., fully informed, free of any defects in their reasoning, and so on, they would play with their children, absent compelling countervailing considerations. In short, the counselors manipulate the parents without thereby bypassing or subverting their rationality.

Cases covered by (3), like those covered by (2), can be understood as a class of actions commonly characterized as being driven by "ulterior motives." The difference between the two forms of manipulation is that manipulation of the form specified in (3) is motivated by what the manipulator takes to be the manipulee's reasons for engaging in the relevant behavior, while manipulation of the form specified in (2) is not motivated by the manipulator's recognition of these reasons. In (2), the manipulator just happens to believe that her manipulee has sufficient reasons to do what the manipulator intends she do. This belief plays no role in explaining why the manipulator does what she does – the reasons for which she acts are independent of her beliefs regarding what her victim has most reason to do. Suppose a professor invites one of her graduate students to have coffee with her, not because she is especially interested in sitting down for a chat, but because she is leaving the country soon and knows that if the graduate student does not complete the next chapter of his dissertation before she leaves he will fall dangerously behind schedule. The professor knows the best way to get her student to do what he has most reason to do – complete the chapter on time – is to put him in a position where if he fails to do this he will feel ashamed to face her. The professor regards her student's shame-based motivation as silly, grounded in his granting undue deference to her authority, and yet she deliberately provides the occasion for this motivation to move him to act. This case falls under (3), while a similar case in which the professor has all the same beliefs about what her student has most reason to do but is motivated instead by her desire to, say, watch her student squirm while describing a rough draft of a chapter falls under (2).

Category (4) covers cases in which a manipulator chooses means of influence that by their natures do not reliably track reasons. I do not have the space here to provide an adequate account of reliable reason-tracking, but the intuitive idea should be clear enough. Such means are not considerations – good or bad – in light of which one acts. For example, if I know you are more likely to do what I want you to do if I play Beethoven on the stereo while I make my request, and you do what I want in the presence of the music but would not have done so in its absence, and yet you do not take the music to be a reason for acting as you do – perhaps you do not even notice the music – then I successfully have employed a means of influence that does not reliably track reasons. The presence of the music is irrelevant with respect to the considerations that speak for or against your acquiescing to my request, and you yourself do not recognize the music as providing any reason for you to do what you do. The methods employed by libertarian paternalists in the construction of "choice architecture" – for example the exploitation of framing effects and the power of defaults – are means that do not reliably track reasons (Sunstein and Thaler 2003). Therefore, on the account of manipulation I favor, libertarian paternalist "nudges" are manipulative.

4 Paternalistic manipulation

Traditionally, paternalism has been thought to involve constraints on or interferences with liberty (Mill 1989 [1859]). It is widely accepted as a conceptual matter that a paternalistic act or policy restricts an agent's liberty or interferes substantially with her ability to act as she wishes and that it does so for the benefit of the agent targeted by the act or policy. For example, a parent who withholds dessert until her child has eaten more nutritious foods behaves paternalistically when this is done for the benefit of the child, and the state acts paternalistically when it coercively enforces seatbelt laws when this is done to protect drivers from injury or death. Objections to paternalism are generally grounded in liberals' commitment to what Joel Feinberg has called "the presumption in favor of liberty" (Feinberg 1984: 9). The presumption in favor of liberty holds that people should be free to do as they wish unless limitations on their liberty can be justified. Theorists who oppose paternalistic policies hold that interferences with an agent's liberty are not justified by the welfare gains (or purported welfare gains) the interferences make possible when those gains accrue not to third parties but to the agent herself, so long as the agent is competent to make her own decisions about how to live her life. By limiting an agent's liberty "for her own good," the paternalist either forcefully imposes the paternalist's judgment about value onto the agent or forcefully imposes the agent's own better judgment onto her when she is unable or unwilling independently to act in accordance with that judgment. Such an imposition, the objection goes, may be an infringement on the victim's autonomy.

It is not clear, however, that paternalistic influence always restricts or interferes with liberty. George Tsai has compellingly argued that rational persuasion can be paternalistic, even though rational persuasion – the offering of reasons, evidence, or argument – does not limit liberty or meaningfully interfere with one's ability to act as one wishes (though it may, of course, alter what one wishes to do). Rational persuasion is paternalistic when

> it is motivated by distrust in the other's capacity to adequately recognize or weigh reasons that bear on her good, when it conveys that she is insufficiently capable of engaging with those reasons, as a competent person is expected to be able to do, and when it occludes an opportunity for her to engage independently with those reasons herself.
> (Tsai 2014: 111)

Imagine you tell a colleague that you plan to take a vacation but do not yet know where you will go. The next day you receive from this colleague an email containing reasons – airfare is cheap this time of year, the beaches are terrific, one cannot find better sunsets anywhere – for why you ought to go to Hawaii. Knowing your colleague to be overbearing, you understand that he is not intending merely to offer you information and suggestions to aid you in your deliberations. Rather, you know he doubts your ability to acquire the relevant information and to weigh the various considerations in the right way. What is best for you, on his view, is that you should go to Hawaii and nowhere else. His email is intended to preempt your making what he is certain will be a foolish decision, coming as it will from someone other than himself, a decision that will fail to promote your best interests, which interests his sagacity allows him to discern. Or, imagine a husband who regularly preempts his wife's deliberations by rationally persuading her as to her best courses of action in the belief that she is incapable of discovering or assessing reasons for herself, of working out for herself what she has most reason to do. Such cases of rational persuasion are paternalistic, though they do not involve in any meaningful way any restrictions of or interferences with anyone's liberty. The recipients of the reasons and arguments are not coerced, nor do they face incentives that are difficult to resist.

Against the background of the commonly held views of paternalism as necessarily liberty-restricting and rational persuasion as autonomy-respecting, rational persuasion is perhaps the most surprising form that paternalistic influence can take. But it is not the only one, for manipulation, too, can be paternalistic. Put succinctly, paternalistic manipulation is a process of influence that deliberately fails to track reasons and that is intended to advance the manipulee's good. It is important to note that my account of manipulation is not the only account on which paternalistic manipulation is possible. Views on which manipulation involves the bypassing or subversion of the manipulee's rationality may also be compatible with paternalism. Even views on which manipulation necessarily leaves the manipulee worse off along one dimension can allow that on balance the manipulation promotes her interests or welfare. For example, according to Jason Hanna, manipulation worsens the manipulee's deliberative position (Hanna 2015). The counselors in *Playtime* may worsen the parents' deliberative position by stressing the health benefits of exercise rather than the psychosocial benefits, but in doing so they promote the parents' interests in improving the welfare of their children. "Manipulative" is a characterization of the means by which one agent influenced another, while "paternalistic" is a characterization of the particular end, namely, the good of the agent targeted by the intervention, at which the influencing agent aims. Just as some, but not all, coercion is motivated by concern for the coerced agent's welfare, so too is some, but not all, manipulation motivated by concern for the manipulated agent's welfare or interests. Only when the manipulation is done to advance the manipulee's good is the manipulation paternalistic. Manipulation that takes the form specified by disjunct (1) above, where the manipulator does not believe the manipulee has good reason to x, is not paternalistic. Nor is manipulation of the form in (2) paternalistic, for there the manipulator is not working to promote the good of the manipulee. Manipulation taking the form of (3) or (4), on the other hand, will be paternalistic when it is done in order to promote the manipulee's good.

5 The ethics of manipulation

On the account I sketched above, a manipulator sometimes deliberately causes her target to behave in ways the manipulator believes are indeed supported by good reasons, but does so not in light of these reasons but rather in light of what the manipulator takes to be some other, weaker reasons or for reasons the influencer regards to be bad. What, if anything, is wrong with this kind of influence, when it is intended to promote the manipulee's good? That is, what, if anything, is wrong with paternalistic manipulation? The following principle can help answer this question.

> *The Reason to Act for Reasons Principle (RARP)*: If an agent has sufficient reason to Φ, then that agent has sufficient reason to-Φ-for-that-reason.

RARP is a modest principle. What it states is that reasons are behavioral guides. Though the principle is a modest one it is not trivial, for it might be thought that so long as a person does what she has sufficient reason to do it does not matter whether or not this reason plays a role in determining her behavior, that so long as one does what one has sufficient reason to do the demands of reason are satisfied. On this view, what matters is that one's behavior accords with reason, irrespective of whether or not the behavior is guided by reason. I will call this view the *Reasons Endorse* (RE) view, as it holds that reasons endorse behavior but do not prescribe it. If RE is true, then RARP cannot help explain the ethical significance of manipulation.

It cannot be right that reasons endorse behavior but do not prescribe it, that is, that a person is rationally unimpeachable so long as she does what she has most reason to do, irrespective of whether those reasons motivate her to act.[2] To see why, consider a person who is not responsive to reasons but who nevertheless exhibits outward behavior that is identical to the behavior of someone in her position who is ideally responsive to reasons. Imagine a mentally ill person – I will call her Deludia – who suffers from all-encompassing delusions. Her mental states uniformly fail to represent reality. However, through sheer happenstance Deludia's mental states correspond (non-representationally) with reality in a way that leads her to behave in reason-supported ways. For instance, Deludia goes to the doctor whenever she has a very high fever but only because this is when she happens to "remember" that medical treatment will cause her hair to turn red, something she welcomes because she believes only red-haired people can avoid detection by CIA agents. There is no causal connection between her high fevers and her "remembering" how to avoid detection by the CIA – the etiology of the psychological processes that lead to this belief and the etiology of the fevers are unrelated. Suppose that Deludia's behavior *always* has this character, so that on the one hand she consistently does what she has good reasons to do but on the other hand she is never motivated by these reasons. An external observer might conclude that she is a robustly reasons-responsive agent, while her own explanations of her behavior make reference to all manner of fantastical entities and events. Deludia's behavior runs counter to what RARP prescribes, for though she behaves in reason-supported ways she does not do so in light of these supporting reasons.

Clearly there is something unfortunate about Deludia's situation. Her relations with the world are deeply defective. She is detached from reality, radically mistaken about various features of her environment and their salience with respect to what she has reason to do. Most obviously, she holds many beliefs that are false, e.g., there are no CIA agents following her and medical treatment will not turn her hair red. But Deludia's situation is an unusual one insofar as her false beliefs about the world do not cause her to behave any differently than she would behave if she were ideally informed and rational. Though her distorted view of the world makes her (or her mental states) defective from an epistemic point of view, Deludia's behavior would be judged by an external observer to be responsive to reasons. Because she does in fact always do what she has good reason to do, it is not a trivial matter to explain what is wrong with Deludia *qua* practical agent. After all, though many of her beliefs are false, she holds true beliefs about what she has most reason to do, e.g., she correctly believes that she has most reason to go to the doctor today, even if she is mistaken about the nature of this reason. Moreover, Deludia does not suffer from weakness of will. When she comes to a settled judgment about what she has most reason to do, she does it. If RE is true, then from a practical point of view Deludia's behavior is beyond reproach, for what she does is endorsed by reasons.

An appeal to RARP can do better because according to that principle Deludia does not do something she has sufficient reason to do, namely, to behave in reason-supported ways *and* to do so in light of those reasons. When Deludia is ill she has reason to see a doctor because medical treatment is likely to promote and protect her health, a claim with which both RE and RARP are consistent. But when Deludia does not recognize and appropriately respond to this feature of a doctor's visit as a guide to her behavior she fails to do what she has sufficient reason to do, for one always has sufficient reason to make one's sufficient reasons action-guiding.

The case of Deludia is analogous to Robert Nozick's well-known "experience machine" thought experiment (Nozick 1974: 42–43). Nozick asks us to imagine a machine that can stimulate our brains in such a way as to bring about any experience we might desire. Someone who has always wanted to climb Mt. Everest, or to have devoted friends, or to write a great book can have the experience of doing or having these things simply by allowing scientists to plug her into a sophisticated machine. Though the subjective states – that is, the experiences – caused by the machine are indistinguishable from the states that would be had were one actually to accomplish the things experienced, Nozick thinks we nonetheless have good reason to reject the invitation to spend our lives plugged into the machine. First, he argues that we want to do certain things and not just to have the experience of doing them and that moreover we want the experience of doing certain things only because we want actually to do them. Second, there is some kind of person that each of us wants to be – for example, we might want to be kind, or courageous, or accomplished – and a human blob plugged into a machine cannot be any of these things. Third, Nozick claims that by plugging in to the machine we "limit ourselves to a man-made reality, to a world no deeper or more important than that which people can construct" (Nozick 1974: 43). Nozick's experience machine argument is meant to show that any moral theory that places primary value on people's having certain subjective mental states conflates the value of the subjective effect of doing or having something with the value of doing or having that thing. The idea is that if we would not choose to live our lives plugged in to the experience machine, then we must not care primarily about experiences. Nozick concludes that we have good reason to reject subjective state theories of well-being like hedonism.

There are obvious differences between Deludia's detachment from reality and that of someone wired up to the experience machine. Nevertheless, the cases are analogous insofar as neither of the "successes" – i.e., behaving in reason-supported ways, having a pleasurable experience – is related in the right way to the features of the world that explain what it is about these things that make them worth having or doing. When the experience machine delivers the experience of winning the Nobel Prize the experience is not caused by winning the Nobel Prize. Deludia's seeking medical care is similarly causally unrelated to the consideration that speaks in favor of her seeking medical care. Deludia's actions as well as the subjective states of the person in the experience machine are defective because they fail to correspond with the relevant features of the world. Nozick maintains that ultimately it is not experiences that people are after. Rather, what people want is actually to do, have, or be the things the purportedly desirable experiences represent them as doing, having, or being. Deludia's case suggests an analogous conclusion, this time with respect to the role reasons play in our behavior.

Reflection on Deludia's situation illuminates an important truth: that an agent's behavior can be defective even when it is supported by good reasons. More specifically, behavior is defective when these reasons fail to play a role in the processes that determine the behavior. Despite always doing what she has good reason to do Deludia is profoundly out of sync with the normatively significant features of her world, totally disengaged from the considerations that ought to govern her behavior. Insofar as we aspire to let our behavior be guided by reason we view Deludia's predicament with sorrow.

Like Deludia, the targets of successful paternalistic manipulation may be left detached from an important aspect of reality, namely, the reasons that ought to govern their behavior. I will now characterize the nature of this detachment. In doing so, I draw upon Nomy Arpaly's and Julia Markovits's work on what makes a morally right action morally worthy or praiseworthy.

Markovits wants to explain what it is about some actions that makes them morally worthy. Kant argued that an action's rightness does not determine its moral worth. One of his famous examples is that of the merchant who does not overcharge an inexperienced customer. The merchant treats his customers honestly but if the merchant behaves this way out of mere prudence, that is, so that he does not develop a bad reputation, then the merchant's motives are selfish and hence not motivated by his recognition of his moral duty (Kant 1997 [1785]). As Markovits reminds us, one of Kant's more controversial claims is that beneficent actions grounded in an agent's selfless desire to promote the happiness or well-being of others are entirely lacking in moral worth. This is so, according to Kant, because such actions are not performed from the motive of duty but by her simple desire to make people better off. Markovits calls this interpretation of Kant's account of moral worth the *Motive of Duty Thesis* (Markovits 2010: 202).

Markovits wants to replace the *Motive of Duty Thesis* – that is, the thesis that an action has moral worth if and only if it is performed because it is right – with an alternative view that can vindicate Kant's emphasis on the role motives play in determining an action's moral worth without entailing some of the less attractive features of his view. Markovits labels her alternative account *The Coincident Reasons Thesis*. According to this thesis,

> *my action is morally worthy if and only if my motivating reasons for acting coincide with the reasons morally justifying the action* – that is, if and only if I perform the action I morally ought to perform, for the (normative) reasons why it morally ought to be performed.
> (Markovits 2010: 205, emphasis in original)

On this account, an agent who performs an action in light of the moral considerations that support that action performs an action with moral worth. It is neither necessary nor sufficient that the motive of the agent refers to the rightness of the action or to her duty to perform it. What matters is that the agent's behavior is guided by the moral considerations that *make* the action right. Markovits illustrates the distinction between the reasons that make an action right and the rightness of the action with an appeal to a scene from Mark Twain's *Huckleberry Finn*. Huck decides to protect his companion, the runaway slave Jim, rather than to turn him in to the slaveholders who are looking for him. Huck settles on this course of action despite his believing that in doing so he is stealing and therefore doing wrong. Huck is not motivated by the rightness of this action because he does not believe it is right. Instead, he is motivated by the considerations that make his action right, namely, that Jim seems to him to display common characteristics of humanity like love, decency, and friendship. According to the *Motive of Duty Thesis*, Huck acted rightly in saving Jim but his action lacked moral worth. According to the *Coincident Reasons Thesis*, Huck's action was both rightful and morally worthy. Clearly the latter view better accounts for our considered judgments about the moral worth of Huck's saving Jim.

Arpaly articulates a principle quite similar to *The Coincident Reasons Thesis*.

> *Praiseworthiness as Responsiveness to Moral Reasons (PRMR)*: For an agent to be morally praiseworthy for doing the right thing is for her to have done the right thing for the relevant moral reasons – that is, the reasons for which she acts are identical to the reasons for which the action is right.
> (Arpaly 2002: 72)

Like Markovits, Arpaly employs the example of Huck to illustrate this principle (Arpaly 2002: 75–79). She argues that it is Huck's responding to the relevant moral reasons – the reasons that make the action right – that makes his action morally praiseworthy even when Huck does not believe these reasons to be moral reasons. Huck does not do what he does because he believes it is the right thing to do. Rather, he does what he does for the reasons that make it the right thing to do.

Markovits's and Arpaly's compelling accounts of an action's moral worth or praiseworthiness can help illuminate what is troubling about paternalistic manipulation. When a manipulator intends a manipulee to behave in ways the manipulator believes to be supported by reasons and yet deliberately fails to make these reasons apparent as action-guides to the manipulee, the manipulator displays an indifference to what I will call the *normative worth* of the manipulee's behavior. Behavior has normative worth to the extent that it is motivated by the right reasons, that is, to the extent that the motivating reasons of the agent coincide with the reasons that justify the behavior. In understanding normative worth in this way, I am extending Markovits's and Arpaly's accounts beyond right actions and to behavior more generally.

The preceding accounts of the conditions that must obtain to render an action morally worthy bear a resemblance to the account of reason-tracking that occupies a central place in my account of manipulation. They too emphasize how the relations between an agent's motivating reasons and her normative reasons can affect our evaluation of the character of the agent's behavior. However, the distinction Arpaly and Markovits draw between morally worthy actions and those lacking moral worth applies to contexts in which the only relevant reasons are those that justify actions morally. The question of normative worth, by contrast, applies to any context in which it is appropriate to explain or justify behavior by an appeal to reasons, irrespective of whether they are moral reasons. As such, I understand behavior to have normative worth *if and only if the motivating reasons that explain the behavior coincide with the reasons that justify the behavior.*

Recall that according to RARP an agent who has sufficient reason to Φ has sufficient reason to Φ-for-that-reason. This principle, when conjoined with the account of normative worth just sketched, entails that the behavior of an agent who has sufficient reason to Φ and yet fails to Φ-for-that-reason (or for any other sufficient reason) lacks normative worth. On this view, Deludia's behavior is lacking in normative worth because the reasons that explain her behavior do not coincide with the reasons that support it.

Manipulators engaged in paternalistic manipulation make use of means of influence that can leave their manipulees detached from the considerations that ought to govern their behavior. As a result, such behavior is defective, lacking in normative worth. Thus, RARP can explain why it is *pro tanto* morally impermissible deliberately to influence a reasons-responsive agent in a way that leaves her in this predicament. But what about agents who are not reasons-responsive or who are generally reasons-responsive but unresponsive to the relevant reason in a particular case? Because such agents are incapable of appropriately recognizing and responding to the relevant reasons – in a case of paternalism the reason will be that x-ing will promote the agent's good – the manipulator cannot be required to limit her modes of influence to those that refer to those justifying reasons. If an agent is incapable of non-defective action there can be no obligation on the part of those wishing to influence her to avoid causing her to behave defectively. In such cases it is sufficient from the moral point of view that the influencer aims to bring about behavior she believes to be endorsed by good reasons and that she believes *would* guide the behavior of the manipulee were the manipulee capable of recognizing and appropriately responding to these reasons. For example, it is permissible to manipulate children for their own good when they lack the resources to act in light of the relevant reasons. Similarly, clinicians may permissibly employ manipulative nudges such as defaults or framing effects in situations where otherwise competent patients cannot discern what they have most reason to do because the options they face are especially difficult to navigate (Gorin et al. 2017). Libertarian paternalist policies in general, premised as they are on the empirically supported view of human beings as "predictably irrational" while exploiting this very irrationality for the benefit of those targeted by the policies, will often paternalistically manipulate their targets via manipulation of the form captured in (4) above without thereby wronging them (Ariely 2008; Sunstein and Thaler 2008).

6 Conclusion

I have argued for a conception of interpersonal manipulation according to which A manipulates B if and only if A deliberately and non-coercively influences B to x via a process that fails to track reasons in one of four distinct ways. When manipulation is *pro tanto* wrongful it is because the manipulator has deliberately caused the manipulee to engage in behavior that is defective insofar as it is detached from an important aspect of reality, namely, the reasons that ought to govern the agent's behavior. Crucially, manipulation is not always even *pro tanto* wrongful, for an influencer cannot be obligated to limit her means of influence to reason-tracking forms of influence when the agent she intends to influence is not responsive to the relevant reasons.

Paternalistic manipulation is manipulation that is done for the purpose of promoting the manipulee's good. When an agent is unable to recognize or respond directly to the reasons that, were she to recognize and respond to them, would promote her good, it is permissible to manipulate her into acting in accordance with these reasons. Thus, there is nothing wrong with paternalistically manipulating young children or even adults who, though otherwise competent, cannot discern or appropriately respond to the relevant good-promoting considerations that motivate the manipulation.[3]

Related topics

Kantian Perspectives on Paternalism; Libertarian Paternalism, Nudging, and Public Policy; Paternalism and the Practitioner/Patient Relationship.

Notes

1 I owe Matthew Flummer (personal communication) my gratitude for his helpful criticism of an earlier formulation of my view.

2 I am grateful to Jason Hanna (personal communication) for pushing me to clarify the Reasons Endorse view.

3 I wish to thank the editors of this volume for their careful, detailed, and incisive comments on an earlier draft of this chapter.

References

Ariely, D. (2008) *Predictably Irrational*, New York: HarperCollins.

Arpaly, N. (2002) *Unprincipled Virtue: An Inquiry into Moral Agency*, New York: Oxford University Press.

Baron, M. (2003) "Manipulativeness," *Proceedings and Addresses of the American Philosophical Association* 77: 37–54.

Cave, E. M. (2007) "What's Wrong With Motive Manipulation?" *Ethical Theory and Moral Practice* 10: 129–144.

Faden, R. R. and Beauchamp, T. L. (1986) *A History and Theory of Informed Consent*, New York: Oxford University Press.

Feinberg, J. (1984) *The Moral Limits of the Criminal Law, Volume 1: Harm to Others*, New York: Oxford University Press.

Gorin, M. (2014) "Do Manipulators Always Threaten Rationality?" *American Philosophical Quarterly* 51: 51–61.

Gorin, M., Joffe, S., Dickert, N. and Halpern, S. (2017) "Justifying Clinical Nudges," Hastings Center Report 47.

Greenspan, P. (2003) "The Problem With Manipulation," *American Philosophical Quarterly* 40: 155–164.

Hanna, J. (2015) "Libertarian Paternalism, Manipulation, and the Shaping of Preferences," *Social Theory and Practice* 41: 618–643.

Hill, T. E. (1984) "Autonomy and Benevolent Lies," *The Journal of Value Inquiry* 18: 251–267.

Kant, I. (1997 [1785]) *Groundwork of the Metaphysics of Morals*, trans. M. Gregor, Cambridge: Cambridge University Press.

Markovits, J. (2010) "Acting for the Right Reasons," *Philosophical Review* 119: 201–242.

Mele, A. R. (2001) *Autonomous Agents: From Self-Control to Autonomy*, New York: Oxford University Press.

Mill, J. S. (1989 [1859]) *'On Liberty' and Other Writings*, ed. S. Collini, Cambridge: Cambridge University Press.

Noggle, R. (1996) "Manipulative Actions: A Conceptual and Moral Analysis," *American Philosophical Quarterly* 33: 43–55.
Nozick, R. (1974) *Anarchy, State, and Utopia*, New York: Basic Books.
Rudinow, J. (1978) "Manipulation," *Ethics* 88: 338–347.
Stern, L. (1974) "Freedom, Blame, and Moral Community," *The Journal of Philosophy* 71: 72–84.
Sunstein, C. R. and Thaler, R. H. (2003) "Libertarian Paternalism Is Not an Oxymoron," *The University of Chicago Law Review* 70: 1159–1202.
Thaler, R. H. and Sunstein, C. R. (2008) *Nudge: Improving Decisions About Health, Wealth, and Happiness*, New Haven, CT: Yale University Press.
Tsai, G. (2014) "Rational Persuasion as Paternalism," *Philosophy & Public Affairs* 42: 78–112.
Wilkinson, T. M. (2013) "Nudging and Manipulation," *Political Studies* 61: 341–355.

20

PATERNALISTIC LYING AND DECEPTION

Andreas Stokke

1 Introduction

Many standard examples of paternalism involve lying.[1] A doctor lies to a patient about her condition because she thinks the patient is likely to get severely depressed by knowing the full extent of her illness. Hoping to avoid unnecessary anguish, the police tell the wife of a man who died in a car crash that he died instantly, although they know that his death was painful and horrible. A teacher hopes to improve a student's dedication and thereby her performance by telling her that she has good philosophical abilities, even though she believes the student's abilities to be poor. A husband is worried that his wife might attempt suicide and tells her that there are no sleeping pills in the house, although he knows there are. Indeed, Gerald Dworkin describes "lying and force" as "the main instruments of paternalistic interference" (Dworkin 2016: sec. 3).

There are also many examples of paternalistic acts that involve deception but not lying. Suppose the husband from above simply hides the sleeping pills from his wife without saying anything to her about it.[2] On any sensible theory of lying, you have not lied unless you have made an utterance that counts as what is variously referred to as a statement, an assertion, or saying something.[3]

This chapter is concerned with paternalistic lying and its relation to deception, but will not consider paternalistic deception that is not lying. We focus on two broad questions concerning paternalistic lying specifically: When is an act of lying an act of paternalism, and where does paternalistic lying fit on the moral spectrum of lying? No attempt is made at giving substantial answers to these questions, nor of surveying all, or even most, existing proposals. Instead this chapter aims at highlighting some of the issues involved and to make explicit some of the relevant ways they interact.

Sections 2–3 locate paternalistic lies within a broader category of altruistic lies. The standard view that paternalistic action involves a certain kind of interference with autonomy is discussed in relation to lying. Sections 4–6 consider the relation between paternalistic lying, deception, and manipulation. It is suggested that paternalistic lying does not always involve deception. Sections 7–10 introduce two main views on the morality of lying and draw attention to three main moral approaches to paternalistic lying.

2 Altruistic lies

It is commonplace to think that lies are often (perhaps even always) harmful at least to some extent.[4] But we should recognize that, while this may be true, arguably most lies are told in order to benefit someone, usually the liar herself. The ways in which a lie can benefit someone are many.[5] A general distinction might be made between cases in which one or more consequences of the lie itself are of positive value to someone and cases in which a lie is used to prevent harm. Here we will speak of a lie benefitting someone to cover both these kinds of situation. So a lie that benefits someone, as we will use this terminology, is a lie that either brings positive value to someone or prevents harm to someone (or both).

Given this very broad understanding of benefitting someone, it is hard to imagine a lie that is told without any hope of benefitting someone in some way or other. Even lies that are told purely for amusement are at least aimed at giving pleasure to the liar and her accomplices. Typical lies are what we might call *selfish lies*, that is, lies people tell in order to benefit themselves. Someone might try to impress their date by bragging that they are an excellent cook, even though they know that their culinary skills are subpar. People sometimes try to gain access to goods and services by lying about things like their health condition, their family history, their financial situation, and so on. Yet selfish lies are also sometimes told in self-defense, and in such cases may be easier to accept.[6]

On the other hand, people often tell *altruistic lies*, that is, lies told to benefit others. The stock example of lying to a murderer who comes to one's door and asks for the whereabouts of her intended victim is a clear instance of an altruistic lie. It is useful to distinguish a number of different forms of altruistic lying. Sometimes people lie in order to benefit others while taking great care to avoid harming anyone in the process. We might call such lying *harmless lying*.

Depending on one's view about the extent to which lying always, mostly, or only sometimes involves harm, it might be thought impossible, hard, or sometimes feasible to succeed in harmless lying, that is, in telling a lie that benefits someone while not harming anyone in the process. For example, if one thinks that even the lie to the murderer at the door harms the murderer to some degree, lying to save someone's life in this case is still not a harmless lie.

Whether or not one thinks that purely harmless lying is possible, it should be clear that there are also many cases in which people try to benefit others by lying while recognizing that doing so will also bring some harm to someone, perhaps the person lied to, perhaps someone else. Such *balanced lies* are arguably common. Indeed, if all lying is harmful at least to some degree, then all altruistic lies are balanced lies.

3 Paternalistic lies

Which of these types of lies can be paternalistic? It is natural to think that, minimally, a paternalistic lie involves some element of altruism. But, just as it is commonplace to observe that not just any kind of altruistic act is an act of paternalism, it is obvious that not all altruistic lies are paternalistic lies.[7] What distinguishes paternalistic lies from altruistic lies more generally? Most writers on paternalism subscribe, more or less explicitly, to a version of the idea that A acts paternalistically toward B only if A's act is done in order to benefit B.[8] Suppose I see you about to shoot someone in the back. If I jump on you, and take away your gun, I am not acting paternalistically if my main concern is to hinder the death of your victim. If, by contrast, I stop you from walking out onto a frozen lake even though you have been warned that the ice is too thin because I am concerned to prevent you from drowning, I may be acting paternalistically.

As an instance of this general idea, it might be held that a lie is paternalistic only if it is told in order to benefit the person lied to. A complication here arises from the difference between two broad ways of characterizing paternalistic acts. One puts weight on consequences, the other on the motives for the act.[9] Accordingly, one characterization of paternalistic lying might hold that a lie is paternalistic only if it benefits the person lied to, while a more motive-oriented view might hold that a lie is paternalistic only if it is told with the motivation of benefitting the addressee. For simplicity, we can put this aside and focus on cases in which a lie fulfills both these criteria, that is, lies that are told in order to benefit the person lied to, and which succeed in doing so.

Further, everyone will agree that paternalism involves more than just acting with the aim of benefitting someone else, even if the person one tries to benefit is also the person toward whom the act is done. A dentist who pulls out a patient's bad tooth is doing something to the patient in order to benefit her, but the dentist is not acting paternalistically in this case. Many theorists will endorse the further claim that paternalistic acts involve a certain kind of *interference* with autonomy. Unsurprisingly, this kind of condition has been spelled out in a great many different ways, and a lot hangs on precisely how it is formulated. (We will consider one in more detail in Section 6.) Regarding paternalistic lying, the relevant question is when, if ever, can lying to someone be considered the kind of interference relevant for paternalism? In the next section this question will be discussed in relation to the involvement of deception in paternalistic lying.

4 Manipulation

Some philosophers think that the moral problem with lying, or at least most lying, stems from its involving *manipulation*. For example, Bernard Williams writes, "In our own time we find it particularly natural to think deceiving people (or at least some people, in some circumstances) is an example of using or manipulating them, and that that is what is wrong with it" (Williams 2002: 93). Similarly, Sissela Bok argues that manipulation is a key factor in the moral status of lying:

> Those who learn that they have been lied to in an important matter . . . see that they were manipulated, that the deceit made them unable to make choices for themselves according to the most adequate information available, unable to act as they would have wanted to act had they known all along.
>
> (Bok 1978: 21–22)

Bok also suggests that paternalism, when it involves deception, is a form of manipulation: "Apart from guidance and persuasion, the paternalist can manipulate in two ways: through force and through deception" (Bok 1978: 216). More generally, it is fair to say that most writers on paternalism will accept that manipulation qualifies as an instance of the kind of interference characteristic of paternalism. So, if one is sympathetic to the idea that lies are typically manipulative, a rough proposal might be that what is distinctive of paternalistic lies is that they manipulate the person lied to with the aim of benefitting them.

One problem with this suggestion is that arguably many examples of paternalistic lying do not seem to trade on manipulation of the person lied to. We have already seen two such examples. The doctor who lies to her patient about her condition in order to stave off depression is naturally seen as engaging in paternalistic lying, but it is less clear that we want to say that she is manipulating the patient. Similarly, if the police lie to a bereaved wife about the death of her husband in order to avoid causing her unnecessary pain, they are arguably telling a paternalistic lie. Yet, again, it is unclear to what extent this kind of lie can be said to manipulate the person lied to.

Some will find it natural to say that, in these cases, those who are being lied to are being deceived, but they are not being manipulated. At the same time, as the remarks by Williams and Bok quoted above suggest, manipulation is typically seen as essentially involving deception, as opposed to other ways of exerting non-rational influence or control over the will of others, such as coercion.[10] So perhaps the right thing to say is that paternalistic lying involves deception, and sometimes this takes the form of manipulation, but sometimes it does not. Accordingly, one proposal is that the hallmark of a paternalistic lie is that it deceives the person lied to with the aim of benefitting them. However, as we will see next, it is arguable that some paternalistic lies do not even involve deception.

5 Paternalism without deception

Against a traditional view of lying as necessarily aiming at deception, it has been argued that there are lies that are told without any intention of deceiving anyone.[11] In the same vein, there are arguably examples of non-deceptive paternalistic lies. Here is one:

> Alan is employed by Burt, and needs to ask Burt's permission to take breaks during his shift. Alan wants to go outside and smoke a cigarette. He asks Burt for permission, but Burt thinks it's better for Alan not to smoke, and so he replies, "There's no time for that," even though he knows that Alan knows that he knows (that Alan knows . . .) there's plenty of time.

Burt lies to Alan, and his lie seems to be a clear case of paternalism. But since Burt knows that Alan will not believe what he says, Burt cannot be trying to deceive Alan.[12] In other words Burt's lie, in this case, is a non-deceptive paternalistic lie.

Consequently, if one thinks that manipulation is a species of deception, one will be reluctant to say that Burt has manipulated Alan in the case above. Indeed, some philosophers who accept that lying does not always involve deception have argued that, for this reason, manipulation cannot, at least in general, be seen as the moral problem with lying. Seana Shiffrin writes,

> Many accounts of the wrong of the lie emphasize the wrong of manipulating or aiming to manipulate the will of the recipient; such accounts implicitly imagine that deception is an aim or product of the lie, for otherwise no such manipulation could occur.
>
> (Shiffrin 2014: 22)

Since Shiffrin agrees that one can lie without intending to deceive, she concludes that the wrong in lying cannot be identified with the wrong in manipulation.

Given this, we might conclude that not all deceptive paternalistic lies are manipulative, as seen from the lies by the doctor and the police above, and moreover, that not all paternalistic lies are deceptive in the first place, as seen from Burt's lie to Alan. Of course, this is not to deny that many paternalistic lies are manipulative. Suppose a doctor thinks that she can make a patient get more exercise and more sleep with benefits to her general state of health if she tells the patient that unless she does these things she will almost certainly go blind, even though she knows that the risk of blindness is minuscule. Arguably the doctor is using the lie to paternalistically manipulate the patient into getting more exercise and sleep in this case.

However, given examples of non-deceptive paternalistic lies, we should ask: Do such lies qualify as interfering in the sense believed to be relevant for paternalism, even if they are not deceptive, and hence not manipulative? One observation here is that Burt's lie arguably restricts Alan's choices, even if it does not do so by means of deception. If Burt had given permission, Alan would have had the choice of smoking or not. But since Burt does not give permission, Alan has no choice in the matter. (Assuming that acting against Burt's instruction is not a live option.) And, moreover, the restriction of Alan's choice was done in order to benefit Alan himself, and not, for example, to benefit Burt or for some other reason. To make this more vivid, one can embellish this story, if one thinks it is necessary. For example, imagine that Burt could not just have said "No," but that, according to rules of the workplace, he must give a reason. In any event, it is plausible that there are cases of this kind in which a paternalistic lie is told without any intention to deceive.

Many writers on paternalism emphasize this kind of influence of choices as a key feature of paternalistic action. For example, on the definition proposed by Simon Clarke (2002: 81), x acts paternalistically toward y only if "x aims to close an option that would otherwise be open to y or x chooses for y in the event that y is unable to choose for himself." And Danny Scoccia (2015: 1) defines paternalism as

> the use of nonrational means (i.e., means other than rational persuasion) to hinder someone's already made choices, preempt the choice that he would make if he could make one (e.g., giving an unconscious Jehovah's Witness a life-saving blood transfusion), or influence his prospective choices, for his own prudential good.

Such ways of understanding paternalism are arguably compatible with non-deceptive paternalistic lies. Burt's lie closes an option for Alan, as Clarke would have it, or preempts the choice that he would have made, as Scoccia might say.

On the other hand, it is not obvious that all paternalistic lies involve influence on choices in this way. Take the original case of the doctor's lie that is aimed at preventing depression, or the police's lie to the wife that is aimed at avoiding excessive grief. These lies do not seem to influence choices, in either Clarke's or Scoccia's sense.

Yet, it might be argued that they nevertheless still interfere with autonomy. That is, one might have a conception of this kind of interference where it is not limited to directly influencing choices in the ways we have been looking at. One might think of autonomy in a more general sense as governing one's own life, or something similar.[13] Arguably, the doctor's lie about the patient's condition violates the patient's autonomy in this kind of broader sense, even if it does not directly influence a particular decision. Similarly, the police's lie to the wife may be seen as diminishing her control of her own life by distorting her perception of it. And, moreover, even if these lies do not influence choices *per se*, they are designed to, at least indirectly, influence behavior, and in this sense they may perhaps be seen as undermining the autonomy of their victims, even in a more narrow sense.

If this is right, then perhaps the best proposal is the general one that what is characteristic of paternalistic lies, as opposed to other forms of altruistic lying, is that they interfere with the autonomy of the person lied to in order to benefit them. On the other hand, some have argued that paternalistic action need not involve interference. Next, we turn to the question of whether paternalistic lies may proceed without interference.

6 Lying and interfering

Even though influence on choices is arguably often a kind of interference, Scoccia (2013, 2015) argues that the fact that paternalism is characterized by influencing choices means that paternalism does not always involve interference with someone's autonomy. He gives the following example:

> Consider a father who pays his teenage son $50 for every random drug test he submits to and passes. The son willingly submits to the arrangement because it rewards him

> for doing what he says he would do anyway. The father believes that his son is sincere when he says this but distrusts his son's strength of will, suspecting that a monetary incentive is needed to help him choose wisely in cases where he faces strong peer pressure to join the fun.
>
> (Scoccia 2013: 75)

Scoccia thinks that "[t]he father's system of rewards for passed tests is surely paternalistic even though it may influence but does not 'interfere with' the son's prospective drug-use choices" (Scoccia 2013: 75). Examples like this motivate the view that paternalism may involve expanding options, creating new alternatives that were not present before. In turn, this is one way of arguing for the possibility of paternalism without interference. That is, examples of this kind can be seen as involving paternalistic actions that do not interfere with autonomy in the sense that they do not block choices or, more generally, inhibit self-governance.

Given that lying is typically thought to be one of the main tools at the paternalist's disposal, it is relevant to ask whether there can be paternalistic lies that do not proceed by means of interference.

Deceptive lies can create the illusion of a situation like the one in Scoccia's example above. One way this may come about is when a lie misinforms about the expected outcome of one or more options. For example, suppose the father in Scoccia's example tells his son that his grandmother has promised to pay him $50 for every successful drug test, even though he knows that she has never done any such thing. At first sight, this might seem, as in the original example, like a case in which a paternalistic lie influences but does not interfere with the son's choices. Indeed, the son has the same choice as in the original example – he can do the drug tests or not. But in the lying case, he is deceived about the outcomes of one of the options.

However, many would probably think that the son is being manipulated in this case, and hence that this is enough to see the case as involving interference of the kind characteristic of standard paternalistic action. Indeed, it is common to remark that one way deceptive lying can manipulate is by misinforming about the available options. That is, not only can deception manipulate by obscuring options, or setting up the illusion of an option that is not in fact available, manipulative deception also sometimes works by misinforming about the consequences of certain (real) options. As Bok says, "the estimates of *costs and benefits* of any action can be endlessly varied through successful deception" (Bok 1978: 20). More generally, then, this may be a reason to think that deceptive paternalistic lying typically works by way of interference.

Can non-deceptive lies be examples of paternalism without interference? It is arguably hard to imagine cases of this kind. We have seen that non-deceptive lies may be used to paternalistically influence choices, as in the case of Burt's lie to Alan. But, as we noted, this kind of case is naturally seen as involving paternalistic interference, since the lie is used to restrict choices. On the other hand, it is difficult to imagine a case in which a non-deceptive lie is used paternalistically to influence choices in the sense of impacting evaluation of the consequences of available options, as in the case of the father and son.

Even if writers like Scoccia are right that paternalistic action generally need not involve interference, then we have some reason to think that it is a feature of paternalistic lying that it involves interference. We have seen that many deceptive paternalistic lies interfere by being manipulative. We also suggested that deceptive paternalistic lies that are not directly manipulative – like the lies told by the doctor or the police – can be construed as interfering given a broader notion of autonomy. And, moreover, even though there are non-deceptive paternalistic lies, such lies appear to turn on interference for their effectiveness.

If paternalistic lying unavoidably, or at least almost always, implies interfering with autonomy, this may be expected to have implications for the moral status of paternalistic lying. The rest of this chapter discusses some of these implications.

7 Absolutism and anti-absolutism

A traditional view on the morality of lying has been that lying is never morally justified. Call this view *absolutism*.[14] As opposed to this, many philosophers have defended the arguably more commonsense *anti-absolutist* view that, although lying is often wrong, lying is at least sometimes morally justified. The most common kind of lying that has been thought by various types of anti-absolutists to be sometimes morally permissible is what we have called altruistic lying. That is, lies told in order to benefit others, in our general sense of either having positive value for someone or preventing harm. Lying to the murderer at the door, for example, is usually thought to be morally permissible – indeed, this kind of case is standardly used to motivate anti-absolutism.

Anti-absolutism about lying can take different forms.[15] One kind of anti-absolutism stems from a view according to which, although lying typically violates particular *rights*, these may be overridden in particular circumstances. For example, one might think that paternalistic lying violates a right to autonomy, and yet at the same time hold that the right to autonomy is not absolute.

Another kind of anti-absolutism is a deontological view that holds that, although there is a standing obligation or duty not to lie, the wrongness of lying can be outweighed by other factors in a given situation, for example, by the danger of someone coming to serious harm. Such a theory might include a principle that says that *pro tanto* one not lie, where a *pro tanto* obligation is, roughly, an obligation that plays a role in determining what one ought to do, but which can be outweighed by other obligations or factors.[16] Lying, according to this kind of anti-absolutist, is sometimes morally permissible, namely in situations where the *pro tanto* wrongness of lying is outweighed by other factors.

Finally, act consequentialists of various kinds will think, roughly, that lying is wrong when and only when there is another available course of action that will lead to a better outcome. Hence, even though one might think that lying often is not the available action with the best consequences, it might sometimes be, and in the latter cases lying is morally right.[17]

For the absolutist, since lying is never permitted, trivially, paternalistic lying is never permitted. To be sure, there might be views on which, even though paternalistic lying is already morally ruled out just by its involving lying, it also has other morally problematic features. Yet, for the absolutist, since lying is always wrong, paternalistic lying is always wrong. On the other hand, for the anti-absolutist, there are particular questions that arise concerning paternalistic lying.

Specifically, for the anti-absolutist, a central question concerns whether there is, or can be, any moral difference between paternalistic lying and altruistic lying more generally. One view is that the kind of interference with autonomy that we have been considering is not special to paternalistic lying. Instead, it might be thought that lying *in general* constitutes a kind of interference with autonomy. For example, as we have suggested, one might think that deceiving someone is a way of interfering with their autonomy. And, moreover, when non-deceptive lies have a point, their point is usually to influence choices or behavior more generally. So there may be ways of arguing that interference with autonomy is a feature of most lying, and not only the paternalistic kind. If one agrees with this line of thinking, one might be inclined to think there is less of a moral difference between paternalistic lying and other forms of lying, and in particular, other forms of altruistic lying.

The rest of this chapter considers some aspects of the issue of to what extent, if at all, paternalistic lying is morally distinct from other kinds of altruistic lying in relation to the three types of anti-absolutist views introduced above.

8 Paternalistic lying and the right to autonomy

On some views the moral problem with paternalistic lying is that, by interfering with the autonomy of its target, paternalistic lying violates a moral right to autonomy. One school of thought considers moral rights to be absolutes in the sense of always overriding other moral considerations.[18] If one thinks of rights in this way, and one also agrees that paternalistic interference invariably violates a right to autonomy, then paternalistic interference is never morally justified. In turn, therefore, paternalistic lying is never permissible.

There are versions of this kind of view that, moreover, commit one to wholesale absolutism about lying. Suppose you agree with the broad idea, mentioned above, that it is not only paternalistic lying, but most kinds of lying, that involve interference with autonomy. Given this, if the kind of interference we are considering constitutes a violation of a right to autonomy, then most lying involves violating such a right. Hence, this kind of approach might be thought to lead to absolutism about lying.[19]

In response one can restrict one's conception of the right to autonomy. For instance, one might appeal to a notion where the right only concerns governance of one's *own* life. If the right to autonomy is understood along these lines, it may be less clear that lying in general violates the right. Lying to the murderer at the door, for example, might be thought not to violate the right to autonomy, if one thinks that the murderer's decision to kill someone else is not a decision about her own life in the relevant sense. Hence, there may be ways of defending a view according to which an absolute right to autonomy rules out paternalistic lying, but does not invariably rule out other kinds of lying. Hence, this kind of view would imply a combination of absolutism about paternalistic lying and anti-absolutism about lying more generally.

On the other hand, others will be more attracted to a more global kind of anti-absolutism. One way of arguing for such a view is to accept that paternalistic lying violates a non-absolute right to autonomy. For example, Thomas Hill suggests that the notion of a right to autonomy be understood as "a moral right against individuals (not the state) (a) to make one's own decisions about matters deeply affecting one's life, (b) without certain sorts of interference by others, (c) provided certain conditions obtain" (Hill 1984: 257). As Hill notes, the type of interference referred to in (b) of his characterization includes the kind of influence on choices we have discussed: "Among these interferences are illegitimate threats, manipulations, and blocking or distorting the perception of options" (Hill 1984: 257). In turn, clause (c) of Hill's characterization implies that the right to autonomy is not absolute, that is, it is not always wrong to violate autonomy. As he says, "The right of autonomy of individuals is also commonly understood to be qualified by a *proviso* that interference is not required to avert a major disaster or to prevent the violation of other, more stringent rights" (Hill 1984: 259). Hence, given this understanding, "Though important, autonomy need not be considered an absolute right" (Hill 1984: 259).

If one thinks of the right to autonomy in this non-absolute way, then paternalistic interference might sometimes be justified. Of course, everything depends on which circumstances one is willing to count among those in which autonomy may be permissibly violated. Hill's suggestion – that the right may be violated in order to "avert a major disaster" – is deliberately vague. Yet it is not unnatural to interpret it such that, for example, lying in order to save someone's life is justified, even if it requires violating their autonomy. For example, consider the husband who lies to his wife about there being sleeping pills in the house because he is worried she might attempt suicide. This kind of paternalistic lying, even though it violates autonomy, might be thought permissible on this kind of view.

On the other hand, if the right to autonomy may only be violated in such acute circumstances, then to the extent that paternalistic lying involves interference, this kind of view will have the consequence that paternalistic lying is rarely morally justified. Hill gives an example in which a roommate lies to her friend in order to prevent her from getting back together with her ex whom the roommate considers to be bad for her friend:

> The roommate manipulates her friend's decision (to call or not to call her "ex"-) by actively concealing pertinent information. If we accept the right of autonomy, this could only be justified if the reunion would have been so great a disaster that the right is over-ridden.
>
> (Hill 1984: 262)

On this understanding of a right to autonomy, then, even though it is not absolute, it is only permissibly overridden in quite narrow circumstances.

9 Paternalistic lying and obligations

From a deontological point of view, anti-absolutism about lying might be grounded in acceptance of *pro tanto* obligations. As we noted, one type of view endorses a *pro tanto* prohibition on lying. As such, it can outweigh other considerations and it can itself be outweighed. Hence, a central issue concerns how this kind of *pro tanto* obligation interacts with other obligations and considerations in various circumstances. In particular, if there is a blanket *pro tanto* obligation not to lie *per se*, it might be asked if paternalistic lying is special morally.

In this connection, it is worth noting that the framework of *pro tanto* obligations can accommodate the idea that interference with autonomy plays a role in moral evaluation. In general, the connection between interference with autonomy and the wrong in lying is independent of any particular moral framework, or at least it can be accounted for by many different frameworks, including one that emphasizes moral rights and one that emphasizes obligations.

Consider again the case of lying to the murderer at the door. Since lying to the murderer at the door is not morally wrong (so we assume), the *pro tanto* obligation not to lie appears to be outweighed, in this case. As we noted earlier, this kind of lie is naturally seen as an example of what we called balanced lies. In particular, the obligation not to lie will usually be seen as outweighed by a consideration concerning the great harm that can be prevented by telling the lie.

Contrast this with Hill's example of the paternalistic roommate lying to her friend in order to prevent her from re-uniting with her ex. Most will think that this lie is harder to justify morally than the lie to the murderer at the door. In terms of weighing different *pro tanto* obligations, there might be two reasons for such an assessment. First, it might be thought that the reason is simply that in the roommate case the harm that can be prevented is less than in the murderer case. Indeed, many will think that, in general, the obligation not to lie is outweighed by considerations concerning harm only if the harm that can be prevented is quite severe.[20] Second, it might be thought that the reason has to do with the paternalistic nature of the roommate's lie.

One argument in favor of the first of these appeals to cases where the harm that can be prevented is more comparable to that in the murderer at the door case. Take the example of the husband lying to his wife about the sleeping pills. If one thinks that, while the roommate's lie to her friend is impermissible, lying to prevent suicide is morally justified, one has some reason to think that the difference between standard lying and paternalistic lying is not significant, when it comes to weighing moral obligations. Rather, the reason the roommate's lie is impermissible is that the harm that can be prevented by lying, in this case, is not great enough to outweigh the obligation not to lie.

Others will think that lying to the suicidal person is harder to justify than lying to the murderer at the door. One can try to account for this by endorsing a principle that says that *pro tanto* one ought not to interfere with autonomy. If there is such a *pro tanto* obligation, in addition to the blanket *pro tanto* obligation not to lie, then it might be thought that together they outweigh the considerations concerning harm in the sleeping pill case. Moreover, as we suggested earlier, if one understands autonomy as concerning decisions about one's own life, it is not obvious that they apply to the murderer at the door, and hence, in that case, the considerations against harm might only have to outweigh the obligation not to lie, without any force coming from considerations about autonomy.

Alternatively, it is sometimes suggested that interfering with someone's autonomy is just another way of harming them.[21] If so, then perhaps those who think that the sleeping pill case and the murderer at the door case differ can argue that we do not need to endorse a specific obligation against interfering with autonomy. Instead, it might be thought that the reason the paternalistic lie in the sleeping pill case is wrong, while lying to the murderer at the door is not, is that in the former case the lie does more harm than good because of its interference with autonomy. Again, this presupposes an understanding of autonomy on which the lie to the murderer at the door does not interfere with autonomy – perhaps because the lie is not about an issue concerning the murderer's own life.

Still, others will feel that it is hard to see how the lie in the sleeping pill case could be said to do more harm than good, especially if it succeeds in preventing suicide. Given this, one might be inclined to either accept that the lie in the sleeping pill case is morally on a par with the lie to the murderer at the door – that is, both are permissible – or one might want to reinstate the idea of a special obligation not to interfere with autonomy.

10 Paternalistic lying and consequences

For the act consequentialist, broadly speaking, paternalistic lying is more often wrong than standard altruistic lying only if, more often than in the case of standard altruistic lying, there is an alternative to paternalistic lying that would lead to a better outcome. This might be the case if paternalistic lying in and of itself carries with it a substantial negative effect.

Why should it do so more than other forms of lying? A possible answer is to say that paternalistic lying tends to leave the person lied to worse off than standard lying does. If one subscribes to the view, mentioned above, that interfering with autonomy is a form of harm, then one can endorse a version of a theory of well-being according to which harming someone typically counts as diminishing their well-being.

This kind of view again assumes that standard lying does not interfere with autonomy – or at least it does so less often than paternalistic lying. But even if this can be made plausible, it might still be doubted whether the act consequentialist can employ this kind of view to argue for a moral difference between paternalistic lying and other forms of altruistic lying.

Contrast again lying to the murderer at the door and the husband's lie about the sleeping pills. On a simple act consequentialist view, the former is right, while the latter is wrong only if in the sleeping pill case there is an alternative course of action to lying with a better outcome profile, whereas in the murderer at the door case there is not. Of course, this might be so due to vagaries of the cases. Supposing we can screen such details off, though, the only way of maintaining a difference of this kind, it might seem, is to argue that it is better not to lie in the sleeping pill case because the harm inflicted by interfering with autonomy is greater than the harm incurred by the suicide (or even attempted suicide). To the extent that this seems implausible, there is some reason to think that the act consequentialist will have a hard time arguing for a moral difference – let alone a principled one – between paternalistic lying and other kinds of altruistic lying.

As has often been observed, though, the act consequentialist nevertheless has many ways of resisting the common complaint that it allows lying whenever it is expedient. Consequentialists typically argue that, although lying is right when it is the course of action with the best outcome, lying tends to have bad outcomes. For example, J.S. Mill (1979 [1861]) famously argued that, even though lying might have good consequences in the short run, it tends to have bad consequences in the long run.[22] In particular, Mill thought that lying tends to make the liar less honest in general and, moreover, undermines trust in others.[23] Similarly, R.M. Hare says,

> For example, it is only too easy to persuade ourselves that the act of telling a lie in order to get ourselves out of a hole does a great deal of good to ourselves at relatively small cost to anybody else; whereas in fact, if we view the situation impartially, the indirect costs are much greater than the total gains.
>
> (Hare 1981: 38)

Carson (in press) argues that this implies that act consequentialism endorses a fairly stringent presumption against lying. This view does not count paternalistic lying as more rarely justified than standard lying, unless a further typical bad consequence of paternalistic lies is identified. If there is no such difference, then paternalistic lying is morally on a par with other kinds of lying, for this kind of theorist.

11 Conclusion

Paternalistic lies are a special case of altruistic lies for which the intended beneficiary of the lie is the person lied to. Further, a common view of paternalistic action implies that paternalistic lies are characterized by involving interference with the autonomy of the intended beneficiary.

One idea is that the kind of interference involved is a form of manipulation. Manipulation is typically thought to be marked by involving deception. Yet paternalistic lying does not necessarily involve deception. One can lie to someone with the aim of thereby benefitting them but without intending to deceive them.

Given this, one conclusion is that paternalistic lying does not necessarily involve interference. However, it is at least hard to imagine cases of paternalistic lying – deceptive or non-deceptive – that parallel the kinds of cases that have been used to motivate the idea that paternalistic actions do not necessarily involve interference with autonomy. There is reason to think, therefore, that at least one key factor in what makes certain altruistic lies paternalistic – deceptive or non-deceptive – is that they involve interference.

According to the anti-absolutist about lying, it is at least sometimes morally permissible to lie. Different moral frameworks can provide different foundations for the further thought that paternalistic lies are among the lies that are sometimes morally permissible. The variations in these explanations depend, among other things, on how individual moral theories view the status of interference with autonomy.

Related topics

Consequentialism, Paternalism, and the Value of Liberty; Epistemic Paternalism; Paternalism and Autonomy; Paternalistic Manipulation; The Concept of Paternalism.

Notes

1 The following are adapted from Dworkin (2016).
2 Since the husband thereby prevents the wife from forming a true belief about whether there are sleeping pills in the house, or perhaps allows her to persist in the false belief that there are not, his act is deceptive, at least on many views. See Chisholm and Feehan (1977).
3 See, e.g., Chisholm and Feehan (1977), Bok (1978), Williams (2002), Carson (2006; 2010), Fallis (2009), Saul (2012), Stokke (2013; 2016).
4 See Kagan (1998: 106–116).
5 See Bok (1978: 84–85).
6 See Bok (1978: 82–83).
7 See, e.g., Shiffrin (2000), Dworkin (2013).
8 An exception is Shiffrin (2000), who holds a view on which A may act paternalistically toward B even though A's act is aimed at benefitting a third party C. See Dworkin (2013) for an overview of different definitions of paternalism.
9 See Dworkin (2013: 26).
10 For discussion, see, e.g., Baron (2003), Todd (2013), Wood (2014).
11 See, e.g., Carson (2006; 2010), Sorensen (2007), Fallis (2009), Saul (2012), Stokke (2013).
12 For discussion of this interpretation of cases of this kind, see Lackey (2013), Fallis (2015), Keiser (2016).
13 See Velleman (1989: chap. 6) for a conception of autonomy of this kind. See also Kagan (1998: 111).
14 Absolutists about lying include Augustine (1952 [395]), Aquinas (1922 [1265–1274]), Kant (1996 [1797]).
15 Defenders of anti-absolutism include Mill (1979 [1861]), Sidgwick (1966 [1907]), Ross (2002 [1930]), Carson (2010), Shiffrin (2014). See also Williams (2002), Carson (in press) for discussion.
16 Ross (2002 [1930]) holds a version of this view. Ross's view is spelled out in terms of what he calls "prima facie" duties. However, as pointed out by Reisner (2013: 4), there are reasons to think that the way he uses this notion is more akin to how the notion of a *pro tanto* obligation is more standardly used. See Reisner (2013) for discussion of the difference.
17 This kind of view is found in Mill (1979 [1861]), Hare (1981).
18 See, e.g., Dworkin (1984).
19 Kant's version of absolutism about lying is often seen as stemming, at least in part, from a view of lies as violating rights. For discussion, see Mahon (2006; 2009).
20 See Ross (2002 [1930]: 41–42) for a suggestion like this.
21 See, e.g., Kagan (1998: 112).
22 On this, see Carson (in press).
23 See Mill (1979 [1861]: 22–23).

References

Aquinas, T. (1922 [1265–1274]) "Of Lying," in *Summa Theologica* (Volume XII, pp. 85–98). London: Burns, Oates, and Washbourne.
Augustine. (1952 [395]) "Against Lying," in R. Deferrari (ed.) *Treaties on Various Subjects* (Vol. 16, pp. 125–179). Washington, DC: Catholic University of America Press.
Baron, M. (2003) "Manipulativeness," *Proceedings and Addresses of the American Philosophical Association* 77: 37–54.
Bok, S. (1978) *Lying: Moral Choice in Private and Public Life*, New York: Random House.
Carson, T. (2006) "The Definition of Lying," *Noûs* 40: 284–306.
———. (2010) *Lying and Deception: Theory and Practice*, Oxford and New York: Oxford University Press.
———. (in press) "What's Wrong With Lying?" forthcoming in E. Michaelson and A. Stokke (eds.) *Lying: Language, Knowledge, Ethics, Politics*, Oxford and New York: Oxford University Press.
Chisholm, R. and Feehan, T. (1977) "The Intent to Deceive," *Journal of Philosophy* 74: 143–159.
Clarke, S. (2002) "A Definition of Paternalism," *Critical Review of International Social and Political Philosophy* 5: 81–91.
Dworkin, G. (2013) "Defining Paternalism," in C. Coons and M. Weber (eds.) *Paternalism: Theory and Practice*, Cambridge: Cambridge University Press, pp. 25–38.
———. (2016) "Paternalism," in E. N. Zalta (ed.) *Stanford Encyclopedia of Philosophy*, http://plato.stanford.edu/entries/paternalism/
Dworkin, R. (1984) "Rights as Trumps," in J. Waldron (ed.) *Theories of Rights*, Oxford and New York: Oxford University Press, pp. 153–167.

Fallis, D. (2009) "What Is Lying?" *Journal of Philosophy* 106: 29–56.
———. (2015) "Are Bald-Faced Lies Deceptive After All?" *Ratio* 28: 81–96.
Hare, R. (1981) *Moral Thinking*, Oxford and New York: Oxford University Press.
Hill, T.E. (1984) "Autonomy and Benevolent Lies," *Journal of Value Inquiry* 18: 251–267.
Kagan, S. (1998) *Normative Ethics*, Boulder, CO: Westview Press.
Kant, I. (1996 [1797]) "On a Supposed Right to Lie from Philanthropy," in M. Gregor (ed.) *Immanuel Kant, Practical Philosophy*, Cambridge and New York: Cambridge University Press, pp. 611–615.
Keiser, J. (2016) "Bald-Faced Lies: How to Make a Move in a Language Game Without Making a Move in a Conversation," *Philosophical Studies* 173: 461–477.
Lackey, J. (2013) "Lies and Deception: An Unhappy Divorce," *Analysis* 73: 236–248.
Mahon, J. (2006) "Kant and the Perfect Duty to Others Not to Lie," *British Journal for the History of Philosophy* 14: 653–685.
———. (2009) "The Truth About Kant on Lies," in C. Martin (ed.) *The Philosophy of Deception*, Oxford and New York: Oxford University Press, pp. 201–224.
Mill, J. S. (1979 [1861]) *Utilitarianism*, Indianapolis: Hackett.
Reisner, A. (2013) "Prima Facie and Pro Tanto Oughts," in H. LaFollette (ed.) *The International Encyclopedia of Ethics*, Oxford: Wiley-Blackwell, pp. 4082–4086.
Ross, W. (2002 [1930]) *The Right and the Good*, ed. P. Stratton-Lake, Oxford and New York: Oxford University Press.
Saul, J. (2012) *Lying, Misleading, and What Is Said: An Exploration in Philosophy of Language and in Ethics*, Oxford and New York: Oxford University Press.
Scoccia, D. (2013) "The Right to Autonomy and the Justification of Hard Paternalism," in C. Coons and D. Weber (eds.) *Paternalism: Theory and Practice*, New York: Cambridge University Press, pp. 74–92.
———. (2015) "Paternalism," in H. LaFollette (ed.) *The International Encyclopedia of Ethics*, Oxford: Wiley-Blackwell, pp. 1–11.
Shiffrin, S. (2000) "Paternalism, Unconscionability Doctrine, and Accommodation," *Philosophy & Public Affairs* 29: 205–250.
———. (2014) *Speech Matters: On Lying, Morality, and the Law*, Princeton, NJ: Princeton University Press.
Sidgwick, H. (1966 [1907]) *The Methods of Ethics*, New York: Dover.
Sorensen, R. (2007) "Bald-Faced Lies! Lying Without the Intent to Deceive," *Pacific Philosophical Quarterly* 88: 251–264.
Stokke, A. (2013) "Lying and Asserting," *Journal of Philosophy* 110: 33–60.
———. (2016) "Lying and Misleading in Discourse," *Philosophical Review* 125: 83–134.
Todd, P. (2013) "Manipulation," in H. LaFollette (ed.) *The International Encyclopedia of Ethics*, Oxford: Wiley-Blackwell, pp. 3139–3145.
Velleman, J. (1989) *Practical Reflection*, Princeton, NJ: Princeton University Press.
Williams, B. (2002) *Truth and Truthfulness: An Essay in Genealogy*, Princeton, NJ: Princeton University Press.
Wood, A. (2014) "Coercion, Manipulation, and Exploitation," in C. Coons and M. Weber (eds.) *Manipulation: Theory and Practice*, Oxford and New York: Oxford University Press, pp. 17–51.

21
EPISTEMIC PATERNALISM

Kristoffer Ahlstrom-Vij

1 The case for external constraints

It's a well-established fact that we often reason by way of *heuristics*, or subconscious rules of thumb, rather than through the systematic application of formal rules or principles of logic, statistics, probability theory, and so forth (Kahneman 2011; Gilovich et al. 2002; Kahneman, Slovic, and Tversky 1982). Such heuristics typically operate by making certain *assumptions*, for example to the effect that all information necessary for making a sound judgment is readily available or otherwise vividly presented to the agent (*availability heuristic*), or that any given sample, even if small, will be representative of the population from which it is drawn (*representativeness heuristic*). In many cases, these assumptions don't present any problem. Indeed, on certain tasks heuristical reasoning outperforms more labor-intensive reasoning strategies by a comfortable margin (Gigerenzer, Todd, and the ABC Research Group 1999), which might in some cases be explained with reference to the adaptive nature of the relevant assumptions (Cosmides and Tooby 1996).

At the same time, the question remains: adapted to *what*? The time span separating the prehistoric from the modern world is too short for any evolutionary pressure to have brought our cognitive apparatus up to speed with a wide variety of modern challenges. Consequently, even if largely adaptive, we have good reason to worry about heuristical reasoning in many contexts making for *bias*, or systematic reasoning mistakes arising when the assumptions that the relevant heuristics are operating on simply don't hold.

Of course, any concern about bias would be greatly diminished were it simply a matter of being more careful and vigilant in our thinking. One way to flesh out this thought is in terms of what we may refer to as *the self-correction strategy*, on which the individual agent corrects for bias on her own accord. However, there are two problems for this strategy. The first problem is one of *motivation*, arising out of the fact that any attempt to deal with bias has to take into account not only that we are biased, but also that we suffer from what Emily Pronin and colleagues (2002) have referred to as a "bias blind spot," on account of which we tend to underestimate the extent to which we are prone to bias. This blind spot should be understood in the context of the well-known psychological fact that, depressed people aside (Taylor and Brown 1988), we tend to rate ourselves as above average on desirable traits (Alicke 1985; Brown 1986). This overconfidence extends to our evaluations of our own epistemic capabilities. As Pronin (2007: 37) notes in an overview, "people tend to recognize (and even overestimate) the operation of bias in human judgment – except when that bias is their own."

Consequently, the first problem facing the self-correction strategy is that, whether or not there are corrective measures available, each and every one of us will tend not to see the point of taking corrective measures, on account of our bias blind spots. But let's assume that there's a way around this problem. Still, merely being *motivated* to correct for bias is not enough – additionally, we need to do so *successfully*. This is the second problem for the self-correction strategy. Solving that problem requires doing two things. First, we need to correct for bias when and only when we are in fact biased. This poses a challenge of *bias identification*. The challenge is that the most obvious way to look for bias is by introspecting, while the bulk of the operations that would require scrutiny simply aren't introspectively accessible (Wilson 2002). And, as pointed out by Wilson and Brekke (1994), while we often have access to the outputs of those operations, bad judgments, unlike bad food, do not smell. But say we find a way to overcome the challenge of bias identification. We now need to correct to and only to the extent needed to remove any bias. This is the challenge of *proper correction*. Such correction is difficult on account of the risks of engaging in *insufficient* correction or *over*correction (Petty and Wegener 1993).

Of course, none of this suggests that it is *impossible* for people to meet the challenges of bias identification and proper correction. Still, it seems clear that, even if we assume that the relevant agents are at all motivated to engage in bias correction – which is far from a trivial assumption, as we've seen – there are substantial challenges they need to meet when it comes to doing so successfully. So what's a more promising strategy?

The strategy to be pursued in what follows focuses not on the correction but on the *prevention* of bias. This is in line with the debiasing literature generally. For example, Wilson and colleagues (2002: 195) note that "[t]he best way to avoid biased judgments and emotions is exposure control." In light of the problem of motivation outlined earlier, the most promising way to achieve such control involves imposing *external* constraints on the agent to shield her from biasing information. For example, consider the practice on the part of US judges to withhold certain types of information from the jurors, such as character evidence or evidence about past crimes, on the assumption that the jurors are likely to systematically overestimate the probative value of such information. Consequently, on the *US Federal Rules of Evidence*, the mere fact that a piece of evidence is *relevant*, in making the hypothesis about guilt more or less likely than it otherwise would have been, is not a sufficient condition for presenting it to a jury. It also matters whether jurors are likely to *gauge* that relevance properly. If not, the presiding judge may withhold the information. This type of evidence control serves to illustrate one particular type of external constraint, namely an external constraint on *information access*, restricting the choices the agent can make when it comes to what information to bring to bear on whatever matter she happens to be considering.

2 Defining epistemic paternalism

The previous section offered some reasons for imposing external constraints on information access as a form of bias exposure control, in light of challenges for the idea of having individual agents correct for bias on their own. In this section, I'll make the case that such external constraints are properly referred to as *epistemically paternalistic*. To make that case, we need to get clearer on what epistemic paternalism is. I'll suggest that there are three jointly sufficient conditions: *the interference condition*, *the non-consultation condition*, and *the improvement condition*. Let's consider each in turn.

One central feature of a paternalistic practice is that it constitutes an interference with the actions of another for her own good. In cases of epistemic paternalism, what's interfered with is someone's *inquiry*, understood here as involving the accessing, collection, and evaluation of information. I will take it that someone is *interfering* with the inquiry of another if the former is compromising the latter's freedom to conduct inquiry in whatever way she sees fit, and that someone is *free* to conduct inquiry thus if there are no constraints imposed by others on her ability to access, collect, and evaluate information. When a practice interferes with someone's inquiry by compromising her freedom in this manner that practice satisfies *the interference condition* on epistemic paternalism.

When judges withhold certain kinds of evidence from the jurors, the jurors' freedom to access whatever information they deem significant in a manner free of constraints imposed by others is compromised. As such, the relevant practice satisfies *the interference condition*. In light of this, it should come as no surprise that this practice has been termed paternalistic in the literature (e.g., Laudan 2006) and, in one instance, *epistemically* paternalistic (Goldman 1991). Whether this practice is in some relevant sense *justified* is, of course, another matter, and one that we'll find reason to return to in Section 5. For now, let's move on to the second condition on epistemic paternalism.

There are reasons why people have found paternalism problematic. For one thing, there is something *arrogant* about paternalistic interference. However, it's not necessary that the person interfered with *objects* to the interference for it to qualify as paternalistic (Dworkin 2010). Consider an example from Shiffrin (2000): if I believe a friend of mine to be financially irresponsible, I might intercept a credit card offer that he gets in the mail. I am not thereby doing something that he objects to, since he is aware neither of the offer nor of my intercepting it. Still, it seems right to say that I am acting paternalistically. This suggests that what makes a practice paternalistic is not that those interfered with are objecting, but that they are not *consulted*. It might be objected that paternalistic interference requires that those interfered with *would* object, had they been consulted. So, assume that my friend actually would have *agreed* that my interference was called for, had I bothered to consult him. Since I don't, it seems that I am still acting paternalistically. As such, the idea that what matters for epistemically paternalistic interference is whether or not those interfered with have been consulted stands. Hence, *the non-consultation condition*.

Do judges withholding biasing evidence from jurors satisfy the non-consultation condition? At no point have they consulted those from whom they are withholding information as to whether the information should be withheld. Consequently, the relevant practice satisfies the non-consultation condition.

We started out by saying that a paternalistic practice involves an interference with the doings of another for her own good. In the case of epistemic paternalism, the relevant good is an *epistemic* good, in turn a function of either succeeding or standing a good chance of succeeding in forming true belief and avoiding false belief (Ahlstrom-Vij 2013a). That said, there are a number of epistemically relevant dimensions along which an agent may be better or worse off in the relevant sense. I will focus on two such dimensions: reliability and (question-answering) power (e.g., Goldman 1992). Reliability is a matter of avoiding error by generating a high *ratio* of true to false belief. Power is the ability to form a large *number* of true beliefs that constitute correct answers to whatever questions are facing the agent.

How are we to weigh improvements in reliability against improvements in power? I will not attempt to answer this question here. Instead, borrowing a term used in contexts of interpersonal comparisons of welfare, I will talk in terms of (intrapersonal) *epistemic Pareto improvements*, that is, improvements along one epistemic dimension that do not entail a deterioration with respect to any other epistemic dimension. In what follows, I will focus on Pareto improvements in *reliability*, the simple reason being that this is a kind of improvement about which we have relevant empirical data. That, moreover, seems a reasonable description of what's going on in cases of evidence control: by withholding biasing information, judges are attempting to increase jurors' reliability, without thereby making them worse off along some other epistemically relevant dimension. The practice thereby satisfies *the improvement condition*.

Does it need to be the case that the relevant agent is interfered with *solely* for the purpose of making her epistemically better off for the practice involved to qualify as epistemically paternalistic? No, and epistemic paternalism is in that respect a *mixed* form of paternalism (Feinberg 1986). By interfering with someone's inquiry for the purpose of making them epistemically better off, we might very well also be looking to make other people better off in *non*-epistemic terms. For example, when withholding information from jurors, judges are motivated not solely by a desire to make the jurors epistemically better off, but also by a desire to protect the defendant's welfare while doing right by those wronged.

However, this might be taken to raise an objection: Aren't we in that case really interfering, *not* for the good of those interfered with, but *rather* for the good of others? Differently put, isn't the *real* reason for interference the good of others? If it is, epistemic paternalism arguably isn't a form of paternalism at all. That, however, seems incorrect. Something is a real reason for someone doing something when it's part of a plausible explanation of *why* that someone did what she did. In the case of epistemically paternalistic practices, the fact that those interfered with are made epistemically better off forms part of the explanation of why we interfere in the manner we do. Part of the reason judges interfere with the information available to the jurors is that it makes them epistemically better off. The judges' *ultimate* aim might not be the jurors' epistemic improvement, but rather pertain to the defendants and those wronged. Still, the fact remains that the judges are interfering *both* for the (epistemic) good of those interfered with *and* for the (non-epistemic) good of others, on the grounds that achieving an improvement in the former is an *instrument* to securing the latter.

However, someone might maintain that paternalistic interference needs to be ultimately concerned with the good of those interfered with. But that's implausible. Another example from Shiffrin helps us see why:

> Suppose a park ranger has the power to refuse permission to climb a steep, dangerous mountain path. If the ranger refuses to allow a person to climb simply because the (fully informed, competent) person might hurt himself and that would be bad for him, that refusal would be paternalist, I think. Suppose the ranger says, "Of course, you may take whatever risks you want with your life, but I refuse permission because you might die and leave your spouse grief-stricken." Such a refusal also seems paternalist. The ranger is substituting his judgment about how the climber should treat her spouse and conduct her marriage. The ranger is taking over the climber's marriage, a bit, on the implicit grounds that his moral judgment of how to conduct the relation is correct and the climber's is incorrect.
>
> (Shiffrin 2000: 217)

In the second case described by Shiffrin, the ultimate aim of the ranger's interference is the welfare of the climber's spouse; his concern for the welfare of the climber is instrumental to this aim. But, according to Shiffrin, the interference is still paternalistic since "it involves a person's aiming to take over or control what is properly within the agent's own legitimate domain of judgment or action" (2000: 216). Something similar is going on in the case of evidence control.

The jurors have a legitimate power to act as the triers of fact. Through her interference, the judge infringes on that power. That, on Shiffrin's account, is what makes her interference paternalistic. But on my account, that's not the whole story. Unlike Shiffrin, I don't find plausible the claim that "the paternalist may be *solely* and directly concerned with the third party's welfare" (2000: 216, my emphasis). Say that the judge, worried about the epistemic caliber of the jury not being high enough to safeguard the defendant and do right by those wronged, dismisses the jury and takes it upon herself to act as the sole trier of fact. In so doing, she certainly takes over the control of the jury's legitimate domain of judgment and action – but she doesn't, I submit, act paternalistically. What's missing is a concern, even an instrumental one, for the epistemic good of the jurors. That concern is present in the original case of evidence control: the judge interferes within the legitimate domain of the jury, but does so with an eye towards making the jury epistemically better off.[1]

So, to sum up, in the case of epistemically paternalistic practices, it's *not* the case that we are interfering for the good of others rather than for the good of those interfered with. We are interfering for the good of both. More specifically, we are interfering on the grounds that making those interfered with (epistemically) better off is an instrument towards making others better off. And, as we have just seen in relation to Shiffrin's account of paternalism, the fact that our *ultimate* aim in so interfering thereby isn't to make those interfered with better off isn't sufficient to show that the relevant form of interference isn't paternalistic.

3 Epistemic paternalism and autonomy

The previous section suggested that we practice epistemic paternalism when interfering with the inquiry of another for her own epistemic good without consulting her on the issue. On this definition, the current practice of evidence control qualifies as a form of epistemic paternalism. It doesn't follow from any of this, of course, that we *should* practice epistemic paternalism. In the present section, we'll consider the strongest form of argument suggesting that we should *not*, on account of considerations about personal autonomy.

The most famous objection of this kind was offered by Joel Feinberg (1986). As Feinberg saw it, "[t]he anti-paternalist . . . must not only argue against particular legislation with apparently paternalistic rationales; he must argue that paternalistic reasons never have any weight on the scales at all" (1986: 25). This is so because "they are morally illegitimate or invalid reasons by their very natures, since they conflict head on with defensible conceptions of personal autonomy" (1986: 26). At the heart of that conception is the idea that

> *respect for a person's autonomy is respect for his unfettered voluntary choice as the sole rightful determinant of his actions except where the interest of others need protection from him.* Whenever a person is compelled to act or not to act on the grounds that he must be protected from his own bad judgment even though no one else is endangered, then his autonomy is infringed.
>
> (Feinberg 1986: 68, emphasis in original)

In other words, someone's autonomy is infringed when the *sole* reason for interference is that it is in his own interest, and it thereby is not the case that the interests of others are being factored in. But notice that this does not imply that paternalistic reasons – that is, reasons in terms of the good of those interfered with – are always invalid. At best, Feinberg's notion of autonomy serves to rule out a certain kind of paternalistic *practice*, namely one motivated *exclusively* with reference to reasons pertaining to the good of those interfered with. As we saw earlier, however, the kind of epistemically paternalistic practices that concern us here are not of this kind. Insofar as they are justified, they are justified with reference to a concern *both* for the epistemic good of those interfered with, and for the non-epistemic good of others. That's why Feinberg's notion of autonomy is compatible with epistemic paternalism.

Notice, however, that Feinberg merely provides a *sufficient* condition on autonomy infringement. Maybe there are other ways to violate people's autonomy, not captured by Feinberg's account. On this point, consider Joseph Raz's notion of autonomy, on which "[o]ne is autonomous if one determines the course of one's life by oneself" (1986: 407). A respect for autonomy rules out coercion or manipulation, both of which constitute ways for one person to impose her will on others (Raz 1986: 378). In the case of coercion, this is done through the reduction or removal of a person's options. Manipulation, by contrast, "perverts the way that [the] person reaches decisions, forms preferences, or adopts goals" (Raz 1986: 377–378).

If a mere manipulation of available options – let alone the removal of such options through coercive means – constitutes a violation of autonomy, it seems that paternalism necessarily violates people's autonomy. Or does it? Not according to Raz:

> [P]aternalism affecting matters which are regarded by all as of merely instrumental value does not interfere with autonomy if its effect is to improve safety, thus making the activities affected more likely to realize their aim. There is a difference between risky sports, e.g., where the risk is part of the point of the activity or an inevitable by-product of its point and purpose, and the use of unsafe common consumer goods. Participation in sporting activities is intrinsically valuable. Consumer goods are normally used for instrumental reasons.
>
> (Raz 1986: 422–423)

As argued above, the epistemic goods promoted through epistemic paternalism are instrumental rather than ultimate or intrinsic goods. In this respect, the practices subject to epistemic paternalism have more in common with the use of consumer goods than with risky sports. For example, unlike in the case of risky sports, it is no essential component of legal proceedings that things might go wrong, and the means involved sometimes fail to secure the ultimate goods. That's why epistemic paternalism also is compatible with Raz's notion of autonomy.

4 Justifying epistemic paternalism

In order to provide a *defense* of epistemic paternalism, it's not sufficient to show that epistemic paternalism isn't inherently objectionable. It also needs to be shown that there are situations in which we are justified in practicing epistemic paternalism. For that purpose, I'll be offering two jointly sufficient conditions for justified epistemically paternalistic interference. *The alignment condition* pertains to the interplay between our reasons and helps us avoid some issues arising when we try to weigh different kinds of reasons against each other. *The burden-of-proof condition* speaks to the circumstances under which one's beliefs about the desired effects are justified. We'll consider these conditions in turn.

Let us return once more to the case of evidence control. The judge withholds biasing information from the jurors in order to make them epistemically better off, because doing so serves the non-epistemic end of protecting the welfare of the defendant and doing right by those wronged. In other words, the paternalistic measure in question involves two serially ordered motivations, picking out means and ends, respectively. But we can certainly imagine cases where epistemic and non-epistemic motivations do not line up so nicely. Consider, for example, a society exercising total control over the minutest details of the epistemic undertakings of their citizens for purposes of making them better off along epistemic dimensions. Even if successful, we might feel that such a regime would be far too intrusive, and that we on that account would have good reason to reject it. Here, we could talk in terms of different kinds of reasons being *weighed* against each other, and say that paternalistic interventions are justified only if the reasons for intervening outweigh those against. This is the strategy employed by Peter de Marneffe (2010) in his discussion of paternalistic restrictions on prostitution. This is not the strategy I'll be employing here, however, the reason being that, while it's fairly easy to make sense of what it is for reasons to have *valence* – being for or against things – it's often more difficult to assign *weights* specific enough to help determine what outweighs what.

What's an alternative strategy, then? The one I'll be pursuing can be framed in terms of the following condition:

> *The Alignment Condition*: The epistemic reasons we have for instituting the relevant epistemically paternalistic practice, on the grounds that it will lead to an epistemic improvement, are *aligned* with our non-epistemic reasons on the issue. Two or more reasons are aligned if and only if they are (*a*) reasons for the same thing, or, failing that, (*b*) silent on the issue, by not constituting reasons either way on the matter.

One benefit of the alignment condition is that it only requires that reasons have valence. The relative weights of reasons do not need to be factored in to determine whether the condition is satisfied. Note, however, that the alignment condition does not provide a *necessary* condition on justified epistemic paternalism. We can imagine epistemically paternalistic practices that fail to satisfy the alignment condition, but nevertheless are justified on weighing grounds. The alignment condition also doesn't offer a *sufficient* condition on epistemically paternalistic interventions. While it guarantees a certain *harmony* among reasons, it doesn't entail that we are justified in believing that the relevant form of interference will actually have the intended effects. As it happens, this gets to a major worry about paternalistic interference, which will motivate our second condition.

The relevant worry was voiced already by John Stuart Mill, who suggested that "the strongest of all arguments" against paternalistic interference is that, if we attempt to interfere, the odds are that we will do so "wrongly and in the wrong place":

> On the question of social morality, of duty to others, the opinion of the public, that is, of an over-ruling majority, though often wrong, is likely to be still oftener right; because on such questions they are only required to judge of their own interests; of the manner in which some mode of conduct, if allowed to be practiced, would affect themselves. But the opinion of a similar majority, imposed as a law on the minority, on questions of self-regarding conduct, is quite as likely to be wrong as right; for in these cases public opinion means, at the best, some people's opinion of what is good or bad for other people; while very often it does not even mean that; the public, with the most perfect indifference, passing over the pleasure or convenience of those whose conduct they censure, and considering only their own preference.
>
> (Mill 1989 [1859]: 83–84)

There are two reasons that this argument does not apply in the case of epistemic paternalism. First, when it comes to epistemic goods, it's *not* the case that each person necessarily knows her own good best. (Young 2008, de Marneffe 2006, and Hart 1963 offer some reasons to think the same goes in non-epistemic cases.) Indeed, given our introspective limitations and tendencies for overconfidence, it cannot be ruled out that the individual agent might in many cases be the *worst* judge on the matter of whether she would undergo an epistemic Pareto improvement on account of some intervention.

Second, the grounds on which it should be judged whether someone is likely to be made better off epistemically through some form of interference are not majority votes, but our best empirical evidence on the issue. More specifically, consider the following:

> *The Burden-of-Proof Condition*: A case can be made that available evidence indicates that it is highly likely that everyone interfered with in the relevant manner is or will be made epistemically better off for being interfered with thus, compared to relevant alternative practices.

What constitutes relevant alternative practices will vary from case to case, but falls into two broad categories. In cases where we are considering implementing a new paternalistic practice to replace the prevailing one, the relevant alternative is the prevailing practice. In cases where we are attempting to justify a practice already in place, the relevant alternatives are whatever practices figure as prominent alternatives, paternalistic or otherwise.

A word is in order on the claim that the evidence needs to indicate that *everyone* interfered with will be made epistemically better off, as it might be taken to raise an objection. Consider a case wherein the burden-of-proof condition is satisfied, and a case consequently can be made that it is highly likely that everyone interfered with will be made epistemically better off for being interfered with. This is compatible with some of those interfered with actually being made worse off. So, let us assume that a small minority actually is. Doesn't this provide an epistemic reason *against* interference, and imply that the alignment condition thereby is not satisfied? And, if it does, can we *ever* assume the alignment condition to hold, given that we typically cannot rule out that *some* people might be made worse off by being interfered with? This is *the problem of the epistemic outlier.*

Let us spell out the imagined scenario in more detail. The scenario is one where the practice satisfies the burden-of-proof condition, while the fact that a minority of those interfered with – the epistemic outliers – will be made worse off is unknown to us. If it were not, the burden-of-proof condition would no longer be satisfied. To determine if the adversely affected minority nevertheless provides us with a reason *not* to interfere, we need to consider on what grounds they are adversely affected. One possibility is that the adverse effect is merely *accidental.* Accidental effects are low probability effects. When it comes to the kind of *ex ante* justifications at issue here, however, we have reason to do whatever is highly likely to generate a good epistemic outcome on our evidence. In the case at hand, our evidence suggests that there is a high likelihood that each person benefits epistemically. Consequently, the mere fact that someone will be affected in unintended and purely accidental ways does *not* provide an epistemic reason against interfering.

Another possibility is that the adverse effect is *not* accidental. Perhaps the epistemic outliers possess some superior epistemic capability that the interference prevents them from relying on. But here, too, it is not clear that this gives us any epistemic reason not to interfere. It is certainly *possible* that there are people who will be systematically disadvantaged by being interfered with in epistemically paternalistic ways. But keep in mind the perspective from which *ex ante* justifications are provided, namely one from which we in effect are placing an empirical bet on what will have the best effect for those interfered with. We could place our bet on the basis of a mere possibility. Or we could acknowledge that it is highly unlikely that people will be disadvantaged, given what we know about our tendencies for bias and overconfidence, as well as the resulting benefits of external constraints, and instead place our bet on the basis of the available evidence. If the burden-of-proof condition is satisfied, that evidence suggests that it is highly likely that the relevant interference will have the intended effect.

Both of these responses assume that we do not *know* that some people are or will be adversely affected. Would finding that out defeat our justification for interfering? Yes, it would, but not on account of the scenario failing to satisfy the alignment condition. As previously noted, if we were to find out that not everyone does or will benefit from the interference, the burden-of-proof condition would no longer hold. But in that case, the response is not necessarily to back away from interference, but to adjust its scope in such a way that those we know to be adversely affected no longer are interfered with in the relevant way.

Hence, the alignment and burden-of-proof are and remain plausible conditions. Of course, no conclusive case has been made to the effect that their combination provides a sufficient condition for justified epistemic paternalism. In the next and final section, we consider whether the epistemically paternalistic practice of evidence control satisfies the conditions. If it does, and it's not obviously unjustified on intuitive grounds, this not only provides evidence for the joint sufficiency of the conditions, but also for the main thesis of this chapter, to the effect that we are sometimes justified in interfering with the inquiry of another without her consent but for her own epistemic good.

5 Epistemic paternalism defended

The question for the present section is whether the type of evidence control discussed above is a *justified* epistemically paternalistic practice. To put the question in terms of the conditions outlined in the previous section: (*a*) Can a case be made that available evidence indicates that it is highly likely that everyone interfered with in the relevant manner is made epistemically better off for being interfered with thus, compared to relevant alternative practices; and (*b*) are the epistemic reasons we have for instituting the relevant practice aligned with our non-epistemic reasons on the issue?

Let us start by considering what the relevant alternatives to prevailing practices would be. According to Larry Laudan, one of the most forceful critics of evidence control, the alternative would be a practice requiring that "the only factor that should determine the admissibility or inadmissibility of a bit of evidence is its relevance to the hypothesis that a crime occurred and that the defendant committed it" (2006: 25). Moreover, Laudan's motivation for this claim is epistemic: "Paternalistically coddling jurors by shielding them from evidence that some judge intuits to be beyond their powers to reason about coherently is not a promising recipe for finding out the truth" (2006: 23).

I think Laudan is mistaken on this point, at least as it pertains to character evidence. To see why, we need to consider the fact that there is a psychological asymmetry in how jurors treat character evidence that favors *negative* character evidence framed in terms of *particular* actions over other kinds of character evidence (Maeder and Hunt 2011). These results are in line with studies suggesting both that anecdotal information in terms of *specific* acts tend to be considered more diagnostic than base rate information (Borgida and Nisbett 1977), and that *immoral* behavior is taken to be more diagnostic of an individual's character than is moral behavior (Lupfer, Weeks, and Dupuis 2000).

Moreover, this psychological asymmetry is a symptom of people *overestimating* the probative value of particular negative character evidence. To see why, consider that what a juror swayed by particular character evidence does is reason (often unconsciously) from evidence about past actions to what social psychologists would call a *personality trait*, consisting in a general disposition to behave in a certain manner. Then, the juror factors in the nature of that trait in her decision about the guilt of the defendant. The problem is that such traits are not particularly predictive of how people behave across different situations, on account of not amounting to what John Doris (2002) calls *robust* traits. The characters we instantiate simply do not manifest a sufficiently high degree of cross-situational consistency. The extent to which we are generous, mean, helpful, or what have you, on any given occasion, owes more to surprisingly small and seemingly irrelevant differences in the situation than to anything like an inherent character.

This is relevant to the epistemic situation of the juror. In cases of negative character evidence framed in terms of particular acts, we seem to invest a significant amount of credence, despite such evidence not providing particularly valuable information about other actions. In that sense, we invest too much credence into negative character evidence, if framed in terms of particular actions. This is sufficient to show that Laudan's (2006: 25) claim – that "the only factor that should determine the admissibility or inadmissibility of a bit of evidence is its relevance to the hypothesis that a crime occurred and that the defendant committed it" – is not epistemically well motivated. Introducing character evidence means introducing a kind of evidence that will make it harder for jurors to evaluate the objective weight of the evidence properly and arrive at an informed verdict. That is why relevance is *not* sufficient for admissibility, even on purely epistemic grounds.

In light of current evidence, it thereby seems a case can be made that it is highly likely that jurors will be made more reliable insofar as there are some restrictions on the admissibility of character evidence, as per the *Federal Rules of Evidence*, compared to there being no such restrictions. But are jurors thereby made worse off along some other epistemic dimension, such as power, by forming fewer true beliefs on the question of guilt?

In one sense, there are only two beliefs available to the jurors: that the defendant is guilty or that the defendant is not guilty. And if so, admitting or not admitting character evidence will make no difference to the number of beliefs formed by the jury. But it might be argued that there are many different kinds of beliefs that jurors might form in relation to the question of guilt, including beliefs about the trustworthiness of witnesses, the probative value of the evidence introduced, and so on. Introducing less evidence might certainly have the jury form fewer such beliefs, for the simple reason that introducing less evidence means that there are fewer things to form beliefs about. Still, if the previous arguments for the idea that introducing character evidence makes jurors less reliable are correct, it is not clear that putting restrictions on the admissibility of such evidence will make them form fewer *true* beliefs.

Consequently, the burden-of-proof condition is met. But what about the alignment condition? To see why our epistemic and *moral* reasons are aligned, remember that, when a judge withholds certain evidence on the grounds that it might bias their judgment, she does this to safeguard the welfare of the defendant while doing right by those wronged. This makes for straightforward alignment: safeguarding the welfare of the defendant and doing right by those wronged requires convicting all and only those who are in fact guilty, and the best way of approximating this ideal is to ensure that those making judgments about guilt evaluate evidence properly.[2]

What about reasons that are neither epistemic nor moral? Say, in particular, that we perform a financial cost-benefit analysis, and it turns out that the costs outweigh the benefits. Does that provide a reason against interference? It is not clear that it does in the kinds of situations that concern us in the present case for epistemic paternalism. The reason is that we have moral reasons for interference, as we saw above, and that moral reasons *silence* countervailing, non-moral reasons.[3] For present purposes, such silencing doesn't require *authoritative* moral reasons, operating independently of desires or interests. After all, insofar as we are inclined to interfere with the inquiry of another in the kind of context that has concerned us, our motivations are most plausibly understood as grounded in a moral concern, such as a moral concern for the welfare of the defendant or for doing right by those wronged, which silences reasons owing to non-moral considerations. That much should be largely uncontroversial. Those critical of "Humean" silencing (e.g., Joyce 2006) are not concerned that moral reasons grounded in *present* desires cannot silence non-moral reasons – they are concerned that this is the only kind of silencing there is, on a Humean picture. But that complaint need not concern us here. That means that, if we can show that the epistemic reasons involved are aligned with our moral reasons on the issue, then no further kinds of reasons need to be considered for us to conclude that the alignment condition is satisfied. Consequently, the aforementioned type of evidence control can be taken to satisfy that condition.

6 Conclusion

We started out by arguing that evidence regarding our dual tendency for bias and overconfidence suggests that our best bet when it comes to counteracting bias and promoting epistemic goods is to have external constraints imposed that restrict our freedom to conduct inquiry in whatever way we see fit. Moreover, it was argued that practices that impose such constraints are properly referred to as epistemically paternalistic when they interfere with our freedom to conduct inquiry in whatever way we see fit, and do so for our own epistemic good without consulting us on the issue. Two objections to such interference framed in terms of autonomy violations were rebutted, and two jointly sufficient conditions for justified epistemic paternalism then defended. Finally, it was argued that the practice of evidence control used throughout the chapter to illustrate the idea of epistemic paternalism satisfies those conditions. As it happens, there are other strong candidates for justified epistemic paternalism, including the practices of relying on experimental randomization in the sciences, and on statistical prediction rules in medical diagnosis and prognosis.[4] But for present purposes, the case of evidence control suffices to demonstrate that we are sometimes justified in interfering with the inquiry of another without her consent but for her own good, and that epistemic paternalism consequently is true.[5]

Related topics

Libertarian Paternalism, Nudging, and Public Policy; Paternalism and Autonomy; Paternalism by and towards Groups; Paternalistic Lying and Deception; The Concept of Paternalism.

Notes

1 That the judge is interfering within the legitimate domain of the jurors might be taken to suggest that her action is at least *pro tanto* morally objectionable (see, e.g., Shiffrin 2000: 220 n. 25). However, whether it's all-things-considered morally objectionable would depend on what other factors are in play. Suffice for present purposes to note that, in the type of case at hand, any legitimate moral claim on the part of the jury not to be interfered with is trumped by considerations about the welfare of others, such as the defendant and those allegedly wronged by the defendant. For more on the moral reasons at work here, see Section 5.

2 What about the fact that the judge is arguably interfering within the legitimate domain of the jurors? See note 1 above.

3 What about situations involving immense financial costs? In such situations, we might hesitate to say that moral reasons silence whatever countervailing reasons we might have (see, e.g., Foot 1978). For present purposes, however, it suffices to note that the kind of situations that will concern us do not involve any such immense costs.

4 See Ahlstrom-Vij (2013b) for an argument to this effect.

5 I am grateful to Kalle Grill and Jason Hanna for extremely helpful comments on an earlier draft of this chapter. Parts of the present chapter are reproduced from Ahlstrom-Vij (2013b) with the permission of Palgrave Macmillan.

References

Ahlstrom-Vij, K. (2013a) "In Defense of Veritistic Value Monism," *Pacific Philosophical Quarterly* 94: 19–40.

———. (2013b) *Epistemic Paternalism: A Defence*, Basingstoke: Palgrave Macmillan.

Alicke, M. D. (1985) "Global Self-Evaluation as Determined by the Desirability and Controllability of Trait Adjectives," *Journal of Personality and Social Psychology* 49: 1621–1630.

Borgida, E. and Nisbett, R. (1977) "The Differential Impact of Abstract vs. Concrete Information on Decisions," *Journal of Applied Social Psychology* 7(3): 258–271.

Brown, J. D. (1986) "Evaluations of Self and Others: Self-Enhancement Biases in Social Judgments," *Social Cognition* 4: 353–375.

Cosmides, L. and Tooby, J. (1996) "Are Humans Good Intuitive Statisticians After All? Rethinking Some Conclusions from the Literature on Judgment under Uncertainty," *Cognition* 58: 1–73.

de Marneffe, P. (2006) "Avoiding Paternalism," *Philosophy & Public Affairs* 34: 68–94.

———. (2010) *Liberalism and Prostitution*, Oxford: Oxford University Press.

Doris, J. M. (2002) *Lack of Character: Personality and Moral Behavior*, Cambridge: Cambridge University Press.

Dworkin, G. (2010) "Paternalism," in E. N. Zalta (ed.) *The Stanford Encyclopedia of Philosophy*, http://plato.stanford.edu/archives/sum2010/entries/paternalism/.

Feinberg, J. (1986) *The Moral Limits of Criminal Law, Volume 3: Harm to Self*, New York and Oxford: Oxford University Press.

Foot, P. (1978) "Are Moral Considerations Overriding?" in her (ed.) *Virtues and Vices*, Oxford: Oxford University Press, pp. 181–188.

Gigerenzer, G., Todd, P. M. and the ABC Research Group. (eds.) (1999) *Simple Heuristics that Make Us Smart*, Oxford: Oxford University Press.

Gilovich, T., Griffin, D. and Kahneman, D. (eds.) (2002) *Heuristics and Biases: The Psychology of Intuitive Judgment*, Cambridge: Cambridge University Press.

Goldman, A. (1991) "Epistemic Paternalism: Communication Control in Law and Society," *The Journal of Philosophy* 88: 113–131.

———. (1992) "Foundations of Social Epistemics," in his (ed.) *Liaisons: Philosophy Meets the Cognitive and Social Sciences*, Cambridge, MA and London: The MIT Press, pp. 179–207.

Hart, H. L. A. (1963) *Law, Liberty, and Morality*, Stanford, CA: Stanford University Press.

Joyce, R. (2006) *The Evolution of Morality*, Cambridge, MA and London: The MIT Press.

Kahneman, D. (2011) *Thinking, Fast and Slow*, London: Penguin Books.

Kahneman, D., Slovic, P. and Tversky, A. (eds.) (1982) *Judgment under Uncertainty: Heuristics and Biases*, Cambridge: Cambridge University Press.

Laudan, L. (2006) *Truth, Error, and Criminal Law*, Cambridge: Cambridge University Press.

Lupfer, M. B., Weeks, M. and Dupuis, S. (2000) "How Pervasive Is the Negativity Bias in Judgments Based on Character Appraisal?" *Personality and Social Psychology Bulletin* 26: 1353–1366.

Maeder, E. and Hunt, J. (2011) "Talking About a Black Man: The Influence of Defendant and Character Witness Race on Jurors' Use of Character Evidence," *Behavioral Sciences and the Law* 29: 608–620.

Mill, J. S. (1989 [1859]) "On Liberty," in S. Collini (ed.) *On Liberty and Other Writings*, Cambridge: Cambridge University Press, pp. 1–116.

Petty, R. and Wegener, D. (1993) "Flexible Correction Processes in Social Judgment: Correcting for Context-Induced Contrast," *Journal of Experimental Social Psychology* 29: 137–165.

Pronin, E. (2007) "Perception and Misperception of Bias in Human Judgment," *Trends in Cognitive Science* 11(1): 37–43.

Pronin, E., Lin, D. and Ross, L. (2002) "The Bias Blind Spot: Perceptions of Bias in Self Versus Others," *Personality and Social Psychology Bulletin* 28: 369–381.

Raz, J. (1986) *The Morality of Freedom*, Oxford: Clarendon Press.

Shiffrin, S. V. (2000) "Paternalism, Unconscionability Doctrine, and Accommodation," *Philosophy & Public Affairs* 29: 205–250.

Taylor, S. E. and Brown, J. D. (1988) "Illusion and Well-being: A Social Psychological Perspective on Mental Health," *Psychological Bulletin* 103: 193–210.

Wilson, T. D. (2002) *Strangers to Ourselves: Discovering the Adaptive Unconscious*, Cambridge, MA: Harvard University Press.
Wilson, T. D. and Brekke, N. (1994) "Mental Contamination and Mental Correction: Unwanted Influences on Judgments and Evaluations," *Psychological Bulletin* 116(1): 117–142.
Wilson, T. D., Centerbar, D. B. and Brekke, N. (2002) "Mental Contamination and the Debiasing Problem," in Gilovich, Griffin, and Kahneman (eds.) 2002, pp. 185–200.
Young, R. (2008) "John Stuart Mill, Ronald Dworkin, and Paternalism," in C. L. Ten (ed.) *Mill's 'On Liberty': A Critical Guide*, Cambridge: Cambridge University Press, pp. 209–227.

Further reading

No one interested in paternalism of any kind can ignore J. Feinberg, *The Moral Limits of Criminal Law, Volume 3: Harm to Self* (New York and Oxford: Oxford University Press, 1986). A more recent and also very insightful discussion of paternalism is P. de Marneffe, *Liberalism and Prostitution* (Oxford: Oxford University Press, 2010). K. Ahlstrom-Vij, *Epistemic Paternalism: A Defence* (Basingstoke: Palgrave Macmillan, 2013) makes a case for the type of epistemic paternalism outlined in this chapter, and defends a variety of epistemically paternalistic practices. A. Goldman, "Epistemic Paternalism: Communication Control in Law and Society," *The Journal of Philosophy* 88(1991): 113–131, introduced the term "epistemic paternalism," and contains a discussion of some epistemically paternalistic practices, including that of evidence control.

PART V

Paternalism in practice

22

PATERNALISM AND THE CRIMINAL LAW

Heidi M. Hurd

A theory of the legitimacy of state action is put to its toughest test when it is employed in the service of determining the kinds of conduct that are eligible for criminal punishment. The purpose of this chapter is to subject paternalism to this test – to explore whether and when the state is morally entitled to use its most coercive means to compel citizens to act in their own best interests when there is reason to believe that they may otherwise fail to do so.

1 Locating paternalism within the larger taxonomy of theories of criminal legislation

A theory of criminal legislation that permits paternalism is one of at least five theories concerning the legitimate scope and limits of the criminal law. These theories are not mutually exclusive; one could embrace one, some, or all of them, so the plausibility of any one of these theories does not undermine the defensibility of others (even as several were originally propounded by theorists who took them to be exclusive justifications for criminalization). Let me begin, then, by locating paternalism within this larger taxonomy.

First, some are tempted by the view that there are no principled limits to a democratic state's authority to criminalize conduct.[1] On this theory, criminal legislation is justified if and only if it has been enacted democratically – namely, through a legislative means that procedurally realizes the value of ensuring that the majority of citizens live under their own laws.[2] Procedural fairness guarantees substantive justice.

There is a significant sense in which this first theory of criminal legislation is a non-theory; or, at best, a theory about why one needs no such theory. Yet it is sadly all too easy in this day and age to imagine electoral enthusiasm for any number of legislative initiatives that would appear morally unjustified despite their democratic pedigree – for example, bans on Muslims who seek refuge within our nation's borders or prohibitions against demonstrators or journalists at political events. For this reason, most legal theorists believe that just as judges are bound by a defensible theory of adjudication, so legislators are bound by a defensible theory of legislation. This must include area-specific constraints that derive not from the whims of an electoral majority, but from the best justification(s) that can be advanced for each area of law in question. For most criminal law scholars, then, the task is to articulate a theory of criminal legislation that harmonizes with persuasive defenses of our criminal justice system, generally, and our practices of punishment, more specifically.

The second theory of criminal legislation answers this call by drawing on John Stuart Mill's famous claim that "the only purpose for which power may be rightfully exercised over any member of a civilized community, against his will, is to prevent harm to others" (1977 [1859]: 223–224). Those who defend this so-called "harm principle" permit the state to eliminate choices from citizens' opportunity sets only when those choices will proximately result in harm to others (for example, by causing physical injury to their persons or property or by exacting unfair advantages through fraudulent means). On this theory, democratically elected legislators are not at liberty to employ their power in any manner that might comport with the will of the majority. Rather, legislators satisfy the obligations of their legislative role if but only if they employ the criminal law and deploy criminal sanctions to prevent conduct that would be harmful to others.

On the third theory of criminal legislation, prohibitions backed by the threat of punishment may be enacted not only as a means of preventing citizens from harming others, but also as a means of preventing them from causing offense to others.[3] Thus we have laws against public nudity, prohibitions on public masturbation, restrictions on noise pollution, sex-segregated restrooms, visible junk collection prohibitions, and so forth. By conceiving of these kinds of offense-generating behaviors as legitimate objects of criminal legislation, this theory expands the criteria that will justify limitations on liberty by making both the harm principle and the so-called offense principle independently sufficient justifications of prohibition.

While those who have sought to supplement the harm principle with the offense principle have clearly found Mill's famous principle of legislation underinclusive, others have framed their criticism of the harm principle in more holistic terms. It is easy to find harms that are not morally wrongful – say, the economic injuries that befall competitors when one is successful in attracting away their customers within a free market or the injuries to self that attend extreme sports; and it is easy to find acts that Mill would have declared harmless to others that nevertheless appear eligible for legal regulation – say, the desecration of a corpse. Inasmuch as some harms are not wrongful and some wrongs are not harmful, it is tempting to conclude that criminal legislation is justified if but only if it prohibits moral wrongdoing. According to "legal moralists" who defend this fourth theory of criminal legislation, the legitimate jurisdiction of the criminal law is co-extensive with actions that constitute (non-*de minimis*) moral wrongs.[4] While some such actions may be harmful or offensive to others, other such actions may be wrong even though they neither harm nor offend anyone. The eradication of species or the destruction of privately owned artistic masterpieces, for example, may constitute "free-standing wrongs," and when sufficiently serious, such actions may justify criminal prohibition in their own right. While there may be both prudential and moral reasons to stay the hand of the state in response to certain sorts of moral wrongdoing – reasons well explored by "liberal perfectionists," for example[5] – these reasons would need to be weighty enough to overcome the *prima facie* legitimacy of extending the reach of criminal prohibitions to all forms of behavior that breach moral duties or infringe moral rights.

There is a fifth category of actions that may be usefully distinguished from those categories that are targeted by the principle of democracy, the harm principle, the offense principle, and the principle of legal moralism. This is the category of actions that are harmful to actors themselves – the category of actions that are of concern to paternalists. Such actions may cause indirect third-party harm or offense, or they may appear independently immoral, but what distinguishes them is the fact that their wrongfulness resides principally in their self-injuring nature. Those who consider such actions eligible for criminal punishment thus defend a fifth theory of criminal legislation, for they add to the menu of reasons that may be independently sufficient to criminalize behavior the fact that such behavior will cause setbacks to the health, welfare, economic well-being, or psychological interests of the actor herself, independently of any harm, offense, or free-standing wrong that it may otherwise cause.

It is worth recognizing that there is logical space for a second version of this theory that effectively collapses the distinction between paternalism and legal moralism. Were one to believe that criminal legislation is an appropriate means of safeguarding the moral well-being of individuals, rather than their health, welfare, economic well-being, or psychological interests, then the demands placed on the criminal law by paternalism would be co-extensive with those of legal moralism.[6] The criminal law would be put to the task of ensuring that citizens fulfill their (weighty) moral obligations, and while legal moralists might claim that this is a good, *tout court*, while moral paternalists might claim that this is a means of saving souls or otherwise ensuring the moral health of individuals, these alternative constructions would mark a philosophical distinction without a legal difference.

Inasmuch as most theorists who are tempted to defend some version of paternalism tend to be concerned with actions that impair citizens' own physical, psychological, or economic well-being, rather than behaviors that negatively affect their moral ledgers, it is to the legitimacy of the first version of the paternalist theory of criminal legislation outlined above that this chapter is devoted. The central question is: Is it legitimate for the state to impose pain, privation, or a loss of liberty on its citizens as a means of preventing them from engaging in conduct that can be predicted to bring them pain, privation, or a loss of liberty? And if not, as a general matter, are there principled exceptions that allow the state to use forcible means to induce citizens to substitute the state's conception of their good for their own?

2 Subjective versus objective conceptions of paternalist theories of legislation

There are two means of formulating the paternalist's essential claim that a just state may properly employ the criminal law as a vehicle for "correcting" the choices that citizens make on their own behalf. One such formulation is subjective, the other objective. On the subjective formulation, paternalism is a theory about the goals, aims, or ends that may subjectively motivate legislators as they pass criminal laws that limit the liberty of citizens. The claim is that limitations on the liberty of citizens are legitimate (other things being equal) when they are imposed by legislators who are subjectively motivated to spare citizens the consequences of self-injuring choices. On the objective formulation, by contrast, paternalism is a theory about external circumstances that justify limitations on the liberty of citizens. According to this formulation, if criminal prohibitions have the effect of curtailing self-injuring choices, this is a basis for considering such provisions justified, regardless of the reasons that subjectively motivated legislators to craft their terms.

These two quite different formulations of legislative paternalism suggest that there are two ways by which critics may formulate their complaints with criminal legislation that curtails the ability of citizens to make certain self-regarding choices. For those who take a theory of legislation to concern the conscious goals that may properly motivate legislators, their thesis is that legislators are not permitted to subjectively aim at substituting their conception of what will be good for citizens for that of citizens themselves.[7] Legislators breach their obligations when they seek to employ the criminal law to improve the lives of citizens who would not themselves choose such "improvements."[8] On this account, legislators may be entitled to act on the subjective motivation to prevent citizens from harming others; and they may be entitled to legislate with the goal of reducing conduct that is offensive to others. But they may not pursue the goal of eliminating conduct that they consider self-injurious by crafting prohibitions that "help" citizens on pain of punishment.

For antipaternalists who believe that a theory of legislation specifies states of affairs that objectively justify liberty-limiting enactments, rather than subjective legislative motivations, their thesis is that criminal legislation is unjustified if its principal effect is neither to prevent harm or offense to others, nor to prevent free-standing moral wrongs, but rather to prevent citizens from making choices that directly harm only themselves. On this account, rational adults cannot be victims of their own informed choices.[9] While the criminal law comports with a defensible theory of criminal legislation when it functions to prevent victimization, actions that harm only (or principally) actors themselves would amount to victimless crimes if prohibited; and victimless crimes are objectively illegitimate within a just regime. Thus, even if the legislation was not crafted by legislators who were subjectively motivated by the goal of preventing self-injuring conduct, the fact that legislation has the effect of reducing such conduct does no justificatory work. If the prohibitions fail to satisfy *other* theories of criminal legislation, they cannot be saved by pointing to the fact that they enhance the well-being of those whose liberty they constrain.

These two distinct means of formulating the paternalist's thesis, and thereby the objections of those who reject paternalism as a defensible theory of criminal legislation, are not mutually exclusive. Indeed, once these claims are teased apart, it is clear that most paternalists embrace both of them: precisely because they believe that restraints on citizens' liberty can be objectively justified by the fact that those restraints enhance citizens' own well-being, they believe that it is legitimate for legislators to subjectively aim at such a goal. And the same is true, in reverse, of critics: precisely because they do not take the increased welfare of citizens to objectively justify constraints on such citizens' liberty, they believe it is illegitimate of legislators to pursue such a goal. But despite the fact that these claims tend to come as a pair, it is useful to maintain their separation. For proponents of paternalism tend to take them as individually sufficient justifications for criminal legislation, while opponents tend to take them as individually sufficient grounds for rejecting legislation. Thus subjectivists who find that a prohibition was paternalistically motivated will not be appeased by the fact that it has the objective effect of eliminating certain kinds of harms or certain sorts of offenses to others. And objectivists who deny that legislation can be justified by the fact that it precludes self-injuring behaviors will not take criminal legislation to be saved by the fact that such a result is an unintended coincidence of provisions that were motivated by other goals.

3 Grounds for rejecting paternalism as a justification for criminal legislation

Why not allow legislators to aim at their citizens' own well-being when criminalizing specific behaviors? Why not allow the justifiability of criminal prohibitions to rest, at least in part, on their success at preventing citizens from engaging in behaviors that risk their own health and welfare?

One set of answers emphasizes the fact that violations of the criminal law are met with punishment and stresses that the conditions under which punishment are appropriate are highly circumscribed. A theory of criminal legislation that is at odds with our most defensible theory of punishment is a theory that itself cannot be justified. Thus, if efforts to punish citizens for their own good are incompatible with the justified conditions of punishment, then paternalistic criminal prohibitions are indefensible. And this is precisely what many punishment theorists would argue.

On a standard utilitarian account, for example, punishment constitutes a harm to the offender that can be justified only by the prevention of greater harm to others. What of punishing an offender in order to prevent the infliction of an even greater harm upon himself? Surely, such a trade-off would be at home within a utilitarian theory of punishment, would it not? Mill's own answer to this suggestion remains a popular response for antipaternalists: rational adults who are eligible for punishment in the first place are also possessed of sufficient self-knowledge to know their interests best and to advance them most efficaciously. "With respect to his own feelings and circumstances, the most ordinary man or woman has means of knowledge, immeasurably surpassing those that can be possessed by any one else" (Mill 1977 [1859]: 277). Each citizen is the one person in the world who is "most interested in his own well-being: the interest which any other person, except in the cases of strong personal attachment, can have in it is trifling, compared to that which he himself has" (Mill 1977 [1859]: 277). It follows, claimed Mill, that "the strongest of all the arguments against the interference of the public with purely personal conduct is that when it does interfere, the odds are that it interferes wrongly and in the wrong place" (1977 [1859]: 283).

If Mill was right to insist that citizens will out-perform the state when it comes to determining what will be in their own best interests, then paternalistically imposed punishments can be expected to produce more harm than good, and thus to offend against utilitarian justifications of punishment. There is, of course, a very considerable literature that takes after Mill on this issue – one that extends from the work of James Fitzjames Stephen,[10] to that of H.L.A. Hart,[11] Gerald Dworkin,[12] and those, like Richard Thaler and Cass Sunstein, who rest contemporary defenses of paternalism on the cognitive and volitional defects that contemporary psychologists have purportedly vindicated.[13] At the very least, this literature suggests that if utility is properly measured by such objective factors as health, wealth, education, life expectancy, and so forth, then there are good reasons to doubt that citizens will inevitably be better at maximizing their own utility than will the state. Yet a more sophisticated utilitarian theory might restore Mill's skepticism on this score. If, for example, the sheer fact that an activity is chosen by a citizen vests that activity with substantial utility – if the autonomy of a choice is itself happiness or satisfaction generating[14] – then it may be that more utility is lost than gained when the state substitutes its choice for that of a citizen, even if in so doing it enhances the citizen's health, wealth, or life expectancy. We need not exhaust this dispute here, however. It is enough to recognize that those who champion a utilitarian theory of punishment may find it to be at odds with paternalistic criminal prohibitions, at least if they conclude that the empirics upon which their theory is inevitably contingent bear out the claim that punishment more often harms than helps those upon whom it is inflicted.

Those who defend, in the alternative, a retributivist theory of punishment may have grounds upon which to object to paternalistic criminal prohibitions that are less empirically contingent. Retributivists take punishment to be justified if, but only if, it is morally deserved. In other words, desert is both a necessary and a sufficient condition for punishment.[15] On the standard view, one is deserving of punishment if one has committed an unexcused (non-*de minimis*) moral wrong. The question then becomes: Can one wrong oneself? Can one have rights against oneself? If yes, then retributivism is consistent with paternalistic prohibitions the violation of which constitute wrongs to self that are eligible for punishment in the same way as are wrongs to others. Yet retributivists are very typically (if not inevitably) deontologists, and deontologists are very typically (if not inevitably) resistant to the claim that there are duties to self (or rights against oneself) that make self-injuring choices into moral wrongs akin to those that that are done to others when others' rights are unjustifiably violated.[16] Self-injurious choices may fall within the province of vice, and so be blameworthy on aretaic grounds; but they do not reflect violations of categorical duties that are blameworthy on deontic grounds. Inasmuch as most retributivists reject the view that the criminal law should be in the business of punishing bad character, rather than culpable actions that violate deontic duties, they do not take expressions of vice to be eligible for punishment. As such, they do not take paternalistic prohibitions to map onto duties that are properly enforceable through the criminal law. Paternalistic criminal laws are thus, for the standard retributivist, at odds with the basis upon which punishment may be justifiably imposed on the citizens of a just state.

Many who theorize about the scope and limits of justified punishment resist both pure utilitarianism and strong retributivism, casting themselves, instead, as "mixed theorists" (or sometimes "weak retributivists").[17] In their view, punishment is unjustified unless it is *both* deserved and achieves a net gain in utility summed across all (unpacked in terms of health, wealth, happiness, preference-satisfaction, etc.).[18] In other words, for mixed theorists, utility-maximization and desert are both necessary and only jointly sufficient conditions for punishment. The same questions will arise for mixed theorists as for utilitarians and retributivists, however, and their answers to these questions may spell an incompatibility between their theory of punishment and the legitimacy of paternalistic criminal prohibitions. If, for example, they conclude that punishment more often hurts than helps those upon whom it is inflicted (because, for example, they take Mill to be right in thinking that citizens do better than the state at advancing their own interests), then their allegiance to the view that punishment is unjustified when it is not utility-enhancing will commit them to antipaternalism. And if they conclude that citizens cannot be victims of their own choices, for there are no duties to self that are akin to the duties we owe to others, then they must conclude that violations of paternalistic prohibitions are not moral wrongs deserving of retribution.

The take-away from this hasty survey is that there are good reasons to think that paternalism is ill at ease with all three of the classic theories of punishment. Those who seek to harmonize paternalistic criminal prohibitions with the justified imposition of punishment have conceptual room to do so, but their success will depend upon contingent and contested empirics or upon philosophical claims that demand considerable heavy lifting. Thus one reason to reject the legitimacy of paternalistic criminal legislation is on the basis that it fails to cohere with our most defensible theory of punishment (whether conceived of as utilitarian, retributivist, or mixed).

There is an independent line of argument, however, against employing the criminal law as a vehicle for paternalistically altering the choices that citizens make for themselves – one that does not rely upon the incompatibility of such laws with the legitimate conditions of punishment. According to this second and quite different thesis, paternalism is an insult to autonomy; it is a denial of what makes us persons at all; it exacts a form of liberty that ought to be unassailable by the state. As Mill put this seemingly non-utilitarian argument, "there is a part of the life of every person who has come to years of discretion, within which the individuality of that person ought to reign uncontrolled either by any other person or by the public collectively" (1965 [1848]: 938).[19] Or in Gerald Dworkin's words,

> To be able to choose is a good that is independent of the wisdom of what is chosen. . . . "It is the privilege and proper condition of a human being, arrived at the maturity of his faculties, to use and interpret experience in his own way."
>
> (1971: 117, quoting from Mill 1977 [1859]: 262)

While the law may tell citizens what they can and cannot do to others, it may not tell them what they can and cannot want for themselves, on pain of usurping their ability to pursue their own theory of the good. For it is in our ability to pursue our own theory of the good – to pick for ourselves what sorts of people we should be, what sorts of ends we should pursue, what sorts of virtues we should cultivate, what sorts of vices we should resist, what sorts of talents we should nurture, what sorts of lifestyle choices we should pursue, what sorts of first-order passions and desires we should indulge or repress – that the autonomy essential to individuality lies.[20] If the state substitutes *its* theory of what will be good for us for our own – if it coerces us into pursuing its theory of what we ought to want for ourselves, what we ought to seek as ends, what we ought to indulge or resist – the state treats us as if we are not persons at all; as if we are not autonomous agents capable of, and entitled to, self-authorship. Inasmuch as a just government presupposes the autonomy of those to whom its laws are applied, it cannot justifiably enact laws that simultaneously deny that autonomy. As such, it is barred from enacting paternalistic prohibitions that purport to punish people for their own good.

4 Ten exceptions to a general rule against criminal paternalism

That there are reasons to think that paternalistic criminal prohibitions are both in tension with the dominant theories of punishment and at risk of violating the independent right of self-authorship that citizens arguably possess does not exhaust the question of whether there are not some exceptional circumstances in which state paternalism is legitimate. While the literature on this question is considerable, it is possible to distill at least ten exceptions that might be defended by those who harbor a general distrust of state paternalism. Not all of these exceptions will provide a sound basis for enacting criminal prohibitions (as opposed to employing other means of state power such as licenses, taxes, or subsidies), but some arguably provide a justification for using punishment as a means of saving citizens from themselves.

First, few deny that paternalism is legitimate when it is true to its name – that is, when it is employed in circumstances in which citizens suffer from cognitive and volitional deficiencies that make them child-like and thus in need of the equivalent of a father's care. Those who cannot autonomously author their lives because they lack the capacities essential to autonomy – the young, the senile, the insane, the intellectually disabled, the comatose, etc. – are properly subject to the state's best theory of what is in their best interests. Inasmuch as those who lack autonomy seem both undeserving of punishment and undeterrable by punishment, this first exception would appear an incongruous basis for criminalizing their conduct. But we can certainly find in the criminal law examples of doctrines that are explicitly designed to protect those who lack rational self-control from the harm that could befall them as a result of their own poor decisions: for example, statutory rape laws that make victims' consent immaterial and criminal prohibitions on under-age drinking, smoking, driving, and prize-fighting.

Second, inasmuch as choices are valued as vehicles of autonomy only when they are fully informed, many believe that the state does not infringe upon its citizens' liberty when it employs its powers to ensure that citizens receive information relevant to important decisions. After all, how can anyone be made worse by having more information? How can autonomy be insulted by efforts to ensure that its informational preconditions are met?[21] Even Mill contemplated the legitimacy of liberty-limiting interventions when their goal and effect was to convey information essential to advancing another's informed choices:

> If anyone saw a person attempting to cross a bridge which had been ascertained to be unsafe, and there were no time to warn him of his danger, they might seize him and turn him back without any real infringement of his liberty; for liberty consists in doing what one desires, and he does not desire to fall into the river.
>
> (1977 [1859]: 294)

Of course, in numerous circumstances people positively desire to remain uninformed – to preserve a kind of ignorance that they take to be essential to bliss: they do not want to know the calorie counts of their favorite fast foods; they do not want to hear about the cruelties that are inflicted on animals in the course of raising the cheap meat and dairy products that they enjoy; they do not want to think about the gas mileage of their sports cars and SUVs; they do not want to know of the ways in which the fetuses they plan to abort have acquired certain attributes of living human beings. And if asked, many report that they feel "coerced" by such information, even as their opportunity sets have remained unchanged (so as to make complaints of "manipulation" more appropriate).[22] Notwithstanding concerns about the liberty-limiting effects of certain kinds of information, there are obvious examples of criminal laws that are crafted to this end; arguably, for example, laws that criminalize the non-prescribed use of prescription medications (the theory being that by having to obtain prescriptions for certain drugs, patients acquire expert judgments concerning the relative safety of their use); and in third-party cases, for example, abortion counseling laws requiring physicians, on pain of punishment, to provide specified information to women seeking abortions, and HIV disclosure laws requiring those with HIV, on pain of punishment, to tell sexual and needle-sharing partners of the risk of HIV transmission.

Implicit in these latter two examples of informational criminal paternalism is a distinction between what Feinberg called "direct" and "indirect paternalism" (1986: 9). This distinction provides a third possible exception to the rule against paternalistic criminal laws motivated in the previous section. According to this exception, while individuals cannot generally be punished for their own good, other parties can be punished as an indirect means of securing that good. Or put differently, while people cannot justifiably be coerced out of making self-injuring choices, they can be justifiably deprived of such choices, for others can be justifiably punished for providing the means by which such choices can be realized. Thus, in the name of the good of prostitutes, "pimps" and "johns" can be punished; in the name of the good of pregnant women (who are thought not to appreciate the lifelong psychological burden of opting for an abortion), abortionists can be punished; in the name of the good of one duelist, the other can be punished; in the name of the good of patients who want to die, euthanasia and assisted suicide can be punished. As many have suggested, all of these deeds might be eligible for punishment under the banner of the harm principle, for in all of these cases, it would seem that someone is punished for harming another. Yet, in all of these cases, the legislative goal is to prevent citizens from purportedly harming themselves by employing their own consent as a means of licensing others to do what they cannot accomplish on their own. The motivation of such laws, in other words, is paternalistic; and the effect of such laws is to thwart the ability of individuals to achieve goals they have set for themselves. As such, indirect paternalism counts as paternalism and requires special justification by virtue of that fact.

Gerald Dworkin (1971: 110–111) tellingly labeled direct and indirect paternalism "pure" and "impure," suggesting, as Feinberg put it, that "the two-party cases are paternalistic in a less genuine, watered-down sort of way" (1986: 9). Whether Dworkin believed that to be the case is hard to know, but many are clearly tempted by the view that it is less objectionable to restrict people's liberty indirectly than to do so directly. Thus, it only took three hours for Donald Trump's 2016 presidential campaign to "clarify" Trump's infamous assertion that women should be punished for obtaining abortions, by announcing that what he "really meant" was that abortionists should be punished for performing abortions upon consenting women. Yet those who maintain that it can be right for the law to thwart actions that citizens *ex hypothesi* have rights to do are committed to the thesis that morality permits what I have called "moral combat" (Hurd 1999). They are wedded to the claim that morality can make it right (indeed, perhaps dutiful) to prevent what another has a right (not just a Hohfeldian liberty) to do.[23] This is not the place to rehearse the book-length argument against the defensibility of such a thesis; instead, suffice it to say that those who seek to build an exception to antipaternalism on the direct–indirect distinction have their philosophical work cut out for them.

A fourth exception to a general rule against paternalistic criminal prohibitions might be extracted from the distinction that John Kleinig draws between "active" and "passive" paternalism (1983: 6–14). Active paternalism requires affirmative actions on the part of citizens – for example, to buckle up seatbelts, to wear motorcycle helmets, or to obtain health insurance – while passive paternalism requires citizens to omit from engaging in certain behaviors – for example, to refrain from using prohibited narcotics, to refrain from selling organs, or to refrain from engaging in unlicensed prize-fighting. If laws that demand affirmative actions are liberty-limiting in a manner that laws prohibiting particular actions are not,[24] then this distinction may provide a sound basis for complaining about some forms of paternalism, but not others. It may be that so long as paternalistic laws are crafted as negative prohibitions that leave citizens with ample alternative choices (for example, drug use prohibitions), they are morally unobjectionable, while paternalistic laws that require positive actions, and are thereby akin to an infinite list of negative prohibitions (for example, helmet requirements), are, for that reason alone, so liberty-limiting as to be illegitimate.

A fifth exception might be derived from another distinction that John Kleinig has insightfully carved; that between "negative" and "positive" paternalism (1983: 13–14). Negative paternalism, as Kleinig uses the term, involves protecting citizens from their own choices when those are likely to cause them harm, while positive paternalism involves bestowing upon citizens positive benefits. It may be that antipaternalists can tolerate, or even celebrate, efforts by the state to create benefits that are paternalistically deemed to be in the best interests of its citizens, particularly when these function as opportunity-expanding options (rather than opportunity-restricting injunctions).[25] Indeed, one might think that the state's provision of many public goods fall within this category – sidewalks, bicycle lanes, museums, zoos, parks, beaches, playgrounds, swimming pools, and so forth – for while these assets expand their opportunity sets, citizens are at liberty to ignore their benefits. But most antipaternalists will regard efforts to restrict opportunity sets by barring choices that may be self-injuring as paradigmatically objectionable.

Sixth, many take self-paternalism to be perfectly rational and perfectly consistent with autonomy, and by extension, they consider consensual state paternalism to be consistent with the obligations of the state to honor, rather than usurp, the liberty of its citizens. We all appreciate the ways in which we sometimes take actions that preclude our future selves from undermining goals we have set for ourselves in the present, as Odysseus famously did when he commanded his men to tie him to the mast and to disregard his future orders to set him free when passing by the Sirens. We hide sweets, chips, and cigarettes so as to remove later sources of temptation; we set our alarm clocks across the room so as to prevent ourselves from going back to sleep; we buy savings bonds and other financial instruments that preclude our immediate liquidation of assets; and we accept invitations to write articles that will compel us to learn new material and force us to sustain vigorous productivity. Those who defend this sixth exception would argue that when we consent to paternalistic strategies by the state – when we license the state to do the equivalent of tying us to the mast – the state cannot be condemned for so doing; for it is simply implementing our will, and thus honoring our autonomy.

One can worry that the state will surreptitiously impose its own conception of the good for its citizens' conceptions by strategically manipulating claims about whose selves are citizens' "real consenting selves" (the Odysseus who wants to be tied to the mast, or the Odysseus who wants to be untied from the mast?; the selves who want to be healthy, or the selves who want to purchase super-sized sodas?; the selves who want to save money for retirement, or the selves who want to engage in heavy future discounting?). And one can worry, in the alternative, that this exception will effectively return us to the very first view articulated in Section 1 – namely, the view that any law that is democratically enacted is morally unobjectionable – for the state has no reliable means of measuring the consent of citizens to (self-)paternalistic prohibitions save by means of the ballot box. But bracketing such concerns, the notion that consensual restrictions are not objectionably paternalistic (whether imposed upon oneself or via the aid of other persons or institutions) is a perfectly sensible one, for such restrictions realize, rather than thwart, citizens' theories of their own good.[26]

A seventh exception that constitutes a natural extension of the sixth one draws on the distinction between "soft" and "hard" paternalism, and is advanced by those who believe that even ardent antipaternalists ought to find soft paternalism unobjectionable. As this distinction is standardly employed in the literature, hard paternalism licenses the state to override the goals or ends that citizens have voluntarily (and with full information) set for themselves, while soft paternalism allows the state simply to ensure that the choices made by citizens are indeed *theirs* – that they are fully voluntary so as to be genuine expressions of autonomous self-governance (Feinberg 1986: 12). As Feinberg puts it, "the soft paternalist points out that the law's concern should not be with the wisdom, prudence, or dangerousness of [the citizen's] choice, but rather with whether or not the choice is truly his" (1986: 12). In Tom Beauchamp's often-cited view, such paternalism "is not paternalism in any interesting sense since it is not a liberty-limiting principle independent of the harm to others principle" (1977: 67). In Beauchamp's view,

> It is not a question of protecting a man *against himself*. . . . He is not *acting* at all in regard to this danger. He needs protection from something which is precisely *not himself*, not his intended action, not in any remote sense of his own making.
>
> (1977: 67, emphasis in original)

The license to engage in soft paternalism is clearly susceptible to abuse, and this is in part because the notion of "voluntariness" upon which it relies is notoriously squishy. If one employs the criminal law's very thin definition of voluntariness and thus thinks of behavior as voluntary so long as it is the product of willed bodily movements, then actions will count as voluntary even though they fail to meet basic conditions of moral agency. Those who are drunk, drugged, senile, confused, mistaken, or grossly ill-informed are nevertheless capable of willing bodily movements; so if it would be a mistake to think of their cognitively impaired (but nevertheless willed) behaviors as fully *theirs*, then we will need a much thicker notion of what voluntariness entails.

One answer might send us back to the first two exceptions with which we began: to act voluntarily is to choose one's conduct under circumstances in which one has exercised ordinary adult capacities for reasoned judgment after assessing available information about the possible outcomes of one's choice. Of course, mistakes are often made by adults of average decision-making abilities who employ reasonably available information, and if such mistaken choices are to be discounted as not fully "theirs" then we will need a still-thicker notion of voluntariness. Such a thicker notion is provided by those who equate a voluntary choice with one that would be made by a perfect reasoner in an idealized choice situation – that is, by a hypothetical version of the citizen in question who is counterfactually unburdened by the cognitive biases, volitional failures, and informational impediments that beset the actual citizen. Of course, having moved this far from the actual choices of the actual citizen, it is then a small step to arrive at hard paternalism. For it is then undoubtedly tempting to conclude that the fully rational and fully informed citizen would never set himself the goal of achieving an evil or self-destructive end – so that the moral deficiencies in his goals would count as evidence of deficiencies in the conditions of his voluntariness. No one could *voluntarily* commit suicide, or take addictive narcotics, or choose to amputate a perfectly good body part, or seek to ride a motorcycle with the wind in his hair, or engage in masochistic sexual activities: so the fact that someone would set such goals for himself is evidence that he isn't "himself" and his choices are not really "his." Such is the slippery slope from soft paternalism to hard paternalism that is licensed by the manipulability of the notion of the "voluntariness" of choice.

A rather different (although often conflated) distinction both seeks to provide a cure for this problem and functions as the basis for the eighth exception to any general rule against paternalistic criminal prohibitions. This is the distinction between "weak" and "strong" paternalism. On this distinction, the state may be weakly paternalistic by intervening in a citizen's choices in order to substitute more efficacious means of achieving the citizen's ends; but it may not intervene to alter those ends, and thus, it may not indulge strong paternalism. The notion is that autonomy resides in the ends that people set for themselves, rather than in the instrumental means that they employ in an effort to achieve those ends. If the state can enable people to achieve their goals more cheaply, more effectively, more directly, less expensively, less dangerously, etc., there can be no objection, for what counts is the fact that people get what they want, not that they get it by any means that they want. Thus if legislators can be confident that people really want to breathe clean air, then they can ban smoking (at least in public places), raise auto emission standards, prohibit polluting activities in areas of high population, and so forth. And if they can be confident that people want to live healthy lives, they can prohibit suicide, ban narcotic drugs, prohibit super-sized sugary drinks, require warnings and nutrition information on tobacco and junk food products, demand that people buy health insurance, and so forth. Again, the worry will be that those enacting criminal prohibitions will confuse ends for means, and thus prohibit, as mere means, activities that people think of as ends-in-themselves (smoking, eating junk food, driving gas-guzzling sports cars, taking drugs, etc.). But at least the distinction between means and ends, and thus between weak and strong paternalism, does not invite the substitution of hypothetical people for real ones, or hypothetical consent for actual consent – a move that manifestly invites the substitution of legislators' theories of the good for those of the citizens whose liberty they curtail.

Ninth, some theorists want to distinguish acceptable from unacceptable forms of paternalism "in accordance with the magnitude of the costs (of whatever kind) imposed on choosers" (Sunstein 2014: 57). In Cass Sunstein's view, for example, all costs should count (both psychic and material), so that graphic warnings about the dangers of certain behaviors that prove psychically traumatic might count as significantly and unacceptably paternalistic, while small fines might function as only modestly, and thus justifiably, paternalistic. As he argues, paternalistic "nudges," which by definition impose no or very small costs on choosers, should be thought of as justifiable, and should be used liberally by "choice architects" (both public and private) as means of enhancing the welfare of citizens whose "biases and blunders" are likely to systematically thwart their success at achieving their own true goals. Sunstein recognizes, of course, that most criminal laws far exceed mere nudges by threatening very significant costs, but he would have no objection to the paternalistic use of the criminal law if its penalties were modest, or if it could achieve its deterrent effects without actually being enforced against any particular offenders (as laws that are on the books but unenforced may do for a long period of time).[27]

Finally, few antipaternalists can resist Mill's own invitation to employ the power of the state to prevent citizens from using their liberty so as to give up liberty altogether.[28] In the name of liberty, choices that will severely and permanently limit the liberty of the chooser are thought by many to be eligible objects of criminal prohibition. Such was the basis upon which Mill was willing to prohibit slavery contracts; and such is the basis upon which some antipaternalists reluctantly admit that the state may employ the criminal law to prevent suicide, bodily maiming, the sale of organs and other body parts, and so forth.[29] While some will insist that the value of liberty is trivialized if it cannot be exercised in ways that ultimately diminish it,[30] others take no exception to the state constraining a citizen's liberty in order to maximize it over the lifetime of that citizen, at least in circumstances in which the citizen's own choice would severely and permanently curtail it.

5 Conclusion

I have suggested that there are two reasons why legislators ought generally to resist the temptation to employ the criminal law (as distinct from other sources of state power) as a means of protecting citizens from their own unfortunate choices. First, legislative paternalism appears ill at ease with each of the major theories of punishment. If our best justification for imposing pain and loss on people (whether utilitarian, retributivist, or mixed) is at odds with doing so as a means of saving them from self-inflicted pain or loss, then there is good reason to refuse paternalists the use of the criminal law. Second, independently of whether paternalism is at home with one or more of the dominant schools of thought concerning the justification of punishment, paternalistic exercises of coercion insult the autonomy of persons in ways that are inconsistent with their fundamental rights as persons. As such, paternalism is inconsistent with criminal legislation.

I have further explored, however, ten possible exceptions to any general rule or presumption against the paternalistic enactment of criminal prohibitions. As I have suggested, those who are distrustful of coercive forms of paternalism might still find room for criminal laws that are designed to (1) protect citizens who do not have the cognitive and volitional capacities for full autonomy; (2) afford citizens the information necessary to make rational decisions; (3) curtail the provision by third parties of opportunities to make self-injuring choices; (4) prevent harmful acts by citizens while permitting their harmful omissions; (5) provide positive benefits to citizens while allowing citizens to suffer the burdens of their own choices; (6) honor citizens' autonomy by curtailing their liberty when and as they have given Odysseus-style consent; (7) interfere with the choices of citizens that are not really "theirs," so as to allow their "true" choices to take effect; (8) interfere with the choices of citizens when doing so will allow them to reach their ends via more effective or efficient means; (9) function as mere "nudges" rather than coercive threats of hefty penalties; and (10) prevent the use of personal liberty to alienate future liberty in a manner that is drastic and permanent. While some of these grounds for criminal paternalism are more appealing than others, those who are persuaded that the state is sometimes permitted, and perhaps sometimes obligated, to save us from ourselves may find among them sufficient grounds to punish us for our own good.[31]

Related topics

Hard and Soft Paternalism; Libertarian Paternalism, Nudging, and Public Policy; Mill's Absolute Ban on Paternalism; Moralism and Moral Paternalism; Paternalism and Autonomy; Paternalism and Contract Law; Paternalism and Duties to Self; Paternalism by and towards Groups; Self-Paternalism.

Notes

1 This is the position that Robert Bork (1971) ultimately took in defending majoritarianism as the only principled basis for settling contested questions of law. Similarly, John Ely (1980: 67) insisted that "as

between courts and legislatures, it is clear that the latter are better situated to reflect consensus," and absent objective moral values that could constrain either individual citizens or legislators, governance by consensus is the best that can be hoped for. On both Bork's and Ely's views, then, the Constitution should be read to guarantee a political process which, if carefully protected, generates necessarily just results, for if the process by which values compete is kept democratic, then whatever the outcome of that competition, it will be fair and just in the only sense in which those terms make sense. For a critique of the "relativist jurisprudence" inherent in these views, see Hurd (1988).

2 This vague formulation clearly disguises large issues concerning how to preserve the consent of the people when direct democracy is replaced with representative democracy; when the views and values of representatives often have little or no direct impact on the formulation of legislation upon which those representatives are asked to vote; when legislators often vote for legislation without any real appreciation of its justification or impact; when legislators have conflicting reasons for, and intentions with regard to, legislation for which they vote so as to defeat any notion that there is any shared majority intention with regard to the legislation in question; and when not all legislators vote for all legislation so as to leave at least some constituencies altogether unrepresented in the legislative process.

3 Recall Joel Feinberg's famous "ride on a bus," during which one encounters fellow passengers who engage in conduct that causes (1) an affront to one's senses (e.g., by dragging their fingernails down a chalkboard); (2) a strong sense of disgust or revulsion (e.g., by eating live insects and raw animal intestines, vomiting on the floor, and eating their vomit); (3) a shock to one's moral and/or religious sensibilities (e.g., by disrobing and engaging in oral sex with one another); (4) the inducement of annoyance or boredom (e.g., by endlessly chattering); (5) a sense of anxiety or anger (e.g., by wearing a t-shirt with openly racist, homophobic, or sexist insults). See Feinberg (1985: chap. 1).

4 See, for example, the trilogy of articles by Michael S. Moore defending legal moralism and its relation both to retributivism and liberalism that comprise Part III of Moore (1997: 637–795). See also his more recent (2014) and (forthcoming).

5 See, for example, Raz (1986: 369–399); Raz (1989); Hurd (2002); Hurd (2004: 37–68); Wall (1998: 145–161); and Moore (1997: 739–795).

6 Gerald Dworkin (2005) labels such a theory "moral paternalism" and very capably explores its implications and problems.

7 This is, for example, how Ten (1980: 40) reads Mill:

> It is not essential to Mill's position that there should be an area of conduct which must always remain completely free from intervention. The absoluteness of Mill's barrier against intervention, or the "theoretical limit" he sets to the power of the state and society to exercise coercion, is of a different kind. There are certain reasons for intervention in the conduct of individuals which must always be ruled out as irrelevant.

8 Michael Moore describes motivational antipaternalism as resting on a commitment to what he calls "the derived right to liberty." As he defines it, "the right is that of every citizen not to have his or her behavior regulated for the wrong reasons by the government." And in his view, at the top of the list of "wrong reasons" for regulation is paternalism:

> One does not, for example, have a right to murder, nor does one have the right that the government not prohibit and punish murder. Rather, one has a right that the government not prohibit murder, because, say, of a view that murder is bad for the murderer's chances of salvation.
>
> (1997: 751)

9 "Over himself, over his own body and mind, the individual is sovereign" (Mill 1977 [1859]: 223–224).

10 See Stephen (1973: 24) (complaining of an absence of proof for the proposition that the "mass of adults are so well acquainted with their own interests and so much disposed to pursue them that no compulsion or restraint put upon them by any others for the purpose of promoting their interests can really promote them").

11 See Hart (1963: 32) (maintaining that "Mill carried his protests against paternalism to lengths that may now appear to us as fantastic." He endows the average individual with "too much of the psychology of a middle-aged man whose desires are relatively fixed, not liable to be artificially stimulated by external influences").

12 See Dworkin (1971: 117) ("To show this [the disutility of paternalistically motivated coercion] is impossible; one reason being that it isn't true").

13 See Thaler and Sunstein (2008: 1–102) (discussing the "biases and blunders" that lead humans to make choices that depart from those that would be made by rational "econs").

14 Mill certainly suggests that he believes just this:

> [I]f it were felt that the free development of individuality is one of the leading essentials of well-being; that it is not only a co-ordinate element with all this is designated by the terms civilization, instruction, education, culture, but is itself a necessary part and condition of all these things; there would be no danger that liberty should be undervalued.
>
> (1977 [1859]: 261)

15 For classic defenses of retributivism, see Moore (1997: 104–152); Morris (1976: 31–63); Murphy (1979); and Sher (1987).

16 Those who are attracted to the notion that people can wrong themselves often think of certain self-injuring actions as "sins." Yet if one presses theology into service here, duties to self quickly become duties to God; and the notion that we wrong ourselves when we make self-injuring choices quickly warps into a "harm to others" argument where the other in question is God.

17 For a defense of the merits of a "weak" theory of retributivism, see Brink (2012).

18 See, for example, Husak (2010) and Husak (2008: chap. 2). See also von Hirsch (1976: chap. 6) (distinguishing theories that take desert to be the accelerator and utility to be the brake from theories that reverse these priorities).

19 It is, of course, famously a matter of dispute whether Mill contradicted his avowed utilitarian commitments in *Utilitarianism* when crafting what appears to be a non-utilitarian defense of liberty in *On Liberty*. For thoughtful reflections on this question, see Donner and Fumerton (2009: chaps. 2–4).

20 "A person whose desires and impulses are his own – are the expression of his own nature, as it has been developed and modified by his own culture – is said to have a character. One whose desires and impulses are not his own, has no character, no more than a steam-engine has a character" (Mill (1977 [1859]: 264). For an excellent discussion of Mill's conceptions of autonomy and individuality, see Donner (2009).

21 For a recent defense of the claim that the availability of a larger rather than a smaller set of valuable options can be inimical to, rather than supportive of, autonomy, see Gordon-Solmon (2017).

22 See, for example, Cass Sunstein's (2014: 65–69) discussion of the outraged responses elicited by the Department of Transportation and the Environmental Protection Agency when they sought to add vivid letter grades to labels on cars as a means of informing consumers about the relative fuel economy of those vehicles. As Sunstein makes clear, even without assigning grades to cars, labels that purport to include "just the facts" inevitably pick out some facts to the exclusion of others, and thereby implicitly convey the government's theory of what consumers ought to care about, rather than, perhaps, what consumers in fact care about.

23 For a recent defense of a deontic logic that precludes moral combat (as opposed to a morality that adheres to Wesley Newcomb Hohfeld's famous definitions of liberties, permissions, privileges, and claim rights, so as to permit moral combat), see Moore and Hurd (draft).

24 The argument for this might rest on the fact that an affirmative requirement precludes all other actions at the time that it applies, while a negative prohibition leaves one with an infinite list of alternative actions that one may pursue.

25 For a fascinating discussion of moral principles that function as side-constraints on the ability of the state to bestow benefits on its citizens – such as, for example, the principle of equality – see Moore (2015).

26 For explorations of the nature and moral force of consent, see Hurd (1996) and Hurd (2017).

27 See Sunstein (2014: 56) (offering no complaints to the idea of "a government that specified its preferred choices with respect to everything in life, and charged everyone one cent for each departure from those preferred choices"). For a sustained critique of the claim that nudging is consistent with a libertarian political philosophy, see Hurd (2016).

28 "By selling himself for a slave, he abdicates his liberty; he forgoes any future use of it beyond the single act. He therefore defeats, in his own case, the very purpose which is the justification of allowing him to dispose of himself. . . . The principle of freedom cannot require that he should be free not to be free. It is not freedom to be allowed to alienate his freedom" (Mill 1977 [1859]: 299).

29 For an excellent discussion of the compatibility of libertarianism with commitments to inalienable rights, see Barnett (1998: 77–82).

30 See, for example, Ten (1980: 117–119) (arguing that slavery contracts should be void on the basis of the harm principle, not on grounds of paternalism).

31 Special thanks to Jason Hanna and Kalle Grill for penetrating comments about this piece, and to Stephanie Davidson for her expert library assistance.

References

Barnett, R. E. (1998) *The Structure of Liberty*, Oxford: Clarendon Press.

Beauchamp, T. L. (1977) "Paternalism and Bio-Behavioral Control," *The Monist* 60: 62–80.

Bork, R. (1971) "Neutral Principles and Some First Amendment Problems," *Indiana Law Journal* 47: 1–35.

Brink, D. O. (2012) "Retributivism and Legal Moralism," *Ratio Juris* 25: 496–512.

Donner, W. (2009) "Liberty," in W. Donner and R. Fumerton (eds.) *Mill*, pp. 56–75.

Donner, W. and Fumerton, R. (2009) *Mill*, Malden, MA: Wiley-Blackwell.

Dworkin, G. (1971) "Paternalism," in R. A. Wasserstrom (ed.) *Morality and the Law*, Belmont, CA: Wadsworth, pp. 107–127.

———. (2005) "Moral Paternalism," *Law and Philosophy* 24: 305–319.

Ely, J. H. (1980) *Democracy and Distrust*, Cambridge, MA: Harvard University Press.

Feinberg, J. (1985) *The Moral Limits of the Criminal Law, Volume 2: Offense to Others*, New York: Oxford University Press.

———. (1986) *The Moral Limits of the Criminal Law, Volume 3: Harm to Self*, New York: Oxford University Press.

Gordon-Solmon, K. (2017) "Why More Choice Is Sometimes Worse than Less," *Law and Philosophy* 36: 25–44.

Hart, H. L. A. (1963) *Law, Liberty and Morality*, Stanford, CA: Stanford University Press.

Hurd, H. M. (1988) "Relativistic Jurisprudence: Skepticism Found on Confusion," *Southern California Law Review* 61: 1417–1509.

———. (1996) "The Moral Magic of Consent," *Legal Theory* 2: 121–146.

———. (1999) *Moral Combat*, New York: Cambridge University Press.

———. (2002) "Liberty in Law," *Law and Philosophy* 21: 385–465.

———. (2004) "When Can We Do What We Want?" *Australian Journal of Legal Philosophy* 29: 37–68.

———. (2016) "Fudging Nudging: Why 'Libertarian Paternalism' Is the Contradiction It Claims It's Not," *Georgetown Journal of Law and Public Policy* 14: 703–734.

———. (2017) "The Normative Force of Consent," in P. Schaber (ed.) *The Routledge Handbook of the Ethics of Consent*, New York: Routledge.

Husak, D. (2008) *Overcriminalization: The Limits of the Criminal Law*, New York: Oxford University Press.

———. (2010) "Why Punish the Deserving?" in *The Philosophy of Criminal Law: Selected Essays*, New York: Oxford University Press, pp. 393–409.

Kleinig, J. (1983) *Paternalism*, Totowa, NJ: Rowman & Allanheld.

Mill, J. S. (1965 [1848]) "Principles of Political Economy," in J. Robson (ed.) *The Collected Works of John Stuart Mill, Vol. 3*, Toronto and London: University of Toronto Press and Routledge.

———. (1977 [1859]) "On Liberty," in J. Robson (ed.) *The Collected Works of John Stuart Mill, Vol. 18*, Toronto and London: University of Toronto Press and Routledge.

Moore, M. S. (1997) *Placing Blame*, Oxford: Clarendon Press.

———. (2014) "Liberty's Constraints on What Should Be Made Criminal," in A. Duff, L. Farmer, S. E. Marshall, M. Renzo and V. Tadros (eds.) *Criminalization: The Political Morality of the Criminal Law*, New York: Oxford University Press, pp. 182–212.

———. (2015) "Liberty and the Constitution," *Legal Theory* 21(3 & 4): 156–241.

———. (forthcoming) "Legal Moralism Revisited," *San Diego Law Review*.

Moore, M. S. and Hurd, H. M. (draft) "Moral Combat and the Hohfeldian Analysis of Active Rights." (available from author).

Morris, H. (1976) "Persons and Punishment," in *On Guilt and Innocence*, Los Angeles: University of California Press, pp. 31–63.

Murphy, J. (1979) *Retribution, Justice, and Therapy*, Dordrecht: Reidel.

Raz, J. (1986) *The Morality of Freedom*, Oxford: Clarendon Press.

———. (1989) "Liberalism, Skepticism, and Democracy," *Iowa Law Review* 74: 761–786.

Sher, G. (1987) *Desert*, Princeton, NJ: Princeton University Press.
Stephen, J. F. (1973) *Liberty Equality, Fraternity [1882]*, New York: Holt and Williams.
Sunstein, C. R. (2014) *Why Nudge?* New Haven, CT: Yale University Press.
Ten, C. L. (1980) *Mill on Liberty*, Oxford: Clarendon Press.
Thaler, R. H. and Sunstein, C. R. (2008) *Nudge: Improving Decisions About Health, Wealth and Happiness*, New York: Penguin Books.
von Hirsch, A. (1976) *Doing Justice*, New York: Hill and Wang.
Wall, S. (1998) *Liberalism, Perfectionism, and Restraint*, New York: Cambridge University Press.

23
PATERNALISM AND CONTRACT LAW

Péter Cserne

Contract regulation is a domain where the conflict between freedom of choice and paternalistic intervention is especially easy to notice. In most countries people are not allowed to sell or buy (although in some cases they can donate and receive) each other's body parts; certain substances deemed dangerous or harmful are not available for sale for the general public; individuals cannot legally commit themselves to servitude or slavery; consumers can access certain unhealthy goods or risky financial products only under strictly regulated terms and after having been confronted with graphic or heavily worded warnings about the risks involved.

In this chapter, contract law is understood as a body of legal rules that pertains to the enforcement and regulation of voluntary private agreements. Contract law provides a normative framework for the social practice of contracting. By enabling, regulating, and selectively enforcing contracts, the state, both through legislation and adjudication, operates as a powerful governance mechanism for private transactions. The law can take various stances towards the social practice of contracting: expressive, constitutive, and regulative. It expresses generally held ideas or aspirations about agency, autonomy, trust, fairness, cooperation, or dependency. Although promises and agreements can operate without law, contract law constitutes categories, conventional forms (e.g., the formalities of offer and acceptance or consideration) in which these cooperative activities can be recognized and carried out. Most conspicuously, the law also regulates the practice of contracting by prohibiting, limiting, or setting terms for various transactions. While paternalism is most visible in the regulatory aspect of contract law, it is traceable in the expressive and constitutive aspects as well.

For the purposes of this chapter, the term "paternalism" will refer to "[t]he interference of a state or an individual with another person, against their will, and defended or motivated by a claim that the person interfered with will be better off or protected from harm" (Dworkin 2017). Given that other chapters of this handbook focus on the meaning, kinds, and normative status of paternalism, this chapter focuses on how the normative problems of paternalism play out in the domain of contract law.

In the following, we shall call a certain rule or doctrine of contract law or a judicial decision paternalistic when it limits the freedom of contract in order to protect from harm or grant benefits to at least one of the contracting parties. The interference of "contract law" typically happens through general rules enacted by legislators or introduced by decisions of state authorities, be they judges or other officials. Paternalism in contract law is easiest to grasp if we look at what legal rules do to contracts *ex ante*, i.e., at the formation of contracts. If we look at legal rules *ex post*, i.e., in case of a dispute, one of the parties willingly invokes a rule, presumably because it is in their interest to do so. So by looking at judicial actions in an individual contract case one would hardly find any intervention paternalistic in the sense of interfering against the current will of the party. The paternalistic character of the interference is clear, however, if we focus on the rule or principle that the court applies to override (some terms of) a binding contract, thus going against the party's will at the time of contracting.

The nature of the interference can vary from refraining from enforcing the contract or some of its terms, to imposing mandatory terms through legislation or various judicial techniques, to introducing default rules which, for various psychological or practical reasons, will be "sticky," i.e., while technically non-mandatory, still difficult to opt out from. Some rules are directly paternalistic, e.g., when they imply mandatory terms favorable to one party. Some rules are indirectly paternalistic, e.g., prohibiting the sale (but not the purchase) and punishing sellers (but not the buyers) of certain goods or services in order to protect potential purchasers from supposedly harmful goods and services.

Even the harshest sanctions of contract law do not amount to coercion (Feinberg 1986: 7–8), so paternalistic interference by contract law is generally seen as less intrusive to autonomy, and in this respect, less morally troubling than coercive paternalistic interventions, e.g., by criminal law.

In sum, contract law is seen as an intelligible and in part systematizable normative practice which can be illuminated by moral and political principles. The main subject of this chapter is to explore the role of paternalism among these principles in the context of a body of legal rules and doctrines which relate, roughly speaking, to judicially enforceable agreements between private parties.

1 The uneasy fit of paternalism with contract law

To make philosophical sense of contract law in terms of moral and political principles, and explicate the role of paternalism among these principles, is not a straightforward exercise. Such an exercise of constructive interpretation faces the difficulty that the doctrines of contract law or judicial arguments in contract cases are rarely openly paternalistic on their face and normative contract theories also rarely endorse paternalistic rules.

In fact, for some, contract law seems inherently antipaternalistic. Libertarian and economic critics of regulative contract law sometimes simply label undesirable regulatory interventions paternalistic. They view contract law as a quintessential domain of private autonomy and the least suitable for paternalism.

The idea of a contract as such may be more rightly associated with autonomy but if one cares to become familiar with contracting practices, there is ample evidence that most contracts do not match those 19th-century models of rational individuals bargaining at arm's length. Most contracts are made on standard terms and conditions, drafted by one party or adopted with reference to general terms drafted by third parties such as trade associations or private standard-setting bodies. Individual negotiation of terms is rather the exception than the rule. In fact, increasingly, contracts are made by algorithms rather than human beings. While these practices make the argument of contracting as an expression of autonomy less plausible, they do not, in themselves, necessarily justify regulation, let alone a paternalistic one.

Still, as the examples mentioned at the outset suggest, both courts and legislators have been interfering with contracts, arguably sometimes from paternalistic motivations. While in classical liberalism, freedom of contract was seen as an institutional guarantee of individual autonomy, for many decades now, contract regulation has been growing in breadth and depth. Contract law is pervaded by *prima facie* paternalistic rules that aim to promote or protect the interest of structurally or situationally "weaker" parties through procedural and substantive limits of freedom of contract.

Yet, paternalism being a problematic justification for both individual and collective (including political and legal) action, most theorists are reluctant to ground contract doctrines in paternalism. In either autonomy-based or welfare-based individualistic approaches, the normative presumption is freedom of contract. Limitations, especially paternalistic ones, are *prima facie* unjustified, and to rebut this presumption, some strong reasons need to be present. In interpreting contract doctrines, theorists are keen on eliminating/avoiding paternalism in the sense of identifying other justifications than benevolence for seemingly paternalistic limitations, based on consent, justice, or fairness, the social protection of autonomy or dignity, third-party effects, or collective action problems.

To be sure, there are exceptions to these attempts of eliminative redefinition and normative reinterpretation. Some current contract theories look at paternalism as a secondary or subsidiary normative principle, one that qualifies or limits those seen as primary in explaining why and to what extent the state is justified or even required to enforce voluntary agreements.

In his 1982 paper, Duncan Kennedy not only endorses paternalism but takes pleasure in unmasking nominally antipaternalistic and non-distributive doctrines as one or the other. In this illuminating exercise in critical legal studies, Kennedy argues that there are genuinely paternalistic doctrines in contract law which should be recognized as such; they are backed by genuine paternalist reasons (or motives) that cannot be meaningfully reduced to those related to efficiency or transaction costs, nor to distributive ones. The underlying reasons or motives for adopting a paternalistic law are concerned with what he calls "false consciousness." False consciousness is claimed to prevent individuals from following their objective interests. A paternalistic law (1) overrides the preferences of the beneficiary and (2) the consequences on third parties are only considered as side effects. Kennedy makes both an interpretative claim (viz., that there are paternalistically motivated doctrines in contract law) and a normative one (acknowledging paternalism as a valid reason to intervene), and distinguishes paternalism from cases where there is "no disagreement as to the values or moral vision on which to act, [in which case] the decision maker is not acting paternalistically" (Kennedy 1982: 572).

Some legal scholars outside the Critical Legal Studies movement, including Kronman (1982), Zamir (1998), Buckley (2005), Eisenberg (see Zamir 2014: 2107), and Mackaay (2013: 444–448) also acknowledge paternalism and assign it an explicit role among the guiding principles of contract law. In German-speaking legal scholarship, which is more philosophical than doctrinal in its starting point, while obviously mindful of Kant's strong antipaternalism as well,[1] there are entire monographs devoted to the topic of paternalism in contract law (see, e.g., Enderlein 1996; Schmolke 2014).

Even more recently, libertarian paternalism has been suggested as a normative political theory or ideological project or platform, pulling together opposed political factions which could, among other things, possibly justify curbing freedom of contract for the benefit of individuals with bounded rationality and willpower. Understanding the background of this argument requires a short detour into legal and economic scholarship. American legal academia has been heavily influenced by economic analysis (law and economics) since the 1980s. The recent rise of behavioral (law and) economics gave currency to what is sometimes called "anti-anti-paternalism" (Jolls, Sunstein, and Thaler 2000: 3) or "new paternalism" (Whitman 2006). In fact, behavioral law and economics is often mentioned in one breath with libertarian paternalism and/or nudges, the last two terms being made popular in joint publications by Cass Sunstein and Richard Thaler (Sunstein and Thaler 2003; Thaler and Sunstein 2008).

A lot of recent discussion of paternalism in contract regulation has been inspired by empirical findings on human behavior concerning bounded rationality, heuristics, and biases.[2] The behavioral (empirical) turn in law and economics is a welcome development as the gap between high-floating normative arguments and particular institutional or practical reality needs to be filled with the help of empirical data. Behavioral economics has also contributed to legal scholarship and public policy by identifying and/or providing evidence for the functioning of various techniques of choice architecture (sticky default rules, options, menus, information provision) which take into account empirical findings on human decision-making and thus can be put to socially beneficial use in contract regulation. Furthermore, empirical findings on how key institutional players operate or how private parties respond to legal rules can be relevant in higher-level policy decisions as to whether particular interventions are feasible or comparatively more or less effective than others, including the status quo. For instance, if judges are prone to hindsight bias (Rachlinski 2000), they might have a biased view on whether a particular loss was foreseeable when the contract was formed and thus whether compensation for this loss is justified or not.

Yet, the idea that behavioral research provides paternalistic arguments of a *new kind* seems misconceived (Cserne 2012: 43–54, 137–139). To be sure, empirical findings give more detail and in this respect further support to the existing argument about the imperfection of the correlation between individual choice and welfare maximization. The increased attention to these findings is expected to lead to more precise and better-founded knowledge about why and when contracting parties make suboptimal choices. Yet, the evidence of various cognitive biases does not, in itself, call for limiting freedom of contract in a different manner than commonsense observations about human frailties do. Knowing more about the likely circumstances or scenarios which justify or even require paternalistic intervention provides valuable information but little in terms of qualitatively new arguments for the justifiability of paternalism.

Related to this, Sunstein and Thaler's (2003) endorsement of "libertarian paternalism" has made the idea of paternalism less inimical to mainstream legal thinking in and beyond contract law. Libertarian paternalism also gives a twist to normative legal and political theorists' overlapping consensus around antipaternalism. Sunstein and Thaler acknowledge the paternalistic character of some of the policy changes they suggest (inducing people, without engaging their deliberative rationality, to make choices that benefit them in terms of their own long-term preferences) and aim at defusing antipaternalist arguments by proposing interventions that at least formally respect freedom of choice (the opportunity to opt out), hence the "libertarianism" of their paternalism.

2 Freedom of contract: meaning and significance

It is hard to talk about contracts without the idea of voluntarily assumed obligations and the state's involvement as enforcer but not source of these obligations. Freedom of contract is both expressing this idea and protecting this domain from intervention. Freedom of contract is an ideologically charged notion which attracts strongly held political views amongst both defenders and critics (Craswell 2000: 82).

Freedom of contract is a fundamental principle enshrined in most modern contract laws, expressing three related ideas: parties should be free to choose their contracting partners ("party or partner freedom"), to agree freely on the terms of their agreement ("term freedom") and "where agreements have been freely made, the parties should be held to their bargains" and contracts should be enforceable by state institutions ("sanctity of contract") (Brownsword 2006: 50). Freedom of contract prevails to "the extent to which the law sanctions the use of contracts as a commitment device," leaving the terms of the agreement to the parties (Hermalin, Katz, and Craswell 2007: 18).

Historical research suggests that the idea of a general enforceability of agreements comes from late medieval and early modern theological and philosophical debates on the moral foundations of contract law. Ancient and medieval laws did not recognize the general enforceability of consensual agreements; only certain types of agreements based on consent were enforceable. Later, both the general principle of freedom of contract and its limits have been systematically discussed in late Scholastic natural law theories, thus providing moral underpinning for the rise of economic freedoms (Gordley 1991; Decock 2013).

In the last few decades, normative theories of contract law have proliferated, especially in the United States (for overviews see, e.g., Hillman 1997; Benson 2001; Klass, Letsas, and Saprai 2014). The aim of these theories is either constructive: to identify a single unifying principle (promise, autonomy, reliance, efficiency, civil recourse, etc.) or a combination of these to make sense of but also to criticize the body of contract law rules and doctrines; or deconstructive: to proclaim the death of contract (law) (Gilmore 1974; Scott 2004). As theoretical propositions, they remain controversial. Even if contract law has no single or essential function, there seems to be a reasonably broad consensus that at least a limited version of freedom of contract may be supported by both autonomy-based and welfarist theories, and perhaps less prominently but also importantly, by aretaic (virtue-based) arguments (Cserne 2012: 82–89).

Autonomy-based, in particular promise-based, theories (e.g., Fried 2015) see contracts as invoking the moral institution of promise-keeping. Consequentialist theories, especially in their welfarist or economic variants, argue that the operation of a modern market economy relies on freely negotiated enforceable contracts. Freedom of contract, sometimes under the label of consumer sovereignty (Persky 1993), has been traditionally supported by its likely benefits in terms of social welfare. This is based on a contingent empirical generalization: "Most people look after their own interests better than anyone else would do for them" (Cooter and Ulen 2012: 342).

Freedom of contract and a competitive market economy seem to simultaneously promote individual autonomy and social welfare, converging towards what could be called a "private ordering paradigm" (Trebilcock 1993). According to this convergence claim, freedom of contract is supported by a combination of autonomy-based and welfare-based arguments (Pincione 2008). Contract is a paradigmatic social institution of liberal individualism, prominent in both of these traditions. This basic agreement dissolves at the margins or when certain implicit assumptions of voluntariness or rationality are questioned.

Aretaic or virtue-based contract theories such as the neo-Aristotelian theory by Gordley (2001) see contract law as embodying and enabling the virtues of commutative justice, prudence, and liberality and also acknowledge freedom of contract as a default position.

While the ultimate justification of freedom of contract is different in these normative accounts, contract doctrines of modern legal systems reflect a sort of overlapping consensus (Rawls 1993). Modern legal systems attach a high value to freedom of contract as a basic legal principle but also set several limits to this freedom. In fact, today's contract law regimes can be seen as long lists of exceptions to the principle of contractual freedom.

In what follows our main focus will be on *prima facie* paternalistic rules, principles, and doctrines of contract law. These can be usefully classified as corresponding to constitutive, procedural, informational, and substantive limits to freedom of contract; four categories that are helpful heuristic devices to systematize the rules of contract law and together probably cover most of the terrain (Cserne 2012: 93–135). We will analyze these limits in turn, indicating (and occasionally discussing) whether and how the legal doctrine or instrument in question can be illuminated, explained, justified, or criticized as an instance of paternalism.

3 Constitutive limits to freedom of contract

Contracting practices implicitly assume some constitutive limits to freedom of contract (Kennedy 1982: 577–578, 581–582). This term refers to those minimal conditions of individual rationality and voluntariness which are necessary for the working of even a libertarian (unregulated) contract regime. Virtually all legal systems impose threshold conditions for making enforceable contracts, requiring capacity, and prohibiting duress and fraud. The key idea is that these "limits," rather than genuinely constraining freedom of contract, constitute our idea of what contracts as voluntary and informed agreements are. These constitutive limits of freedom of contract not only guarantee contracting as a domain of individual autonomy but are also valuable for instrumental reasons, selectively enforcing transactions that increase social welfare (Cserne 2012: 93–106). Let me briefly discuss the most important constitutive limits in more detail.

Capacity

As Ganuza and Gomez Pomar write,

> [M]ost legal systems establish some limitations to consent by individuals whose age, maturity, or mental or physical condition pose them in danger of taking contractual decisions that are not in line with their best interest. This is why the legal system, paternalistically, restricts contracting by minors and severely mentally or physically handicapped individuals, who will be essentially represented in the contractual sphere by their parents or legal guardians.
>
> (2010: 52)

The rules of capacity are supposed to regulate transactions in such a way that only those who are capable of making choices in the sense of being autarchic (minimally autonomous) agents can conclude a valid contract. In modern legal systems, "capacity" is the principal juridical mechanism by which individuals and entities are empowered to enter into legally binding agreements and, more generally, to arrange their affairs using the instruments of private law. Legal capacity is the gateway to involvement in the operations of a market economy (Deakin 2006).

Incapacity rules are clear instances of soft paternalism (Feinberg 1986: 11–12): they protect presumably not fully autonomous persons or persons in situations when their capacity for autonomous decisions is temporarily impaired, "from themselves," by disqualifying their contractual consent as not genuinely voluntary (Feinberg 1986: chap. 26). Those who lack actual capacity to voluntarily consent are protected by being deprived of their legal capacity to participate in potentially self-harming transactions. As long as the contracts concluded by minors or mentally incompetent adults are void or voidable, incapacity rules also defend these vulnerable persons from being exploited by others. The rules operate to this effect indirectly, by changing the *ex ante* incentives of potential contractual partners. Paternalism also has a more active dimension with regard to incapacitated persons. The law has installed guardianship and provides rules for statutory representation as ways to directly take care of the interests of the incapacitated.

Factors vitiating consent

In modern doctrinal understanding, a contract is founded on the agreement ("the will") of the parties. In the context of contract formation, the law is concerned with the voluntariness of agreements and takes care of situations where individuals should be protected from their not fully voluntarily undertaken obligations. A set of legal doctrines, technically called formation defenses, are meant to guarantee the lack of certain substantial external controlling influences that would vitiate genuine consent in the formation of contracts. Formation defenses include duress, fraud, misrepresentation, and undue influence (Mehren 1992; Zweigert and Kötz 1998: 410–430). Like incapacity rules, formation defenses also represent soft paternalism insofar as they protect individuals from forming less than substantially voluntary contracts (see Feinberg 1986: chaps. 23–25 for analogous doctrines in criminal law).

While the justification of the core of these doctrines is relatively uncontroversial, the degree of voluntariness that the law deems (or should deem) necessary in a given contract formation setting is not easy to discern. Voluntariness comes in degrees, and the legal rules on what constitutes voluntary consent are fuzzy. There are many kinds of pressure on a person entering into a contract, and the law has to determine which of these is serious enough to make the contract either (automatically) void or voidable (allowing the party whose consent is problematic to make a claim to this effect). One must also think of categorizing different kinds and degrees of pressure or advantage-taking by the other party as to their vitiating effect. Suppose there is full information, neither party is subject to significant cognitive deficiencies, and the contract is complete. In such a case the question is whether the *constrained choice set* of one party renders his consent "involuntary." If it does then practically every contract can be seen as "coerced" because of the relative scarcity of resources and opportunities. This would set the threshold of voluntariness unreasonably high. Alternatively, one could say that, except for extreme cases such as actual physical force, psychic torture, or hypnotic trance, almost every exchange is "voluntary" in the weak psychological sense of reflecting the choice of the individual. The victim of coercion is acting voluntarily in the sense of evaluating his reasons for action and choosing the least painful one. Yet, setting the threshold this low is not convincing either. In fact, any decent court of law will find a contract signed at gunpoint, although psychologically voluntary, void, in reason of a *normative judgment* about the quality of the choice set available to the coerced party. In such a situation consent does not carry moral weight (Feinberg 1986: 188).

Most coerced transactions are instances when either consent is obtained improperly or the alternatives available for the coercee are such that consent has a different moral weight than under normal circumstances. Arguably, if the choice set is too constrained, so that there are no morally acceptable options, then the choice is not voluntary (Olsaretti 2004, 2008). While this is a sensible general point, the details, such as the meaning and criteria of moral acceptability, are open for discussion. In consequence, the extent of this set of constitutive limits is controversial, both in theory and in practice.

Philosophers concentrate their efforts on identifying those proposals or directives which count as coercive and are, on this account, morally problematic (Wertheimer 1987). Starting with the classic article by Nozick (1997 [1969]), philosophical analysis has been concerned with drawing a distinction between (illegitimate) *threats* and (legitimate) *offers*. While threats reduce the possibilities open to the recipient of the proposal, offers expand them. One of the difficulties for this approach is to specify a baseline against which the proposal is to be measured. The position of this baseline is not natural or self-evident. It may be statistical (what the offeree can reasonably expect), empirical, phenomenological (what he in fact expects), or moral (what he is entitled to expect).

As for disagreements in practice, while it is often said that the law needs to draw clear distinctions, the interpretation of contract formation doctrines usually allows a wide margin of discretion to courts to decide on a case-by-case basis. Courts sometimes expand the scope of formation doctrines to cases where the original justification does not apply. It has been discussed in the legal literature for a long time whether and under what circumstances the law should void contracts for economic duress (Hale 1943; Dawson 1947). The justification of the duress doctrine is less problematic within its traditional narrow boundaries, i.e., when a threat or actual coercion was used. But if one argues that the consequences of an exercise of autonomy depend on the opportunities available to the individual, then the range of circumstances when consent is vitiated gets much broader. It may also be considered a case of coercion vitiating consent if a contract is formed under what is called "economic duress" or "necessity," i.e., when a contracting party lacks alternative ways to procure income sufficient for a decent living.

When an offer is made or an agreement is reached in a situation when one of the parties is in "economic duress" or "necessity" in the sense of having inadequate alternatives, the question arises whether courts should consider the agreement as a valid contract, modify it, or invalidate it. In many national contract laws, promises made under necessity are either void or voidable, or the party claiming that his situational vulnerability was exploited may seek a judicial modification of the contract. Should the "non-coercive exploitation" (Feinberg 1983) of the economic necessity of another party trigger any of these remedies? Or should the lack of alternatives for one contracting party, rather, count among those "legitimate inequalities of fortune" (Feinberg 1986: 196–197) that contract law is not supposed to care about? Such questions are doctrinally linked to contract formation but they are beyond its core. In fact, as we shall see below in the discussion of unconscionability, constitutive and substantive limits to freedom of contract are sometimes hard to distinguish and the answer may also depend on epistemic and institutional capacities of courts to correctly identify cases when interference is justified.

Note also that voiding unequal bargains or non-coercive but exploitative contracts is not necessarily paternalistic: besides concerns about the relevant party's well-being, interference may be motivated by ideas of substantive fairness or guided by ambitious redistributive policies.

4 Procedural limits to freedom of contract

Procedural limits to freedom of contract do not constrain the parties' agreement on any substantive term they choose (accept or bargain for); they merely require certain actions to be taken (or not taken) before contracting or during the contractual relationship. These relate to the process of securing, evidencing, recording, or authenticating the agreement. For instance, the law may prescribe a written form for the contract or a waiting period before it becomes valid, impose a requirement of mandatory advice, or grant a right of withdrawal (cooling-off period).

These procedural requirements of contract validity may be interpreted as legal responses to bounded rationality of a contracting party or some shortcoming of the contracting process. At first sight, the link between procedural rules and paternalism is relatively straightforward. As policy instruments, their function is to make the conclusion of contracts more difficult and in this way more deliberate (rational or autonomous) and eventually more welfare-enhancing. Formal requirements may serve various other functions as well; not all are related to the protection of individuals from self-harming contracts.

While institutional arrangements vary from jurisdiction to jurisdiction, most legal systems require relatively strict formalities (certificate by courts, notarial document, witnesses, etc.) for certain types of transactions, where special importance is attached to the autonomy and deliberateness of the choice. These usually include unilateral obligations (gifts), dispositions over real estate, certain transactions of a highly personal nature such as surrogate motherhood or marriage, or with large economic impact such as various transactions related to inheritance. Many other contracts need to be formed or evidenced in writing.

Cooling-off periods

In the context of consumer contracts, one of the best-known formal requirements is the mandatory provision of a withdrawal right, often called a cooling-off period. This is a short time period within which one of the parties (usually the consumer) can withdraw from the contract unilaterally, without further explanation. This is usually justified by emotional and cognitive biases which are typically present in doorstep or distance-selling situations or other contexts where aggressive sales techniques tend to be used. Yet, mandatory withdrawal or confirmation periods are granted in a much broader set of contexts. Often, they are related to situations where emotional biases are strong and the stake is large enough to make *ex post* regret weigh heavily. Marriage, divorce, or surrogate motherhood contracts belong to this category.[3]

Generally, a withdrawal period can be helpful when, for reasons related to the contracting situation, it is not to be expected that a party is capable of making a deliberate choice. The scarce empirical evidence available suggests, however, that actual withdrawal from contracts is rare.[4] Even if consumers regret their choice afterwards, they tend to keep the purchased good. It is plausible that the paternalistic rule allowing people to "cool-off" does not work effectively because there are emotional and cognitive mechanisms that have not been taken into account. As some commentators suggest, this would provide an argument for changing withdrawal periods into confirmation periods and requiring a confirmatory action subsequent to the agreement before it can take effect (Eidenmüller 2005).

Mandatory advice

In practice, contracting parties often use the services of advisors and experts on their own initiative. Making this assistance mandatory is potentially paternalistic; it needs further justification. A typical case when contract law prescribes mandatory advice is related to suretyship and guarantee contracts where family members agree to provide security to a bank loan of their spouse, child, or parent. A number of controversial judicial decisions in Germany, England, and elsewhere have attracted some public attention to this kind of contract. In some cases, court released the guarantors from their contractual obligations with the argument that the family member agreed to guarantee the loan under undue influence and the bank did not take sufficient reasonable steps to ensure that the guarantor agreed voluntarily, in particular free from undue influence by a family member.[5] The typical banking practice to fulfill this requirement is to require guarantors to seek independent advice on the nature of obligations they are about to undertake. As these financial arrangements concern close family members, contract law is supposed to play a somewhat different role than in arm's-length transactions among strangers. Within the family, calculative behavior is typically less prominent, while altruism is frequent.[6] Strong personal ties often imply power relations of such intensity that would, in an extra-familial context, require intervention and a strict legal response. Yet, courts are reluctant to use heavy-handed contract doctrines for interfering with family finances. By adding a procedural limit of mandatory advice on top of the constitutive limit of undue influence, courts are arguably striking a reasonable balance between these various considerations (Trebilcock and Elliott 2001).

In sum, from the perspective of soft paternalism, formalities work as "indices of seriousness" (Kötz and Flessner 1997: 77), aimed at preventing myopic or impulsive decisions. The higher the potential risks, the more cumbersome the formalities they can justify. Note, however, that this protective function of formalities is ambivalent. While they can indeed protect an unsophisticated party from being bound by an ill-considered decision, if that party is uninformed about these very formal requirements, he may be later disappointed when it turns out that, contrary to his intentions and expectations, he did not enter into a binding agreement. Courts may feel inclined to exempt unsophisticated parties from strict formalities and disregard minor formal defects, if they think that upholding the contract would benefit the relatively unsophisticated or uninformed party. Thus, paternalism may provide an anti-formalist argument as well.

5 Informational limits to freedom of contract

These limitations regulate the information flow between the parties before or during the contract. Contract law deals with situations where at least one contractual party typically has imperfect information through various judicially enforced doctrines, such as fraud, misrepresentation, mistake, or pre-contractual duties. Government regulation of the content, presentation, location, structure, and format of advertising, labeling, and other information provision is also increasingly important. This includes rules that mandate the pre-contractual furnishing of information and prohibit the provision of fraudulent, misleading, or irrelevant information.

There are important differences among these measures as to their links to paternalism. As argued above, prohibiting fraud is necessary to ensure that people enter into the contract voluntarily (it is a constitutive limit to freedom of contract) and is at most an instance of soft paternalism. The regulation of product labeling may be paternalistic and less easy to justify, depending on how it is carried out.

The regulation of information provision and disclosure is often seen as an optimal method of consumer protection. Mandatory information disclosure assumes that when consumers are presented with sufficient relevant information, they can make sufficiently autonomous and rational decisions on the market. However, this assumption is often false. For information-based contract regulation to work effectively, it is crucial for the regulator to understand how contracting parties (in particular consumers) process information and how disclosure rules are likely to impact on this process.

Both theoretical considerations and empirical findings suggest that the mandatory provision of more information or more detailed information does not necessarily make the consumers' choice more rational or more autonomous. More information is not an unqualified benefit for contracting parties. Disclosed either by central agencies or contracting partners (or producers), information is often technically complex, and the non-professional party (consumer) has limited time, motivation, and abilities to process it. For instance, "providing too much information on a product label can be counterproductive in the sense that label clutter reduces total recall of the information on it" (Magat 1998: 310).

Information provision is an imperfect, mostly ineffective, and arguably even counterproductive tool for paternalistic intervention. Its wide use in consumer protection and beyond is hard to justify but can be explained by political economy reasons: mandatory disclosure rules often represent a viable political compromise between powerful interest groups and persistent ideological oppositions (such as controversial laissez-faire and heavy-handed substantive regulation) (Ben-Shahar and Schneider 2014).

Information about product safety provides a good example of the behaviorally inspired idea of debiasing through law (Jolls and Sunstein 2006), a more recent sophisticated information provision mechanism that takes into account empirical research on human decision-making and judgment. It is meant to remedy certain biases in information processing by harnessing other countervailing biases. People tend not to process information in a rational way, not only in the sense of deviating from Bayesian theory in estimating future risks. Even if their estimations of the general probabilities were correct in a statistical sense, they may be too optimistic about the occurrence of safety risks in their own specific case. In the psychological literature, this is called over-optimism bias. The tendency to underestimate the risk of a personal injury occurring to the buyer or user (over-optimism) can be counteracted by using vivid and personified examples in the information provided about the product (availability heuristic). Confronted with a story about a recent real-life case of harm caused by a defective product, consumers become more aware, perhaps even overly aware of the risks involved. This rudimentary sketch of the idea behind debiasing through law should be refined in many ways before making actual use of it in information regulation. Still, the basic notion behind it is plausible, some would even say trivial. The manner in which information is presented and communicated matters.

6 Substantive limits to freedom of contract

Substantive contract law rules can take many forms, but they all set limits to "term freedom." They mandate or prohibit terms of contracts either *ex ante*, e.g., by statutory rules regulating interest rates and other terms of consumer credit contracts, or *ex post*, e.g., by non-enforcement of terms that courts find unconscionable, unreasonable, or unfair. They regulate not only *how* but *what* individuals can legally (enforceably) promise to another. In modern legal systems substantive limits to contracting are numerous and extremely varied. Many rules of labor, housing, or consumer protection law belong here. Because of the volume and heterogeneity of such rules, this section focuses on a single doctrine of contract law, characteristic for its paternalistic uses.

Most legal systems allow for the judicial control of the fairness of certain or all contract terms, though under different doctrinal names and with substantive differences. When a contract (term) is denied enforcement for reason of unfairness, unconscionability, or gross disparity, the reasons evoked by courts often combine procedural and substantive elements (Leff 1967). Procedural unconscionability concerns the circumstances of contract formation and refers to the "absence of a meaningful choice" for one party. Substantive unconscionability refers to the allocation of gains, burdens, and risks of the contractual bargain. This is deemed unconscionable if it is considered "unexpected" or "objectively unreasonable."

It may be interesting to identify and analyze the case groups where unconscionability stands as a legal *façon de parler* for paternalism and distinguish it from other uses of the doctrine that are better understood as justified differently. Sometimes courts invoke the doctrine of unconscionability, implicitly or explicitly, as a proxy for involuntariness. If consent was not voluntary or sufficiently informed but the evidentiary requirements for a formation defense are difficult to meet or some other legal technicalities raise practical difficulties in applying a formation defense, unconscionability may provide relief. As an empirical generalization, a very unequal distribution of the contractual cooperative surplus may signal that the disadvantaged party did not agree fully voluntarily. Based on this generalization, voluntariness might be ascertained indirectly through a combination of procedural and substantive fairness rules (Epstein 1975). This interpretation would reduce unconscionability to a constitutive limit of freedom of contract.

At the other extreme, we find procedurally fair risky exchanges. Can purely substantive limits on contractual freedom be justifiably used to protect the interests of a contracting party who fully voluntarily agreed to a high-risk contract which turns out to be unfavorable to him? This would be a clear-cut case of hard paternalism or perhaps legal moralism. Suppose the contract was formed fully rationally but under uncertainty, such that the value of the contractual surplus is unknown to the parties at the time of agreement. If unconscionability is applied in case of the transfer of a good of uncertain value, this allows opportunism by the buyer. After finding out that the low-value case materialized, he refuses to pay and asks the court for assistance to rescind from the contract. To grant relief in such cases provides perverse incentives for opportunism.

Sometimes paternalistic and redistributive goals are combined in support of certain policies, for example, the protection of so-called weak or structurally unequal parties in contract law. However, unless both contract price and all other contractual terms are regulated, contract law tends to be largely ineffective for redistributing wealth between the two sides of the transaction. The contract price or unregulated terms are changed accordingly so as to reflect and pass on increased costs. This implies that the non-enforcement of certain odious terms does not lead to redistribution from the contracting partners to the class of paternalized individuals.

A further troublesome counter-intentional consequence is when the interference is, on balance, harmful to the targeted beneficiaries. This may take different forms; one of them is the so-called double-bind effect. If certain transactions are prohibited for paternalistic or moralistic reasons, this may hurt those who are engaged in those transactions more than if the activity were tolerated. If individuals are prohibited from voluntarily engaging in dangerous, degrading, or self-harming yet voluntary and at least minimally beneficial activities, this may deprive them from opportunities and benefits which, given their circumstances, are the best options available to them. Economists often argue against certain types of market control in this way.

In certain market segments, harsh contract terms may in fact be efficient, reflecting customers' preferences and/or higher credit risk, collecting costs, or marketing costs. These contracts may be expensive for consumers but justified by costs and risks. The non-enforcement of such contracts would "change sellers' incentives to seek out certain kinds of customers, or to raise the price charged to an entire class of customers who might later be released by courts" (Craswell 2001: 38). If such terms are deemed unenforceable, those market segments are likely to be left unserved. The ready availability of unconscionability in such cases is expected to have unintended systemic consequences.

Even without prohibition, overly protective regulation runs the risk that certain transactions become unprofitable and lead to the collapse of the entire market or of the segment that was the object of protective regulation. For instance, limits on interest rates may make legal credit unavailable to certain groups of borrowers. In such cases, paternalism becomes problematic also because it backfires and makes the very group of persons it intended to protect worse-off on balance.

The American case *Williams v. Walker-Thomas Furniture Co.*[7] provides a good example. The case involved Walker-Thomas extending credit from 1957 to 1962 to Williams, a low-income customer living on government benefits, for a series of furniture purchases. The contract was written in such a way that no furniture could be paid off until all of it was (cross-collateralization). When Williams defaulted on the contract in 1962, Walker-Thomas tried to repossess all the furniture sold since 1957. The court found that the cross-collateralization clause was unconscionable. After the case had been decided, the federal government commissioned sociological and economic research on the characteristics of consumer durable markets where cross-collateralization and other apparently "unfair" clauses were prevalent. It turned out that both in economic and in sociological terms, there are significant differences between the low-income and middle-income segments of these markets (Collins 1999: 106–107, 262–265; Korobkin 2004). One of the differences was this:

> Low income customers may be purchasing with a particular sales method more amenable to their backgrounds and traditions. Immigrants from a rural world may seek not only the physical good but also a personalized sales method. Prices to customers will rise with the increased costs of buying from a sales person who knows one's name, one's family, and one's interests and who consents to talk about them.
>
> (Kornhauser 1976: 1171)

If oppressive terms or exorbitant prices are not due to fraud, duress, or misrepresentation, they may be related to market imperfections. These may justify intervention although not necessarily a judicial one. Even when it can be established that the contract price or a particular pricing practice is not efficient, the remedy for unconscionability may not ideally lie with private law courts.

One relevant question is whether, on balance, courts are better at identifying such cases than legislators or the parties themselves. There can be cases when a party voluntarily agrees to a clearly inefficient contract (with negative consumer surplus) and courts are better at judging the value of the transaction than the paternalized party:

> If an expensive English-language encyclopedia has been sold to a childless couple that does not even speak English, or if fifty years' worth of dance lessons have been sold to an eighty-year-old widow, it is difficult to argue that – at least in these cases – a court could not make a better judgment than that made by the individuals involved.
>
> (Craswell 2001: 37–38)

But courts also make errors in identifying inefficiency. It is an unsettled empirical question whether the total loss from errors can be minimized by granting courts discretion for such interventions.

Furthermore, in many cases private litigation is an ineffective and potentially counter-effective remedy. Even with the use of class actions, private adjudication is unable to compensate every consumer for his losses. "Persons who failed to purchase because of too high prices or too low quality have no cause of action though these unrealized sales represent the efficiency loss of the market" (Kornhauser 1976: 1180).

Not all seemingly paternalistic rules are in fact such. They can be justified by nonpaternalistic arguments as well. Seemingly paternalistic doctrines can be reinterpreted such that some substantive limits to freedom of contract may be justified on other grounds. To conclude this section, I now discuss three such suggestions.

An alternative nonpaternalistic justification for striking down unconscionable contract terms has been suggested by Seana Shiffrin (2000). Shiffrin explicitly aims at a nonpaternalistic justice-based argument for the unconscionability doctrine. She suggests that the best justification for the non-enforcement of certain exploitative or otherwise problematic contracts is based on justice. By enforcing grossly unfair contract terms, the state would be complicit in a moral wrong. Non-enforcement of unconscionable contracts allows the state to withdraw its assistance from objectionable private practices.

While this argument has been widely seen as a sound one, in a recent paper Rebecca Stone (2014) added an important qualification to Shiffrin's argument. Stone claims that in most cases it is hypocritical for the state to argue along Shiffrin's lines. As the state is "in the business of enforcing property rights and determining the distribution of resources through taxation and spending" (Stone 2014: 36), insofar as it upholds (does not start reforming) an unjust initial allocation of entitlements which provide the basis for exploitative contracts, the Shiffrinian non-paternalistic justification of non-enforcement of these contracts is hypocritical. When courts refuse the enforcement of unequal bargains, unless these are based on fraud or coercion (in which case, as we have seen, unconscionability serves as a proxy for the lack of voluntariness), they hold contracting parties to a standard that the state itself does not respect. This is hypocritical and as such *pro tanto* wrong. This may bring us back to paternalism:

> if the weaker party's assent really was genuine, the court may find it difficult to appeal non-hypocritically to the necessities and distress of the party to justify any refusal to

> enforce without resorting to paternalism, unless the stronger party has violated some independent moral duty to the weaker party.
>
> (Stone 2014: 47)

Even if Shiffrin's argument is much weakened, there are at least two other ways of reinterpreting seemingly paternalistic (and therefore *prima facie* controversial) limits on freedom of contract as nonpaternalistic (and potentially less controversial) regulatory interventions: as mechanisms for protecting cooperative schemes against collective action problems and harmful externalities (third-party effects). Examples include regulations that maximize daily or weekly working hours or the recent French law that gives employees the right not to read their email out of set business hours.[8]

In his *Principles of Political Economy* (Mill 1997 [1848], Book V chapter XI §12), John Stuart Mill analyzed and argued for a number of limitations on freedom of contract in surprisingly modern (and, in contrast to *On Liberty*, entirely consequentialist) terms. He made a case for prohibiting child labor based on a version of soft paternalism towards non-competent agents, and discussed women's labor as a matter of removing existing injustices in their social position. Most interesting for our purposes is the way he argued for the limitation of working hours. He referred to the statutory limitation of working hours as an example of what we would now call a governmental solution to a collective action problem:

> classes of persons may need the assistance of law, to give effect to their deliberate collective opinion of their own interest, by affording to every individual a guarantee that his competitors will pursue the same course, without which he cannot safely adopt it himself.
>
> (Mill 1997 [1848]: 964–965)

He argued that even if workers as a class would prefer to work for shorter periods, they cannot achieve it without mandatory rules limiting working hours, because each would have an individual interest in working longer.

In general terms, Mill's argument was this. The role of government is not to overrule the judgment of individuals but to give effect to that judgment. But sometimes this is not possible in a decentralized manner: "large number" or "free rider" problems lead to a market failure. It is difficult to take collective action in the group's interest (Mill simply assumes there is such a common collective interest) without selective incentives. Gerald Dworkin's influential paper on paternalism makes a very similar argument for the interference being potentially nonpaternalistic (Dworkin 1983a: 23).

While a full-fledged analysis of such problems is rather complex, the economist Kaushik Basu argued that even in a competitive market there may be cases when such interference is Pareto improving – generally a strong justification (Basu 2007). If this is so, then in this category of cases, overriding contract terms can be justified within a welfarist framework, without reference to paternalism, moralism, or even autonomy.

Still, I hesitate to accept that Mill's argument provides a genuinely nonpaternalistic justification for interference. Even though the regulator might know that long hours are against the objective interest of the workers, they may not be against their will. The regulator just assumes that the majority of the workers actually would want to work less; there may not be a clear expression of such a will. The workers might decide to work long or read their emails on vacation, against their own interest, but for some other valid reason. Apparently there is no clear mechanism available for them to express this will one way or another, other than in agreeing to a contract; *ex hypothesi* they are not able to express their will through collective private or political action.

Another way of reinterpreting a limitation on contractual freedom as nonpaternalistic refers to the economic costs imposed on third parties to justify limiting certain risky contractual choices. In a thoughtful article, Eric Posner suggested that many protective laws of modern welfare states serve to redress imbalances created by social security and welfare laws (Posner 1995). By providing a social safety net, welfare states effectively truncate the downside of financial and other risks to citizens. The regulatory environment of a welfare state has the unintended effect of encouraging socially harmful behavior, such as irresponsible spending, risky borrowing, and over-indebtedness. This suggests that many seemingly paternalistic limitations on freedom of contract may be best justified as a mechanism to prevent harmful externalities. When individuals take on too much risk in reliance on the welfare state, they impose external costs on society. Thus, what at first sight looks like a rule protecting vulnerable groups may in fact be protecting the public budget. In the philosophical literature, a similar argument has been discussed by Dworkin (1983b: 108–110) and Feinberg (1986: 138–141), in non-contractual contexts with reference to cases such as mandatory motorcycle helmets.

7 Conclusion

Paternalism seems antithetical to freedom of contract, yet contract law provides a wide range of doctrinal instruments for judicial intervention and often includes regulatory rules which, at first glance, look paternalistic: either in the soft sense of protecting or promoting the interests of not fully voluntary contracting parties – formation defenses, procedural requirements (formalities, cooling-off periods), information disclosure rules – or in the hard sense of overriding contracting parties' will, as expressed in a fully voluntary contract, by mandatory substantive rules. Even when such interference is justified, contract law is not the only, and often not the best, policy instrument in the service of paternalism. Thus, two general *kinds* of objections can be raised against *prima facie* paternalistic contract law rules. The interference may not be justified in the first place. Or the interference may be justified, but the particular regulatory technique does not achieve its goal or achieves it at a higher price, in terms of direct costs, unintended consequences, or potentially undesirable redistributive effects, than an alternative policy instrument. For instance, some problematic contracting practices and market failures cannot be appropriately addressed by piecemeal judicial decisions or through private law rules. This suggests that situations and environments which, for instance, lead to fully voluntary exploitative contracts or unequal bargains, may be better regulated by a mix of policy instruments and regulatory tools beyond contract law. For instance, Michael Trebilcock argues for a "relative institutional division of labor" in which "the common law of contracts will be principally concerned with autonomy issues in evaluating claims of coercion, antitrust and regulatory law [with] issues of consumer welfare, and the social welfare system [with] issues of distributive justice" (Trebilcock 1993: 101). This raises further questions of legal and institutional design.

Related topics

Epistemic Paternalism; Libertarian Paternalism, Nudging, and Public Policy; Hard and Soft Paternalism; Paternalism and the Criminal Law; Paternalism by and towards Groups; Self-Paternalism.

Notes

1 Kant referred to "imperium paternale" (paternal government) as "the greatest conceivable despotism" (1991: 74).

2 For a contribution by a philosopher outside the law and economics tradition, see Trout (2005).
3 As to divorce, Trebilcock (1993) argues for a 60-day cooling-off period for separation agreements. For surrogate motherhood contracts he proposes to allow withdrawal for the biological mother after delivery. Contracts for prostitution in Germany provide another example. Since 2002, contracts for prostitution have been enforceable under certain conditions. The client has limited rights to remedies for nonperformance. The prostitute as service provider enjoys a strong right to withdraw from the contract at any point of time, even after payment by the client. Such an easy withdrawal may be a sensible or even humane legal instrument in this very special context. But its justifiability is much weaker beyond this context. Contract law as we currently know it, i.e., as a basis of cooperation and reliance, would be at an end if withdrawal were possible in every case of regret.
4 Eidenmüller (2005) cites data about withdrawal rates between 0.8 to 1.8%. Epstein (2006: 129) cites anecdotal evidence for the irrelevance of the rule in practice.
5 Royal Bank of Scotland plc v Etridge (No 2) [2001] UKHL 44.
6 The code of "Community" is used; see Pinker (2007: 409).
7 121 U.S. App. D.C. 315, 350F. 2d 445 (1965).
8 According to a BBC report, since 1 January 2017, "Companies with more than 50 workers [are] obliged to draw up a charter of good conduct, setting out the hours when staff are not supposed to send or answer emails." www.bbc.co.uk/news/world-europe-38479439.

References

Basu, K. (2007) "Coercion, Contract and the Limits of the Market," *Social Choice and Welfare* 29: 559–579.

Ben-Shahar, O. and Schneider, C.E. (2014) *More Than You Wanted to Know: The Failure of Mandated Disclosure*, Princeton, NJ: Princeton University Press.

Benson, P. (ed.) (2001) *The Theory of Contract Law: New Essays*, Cambridge: Cambridge University Press.

Brownsword, R. (2006) "Freedom of Contract," in R. Brownsword *Contract Law: Themes for the Twenty-First Century*, 2nd edition, Oxford: Oxford University Press, pp. 46–70.

Buckley, F. H. (2005) *Just Exchange: A Theory of Contract*, London and New York: Routledge.

Collins, H. (1999) *Regulating Contracts*, Oxford: Oxford University Press.

Cooter, R. and Ulen, T. (2012) *Law and Economics*, 6th edition., Boston: Pearson Education.

Craswell, R. (2000) "Freedom of Contract," in E. Posner (ed.) *Chicago Lectures in Law and Economics*, New York: Foundation Press, pp. 81–103.

———. (2001) "Two Economic Theories of Enforcing Promises," in P. Benson (ed.) *The Theory of Contract Law: New Essays*, Cambridge: Cambridge University Press, pp. 19–44.

Cserne, P. (2012) *Freedom of Contract and Paternalism: Prospects and Limits of an Economic Approach*, New York: Palgrave Macmillan.

Dawson, J. P. (1947) "Economic Duress – An Essay in Perspective," *Michigan Law Review* 45: 253–290.

Deakin, S. (2006) "'Capacitas': Contract Law and the Institutional Preconditions of a Market Economy," *European Review of Contract Law* 2: 317–341.

Decock, W. (2013) *Theologians and Contract Law: The Moral Transformation of the Ius Commune (ca. 1500–1650)*, Leiden: Martinus Nijhoff.

Dworkin, G. (1983a) "Paternalism," in R. Sartorius (ed.) *Paternalism*, Minneapolis: University of Minnesota Press, pp. 19–34.

———. (1983b) "Paternalism: Some Second Thoughts," in R. Sartorius (ed.) *Paternalism*, Minneapolis: University of Minnesota Press, pp. 105–111.

———. (2017) "Paternalism," in E. N. Zalta (ed.) *The Stanford Encyclopedia of Philosophy* (Spring 2017 edition), https://plato.stanford.edu/archives/spr2017/entries/paternalism/.

Eidenmüller, H. (2005) "Der Homo Oeconomicus und das Schuldrecht. Herausforderungen durch Behavioural Law and Economics," in R. Alexy (ed.) *Juristische Grundlagenforschung – Tagung der Deutschen Sektion der Internationalen Vereinigung für Rechts- und Sozialphilosophie (IVR) vom 23. bis 25. September 2004 in Kiel*, Stuttgart: Franz Steiner, pp. 13–28.

Enderlein, W. (1996) *Rechtspaternalismus und Vertragsrecht*, München: C. H. Beck.

Epstein, R. A. (1975) "Unconscionability: A Critical Appraisal," *Journal of Law and Economics* 18: 293–315.

———. (2006) "Behavioral Economics: Human Errors and Market Corrections," *University of Chicago Law Review* 73: 111–132.

Feinberg, J. (1983) "Noncoercive Exploitation," in R. Sartorius (ed.) *Paternalism*, Minneapolis: University of Minnesota Press, pp. 201–235.

———. (1986) *The Moral Limits of the Criminal Law, Volume 3: Harm to Self*, New York: Oxford University Press.

Fried, C. (2015) *Contract as Promise: A Theory of Contractual Obligation*, 2nd edition, Oxford: Oxford University Press.

Ganuza, J. and Gomez Pomar, F. (2010) *The Strategic Structure of Contract Law*, Unpublished Book Manuscript, Universitat Pompeu Fabra, Barcelona, https://www.scribd.com/document/228415029/Strategic-Structure-Contract-Law.

Gilmore, G. (1974) *The Death of Contract*, Columbus, OH: Ohio State University Press.

Gordley, J. (1991) The Philosophical Origins of Modern Contract Doctrine, Oxford: Clarendon Press.

———. (2001) "Contract Law in the Aristotelian Tradition," in P. Benson (ed.) *The Theory of Contract Law: New Essays*, Cambridge: Cambridge University Press, pp. 265–334.

Hale, R. L. (1943) "Bargaining, Duress, and Economic Liberty," *Columbia Law Review* 43: 603–628.

Hermalin, B. E., Katz, A. W. and Craswell, R. (2007) "Contract Law," in A. M. Polinsky and S. Shavell (eds.) *The Handbook of Law & Economics vol. I*, Amsterdam: Elsevier, pp. 3–136.

Hillman, R. A. (1997) *The Richness of Contract Law: An Analysis and Critique of Contemporary Theories of Contract Law*, Dordrecht, Boston and London: Kluwer Academic Publishers.

Jolls, C. and Sunstein, C. R. (2006) "Debiasing Through Law," *Journal of Legal Studies* 35: 199–242.

Jolls, C., Sunstein, C. R. and Thaler, R. H. (2000) "A Behavioral Approach to Law and Economics," in C.R. Sunstein (ed.) *Behavioral Law and Economics*, Cambridge: Cambridge University Press, pp. 13–58.

Kant, I. (1991) *Political Writings*, 2nd edition, ed. H.S. Reiss, Cambridge: Cambridge University Press.

Kennedy, D. (1982) "Distributive and Paternalist Motives in Contract and Tort Law, With Special Reference to Compulsory Terms and Unequal Bargaining Power," *Maryland Law Review* 41: 563–658.

Klass, G., Letsas, G. and Saprai, P. (eds.) (2014) *Philosophical Foundations of Contract Law*, Oxford: Oxford University Press.

Kornhauser, L. A. (1976) "Unconscionability in Standard Forms," *California Law Review* 64: 1151–1183.

Korobkin, R. (2004) "A Comparison of 'Traditional' and 'Behavioral' Law-and-Economics Analysis of Williams v. Walker – Thomas Furniture Company," *University of Hawaii Law Review* 26: 441–468.

Kötz, H. and Flessner, A. (1997) *European Contract Law Vol. 1: Formation, Validity, and Content of Contracts: Contract and Third Parties*, Oxford: Clarendon Press.

Kronman, A. T. (1982) "Paternalism and the Law of Contracts," *Yale Law Journal* 92: 763–798.

Leff, A. A. (1967) "Unconscionability and the Code – The Emperor's New Clause," *University of Pennsylvania Law Review* 115: 485–559.

Mackaay, E. (2013) *Law and Economics for Civil Law System*, Cheltenham: Elgar.

Magat, W. A. (1998) "Information Regulation," in P. Newman (ed.) *The New Palgrave Dictionary of Economics and the Law, vol. 2*, London and New York: Palgrave Macmillan, pp. 307–311.

Mehren, A. T. von. (1992) "Formation of Contracts," in A.T. von Mehren (ed.) *International Encyclopedia of Comparative Law vol. 7: Contracts in General*, Tübingen: Mohr Siebeck and Amsterdam: Martinus Nijhoff, chap. 9.

Mill, J. S. (1997 [1848]) *Principles of Political Economy With Some of Their Applications to Social Philosophy*, London: Routledge.

Nozick, R. (1997 [1969]) "Coercion," in Nozick, *Socratic Puzzles*, Cambridge, MA: Harvard University Press.

Olsaretti, S. (2004) *Liberty, Desert and the Market: A Philosophical Study*, Oxford: Oxford University Press.

———. (2008) "Debate: The Concept of Voluntariness – A Reply," *Journal of Political Philosophy* 16: 112–121.

Persky, J. (1993) "Consumer Sovereignty," *Journal of Economic Perspectives* 7: 183–191.

Pincione, G. (2008) "Welfare, Autonomy, and Contractual Freedom," in M.D. White (ed.) *Theoretical Foundations of Law and Economics*, Cambridge: Cambridge University Press, pp. 214–233.

Pinker, S. (2007) *The Stuff of Thought: Language as a Window into Human Nature*, London: Penguin Books.

Posner, E. A. (1995) "Contract Law in the Welfare State: A Defense of the Unconscionability Doctrine, Usury Laws, and Related Limitations on the Freedom to Contract," *Journal of Legal Studies* 24: 283–319.

Rachlinski, J. J. (2000) "A Positive Psychological Theory of Judging in Hindsight," in C.R. Sunstein (ed.) *Behavioral Law and Economics*, Cambridge: Cambridge University Press, pp. 95–115.

Rawls, J. B. (1993) *Political Liberalism*, Cambridge, MA: Harvard University Press.

Raz, J. (1986) *The Morality of Freedom*, Oxford: Clarendon Press.

Schmolke, K. U. (2014) *Grenzen der Selbstbindung im Privatrecht: Rechtspaternalismus und Verhaltensökonomik im Familien-, Gesellschafts- und Verbraucherrecht*, Tübingen: Mohr Siebeck.

Scott, R. E. (2004) "The Death of Contract Law," *University of Toronto Law Journal* 54: 369–390.
Shiffrin, S. V. (2000) "Paternalism, Unconscionability Doctrine, and Accommodation," *Philosophy & Public Affairs* 29: 205–250.
Stone, R. (2014) "Unconscionability, Exploitation, and Hypocrisy," *Journal of Political Philosophy* 22: 27–47.
Sunstein, C. R. and Thaler, R. H. (2003) "Libertarian Paternalism Is Not an Oxymoron," *University of Chicago Law Review* 70: 1159–1202.
Thaler, R. H. and Sunstein, C. R. (2008) *Nudge: Improving Decisions About Health, Wealth, and Happiness*, New Haven, CT: Yale University Press.
Trebilcock, M. J. (1993) *The Limits of Freedom of Contract*, Cambridge, MA: Harvard University Press.
Trebilcock, M. J. and Elliott, S. (2001) "The Scope and Limits of Legal Paternalism: Altruism and Coercion in Family Financial Arrangements," in P. Benson (ed.) *The Theory of Contract Law: New Essays*, Cambridge: Cambridge University Press, pp. 45–85.
Trout, J. D. (2005) "Paternalism and Cognitive Bias," *Law and Philosophy* 24: 393–434.
Wertheimer, A. (1987) *Coercion*, Princeton, NJ: Princeton University Press.
Whitman, D. G. (2006) "Against the New Paternalism: Internalities and the Economics of Self-Control," Policy Analysis No.563 (Cato Institute, 22 February), www.cato.org/pubs/pas/pa563.pdf.
Zamir, E. (1998) "The Efficiency of Paternalism," *Virginia Law Review* 84: 229–286.
———. (2014) "Contract Law and Theory: Three Views of the Cathedral," *University of Chicago Law Review* 81: 2077–2123.
Zweigert, K. and Kötz, H. (1998) *An Introduction to Comparative Law*, 3rd edition, Oxford: Oxford University Press.

Further reading

R. Bigwood, *Exploitative Contracts* (Oxford and New York: Oxford University Press, 2004) is a thorough monograph on exploitative contracts. The separate but related philosophical discussion on the ethical limits of markets has obvious relevance for paternalism and contract law. Some key literature includes M.J. Radin, *Contested Commodities* (Cambridge, MA: Harvard University Press, 1996), D. Satz, *Why Some Things Should Not Be for Sale: The Moral Limits of Markets* (Oxford: Oxford University Press, 2010), and M.J. Sandel, *What Money Can't Buy: The Moral Limits of Markets* (New York: Farrar, Straus and Giroux, 2012).

24

PATERNALISM AND THE PRACTITIONER/PATIENT RELATIONSHIP

Emma C. Bullock

1 Introduction

In therapeutic practice, medical paternalism typically involves a medical practitioner interfering with a patient's choices regarding her medical treatment for the sake of protecting or promoting that patient's best interests. There are a number of different ways a medical practitioner can behave paternalistically towards her patient.[1] A "soft" medical paternalistic interference will occur when the patient is judged to be incapable of making her own medical decisions. Soft medical paternalistic interferences are usually directed towards young children, the profoundly mentally impaired, or the comatose and are generally considered to be morally unproblematic (Grisso and Appelbaum 1998; Beauchamp and Childress 2009; Feinberg 1986).[2] A "hard" medical paternalistic interference, on the other hand, involves interfering with a patient who is judged to be competent to make her own medical decisions. A further distinction can be made between "direct" and "indirect" forms of medical paternalism. With "direct" medical paternalism the medical practitioner aims to benefit the patient interfered with, while "indirect" medical paternalistic interferences aim at benefitting some other patient(s).[3] This chapter focuses exclusively on the moral justification of hard and direct medical paternalistic interferences.

Direct medical paternalism with competent patients is typically composed of the following three features:

(a) *Interference.* The medical practitioner interferes with the patient's decision about her therapeutic treatment.
(b) *Lack of valid consent.* The patient does not validly consent to the interference, or in cases of ignorance, would not validly consent to the medical practitioner's interference were she to be made aware of its occurrence.
(c) *Beneficence.* The medical practitioner believes that the interference will promote the patient's best interests, and this features as a reason for the medical practitioner's behavior.[4]

There are a range of ways in which a medical practitioner might interfere with a patient's decision regarding her medical treatment, (a): the practitioner might pressure a patient into consenting to an unwanted treatment, withhold pertinent information about the nature and risks of various treatment options, lie to or deceive the patient about the state of her health, divulge unwanted medical information, or submit the patient to treatments she does not want (Häyry 1991: 141). There are also a number of ways in which a patient might not validly consent to this interference, (b): the patient might explicitly refuse the treatment or medical information, her consent might be invalidated because it is based on inadequate information, there might be coercive or manipulative pressures from her employer, her family, or the medical practitioner to consent, or her consent might not be sought in the first place. Given the vagueness of the concept of "best interests," the third feature of medical paternalism can also be met in different ways, (c). In the medical context a patient's best interests are most commonly reduced to her physical or mental health, or her overall quality of life, but have also been identified in terms of the patient having true beliefs about the state of her health[5] or in terms of promoting the patient's actual or future autonomy (Sullivan 2016; Komrad 1983).[6] In this chapter I will begin by providing some background to the debate over medical paternalism in western therapeutic practice (Section 2), before moving on to evaluate two major considerations in favor of medical paternalism (Sections 3 and 4).

2 Background

Various forms of medical paternalism have been commonplace throughout much of the history of medical practice.[7] The pervasiveness of medical paternalism is underpinned by the Hippocratic Oath (300 BC), which not only requires medical practitioners to "benefit the sick according to [their] ability and judgement," but to also "keep them from harm" by, for instance, refusing to give their patients deadly drugs when the patient requests them to do so (Hippocratic Oath, Classical Version). Since the Hippocratic Oath requires that the medical practitioner refrain from doing what she perceives to be harmful to the patient even at the patient's request, adhering to the duty may pre-dispose medical practitioners to behave paternalistically (Gordon 2014: 73–74).

The requirements to "do no harm" and to "benefit the sick" persist in medical ethical codes to this day and medical practitioners remain accountable to a "duty of care." According to the General Medical Council, for instance, doctors must make the care of the patient their concern (English et al. 2004: 15). The Nursing and Midwifery Council (2008) states that the "first concern" of nursing is the provision of care and that this is achieved by protecting and promoting the health of the patient. Likewise, the International Code of Ethics for Nurses (2012) states that one of the fundamental responsibilities of a nurse are "to promote health, to prevent illness, to restore health and to alleviate suffering."

Over the last 50 years or so medical paternalism in therapeutic practice has been considered to be both morally and legally problematic (Groll 2014: 195; Veatch 2009: 92–93). Some forms of medical paternalism are argued to be legally problematic because the sorts of actions involved in carrying out a medical procedure would normally infringe upon an individual's human rights when undertaken without valid consent (Manson and O'Neill 2007: 75). Under ordinary circumstances, for instance, an individual has a legal right against having her blood extracted: this would normally constitute a battery or an assault. In the medical context, however, it is sometimes necessary for medical practitioners to extract blood for diagnostic purposes. In order to prevent such medical procedures being classified as battery or assault, the patient can consent to her blood being taken, thereby protecting the medical practitioner from criminal repercussions (Davis 2003: 368; Manson and O'Neill 2007: 76).[8] However, if the medical practitioner proceeds to act paternalistically, and so without the patient's consent, then in some cases her actions may legally count as assault.

The central moral justification for the shift away from medical paternalism has been the importance of respecting patient autonomy (Manson and O'Neill 2007: 17). There are two main ways in which patient autonomy is said to be morally important in this context. First, it has been argued that the patient has sovereign authority over her "bodily domain" and may freely choose what happens to and what she does with her body even if such behavior is detrimental to her overall well-being: when such choices are autonomous they morally ought never to be overridden (Faden and Beauchamp 1986: 19; Groll 2014: 200). Second, the move away from medical paternalism holds that respect for patient autonomy is the best way to protect and promote the patient's overall best interests (Groll 2014: 199). Respecting patient autonomy allows for the maximization of patient best interests either because patients are taken to be best placed to judge the medical treatment most compatible with their broader interests (Tännsjö 1999: 16; Veatch 2000: 704) or because making one's own medical decisions is directly therapeutically beneficial, improving both psychological and physical health (Schneider 1998: 18–19).

In medical ethics and law the demand to respect patient autonomy is implemented by respecting the informed choices of patients to consent to or refuse treatment (Appelbaum, Lidz, and Meisel 1987: 22; Dworkin 1988: 5; Manson and O'Neill 2007: 185; O'Neill 2002: 4). The "doctrine of informed consent"[9] is premised on the plausible idea that a patient cannot make autonomous treatment decisions without adequate and relevant information about her diagnosis, prognosis, and treatment options (Walker 2013).[10] Since it is impossible to give patients "full" or "complete" information about their condition and the treatments available to them, the demand is that the patient receives an *adequate* amount of relevant information in order for her consent to be valid (O'Neill 2003). Patients who are adequately informed (and otherwise competent) are thought to be able to meaningfully consent in a way that is respectful of their autonomy.

Despite the movement of medical ethics away from a focus on a duty of care to an ethic that explicitly demands respect for patient autonomy, the implementation of the doctrine of informed consent has been met with pessimism. Some have argued, for instance, that medical paternalism remains prominent in therapeutic practice because of a deeply imbedded professional authoritarianism (Katz 2002). Others have argued that the high standards of informed consent can never be reasonably met and may undermine trust in medical institutions (O'Neill 2002). There also remains a deep conflict within medical ethics between the duty of care and respect for patient autonomy, since a patient may freely choose or refuse treatment that may lead to a decline in her health (Groll 2014: 197). A Jehovah's Witness, for instance, might autonomously refuse a certain medical treatment on the basis of her religious values; a cancer patient might refuse chemotherapy in favor of homeopathic treatment; and a patient might request information about her medical condition that would likely cause her to become clinically depressed. While the doctrine of informed consent will demand that the medical practitioner respect the patient's autonomous decision, the duty of care may pull the medical practitioner in the direction of behaving paternalistically, either by directing her to administer the relevant medical treatment without the patient's consent or to withhold medical information from her. Plausibly the conflict between the duty of care and the doctrine of informed consent will not be equally strong in all cases: a medical practitioner might be more comfortable paternalistically withholding information than paternalistically subjecting a patient to treatment. It nonetheless remains important to determine a method for resolving the conflict between the duty of care and respect for patient autonomy when these conflicts arise (Bullock 2014: 376).

One way of resolving the tension between the duty of care and the demand to respect patient autonomy is to prioritize one or the other. Typically, the conflict is resolved by prioritizing the doctrine of informed consent whenever the patient is recognized to be a competent decision-maker (Beauchamp 2003: 271; Cassell 1977: 18; Faden and Beauchamp 1986: 19). On this blanket antipaternalistic approach medical practitioners are required to respect the free choices of their patients even if they think that respecting the decision will be harmful to the patient's health (Bullock 2014: 377). There are a number of reasons, however, to be dissatisfied with this antipaternalistic resolution to the conflict, and so some motivation to retain an element of medical paternalism in the practitioner/patient relationship. Key arguments against the prioritization of informed consent are essentially (1) well-being-based or (2) autonomy-based. In the remainder of this chapter I present each of these considerations in favor of medical paternalism in turn and outline some of their limitations.

3 Well-being considerations

A central rationale for medical paternalism is that absolute respect for patient autonomy effectively abandons patients to the harmful consequences of their decisions, and this is at odds with the medical practitioner's duty of care. Take the following case:

> Imagine that Frank suffered a shoulder injury a year ago that will prevent him from ever again playing competitive golf, his life's passion. He is no longer depressed about his situation but feels certain that he has nothing to live for and would be better off dead.
>
> (Scoccia 2008: 367)

Were Frank to require a life-saving treatment the medical paternalist might try to justify overriding Frank's refusal of treatment by claiming that it is in Frank's best interests to continue living even though he can no longer play golf.[11]

Recall that one of the justifications for the doctrine of informed consent was that respect for patient autonomy is the best means of maximizing patient best interests, because patients are best placed to determine what is in their best interests. One reason we might think this is if we adopt a desire-satisfaction view of well-being that identifies individual best interests with the satisfaction of that individual's desires. On this view, it might be tempting to suppose that since Frank prefers not to continue living it really is in Frank's best interests to refuse the life-saving treatment. As such we ought to respect his decision: it would be self-defeating to paternalistically impose treatment upon him in the sense that blocking the satisfaction of his desire to die would necessarily make him worse off.

However, there are reasons to doubt that respecting patient autonomy maximizes patient best interests, even if a desire-satisfaction account of well-being is true. First, an individual's well-being is plausibly not only dependent on the satisfaction of his current desires, but on the satisfaction of the desires he has over the course of a lifetime. As such, if the satisfaction of a patient's strongest current desire is likely to prevent that patient from satisfying his future desires – such as is the case with Frank's current desire to die – then even on the desire-satisfaction theory of well-being we have a reason not to respect the patient's current decision (Scoccia 2008: 362).

A second reason to doubt that respecting patient autonomy maximizes patient best interests is that a patient might be mistaken about the best way to satisfy her current desires – even when she is adequately informed of the medical aspects of her decision – if she is ignorant about a number of non-medical factors that bear on whether or not the decision she makes will likely promote or protect her best interests. Suppose, for instance, that a patient misinterprets the central tenets of her religion and refuses a life-saving treatment, but would not have refused it had she known that it was permitted (Bullock 2014: 379–381). Or consider a patient who has been deceived by her spouse about the state of their finances and decides to continue a lengthy and expensive course of cancer treatment despite there being a low chance of recovery, and would have chosen an alternative course of treatment had she known that they were heavily in debt.

In both cases the outcome of the decision is detrimental to the patient's best interests by "her own lights." A justification for medical paternalism might thus arise whenever we have reason to suspect that respect for patient autonomy is likely to fail to maximize patient best interests.

It is unclear, however, whether the medical practitioner is any better placed to make a medical decision that will maximize patient best interests when the patient is ignorant about pertinent non-medical information, because we have no good reason to suppose that the medical practitioner will have access to this information any better than the patient herself. Unless the medical practitioner happens to know the details of her patient's religion then it is unlikely that she herself will know which medical treatments the religion permits or forbids. It is even less likely, moreover, that the medical practitioner will know the true state of her patient's financial affairs or whether her spouse is deceiving her about them. One possible solution to this knowledge-gap would be to demand a rich two-way communication between the medical practitioner and her patient. Rather than simply disclosing medical information to the patient, the patient would also be expected to share information with her medical practitioner about her values and personal circumstances. Such models of medical decision-making are increasingly supported (Birchley 2014; Maclean 2006; Sandman and Munthe 2010). This solution is, however, undesirable insofar as it is intrusive: it is not unreasonable for a patient to want to keep certain features of her lifestyle – such as her financial affairs – private. Moreover, to expect the medical practitioner to reach a judgment on the best course of action by taking into account as many of the patient's medical and non-medical circumstances as possible is exceedingly demanding on her time and mental energy (Veatch 2009: 34).

A third way in which the claim that respect for patient autonomy is the best way to maximize patient best interests might be undermined is if an alternative theory of well-being is true. Advocates of objective list theories of well-being will typically identify patient best interests in a way that is not reducible to the patient's desires and values (Arneson 1999: 118–119). On an objective list theory of well-being we could determine that it is in Frank's objective best interests to continue living (despite his current aversion to living a life without being able to play golf) because his best interests are not reducible to his desires.

There are, however, several reasons to continue to respect patient autonomy even if an objective list account of well-being is true. First, a patient's preferences will often provide a good indication of the treatment option that is most compatible with her best interests (Mill 2008 [1859]). Moreover, if health is objectively valuable, and achieving good health requires that the patient undergo some form of treatment, then the treatment that the patient prefers will be important for ensuring that she follows that particular treatment plan. Indeed, some research indicates that patients comply more readily with treatment recommendations when they have played an active role in medical decision-making (Ghane et al. 2014). Objective accounts of patient well-being thus need not deny that typically respecting patient autonomy facilitates the protection and promotion of that patient's best interests (Arneson 1999: 117; Bullock 2015).

Secondly, a paternalistic interference might itself have a negative effect on the patient: having your choice paternalistically usurped can be offensive and upsetting (Mill 2008 [1859]). Justified medical paternalism will need to take into account the possible negative ramifications that having a decision overridden will have on the patient's overall best interests. It might turn out that these negative ramifications always outweigh any possible benefit to be gained from the paternalistic interference. In other words, medical paternalism might turn out to be self-defeating.

Third, autonomous decision-making might itself be one of the goods that contributes to objective well-being. If so, then paternalistic interferences will necessarily reduce objective well-being insofar as they impede autonomous decision-making. Moreover, it might be argued that the objective good of autonomous decision-making is the most important element of well-being. On this view, the negative impact that disrespect for individual autonomy has on objective well-being will always outweigh any positive impact the interference would have on other elements of a patient's objective well-being, thereby rendering medical paternalism as necessarily unjustified.

4 Autonomy considerations

A second type of consideration in favor of medical paternalism is the thought that patients are in fact unable to autonomously consent to medical treatment. There are at least two reasons we might have to doubt that generally competent adult patients are capable of autonomous medical decision-making. First, patient autonomy might be undermined because the patient does not understand complex medical information. Alternatively, patient autonomy may be undermined because the patient's decision is affected in some way by a cognitive bias. Ultimately, if it turns out that patients are unable to make autonomous decisions then medical practitioners ought to instead operate under the duty of care, since the doctrine of informed consent applies only if the patient is capable of making an autonomous decision.

Medical practitioners are medical experts; they are more knowledgeable about diseases, their alleviation, and the proposed risks and benefits of treatments or the failure to treat than the typical patient (Häyry 1991: 2). One justification for medical paternalism, then, is that the medical practitioner is epistemically privileged when it comes to determining the best course of medical treatment. As such, the medical practitioner is justified in interfering with a patient's decision whenever her decision is at odds with what is medically recommended.

There are at least two reasons to resist this simplistic argument. First, one of the motivations for the doctrine of informed consent is to address this imbalance of knowledge: since the patient is not a medical expert then the medical practitioner is obliged to communicate pertinent medical information to the patient in a way that meets her communicative needs. Second, as noted in the previous section, medical decisions will impact the patient in non-medical ways. The side effects from a course of chemotherapy could, for instance, prevent a patient from continuing paid employment; a major limb amputation might put an end to a young athlete's career; antidepressants might cloud a PhD candidate's ability to continue her research, and a blood transfusion might cause a Jehovah's Witness to be ostracized from her community. In these ways the medically best outcome might not be the best overall outcome for the patient (Veatch 2009: 36). Once again, there is no reason to think that a medical practitioner is in any way privileged when it comes to making a medical decision that takes a patient's culture, lifestyle, religion, personal values, and worldview into account (Buchanan 1978: 381–383; Gordon 2014: 74).

A different autonomy-based consideration that can motivate medical paternalism is the idea that the patient is unable to make an autonomous decision because her decision-making process is in some way non-rationally influenced. Empirical studies have indicated that often a patient's decision can be non-rationally influenced by one or more cognitive biases.[12] Patients awaiting a kidney transplant, for example, have been shown to be susceptible to "impact bias," significantly overestimating the improvement to their quality of life with a transplant (Smith et al. 2008). This is problematic insofar as this overestimation non-rationally influences their decision to undergo treatment such that they would not have pursued the treatment had they known that it would not improve their quality of life very much. Patients have also been shown to be bad at making judgments about the riskiness of various treatment options (Gilovich, Griffin, and Kahneman 2002; Lloyd 2001) and other studies have indicated that whether or not a patient consents to or refuses treatment can be non-rationally influenced by the way in which the information about her treatment options is presented to her. Individuals react differently, for instance, to being told that a medical procedure carries a 10% chance of death than they do to being told there is a 90% chance of survival (McNeil et al. 1982).

The presence of a non-rational influence on a patient's consent may invalidate her consent by way of undermining patient autonomy. Take the following example:

> *Framing-Induced Consent*: Patient has lung cancer. Doctor explains the treatment options in terms of the survival rate description, and Patient consents to surgery. Had Patient been given the mortality rate description, however, she would *not* have consented to surgery.
>
> (Hanna 2011: 520 [original emphasis])

Since the influence on the patient's consent in this case is non-rational we have reason to doubt that her resulting decision to consent to or refuse the surgery is autonomous in any deep sense. Indeed, "having different [preferences] depending only on how information has been framed is surely irrational, and therefore incompatible with autonomous choice" (Chwang 2016: 275). As such, whenever a patient's consent is influenced by the way in which the medical information has been framed it looks as though her consent is morally irrelevant to how she may be treated (Hanna 2011: 518).[13] Since framing effects are endemic (Thaler and Sunstein 2008: 23–24), it looks as though patient consent might often fail to be morally decisive when it comes to making treatment decisions. If so, this gives us a moral justification for privileging the duty of care and adopting a blanket policy of soft medical paternalism.

This conclusion may be resisted, however, in at least three ways. First, instead of treating the patient paternalistically we could try to rid the patient of her cognitive biases and so restore patient autonomy. It is, however, difficult to "debias" patients because cognitive influences are not always easily avoided even when a patient is aware that her decision is subject to them (Schiavone et al. 2014; Thaler and Sunstein 2008). In the context of framing effects this is because the different ways of providing information about framing effects might themselves generate different preferences on the part of the patient. While this might be avoided with meta-level debiasing, i.e., by informing the patient that the information provided about framing effects might itself non-rationally influence their decision (Chwang 2016: 281–282), it is difficult to see how this disclosure of debiasing information could avoid an infinite regress.

A second response would be to try to mitigate (rather than eliminate) the effects of a patient's cognitive biases by "nudging" the patient toward the option they would choose were they free from the non-rational influence. This could be achieved by framing the information communicated to the patient during the informed consent process in a particular way. If patients are liable to, say, overestimate or underestimate the riskiness of a certain treatment option then information framing could be harnessed to help them choose what they themselves would judge to be the better treatment option, were it not for their cognitive biases. We should, however, be cautious about nudging patients in this way since it is difficult (if not impossible) to identify what the patient would choose had she not been non-rationally influenced. This could either be because there is no such thing as her "real" preference (Sunstein and Thaler 2003: 1164) or because a patient's preference cannot be reconstructed without relying on an independent judgment about what the "best" decision would be (Grüne-Yanoff 2012: 641–644). This is problematic because we have some reason to doubt that the medical practitioner is able to determine what the best overall decision would be: as previously mentioned, medical practitioners are at an epistemic disadvantage when it comes to knowing the patient's values and lifestyle preferences, and we have no special reason to think that medical practitioners can determine what is objectively valuable. Moreover, some empirical evidence indicates that the decisions of medical practitioners are themselves subject to cognitive biases (Wright, Bolger, and Rowe 2002).[14] As such their judgments about what counts as the best medical decision could equally be non-rationally influenced.

A third response to the autonomy-based justification for medical paternalism gives up on the claim that patient consent ought to be respected because patients have a sovereign right to autonomy and instead insists that respecting a patient's cognitively biased consent might nonetheless be the best way of promoting patient best interests. As suggested in the previous section this might be because respect for patient consent facilitates treatment compliance or because having one's consent overridden has a negative effect on well-being. A blanket policy of respect for patient consent, regardless as to whether it is cognitively biased, might thus be justified if it would generally protect patient best interests. Since what is now at stake is securing the best outcome, rather than respect for a patient's sovereign right to autonomy, one might be tempted to adopt a policy of respecting patient consent on a case-by-case basis depending on whether or not respect for patient consent would produce the best outcome in a given situation (Bullock 2014: 388; Bullock 2016).

5 Conclusion

While medical paternalism has been historically pervasive, western medical ethical standards have changed over the last 50 years to include respect for patient autonomy as a governing ethical principle. The moral justifiability of medical paternalism remains a pressing issue, however, given that medical practitioners remain accountable to a duty of care. Since respect for patient autonomy and exercising the duty of care can sometimes conflict, an ethical rationale needs to be given as to which has priority. It is typically argued that respect for patient autonomy takes precedence in cases of conflict, either because patients have a sovereign right to determine their medical treatment, or because respecting their autonomous preferences is the best way to maximize their best interests.

There are, however, at least two broad motivations for instead prioritizing the duty of care and permitting paternalistic interferences with a patient's medical decision whenever it appears to conflict with it. First, medical paternalism can be motivated out of a concern for individual well-being. On a desire-satisfaction account of well-being considerations of patient well-being might motivate medical paternalism because respect for a patient's autonomy could be in tension with the satisfaction of the patient's future desires. Patients can also sometimes be mistaken about the decision that will most likely satisfy their desires when they lack relevant medical and non-medical information about the likely ramifications of their decision. Alternatively, on an objective account of well-being the patient might simply be wrong about what her best interests are.

These considerations alone, however, do not straightforwardly justify medical paternalism since the medical practitioner is unlikely to be well placed to determine whether or not the patient is making a mistaken decision. While this might be rectified by requiring a communicative, open, and truthful relationship between the medical practitioner and patient, this could be overly intrusive for the patient and overly demanding on the medical practitioner. Moreover, there might be good reason to continue to respect patient autonomy even if the decision that would be in her best interests can be independently identified: paternalistic interferences can have negative consequences that potentially outweigh any net benefit.

The second major consideration in favor of medical paternalism focuses on the limits of patient autonomy. In particular, empirical research suggests that patients are susceptible to a number of cognitive biases. Since a patient's decisions might turn out to be non-autonomous there is seemingly no autonomy-based reason to respect them. The mere fact that a patient's medical decisions are typically cognitively biased, however, does not provide a sufficient justification for implementing medical paternalism: instead, respecting cognitively biased patient consent might still be the best way to protect patient best interests. We might thus adopt a policy of (paternalistically) respecting a patient's cognitively biased consent, either generally or on a case-by-case basis, if this is likely to produce the best outcomes.

Balancing the different goods that contribute to a patient's best interests is extremely difficult. The advantage of prioritizing the doctrine of informed consent when it comes into conflict with the duty of care is that it removes responsibility for this difficult task from the medical practitioner and places it squarely on the shoulders of the patient. However, given that patients are typically in need of care because they are ill or suffering, medical practitioners might have an ethical responsibility to take up some of this burden under the duty of care. As such, while medical practitioners cannot make the best medical decisions for their patients (as was traditionally thought), there is at least some reason to think that they have an obligation to help patients make better ones. Given the patient's epistemic limitations and her vulnerability this will plausibly not be achieved with an absolute respect for patient autonomy.[15]

Related topics

Deciding for the Incompetent; Paternalism and Autonomy; Paternalism and Well-Being; Paternalistic Lying and Deception.

Notes

1 The following distinctions are adapted from Feinberg (1986).
2 Although see Fateh-Moghadam and Gutmann (2014), and Howard, this volume.
3 I leave it open as to whether indirect medical paternalism must involve an interference with one patient for the sake of another patient, and allow the possibility that interferences with a non-patient for the sake of a patient could also count as specifically medically paternalistic.
4 For alternative formulations of medical paternalism that omit the beneficence feature see Groll (2014).
5 This account of best interests would make an interference a case of epistemic paternalism (Bullock 2016).
6 Sullivan (2016) refers to this sort of paternalism as maternalism, while Komrad (1983) refers to it as limited paternalism.
7 For the opposed view see McCullough (2011).
8 That is, unless they are negligent.
9 For various formulations of the doctrine of informed consent see the Nuremberg Code (1949), the Belmont Report (1979: Section C1), and the Declaration of Helsinki (1964–2008).
10 The implementation of a policy of free and informed consent in therapeutic contexts is partly a consequence of its instantiation in medical research (Groll 2014: 195; Manson and O'Neill 2007: 4; O'Neill 2002: 19). It can also be viewed as a response to controversies within case law (Faden and Beauchamp 1986: 101) and the civil-rights movement during the middle of the 20th century (Schmidt 2004: 281). It is worth noting too that the move away from medical paternalism is for the most part a project of western societies, with countries such as China, Japan, and most African countries rejecting the ideal of informed consent (Gordon 2014: 77).
11 In Scoccia's use of the example he takes it to be the case that Frank should be denied suicide assistance because he is better off alive than dead (Scoccia 2008: 367).
12 For a comprehensive overview of the varieties of cognitive biases affecting patient decision-making see Aggarwal, Davies, and Sullivan (2014: 32).
13 A related issue is whether or not a patient can ever be adequately informed if her decision is influenced by framing effects (Hanna 2011: 523–525). A separate motivation for medical paternalism might thus be that medical practitioners ought to operate under the duty of care since patient consent can never be adequately informed in the way the doctrine of informed consent requires.
14 For an overview see White 2013.
15 I would like to thank Kalle Grill and Jason Hanna for their helpful feedback on previous drafts of this paper.

References

Aggarwal, A., Davies, J. and Sullivan, R. (2014) "'Nudge' in the Clinical Consultation – An Acceptable Form of Medical Paternalism?" *BMC Medical Ethics* 15(1): 31–36.

Appelbaum, P. S., Lidz, C. W. and Meisel, A. (1987) *Informed Consent: Legal Theory and Clinical Practice*, New York: Oxford University Press.

Arneson, R. J. (1999) "Human Flourishing Versus Desire Satisfaction," *Social Philosophy and Policy* 16(1): 113–142.

Beauchamp, T. L. (2003) "Methods and Principles in Biomedical Ethics," *Journal of Medical Ethics* 29(5): 269–274.

Beauchamp, T. L. and Childress, J. F. (2009) *Principles of Biomedical Ethics*, 6th edition, New York: Oxford University Press.

"The Belmont Report: Ethical Principles and Guidelines for the Protection of Human Subjects of Research," (1979) www.hhs.gov/ohrp/regulations-and-policy/belmont-report/ [Accessed 28 Jul. 2016].

Birchley, G. (2014) "Deciding Together? Best Interests and Shared Decision-Making in Paediatric Intensive Care," *Health Care Analysis* 22(3): 203–222.

Buchanan, A. (1978) "Medical Paternalism," *Philosophy and Public Affairs* 7(4): 370–390.

Bullock, E. C. (2014) "Free Choice and Patient Best Interests," *Health Care Analysis*, 24(4): 374–392.

———. (2015) "Assisted Suicide and the Proper Role of Patient Autonomy," in J. Varelius and M. Cholbi (eds.) *New Directions in the Ethics of Assisted Suicide and Euthanasia*, Switzerland: Springer International Publishing, pp. 11–25.

———. (2016) "Mandatory Disclosure and Medical Paternalism," *Ethical Theory and Moral Practice* 19(2): 409–424.

Cassell, E. J. (1977) "The Function of Medicine," *The Hastings Center Report* 7(6): 16–19.

Chwang, E. (2016) "Consent's Been Framed: When Framing Effects Invalidate Consent and How to Validate It Again," *Journal of Applied Philosophy* 33(3): 270–285.

Davis, H. (2003) *Human Rights and Civil Liberties*, Devon: Willan.

"Declaration of Helsinki: Ethical Principles for Research Involving Human Subjects," (1964; amended 2008) http://bit.ly/1dWyRro [Accessed 27 Jul 2016].

Dworkin, G. (1988) *The Theory and Practice of Autonomy*, Cambridge: Cambridge University Press.

English, V., Romano-Critchley G., Sheather, J. and Sommerville, A. (2004) *Medical Ethics Today: The BMA's Handbook of Ethics and* Law, 2nd edition, London: BMJ Books.

Faden, R. R. and Beauchamp, T. L. (1986) *A History and Theory of Informed Consent*, Oxford: Oxford University Press.

Fateh-Moghadam, B. and Gutmann, T. (2014) "Governing [Through] Autonomy: The Moral and Legal Limits of 'Soft Paternalism'," *Ethical Theory and Moral Practice* 17(3): 383–397.

Feinberg, J. (1986) *The Moral Limits of the Criminal Law, Volume 3: Harm to Self*, Oxford: Oxford University Press.

Ghane, A., Huynh, H. P., Andrews, S. E., Legg, A. M., Tabuenca, A. and Sweeny, K. (2014) "The Relative Importance of Patients' Decisional Control Preferences and Experiences," *Psychology and Health* 29(10): 1–29.

Gilovich, T., Griffin, D. and Kahneman, D. (2002) *Heuristics and Biases: The Psychology of Intuitive Judgment*, Cambridge: Cambridge University Press.

Gordon, J-S. (2014) "Medical Paternalism and Patient Autonomy," in M. Boylan (ed.) *Medical Ethics*, 2nd edition, West Sussex: Wiley-Blackwell, pp. 72–82.

Grisso, T. and Appelbaum, P. S. (1998) *Assessing Competence to Consent to Treatment: A Guide for Physicians and Other Health Professionals*, Oxford: Oxford University Press.

Groll, D. (2014) "Medical Paternalism- Part 2," *Philosophy Compass* 9(3): 194–203.

Grüne-Yanoff, T. (2012) "Old Wine in New Casks: Libertarian Paternalism Still Violates Liberal Principles," *Social Choice and Welfare* 38(4): 635–645.

Hanna, J. (2011) "Consent and the Problem of Framing Effects," *Ethical Theory and Moral Practice* 14(5): 517–531.

Häyry, H. (1991) *The Limits of Medical Paternalism*, London: Routledge.

"Hippocratic Oath (Classic version)," www.nlm.nih.gov/hmd/greek/greek_oath.html [Accessed 8 Feb. 2012].

International Code of Ethics for Nurses. (2012) www.icn.ch/images/stories/documents/about/icncode_english.pdf [Accessed 27 Jul. 2016].
Katz, J. (2002) *The Silent World of Doctor and Patient*, London: The Johns Hopkins University Press.
Komrad, M. S. (1983) "A Defense of Medical Paternalism: Maximizing Patients' Autonomy," *Journal of Medical Ethics* 9(1): 38–44.
Lloyd, A. J. (2001) "The Extent of Patients' Understanding of the Risk of Treatments," *Quality in Health Care* 10(I): i14–i18.
Maclean, A. (2006) "Autonomy, Consent and Persuasion," *European Journal of Health Law* 13: 321–338.
Manson, N. C. and O'Neill, O. (2007) *Rethinking Informed Consent in Bioethics*, Cambridge: Cambridge University Press.
McCullough, L. B. (2011) "Was Bioethics Founded on Historical and Conceptual Mistakes About Medical Paternalism?" *Bioethics* 25(2): 66–74.
McNeil, B. J., Pauker, S. G., Sox, H. C. Jr. and Tversky, A. (1982) "On the Elicitation of Preferences for Alternative Therapies," *New England Journal of Medicine* 306(21): 1259–1262.
Mill, J. S. (2008 [1859]) *On Liberty*, Oxford: Oxford University Press.
Nursing and Midwifery Council. (2008) "The Code," www.nmc.org.uk/standards/code/read-the-code-online/ [Accessed 18 Oct. 2011].
"Nuremberg Code: Directives for Human Experimentation: (1949)," https://history.nih.gov/research/downloads/nuremberg.pdf [Accessed 1 Dec. 2016].
O'Neill, O. (2002) *Autonomy and Trust in Bioethics*, Cambridge: Cambridge University Press.
———. (2003) "Some Limits of Informed Consent," *Journal of Medical Ethics* 29(1): 4–7.
Sandman, L. and Munthe, C. (2010) "Shared Decision Making, Paternalism and Patient Choice," *Health Care Analysis* 18(1): 60–84.
Schiavone, G., De Anna, G., Mameli, M., Rebba, V. and Boniolo, G. (2014) "Libertarian Paternalism and Health Care Policy: A Deliberative Proposal," *Medicine, Health Care and Philosophy* 17(1): 103–113.
Schmidt, U. (2004) *Justice at Nuremberg: Leo Alexander and the Nazi Doctor's Trial*, New York: Palgrave Macmillan.
Schneider, C. E. (1998) *The Practice of Autonomy; Patients, Doctors and Medical Decisions*, New York: Oxford University Press.
Scoccia, D. (2008) "In Defense of Hard Paternalism," *Law and Philosophy* 27(4): 351–381.
Smith, D., Loewenstein, G., Jepson, C., Jankovich, A., Feldman, H. and Ubel, P. (2008) "Mispredicting and Misremembering: Patients With Renal Failure Overestimate Improvements in Quality of Life After a Kidney Transplant," *Health Psychology* 27(5): 653–658.
Sullivan, L. S. (2016) "Medical Maternalism: Beyond Paternalism and Antipaternalism," *Journal of Medical Ethics* 42(7): 439–444.
Sunstein, C. S. and Thaler, R. H. (2003) "Libertarian Paternalism Is Not an Oxymoron," *University of Chicago Law Review* 70(4): 1159–1162.
Tännsjö, T. (1999) *Coercive Care: The Ethics of Choice in Health and Medicine*, London: Routledge.
Thaler, R. H. and Sunstein, C. R. (2008) *Nudge*, London: Yale University Press.
Veatch, R. M. (2000) "Doctor Does Not Know Best: Why in The New Century Physicians Must Stop Trying to Benefit Patients," *Journal of Medicine and Philosophy* 25(6): 701–721.
———. (2009) *Patient, Heal Thyself: How the New Medicine Puts the Patient in Charge*, New York: Oxford University Press.
Walker, T. (2013) "Respecting Autonomy Without Disclosing Information," *Bioethics* 27(7): 388–394.
White, M. D. (2013) *The Manipulation of Choice: Ethics and Libertarian Paternalism*, New York: Palgrave Macmillan.
Wright, G., Bolger, F. and Rowe, G. (2002) "An Empirical Test of the Relative Validity of Expert and Lay Judgments of Risk," *Risk Analysis* 22(6): 1107–1122.

Further reading

J. Jackson, *Truth, Trust and Medicine* (London: Routledge, 2001) investigates the ethical and epistemic repercussions of epistemic medical paternalism. E.D. Pellegrino and D.C. Thomasma, *For the Patient's Good: A Restoration of Beneficence in Health Care* (New York: Oxford University Press, 1988) provides an account of patient best interests in the context of a fully developed philosophy of medicine. D.J.

Rothman *Strangers at the Bedside: A History of How Law and Bioethics Transformed Medical Decision Making* (New York: Basic Books, 1991) is a comprehensive overview of the history of medical paternalism and the development of the doctrine of informed consent in America. H. Marsh, *Do No Harm: Stories of Life, Death and Brain Surgery* (London: Weidenfeld and Nicolson, 2014) is an insightful autobiographical account of the difficulties of adhering to the duty of care, while respecting patient autonomy, in the context of neurosurgery.

25
DECIDING FOR THE INCOMPETENT

Dana Howard

What is often seen to be wrong about acting paternalistically towards others is that we are treating them *as if* they cannot make their own decisions about their own good. So how should we think about situations where it is incumbent upon us to make decisions on behalf of people who *indeed* cannot make their own decisions about their own good – especially if these people are still to some extent capable of expressing their desires and concerns about what happens to them? This is often the case in the medical setting, when a patient has lost decisional capacity or has not developed such a capacity in the first place. Treatment decisions must nevertheless be made, and so family members, friends, and other third parties are tasked with the role of being the patient's surrogate decision-makers. How should these surrogates decide? Is there a risk that they treat incompetent patients paternalistically? If so, how can this be avoided?

Thinking about paternalism in the context of making decisions on behalf of incompetent others is important for a number of reasons. First, it is a topic that is often overlooked in the literature since many assume that insofar as a person is decisionally incompetent, it is either impossible to act paternalistically or there is nothing ethically suspect in treating such a person paternalistically. Yet in the actual medical context, surrogates are often confronted with difficult choices, which can have lasting and irreversible effects on the patients; to disregard these patients' concerns and wishes about such important matters can feel paternalistic. Deeper philosophical investigation would help mark out the difference between unjustified paternalism and deciding in a manner that is not overly deferential to the preferences of incompetent patients.

Second, when we examine what it means to respect persons who are at the margins of agency, we can refine our understanding of what is indeed problematic about treating others paternalistically more generally. That is, if we are capable of treating the decisionally incompetent paternalistically and if such treatment can constitute a moral offense of some kind, then what is wrong about treating someone paternalistically does not solely consist in treating them as if they can't make their own decisions. Seana Shiffrin has argued that, at its heart, paternalistic treatment conveys an insult to the agency of another (Shiffrin 2000). It is easy to see how treating someone who has decisional capacity *as if* they cannot make their own decision insults their agency. I want to suggest, more controversially, that there are ways to lodge an analogous insult to the agency of those who cannot make their own decisions and for whom it would be inappropriate to pretend that they had such capacities. Accordingly, I will argue that insofar as the decisionally incompetent possess some elements of their agency, they can indeed be wronged by being the target of paternalistic treatment.

This chapter takes steps at uncovering the contours and limits of this type of moral insult. To do so, we need a better understanding of the ways in which patients may lack the capacity to make their own medical decisions and whether such patients nevertheless retain a sense of agency that can be insulted. We can then examine whether, on some plausible accounts of what is wrong with treating the decisionally competent paternalistically, the same treatment can constitute an analogous insult to the agency of the decisionally incompetent.

1 Varieties of decisional incompetence

When it comes to medical decision-making, much philosophical thinking is premised on there being two sets of rules: one concerning treatment of competent people, the other concerning treatment of the decisionally incompetent. Although people concede that there may be some epistemic challenges in figuring out the marginal cases, the general idea is that there exists some important threshold delineating between these two categories of people. Once we determine on which side of the capacity divide someone lies, we then simply apply the rules and treat that person accordingly – deference is owed to the expressed preferences of the competent, whereas the preferences of the incompetent may be overridden or remain unsolicited.

This threshold setting plays out in how tools such as advance directives have been defended for the purposes of surrogate decision-making. It is often asserted that if a person was a competent decision-maker when she wrote her advance directive and has now crossed over the divide and can no longer make her own decisions, then the only way to respect her will is to do what the advance directive tells us to do regardless of whether we agree with its prescriptions (Groll 2014: 190). Notably, this standard view can still allow for scenarios in which it would be best, all things considered, to go against the dictates of the advance directive in light of the incompetent patient's current preferences. That is, in some cases, the level of distress incompetent patients can experience when their wishes are ignored is so high that it may turn out that what would be best for them, from our perspective, is to accede to their wishes, even if we take these wishes to be misguided. Important for this standard view, however, is that acceding to the incompetent patient's wishes in such a scenario would be justified *not* on the grounds of respect for the patient's will but rather on the need to safeguard the patient's well-being. Indeed, defenders of this view have argued that acceding to the contemporaneous wishes of the incompetent violates rather than respects patient autonomy (Dworkin 1993: 229). They argue that insofar as patients presently lack decisional capacity, they have no active will or capacity for autonomy to respect.

The motivation to maintain some clear threshold between the competent and incompetent and then to treat all cases alike in each category is an understandable one – especially when we focus on what is entailed in respecting competent decision-makers. Once we determine that people are generally capable of deciding for themselves, there is a strong presumption to defer to their informed choice regarding how to exercise that capacity. We can still try to offer them more information or to persuade them that they are making a mistake. Ultimately, though, respect for persons generally requires us to acknowledge that it is their choice to make.[1]

On the other hand, treating all cases of incompetent decision-making alike is less justifiable. That there should be a presumption of respect for the self-regarding choices made by competent people doesn't imply that there should conversely be a presumption of disrespect for the self-regarding choices made by the incompetent – even those self-regarding choices that may leave such persons less well off. When it comes to showing respect for persons who generally fall short of being fully able to decide for themselves, context and capacity continue to matter.

Consider the following three cases:

> *Precompetent*: Jesse is a pre-pubescent girl with sarcoma. Before she undergoes chemotherapy, which carries with it a significant risk of fertility loss, her parents are considering an optional and experimental procedure, Ovarian Tissue Cryopreservation (OTC), to preserve her future reproductive potential (Resetkova et al. 2013). Jesse's parents don't think that it is appropriate to discuss complicated matters of her future fertility with her at present. Given this incredibly trying episode in her life, they want to retain some semblance of a childhood for Jesse in whatever way they can. On their view, to seek her input about maybe becoming a mom someday would just add further anxiety and confusion to an already dispiriting situation. Is Jesse being treated paternalistically since her views are not being solicited? If so, is such paternalistic treatment objectionable?
>
> *Never Competent*: Jerry is a cognitively impaired 27-year-old. His brother, Tom, has fatal kidney disease and will soon die without a kidney transplant. Jerry is Tom's only acceptable match for organ donation among relatives. He knows Tom is sick and keeps on saying he wants to give Tom his kidney. However, given his cognitive deficits, he does not really know what a kidney is and cannot fully understand what is involved in such a donation. He thus cannot provide informed consent to the procedure. Members of his family are trying to figure out what to do on Jerry's behalf. While they believe that Jerry donating his kidney to his brother would be best all things considered, they do not think this course of action is best for *Jerry*. Consequently, they are not comfortable authorizing the donation on his behalf. Are Jerry's family members treating him paternalistically in the way they are making this decision? If so, is such paternalistic treatment objectionable?[2]
>
> *Previously Competent Incompetent*:
>
>> Mr. O'Connor was a deeply religious man for whom thoughts of taking his own life or of withholding lifesaving measures for whatever reason were completely unacceptable. In his seventies he developed Alzheimer's disease. He lost his ability to do many things he used to enjoy, such as playing the piano; soon he could no longer take care of himself. With the loss of capacity for complex reasoning, most of his religious beliefs gradually faded away. Then came a terrible emotional blow: the death of his wife. He has now begun saying that he does not want to go on, that he does not want to live.
>>
>> (Jaworska 1999: 107)
>
> Imagine that upon first being diagnosed with Alzheimer's disease, Mr. O'Connor wrote an advance directive communicating his wish to have all life-saving measures attempted on his behalf. Imagine further that he now has metastatic cancer that will require chemotherapy treatments, which have the prospect of extending Mr. O'Connor's life (for about six to 12 months). His daughters must decide whether to make a treatment decision based on the dictates of his advance directive that reflect his earlier values or to decide in a manner that respects his current wishes. How should they go about making this choice? Is either choice invulnerable to the charge of being objectionably paternalistic?

Each of these cases depicts a distinct category of decisional incapacity in a different period of the patient's life. Jesse is what Donald VanDeVeer has called *Precompetent*; she is not yet capable of making her own decisions but she will likely develop these capacities as she grows up (VanDeVeer 1986: 345). Moreover, the way in which her parents presently make decisions on her behalf can affect the nature of this development and can affect the values she will grow up to have. Jerry, on the other hand, is best regarded as *Never Competent* since he has never been nor ever will he be capable of making his own decisions about such complex medical matters. He may never have the capacity to fully understand or evaluate the decision that his family must make on his behalf. Finally, Mr. O'Connor can be understood as *Previously Competent Incompetent*; in the past he had the full use of his faculties, and he actually exercised them by writing an advance directive in anticipation of finding himself in a condition in which others must make decisions for him.

These three cases also share some common features. Each case depicts a patient who cannot make an independent decision regarding the medical choice that he or she faces. We can presume then that in each of these cases it is appropriate for a surrogate to be making a decision on behalf of the patient. But just because the authority for decision-making lies with the surrogates and not with these patients, it does not follow that the patients can have no say in the medical decisions made on their behalf. On this point, it is important to distinguish these three cases from another category of persons who cannot decide for themselves: the *Unconscious*. People who are either temporarily or persistently in an unconscious state cannot be consulted and do not generate any new wishes or commitments about their current condition. As Agnieszka Jaworska has argued, when we are making decisions on behalf of unconscious patients, "there is no active agent whose interests need to be taken into account" (1999: 137). Of course, it may be the case that the current interests of unconscious patients would not be best met by adhering to the wishes and commitments they expressed when they were active agents. However, it would be mistaken to regard these current interests as tracking any new wishes or commitments that the patients have adopted since becoming unconscious. Unconscious patients lack all agential capacities (at least temporarily).

This is not the case with Jesse, Jerry, or Mr. O'Connor. Although they lack certain decisional capacities, they still retain some constituent components of their agency that others are in a position to respect and support. They retain, for instance, the capacity to express their wishes about a given choice even if they do not fully understand its implications. They have the capacity to value and uphold certain commitments: such as a deep commitment to one's family or the value of helping others. The decisional incompetency of these three patients lies in their incapacity to successfully translate these values and commitments into an informed decision when it concerns complex medical considerations.

Accordingly, when it comes to medical matters, Jesse, Jerry, and Mr. O'Connor are not autonomous decision-makers, but it would be wrong to treat them as non-agents in every respect. For one thing, it is possible that in other domains of their lives, they are capable of independent decision-making, and that what is decided upon in the medical setting can have important implications in these other domains.[3] For instance, imagine that Jerry works as a stock clerk at a local market that requires heavy lifting. He finds the work satisfying, and donating a kidney would require him to take a leave of absence from work for recovery. While Jerry may not be able to independently weigh all the risks and benefits of surgery, he likely has the capacity to independently decide whether to take time off work. Second, it is possible that even within the medical domain itself, there are some choices that these patients still have the capacity to make independently. Even as Mr. O'Connor does not have the capacity to choose whether to undergo chemotherapy, he can have a good sense of whether his daughters are taking his concerns seriously, treating him with dignity, and acting in good faith on his behalf. Accordingly, Mr. O'Conner may still retain the capacity to appoint a specific person whom he trusts to be his proxy decision-maker (Kim et al. 2011).

Jesse, Jerry, and Mr. O'Connor are also not merely *temporarily incompetent*. When we find ourselves in the position to make decisions on behalf of persons who are just undergoing temporary episodes of incompetence, two strategies are available to us. First, we may choose to delay the decision-making until the patient possesses the capacity to decide for him- or herself. Sometimes, however, these decisions are pressing and cannot wait until the patients regain their competency – e.g., deciding whether to authorize intravenous hydration for a person who is delirious due to high fever. A second strategy then is to disregard the incompetent persons' current wishes as transitory lapses in judgment and instead decide on the basis of what these people would have chosen were they to momentarily regain the capacity to see matters clearly and decide for themselves (Feinberg 1986; Kuflik 2009).

When it comes to the cases above, however, these two strategies do not apply. Delaying decision-making won't do; Jerry and Mr. O'Connor will never gain the capacity to decide these matters for themselves and the decision that Jesse's parents face cannot be delayed until she develops decisional competency. Additionally, it is problematic to discount these patients' preferences as transitory. For all three patients, their currently expressed wishes, values, and commitments may be relatively stable features of their personality, reflective of and informed by their present condition. This is clearly the case for Jesse and Jerry, who have never had the capacity to make their own decisions; imagining what they would choose were they temporarily to gain such a capacity would be imagining very different people who may never come to exist.

Mr. O'Connor's case is somewhat different since the daughters could imagine the values and preferences he would have held had he never developed dementia. However, even in this case, the cognitive condition that leads to Mr. O'Connor's decisional incapacity may not be so easily disentangled from the features that ground his current values and commitments. The development of dementia has the potential to permanently change both one's capacities as well as one's values. For instance, with the loss of complex reasoning, some of Mr. O'Connor's religious commitments have lost their grip on him and will not likely return. If his daughters were to make a decision on the basis of what Mr. O'Connor would have chosen were he to regain his capacities, their decisions would be based on values that he may no longer hold. One could still argue that Mr. O'Connor's daughters should nonetheless defer to the values and preferences that he expressed when he was competent – that these past values and preferences take priority. But such an argument cannot rely on the assumption that past expressions are automatically representative of Mr. O'Connor's current values and preferences, especially when he is presently communicating that the situation is otherwise.

It follows that while each of these cases represents a class of decisional incompetence at different developmental stages of a patient's life, there are some similarities that make decision-making on their behalf particularly challenging and open to the charge of unjustified paternalism. First, the patients in each case are at the margins of agency: while they cannot make independent decisions regarding complex medical procedures, they do possess some constituent components of agency, which can be respected and supported. Second, one of the constituent components of their agency is the capacity to express current preferences that are potentially at odds with their past or future preferences. Third, these current preferences are not necessarily the result of capricious whims and urges; they could be tracking relatively stable values and commitments on the part of the patients.

Taken together, these features open up a space in which patients exhibit enough agency to be treated paternalistically, even if they lack the autonomy necessary to be left fully in charge of their lives. It is at least possible that in ignoring the current preferences of such patients or in disregarding the components of their agential capacities, we end up treating them paternalistically in an objectionable way. The rest of this chapter will explore how a number of leading accounts of paternalism would make sense of these cases. The aim is not to offer a defense of one specific account of paternalism that best captures how we ought to treat those with decisional incapacities; rather it is to highlight how such persons can also be the targets of unjustified paternalism.

2 What is objectionable about paternalistic treatment?

What constitutes the core wrong of treating someone paternalistically has long been a contested topic. Paternalism is often taken to consist in inappropriate interferences in the autonomous choices of others for their own good (Dworkin 1983 [1971]: 20–21). If having one's autonomy interfered with is a necessary condition for being treated paternalistically, then worries about paternalism are unwarranted in the above cases. The surrogates are not interfering in the patient's affairs, and the patients do not have the capacity for autonomy.

Yet as philosophers have attempted to flesh out what is distinctively objectionable about paternalistic behavior, they have cast a wider net than looking solely at interferences. Bernard Gert and Charles Culver, for instance, argue that what is essential to cases of paternalism is that someone is violating a moral rule in relation to another for that person's own sake. This may include interfering or restricting another's freedom, but it can also include failing to keep one's promise, deceiving another, or causing another pain (Gert and Culver 1976). Others have argued that even the violation of a moral rule is not a necessary component of paternalistic behavior; morally innocuous acts such as attempting to exorcize someone's demons can be paternalistic when one does so explicitly against that person's operative preferences (VanDeVeer 1986: 37–38). Moreover, some have argued that we may treat others paternalistically through behavior that respects (and even promotes) their freedom – e.g., by expanding their options even as they worry about unnecessary temptations, by refusing to assist them so they develop independence, or by deliberating with them about an important decision rather than letting them flounder on their own (Savulescu 1995; Shiffrin 2000; Tsai 2014). Once interferences or restrictions of freedom are no longer understood as necessary components of objectionably paternalistic behavior, then decisionally incompetent patients become possible targets of objectionable paternalism.

If it isn't all beneficent interference or some violation of a moral rule, what is it about paternalistic behaviors that makes them objectionable? Seana Shiffrin has argued that paternalism conveys a special kind of insult to the agency of another. Being the target of paternalism differs from being disrespected in other ways. When people coerce us or manipulate us for their own gain, or when they inadvertently violate our rights due to negligence, they are disrespecting our moral status as agents but not our capacity for agency itself. Shiffrin argues that what is distinctively problematic about paternalistic behavior is that it "directly expresses insufficient respect for the underlying valuable capacities, powers, and entitlements of the autonomous agent" (2000: 220). One is treated paternalistically when one's capacity to judge or to act or when one's decisional authority in a specific domain is not sufficiently respected. On this view, beneficent interferences are prime examples of the distinctive insult of paternalistic treatment in that they express insufficient respect for a person's agential capacity to judge for herself, but a person's agential capacities can be insulted in other ways as well.

Of course, Shiffrin's focus is specifically on the way that paternalistic treatment disrespects the agency of *autonomous* others. It is insulting to one's agential capacities when other people, just because they take themselves to know better, insert their wills into realms over which one has ultimate decisional authority. However, I want to suggest that it is the attitude of disrespect towards the agency of another, rather than the fact that the target of the attitude is fully autonomous, that makes paternalistic treatment problematic. This is a departure from Shiffrin's explicit account.[4] However, a deeper investigation into the cases of Jesse, Jerry, and Mr. O'Connor will illuminate the fact that insofar as a person has elements of valuable agential capacities and powers (albeit not fully formed or well integrated), these capacities can also be disrespected.

Let us then examine some common characterizations of paternalistic behavior to see whether any of these behaviors may end up conveying an insult to the agency of people who lack full decisional competence. When it comes to the role played by the target's will, philosophers have defined paternalistic behavior in a number of ways:

A *Contrary to Will Account*: A treats B paternalistically when A acts against B's will to promote B's good (Dworkin 2010; Arneson 1980; Arneson 2015).

B *Disregard of Will Account*: A treats B paternalistically when A acts to promote B's good without regard to B's present attitudes or operative preferences regarding the action (Feinberg 1986: 10; Archard 1990).

C *Disrespect Authority of Will Account*: A treats B paternalistically when A acts to promote B's good and in doing so fails to treat B's will as authoritative in determining what to do (Groll 2012; Cholbi 2017).

These various characterizations highlight normatively distinct ways we can insult someone's capacity for agency even as we aim to act for their benefit. It is still a matter of debate how these different behaviors relate to each other and whether one of these characterizations gets at the heart of why paternalistic treatment poses a unique *prima facie* offense. The question posed here, however, is whether any of these characterizations constitute objectionable ways of treating those who are not competent to independently decide for themselves.

3 Contrary to Will Account

Jesse has no grounds for complaint on the Contrary to Will Account. Her parents are not choosing against her will; rather they are doing what they can to safeguard Jesse from having to exercise her will in the first place. Jerry, on the other hand, seems to have strong grounds for complaint. His family refuses to authorize the treatment option that would be consistent with his stated and stably held preferences, all in an effort to safeguard his interests. Such a choice may ultimately be justified on the grounds that it would in fact be best for Jerry to forgo donating his kidney, but this justification would not absolve the paternalistic nature of the family's decision.

It is less clear how the Contrary to Will Account would apply in Mr. O'Connor's case. His daughters must choose between authorizing or forgoing chemotherapy on behalf of their father, but each of these options conflicts with a set of preferences articulated by Mr. O'Connor at some point in his life. Would this mean that Mr. O'Connor invariably has grounds for complaint that his agential capacities are being disrespected? It depends on his daughters' reasoning for their decision (Grill 2007). Insofar as his daughters justify their decision by appeal to Mr. O'Connor's attitudes at some point in his life, rather than by appeal to what would promote Mr. O'Connor's good, there is no objectionable paternalism according to the Contrary to the Will Account. The Contrary to Will Account thus offers little practical guidance for his daughters in determining which of the two options would be less objectionable. Choosing an option that conflicts with the preferences Mr. O'Connor had communicated at one point in his life in order to adhere to an expression of his will at another point in his life does not in itself constitute an insult to his overall agential capacities. As long as they are deciding based on Mr. O'Connor's explicit preferences rather than what they think is in his interest, then they can avoid the charge of paternalism.

This account could however be supplemented with the view that only those preferences expressed by Mr. O'Connor when he had full decisional capacity should be counted as genuine exercises of his will.[5] On this view, Mr. O'Connor's daughters should avoid making decisions that conflict with the dictates of the advance directive if they want to respect his agency. They should thus authorize the chemotherapy. This is a sensible position to hold, and I will return to its merits in Section 5. At present, we can conclude that for all of these cases, it is only Jerry's case where the Contrary to Will Account can *on its own* pick out any behaviors on the part of surrogate decision-makers that clearly constitute an insult to these patients' agency.

We have two options for what to make of Jesse and Mr. O'Connor's cases. We can either conclude that there is nothing objectionably paternalistic in these two cases or that the Contrary to Will Account is inadequate for the purposes of picking out all that is objectionable about paternalistic treatment. I am partial to the second option. Even when it comes to autonomous adults, acting contrary to the will of another is not a necessary condition for treating them paternalistically. Many cases exist in which we are clearly treating others paternalistically and insulting their agency, even though we never explicitly go against their will. Imagine the following scenario:

> *Fully Competent*: Robert is having lunch with an old fraternity brother who also happens to manage a firm for which his daughter, Miriam, is applying to work. As he catches up with his old friend, Robert makes sure to mention his daughter's application and talks up her professional achievements – as he thinks any proud father would. Robert didn't tell Miriam about the lunch meeting. He did not want to risk finding out what she thought about his advocating on her behalf. Knowing his daughter, Robert has reason to suspect that she may disapprove of his dealings and communicate as much. This would put Robert in an awkward position. He would either end up acting against her stated will or oblige her preferences and lose out on an opportunity to give her a competitive edge.

Miriam appears to have grounds for the complaint that she is being treated paternalistically in this situation, but not in virtue of her father acting against her will. By not informing his daughter about his decision, Robert is completely bypassing Miriam's own deliberation about the matter. He is using his own judgment of what he thinks is in her best interest rather than giving her the opportunity to figure it out for herself. What is potentially insulting about Robert's actions is that he is not giving Miriam an opportunity to exercise her will; his fidelity to what Miriam has willed up to this point remains intact.

Put most broadly, what makes Robert's treatment paternalistic is that he doesn't appropriately take his daughter's will into consideration on the matter even though she is the intended beneficiary of his action. The Disregard of Will Account or Disrespect Authority of Will Account thus seem better equipped to offer a more inclusive characterization of the sorts of behaviors that are paternalistic and what it is about these behaviors that is insulting. Does either account offer any insight about decision-making on behalf of the decisionally incompetent?

4 Disregarding the will of another

Another potentially insulting feature of paternalistic behavior is that it sometimes involves acting in a way that promotes another's good without regard to that person's present attitudes or operative preferences. This seems to be what is problematic about the way that Robert treats Miriam. He is intentionally avoiding any effort to consider Miriam's will on whether he should advocate on her behalf. He can thus be understood as failing to give Miriam an opportunity to exercise her will rather than as doing anything that directly contradicts her will.

Robert's disregard of Miriam's wishes and preferences is a clear case of insulting her agential capacities. Among our three incompetent patients, Jesse's case is the only one in which the surrogates are wholly disregarding the will of the patient. Does a similar avoidance on the part of Jesse's parents convey the same sort of insult to her budding agential capacities and powers? Of course, Miriam is a competent adult, while Jesse is still developing her agential capacities and requires the support and guidance of her parents to do so. Jesse's parents cannot help but insert their will in Jesse's affairs. This does not mean that Jesse's parents may bypass her will on all matters. Respect for the inchoate agential capacities of children requires parents to create opportunities for their children to develop and exercise these capacities. However, such a requirement cannot entail that parents must on every occasion create such an opportunity. Sometimes the stakes of the decision are too high; other times the factors that go into the decision are too complicated. It therefore cannot be the case that Jesse's parents insult her agential capacities each time they refrain from giving her the opportunity to figure things out for herself.

A less obvious difference between Jesse's and Miriam's situations lies in the motivation behind their parents' reluctance to solicit their views. Robert doesn't bother consulting with his daughter on the matter because he takes himself to be a better judge and executor of what is best for her. Jesse's parents may also take themselves to be better judges about what is best for her when it comes to issues such as fertility. But that is not the sole reason for bypassing her will on the matter. Rather, they refrain from consulting Jesse because they want to shield her from needing to exercise her will in the matter. They may very well believe that were she consulted on the matter (and given information at a level that is appropriate to her age), Jesse would have the judgment necessary to make a thoughtful decision about whether to undergo OTC; a decision they would be disposed to respect. However, Jesse's parents don't think it is in her interest to have to make such a decision given everything else that she is dealing with. In this way, Jesse's parents are surely acting paternalistically; however, their behavior does not automatically convey an insult to her agential capacities.

Accordingly, context and motivation matter in how we are to evaluate Jesse's parents' treatment of her agency. Were her parents to refuse consulting with her as a matter of course, never giving her the opportunity to develop her judgment or exert any control over her medical care (perhaps because on their view *any* choice would be too difficult for Jesse to deal with given her disease), then such behavior would convey a general disrespect of her limited agential capacities. Moreover, were her parents' motivated to refrain from soliciting Jesse's preferences for reasons closer in kind to Robert's – i.e., they don't want to deal with the prospect of Jesse getting things wrong and thus complicating their efforts to do what is best for her – then such a disregard for her will could very well constitute an insult to her agency.

5 Disrespecting the authority of another's will

Let us next examine one final characterization of paternalism recently developed by Daniel Groll, which I will call the Disrespect Authority of Will Account. Groll argues that it is possible to disrespect the agency of another even as we act in accordance with that person's will; in fact, it is possible to disrespect another's agency even as we treat that person's will as a decisive factor in our deliberation (Groll 2012). To see what this means and how it relates to respectful decision-making on behalf of the decisionally incompetent, let us return to Jerry's family but imagine a different course of deliberation about whether to authorize the kidney donation.

> *Never Competent**: Because Jerry cannot fully understand the risks and benefits of the procedure, his family doesn't want to give too much weight to his explicit wish to

donate his kidney. Instead, as they deliberate, family members focus on the devastating emotional effect that Tom's death would have on Jerry, the value of having a living sibling in the event that Jerry's parents pass away, and the need to shield Jerry from experiencing any guilty feelings related to Tom's death. In light of these considerations, they determine that donating a kidney is in Jerry's best interest and so they decide to authorize the treatment. While this decision happens to accord with Jerry's preferences, it was not made in deference to those preferences.

In *Never Competent**, we can see that Jerry's family is definitely not disregarding or acting contrary to his will in their decision-making. In fact they may even be treating his will as decisive in his deliberation. As family members consider Jerry's expressed willingness to help, they reckon that Jerry would be inconsolable were he not allowed to donate his kidney. So they decide to authorize the donation in light of Jerry's preferences – not out of respect for his wishes but rather because of the suffering he would feel were his family to refuse his wishes. This does not mean that their decision is not paternalistic according to the Disrespect Authority of Will Account.

Groll contrasts two different ways in which Jerry's will can be decisive in his family's deliberation: it can be *substantively decisive* or *structurally decisive*. We treat another person's will as *structurally decisive* when their say-so acts as an "authoritative demand" in our decision-making. The expression of their will determines what we do "not because it outweighs other considerations but because it is meant to silence or exclude those other considerations" in our decision-making (Groll 2012: 701). When a person's capacity to make his or her own decision is diminished, as is the case with Jerry, they may not have the cognitive powers to make an authoritative demand on others, but they may have enough sense of their situation and of their life as a whole that acting according to their will is still what is best for them. This is because acting against their will could be deeply distressing. In these sorts of situations, we treat the will of another as *substantively decisive*. We decide to act according to their will not because it is an authoritative demand, but rather because it is an important constituent of their well-being.

While Groll argues that paternalism involves refraining from treating the will of another person as structurally decisive, on his view there is nothing impermissible or disrespectful about treating the decisionally incompetent paternalistically. Insofar as a person does not have the sort of will that ought to be treated as structurally decisive, it is impossible to disrespect their agency by treating them paternalistically (Groll 2012: 710). It follows that there is no normative difference between *Never Competent* and *Never Competent**. On both versions of the case, Jerry would be treated paternalistically on Groll's account, but not in a manner that is objectionable.

In closing, I want to return to Mr. O'Connor's case and ask whether his daughters can escape the charge of paternalism if they choose to go against the dictates of the advance directive. It is tempting to hold that the only way for Mr. O'Connor's daughters to treat his will as structurally decisive is by strictly adhering to the dictates of his advance directive. Groll defends such a position. He argues that to the extent that people like Mr. O'Connor are decisionally incompetent at present, we cannot treat their occurrent wills as structurally decisive. However, since Mr. O'Connor previously made his wishes clear in an advance directive, then "the surrogate decision-maker is subject to competent-[Mr. O'Connor]'s authoritative demand" (Groll 2012: 702). On Groll's view, we should respect the prescriptions of an advance directive in precisely the same manner as we respect the explicit wishes of a competent patient at present; we should treat both as structurally decisive in our decision-making. Failing to do so in either scenario constitutes the same insult to the agency of the patient.

While I agree with Groll that failing to treat an incompetent patient's will as structurally decisive does not in itself constitute any objectionable form of paternalism, I do not think that this claim leads to any obvious conclusions about how to adjudicate between Mr. O'Connor's past and current stated preferences. It would be a mistake to hold that anytime a surrogate acts counter to the dictates of an advance directive, she is insulting the patient's agency. This is because it is a confusion to hold that the only way for advance directives to be authoritative at all, their prescriptions must be decisively binding.

In equating the authority of advance directives with the authority of expressed wishes of presently competent patients, Groll relies on a common view that an advance directive should be respected as a strategy of self-binding: what patients are doing in drafting an advance directive is using the window of opportunity available to them now – when they still have decisional capacity – to exercise their will so as to direct their medical treatment in the future. Drafting such a directive is the only way, then, for patients to even have *the opportunity* to have their will carried out by others (Robertson 2003; Buchanan and Brock 1990: 99).

However, when we look at how advance directives are actually used, we come to realize that advance directives are not always drafted for the purposes of self-binding. A majority of patients – when asked whether they want their directives to be strictly followed or whether the surrogate should be left some leeway in the treatment decision – prefer surrogates to exercise some measure of discretion (Ashwini et al. 1992). These patients thus write advance directives with the knowledge that someone else may have to step in and make difficult decisions on their behalf; this decision-maker is often a loved one for whom the patient may be deeply concerned and for whom the patient may wish to support in whatever ways possible. As they draft their advance directive, they may have preferences both about the course of treatment and about how the course of treatment will be decided on their behalf were they no longer in a position to choose on their own. An advance directive may thus be written as an act of care for that person or as a way to legitimate certain choices that would have otherwise been either ethically impermissible or motivationally impossible. Accordingly, in many cases it is more accurate to view advance directives as tools for interpersonal co-deliberation rather than as tools for intrapersonal self-binding.

Mr. O'Connor's daughters can thus respect their father's past will, and his advance directive could maintain its authoritative force without requiring anyone to do exactly as the advance directive demands. Many sorts of expressions of one's will are of this sort. They are acts that make an authoritative difference in the decision-making structure of another person but are also not binding of that structure. When we make requests or grant permissions, for instance, we change the normative status of our relationship with another person and how that person's decisions are going to be made. But we don't force that person's hand according to our will.

When a patient that is confronting the decision of how to have others decide on her behalf, advance directives can be used for many purposes. They can be used to ensure that others have clear guidelines to implement her wishes. They can be used to relieve the burden the surrogates must face in making decisions without the patient's input. They can also be used to recruit a specific person as their surrogate to share in the burden of making such decisions on one's own behalf. All these ends of advance directives are markedly different from binding the surrogate to one's will or even making the surrogate better at accurately reflecting one's will. Insofar as Mr. O'Connor's daughters are attempting to decide in a way that accurately reflects his aims in writing an advance directive, they need not be charged with treating him paternalistically even if they opt for a treatment preference that conflicts with the explicit dictates of the directive.[6]

6 Conclusion

This chapter has asked whether surrogate decision-makers run the risk of treating incompetent patients paternalistically. I have argued that insofar as someone has the component elements of agency, they can be the targets of unjustified paternalism. As we think about the ways in which the decisional incompetent can still have their agential capacities insulted, we have arrived at some conclusions that may be surprising. In particular, I have suggested that acting according to the explicit preferences of the patient (either past or present) can still be understood as paternalistic in some situations. Inversely, choosing a treatment option that conflicts with the preferences of the patient (again, either past or present), may turn out to not only be justified but also a decision altogether devoid of paternalism.[7]

Related topics

Paternalism and the Practitioner/Patient Relationship; Self-Paternalism; The Concept of Paternalism.

Notes

1 There are, of course, exceptions to this presumption. When the stakes are high enough or when it is not clear whether a person is acting with full information, the presumption against paternalism may be overridden (Feinberg 1986: 12; Hanna 2012).
2 Adapted from Strunk vs. Strunk, Court of Appeals of Kentucky, 445 S. W. 2d 145, Sept. 26, 1969.
3 Modern methods of capacity assessment take a domain-specific and risk-sensitive approach rather than appeal to a diagnosis or a label ("unsound mind") of global decisional incapacity (National Bioethics Advisory Commission (United States) 2002; Kim 2010).
4 Shiffrin argues that behavior is paternalistic only if it is "directed at matters that lie legitimately within" the control of the paternalized agent (2000: 218–219). It follows that no behavior described by the above cases would be characterized as paternalistic since the decisions to be made do not lie in the incompetent patient's legitimate control.
5 Alternatively, one could argue that an advance directive can serve as a way for a patient to communicate his "resolution preference" for how surrogates should adjudicate between two conflicting preferences (Davis 2004).
6 Though, of course, they could still be making the decision irresponsibly or negligently.
7 The views expressed in this chapter are the author's own and do not reflect the view of the National Institutes of Health, the Department of Health and Human Services, or the United States government.

References

Arneson, R. (1980) "Mill Versus Paternalism," *Ethics* 90: 470–489.
———. (2015) "Nudge and Shove," *Social Theory and Practice* 41: 668–691.
Archard, D. (1990) "Paternalism Defined," *Analysis* 50: 36–42.
Ashwini, S. et al. (1992) "How Strictly Do Dialysis Patients Want Their Advance Directives Followed?" *JAMA* 267(1): 59–63.
Buchanan, A. and Brock, D. (1990) *Deciding For Others*, Cambridge: Cambridge University Press.
Cholbi, M. (2017) "Paternalism and Our Rational Powers," *Mind* 126: 123–153.
Davis, J. (2004) "Precedent Autonomy and Subsequent Consent," *Ethical Theory and Moral Practice* 7: 267–291.
Dworkin, G. (1983 [1971]) "Paternalism," in R. Sartorius (ed.) *Paternalism*, Minneapolis: University of Minnesota Press.
———. (2010) "Paternalism," in E. Zalta (ed.) *Stanford Encyclopedia of Philosophy*, https://plato.stanford.edu/entries/paternalism/.
Dworkin, R. (1993) *Life's Dominion*, New York: Vintage Books.
Feinberg, J. (1986) *The Moral Limits of the Criminal Law, Volume 3: Harm to Self*, Oxford: Oxford University Press.
Gert, B. and Culver, C. (1976) "Paternalistic Behavior," *Philosophy & Public Affairs* 6: 45–57.
Grill, K. (2007) "The Normative Core of Paternalism," *Res Publica* 13: 441–458.
Groll, D. (2012) "Paternalism, Respect, and the Will," *Ethics* 122: 692–720.
———. (2014) "Medical Paternalism, Part 1," *Philosophy Compass* 9: 186–193.

Hanna, J. (2012) "Paternalism and the Ill-Informed Agent," *Journal of Ethics* 16: 421–439.
Jaworska, A. (1999) "Respecting the Margins of Agency: Alzheimer's Patients and the Capacity to Value," *Philosophy & Public Affairs* 28: 105–138.
Kim, S. (2010) *Evaluation of Capacity to Consent to Treatment and Research*, New York: Oxford University Press.
Kim, S. et al. (2011) "Preservation of Capacity to Appoint a Proxy Decision Maker: Implications for Dementia Research," *JAMA: Archives of General Psychiatry* 68(2): 214–220.
Kuflik, A. (2009) "Hypothetical Consent," in F. Miller and A. Wertheimer (eds.) *The Ethics of Consent: Theory and Practice*, Oxford: Oxford University Press.
National Bioethics Advisory Commission (United States). (2002) "Research Involving Persons With Mental Disorders That May Affect Decision-Making Capacity," *Journal of International Bioethique* 13(3–4): 173–179.
Resetkova, N., Hayashi, M., Kolp, L. A. and Christianson, M. S. (2013) "Fertility Preservation for Prepubertal Girls: Update and Current Challenges," *Current Obstetrics and Gynecology Reports* 2(4): 218–225.
Robertson, J. (2003) "Precommitment Issues in Bioethics," *Texas Law Review* 81: 805–821.
Savulescu, J. (1995) "Rational Non-Interventional Paternalism: Why Doctors Ought to Make Judgments of What Is Best for their Patients," *Journal of Medical Ethics* 21: 327–331.
Shiffrin, S. (2000) "Paternalism, Unconscionability Doctrine, and Accommodation," *Philosophy & Public Affairs* 29: 205–250.
Strunk vs. Strunk. (1969) Court of Appeals of Kentucky, 445 S. W. 2d 145.
Tsai, G. (2014) "Rational Persuasion as Paternalism," *Philosophy & Public Affairs* 42: 78–112.
VanDeVeer, D. (1986) *Paternalistic Intervention: The Moral Bounds on Benevolence*, Princeton, NJ: Princeton University Press.

26

PATERNALISM AND EDUCATION

Gina Schouten

Many people think that paternalism is generally at least *prima facie* wrong. But in some cases, paternalism seems intuitively justified. The challenge of paternalism is to "reconcile our general repugnance for paternalism with the seeming reasonableness of some apparently paternalistic regulations" (Feinberg 1986: 25).

Paternalism is standardly understood as the "interference of a state or an individual with another person against their will," where that interference is "defended or motivated by the claim that the person interfered with will be better off or protected from harm" (Dworkin 2014).[1] In one obvious sense, education is pervasively paternalistic, because it involves various actors – teachers, school boards, policy-makers, parents – acting on judgments about what is good for students: judgments that students will be better off if they are educated according to the program those actors administer. We might wonder, though, whether the *challenge* of paternalism arises within the domain of education. Most students are children, and the repugnance generally provoked by paternalism seems misplaced in the case of children. In fact, paternalism is often taken to be repugnant *precisely because* it involves treating someone like a child (de Marneffe 2006). What could be wrong – even *presumptively* wrong – about treating a child like a child?

As we shall see, education does raise philosophical questions about paternalism. But while paternalism toward adults raises instances of the general challenge expressed above, the paternalism involved in educating children raises a distinctive set of questions: When do children *become* the kinds of people toward whom acting paternalistically is presumptively repugnant? *Whose judgment* should be taken authoritatively to reflect a child's best interest for the purposes of informing educational decisions? When parents' judgments conflict with the state's judgments, whose should be decisive?

This chapter unfolds as follows: I first briefly consider a definitional maneuver that would rule out paternalism toward children on conceptual grounds, but suggest that the most plausible way of understanding the concept *does* allow for paternalism toward children and thus for educational paternalism. I go on to argue, however, that the reasons for thinking that paternalism is at least presumptively wrong do not apply to educational paternalism toward children: it is not even presumptively wrong to treat children as children for the purposes of educating them. In the final two sections, I explore two sets of important philosophical questions about educational paternalism: First, at what stage does educational paternalism *become* presumptively repugnant? When do children become adults toward whom it is presumptively wrong to act paternalistically? Second, how are we to distribute authority paternalistically to educate children? In particular, how should such authority be shared between parents and the state? I don't defend decisive answers to either set of questions, but I explore approaches that seem plausible.

1 Definitional matters

Education *seems* paternalistic because we compel children to consume it, for their own good, without regard to what *they* take their good to be. But for most of us, educating children does not provoke repugnance. How to account for this? Perhaps the concept of paternalism simply doesn't apply to relationships with children. Does the right account of paternalism conceptually exclude the possibility of paternalism toward children, and thus of educational paternalism? In this section, I argue that it does not.

Roughly, paternalism involves interference with another person, defended or motivated by the claim that the person interfered with will be better off in virtue of that interference. Educational paternalism certainly has all these characteristics, but many would argue that these characteristics do not *suffice* for paternalism to occur – that some additional conditions are necessary.[2] What additional conditions might rule out paternalism toward children on conceptual grounds? Perhaps, by definition, paternalism is *morally objectionable*. On this basis, we might argue that we cannot act paternalistically toward children because while paternalism is by definition morally problematic, treating children like children is morally *un*problematic. But such a definition should be avoided; defining paternalism in this way would be artificially to settle, rather than philosophically to engage with, the permissibility of paternalism.

Alternatively, we might think that educational paternalism is conceptually incoherent because an act counts as paternalistic only if the substitution of judgment occurs *against the will* of its target. We might think it conceptually impossible to act *against* the will of a child in the relevant way, if their will is not morally authoritative or does not carry authorizing power.

There are two problems with this move. First, acting *against* one's will seems an implausibly demanding requirement for paternalism to occur.[3] Seatbelt laws would strike us as paternalistic, I think, even if we were to learn that, by some extraordinary coincidence, all those to whom they apply at some particular time would have been independently sufficiently motivated to wear seatbelts, or to consent to a law requiring that they do so.[4] A better way to understand the relevant condition, I suggest, is that the substitution of my judgment for yours must occur *without regard for* your consent or authorization. This revised condition does not rule out paternalism toward children. Whether or not we can act *against* their consent, we can surely act *without regard* for it.

More importantly, even if we impose the strong requirement that paternalistic actions must go *against* the will of their target, we should avoid defining "against the will" in a moralized way that rules out children. We can surely act against the will of a child in a non-moralized sense; we do it all the time, and often in the context of education: we make children go to school even when they do not want to, and on our terms, even when they would choose others. We may not act against their will *in a morally relevant way* – their will might not be morally authoritative in certain contexts – but this possibility is not unique to children. Under some circumstances, the will of adults may be morally non-authoritative. These are questions about the *justifiability* of paternalism, and, as I argue above, we should avoid building their answers into the definition of paternalism.

We can proceed, on the basis of the above considerations, with the characterization of paternalism with which we began: "interference of a state or an individual with another person against [or without regard for] their will . . . defended or motivated by the claim that the person interfered with will be better off or protected from harm" (Dworkin 2014).[5] This working characterization leaves open the possibility that other elements are required for an action to constitute paternalism. Even so, we should reject the definitional maneuvers canvassed above that rule out paternalism toward children on conceptual grounds. Assuming I am right to dispense with moralized analyses of paternalism, there is nothing incoherent in the idea of educational paternalism.

2 Presumptive wrongness

An adequate definition of "paternalism" will not rule out the possibility of paternalism toward children – and so will not rule out educational paternalism – on conceptual grounds. I now want to suggest, however, that the best reasons for thinking that paternalism is *presumptively wrong* do not apply to paternalistic acts toward children. Although we should not regard paternalism as unjustified by definition, the repugnance that motivates the general challenge of paternalism should not arise in response to cases of educational paternalism toward children. What accounts for that repugnance when it does arise?

According to Peter de Marneffe, "paternalism seems repugnant because it seems infantilizing. In limiting our liberty for our own good, it seems that the government treats us like children or that it impedes our development into fully mature adults" (2006: 68). Seana Shiffrin elaborates on the way in which paternalistic actions treat adults as if they were children: paternalism conveys "a special, generally impermissible, insult to autonomous agents," an "intrusion into and insult to a person's range of agency" (2000: 207, 218). For Elizabeth Anderson, certain paternalistic intrusions amount to "telling citizens that they are too stupid to run their lives," and thus threaten citizens' self-respect (1999: 301).

Despite disagreement over precisely how to understand the insult, a common thread here is that paternalism is presumptively wrong because it is *insulting* to be *infantilized*, to be *treated like a child*.

If this is what accounts for our repugnance toward paternalism, how might we reconcile that repugnance with the apparent acceptability of some paternalistic policies? We could either offer nonpaternalistic justifications for those policies (e.g., Shiffrin 2000) or show that some paternalistic policies are not objectionably infantilizing after all (e.g., de Marneffe 2006). But in the case of educational paternalism, I suggest, there is nothing to reconcile. Insofar as students are children, paternalistic educational policies should not provoke repugnance in the first place. There is no insult in treating children like children.[6]

Of course, there may be objectionable *ways* of treating children like children, and here we can see the import of a distinct source of repugnance that we might draw out of the de Marneffe quote above: in acting paternalistically, the government "impedes our development into fully mature adults." Because it is in children's interest, generally, to develop into mature adults, we should often empower children to be self-directing. We might foster their expressive agency by letting them choose their own clothes, for example, or their moral agency by allowing them to choose a charity to support. We might encourage children to be self-directed in these ways for many reasons: to promote their childhood flourishing; to help them grow into ethically sensitive adults; to equip them to be self-reliant so that they are less burdensome to parents; to help them become the kinds of people whom it *will* be presumptively insulting to treat paternalistically.

These reasons, and particularly the last, might justify giving children a great deal of space for self-direction, and tell against many of the ways in which parents typically constrain children's options. It might be developmentally beneficial, for example, to permit children to refuse the piano lessons and other activities that parents often push upon them. But in such cases, *the preferences themselves* are not morally decisive; rather, we defer to children's preferences to foster agency for their developmental benefit. In the case of adults, in contrast, we defer to preferences because those preferences are themselves presumptively morally authoritative.

While there are clearly objectionable and repugnant ways of acting paternalistically toward children, then, we must bear in mind that there is no insult in the *mere fact* of treating them like children. In the case of adults, the *mere fact* of treating them like a child can be sufficient to insult. In the case of children, many particular ways of acting paternalistically may be problematic, but there is no insult in the mere substitution of our judgment for theirs. Paternalism toward children, *as such*, is not presumptively wrong.

All this suggests that thinking clearly about educational paternalism requires developing some account of *when* students become the kinds of people whom it is presumptively wrong to treat paternalistically. At the beginning of the process of education, those toward whom we act paternalistically are not the kinds of people toward whom it is insulting so to act. Presumably, at some point prior to the *end* of the process of education, students *become* the kinds of people – adults – toward whom paternalistic behavior is presumptively wrong. And, presumably, there is no bright line indicating a point at which this transformation occurs; rather, children gradually become adults, with gradually greater capacity and entitlement to chart the course of their own lives.[7] If this is right, then it *becomes* presumptively insulting to substitute our judgment for theirs, in proportion to the stage at which their capacity and worthiness to act on their own judgment have developed. One urgent question about educational paternalism, then, concerns how and when students transition from children to adults. I address this question in the next section; in the final section, I turn to the question of who has the authority to make judgments about where students' educational interests lie. Answering these questions will be crucially important to designing educational systems and policies that are paternalistic *in the right ways*.

3 Educational paternalism and the agency of children

In early stages of education, those whom we treat paternalistically are children. As such, they have not yet developed the self-regulation mechanisms that would make their behavior truly autonomous.[8] The precise requirements for autonomous agency are subject to much dispute, but on any plausible account, the necessary cognitive and regulative capacities are absent or not fully developed during childhood. Indeed, one of the defining features of childhood is that these capacities are not yet sufficiently intact to render children fully responsible for their choices. We often *treat* children as if they are responsible, for example when we punish them for bad behavior. But when we do so, we do not think that we are responding to their autonomous exercise of agency so much as facilitating their *development* of agency. Our punishments are most effective, in this regard, when they don't merely discourage future instances of the bad behavior in question, but encourage reflection on what makes it bad and how best to avoid it, thus facilitating the development of the cognitive and regulative capacities necessary for agency.

How does this bear on the right account of permissible educational paternalism? Such an account must ultimately be responsive to questions about how and when children become agents in the relevant sense. These are important questions,[9] but I will set them aside here. I focus instead on how their answers *should inform policy questions* about justified paternalistic education. We might think that, at whatever point students become autonomous agents, the presumption in favor of deferring to their own judgment applies immediately. I will argue that, for the purposes of *education policy*, things are not so simple. Our judgment of the status of a particular student – whether she is more child or more adult – should be moderated in certain ways based on the ends that judgment will serve.

Presumably, the degree to which paternalism is insulting depends on the degree to which its target has developed their capacities for agency, and the degree to which paternalism restricts their exercise. We are epistemically limited in discerning these considerations about any particular target of paternalism, but profoundly so in the case of children, who presumably attain the capacities gradually and on their own schedule. Notice, though, that educating children is at least in part a *social* project. Insofar as we are obligated to ensure that all children are educated, we must craft education *policy* to ensure that this obligation is reliably discharged. Education policy systematizes instances of educational paternalism. But policy is a blunt instrument, and even the best policy is limited in its responsiveness to fine-grained developmental differences among students. For example, we allow students gradually greater self-direction as they proceed through their schooling by gradually increasing their control over the subjects they study. But we are limited, for the purposes of schooling policy, in our ability to tailor this aspect of schooling to account for differences among students in their development of the capacities for self-direction. We rely largely on rough projections instead of nuanced, individualized judgments.

Because of our epistemic limitations, and because of the unavoidable bluntness of social policy, our judgments about educational paternalism cannot be responsive *only* to the facts about when students become the kinds of people toward whom it is insulting to act paternalistically. Our epistemic limitations make these facts impossible to discern with confidence, and the bluntness of policy makes them impossible to respond to with precision. We must incorporate strategies for how best to discern and respond to students' developing agency, given the unavoidable limitations in both discernment and response. I propose that compulsory primary and secondary schooling should be guided by a strategy of *conservatism* in delineating students' space for self-direction: education policy should take students' capacities for self-direction to be *developing* rather than *developed*. It should err on the side of promoting students' development of agential capacities, rather than on the side of respecting agential capacities taken already to be well-developed. We should allow increased scope for self-direction as students move through schooling, because "children at different stages of development differ from one another in the extent of their hegemony over themselves" (Schapiro 1999: 734). But at each step, we should err on the side of conservatism in our judgments about how fully formed children's agency is.

Epistemic limitations and the unavoidable rough-grained-ness of policy necessitate the use of *some* strategies for making judgments about children's development; but what favors strategies of *conservatism*?

We should be *highly risk averse* about mistakenly regarding students as autonomous for the purposes of educational policy. The appropriate level of risk aversion regarding responsibility attributions depends on what the stakes are – on how severe the consequences are of attributing responsibility. In the case of education, the stakes can be tremendously high, because education serves as a gateway to social goods like income, meaningful employment, power, leisure, and self-efficacy. By designing education policy that affords students more scope for self-direction than we are *highly confident* their agency warrants, we would risk forfeiting their attainment of important educational goods. Because educational goods position students to attain so many other goods that constitute a flourishing life, we should be highly risk averse about forfeiting them based on mistaken judgments about students' development of agency.

A parallel argument might support conservatism with regard to the sentencing of juvenile offenders. In sentencing juvenile offenders, we take the most debilitating punishments off the table from the start. Because the capacities on which culpability is predicated may not be fully intact in juvenile offenders, and because the cost imposed by certain punishments is so severe, we should be highly risk averse about mistakenly imposing such penalties on juvenile offenders.

Of course, with respect to *some* educational decisions, the stakes are not so high. Our reasons to be concerned that a first grader learn to read are more decisive than are our reasons for wanting a high school sophomore to take a second year of calculus. Part of the appropriateness of granting more control to older students over their course of study is based on a judgment that they are relatively closer to adulthood than younger students. But another part is our judgment that the stakes are less high with respect to some curricular decisions than with respect to others. Those with the highest stakes – as measured by their import for ultimate life prospects – tend to congregate in the early years of a child's education, because early educational experiences are often prerequisite for subsequent educational success. This helps justify – even beyond judgments about students' capacities for mature choice – a system that is more directive early on.

A policy of conservatism is not without costs. Adults are generally best positioned to know what is good for them; if advanced students are adults in this sense but we treat them as children, we not only risk being insulting; we also risk forfeiting opportunities to advance interests that they authoritatively and correctly take themselves to have. These latter might be regarded as epistemic costs of paternalism, because they are costs we incur when we act without regard to the first-person knowledge that an agent generally has of their own good. These are genuine costs. But we can reduce them by ensuring that education aims to help students acquire "all-purpose" goods: goods that equip them to pursue their *own* ends, whatever they ultimately take those ends to be.[10] All-purpose goods include skills for economic independence, meaningful employment, and critical reflection.

To be sure, not all educational goods are all-purpose goods, and even the best education policy invariably forecloses some ends that students might come to have. But insofar as the goods for which education aims are all-purpose goods that are instrumentally valuable in pursuing a wide diversity of ends, there is less to be lost by a strategy of conservatism in assessing students' development into adults. Equipped with all-purpose educational goods, those students will subsequently enjoy a broader range of possibilities for pursuing the ends they take themselves to have. The cost of disregarding the epistemic advantage students have in judging their own good can be reduced by making educational decisions with an eye toward enlarging students' scope for judgment moving forward.

Paternalism is *prima facie* objectionable only when the individual in question has acquired the capacities that make it insulting for us to substitute our judgment for theirs. If this is right, then for educational paternalism, we will want to know when students have reached that threshold – or, likelier, where they fall on a spectrum of progress toward it. I have not weighed in on the question of when or how students develop into adults. I have argued, however, that we should be strategically conservative in assessing students' place on the developmental spectrum. At each stage, we should err on the side of *facilitating students' development* of the capacities that confer adulthood, even when this licenses intrusion into the domain of self-direction to which we would defer were we to assume those capacities to be fully present.

One important project of education is to facilitate students' transition to adulthood by gradually expanding their scope for self-direction and by equipping them with the tools to choose well within it. These tools for choosing well include the skills associated with autonomy. But some parents reject autonomy as a worthy educational goal for their children. The debate over autonomy education nicely illustrates a second fundamental question about educational paternalism: How is paternalistic authority over children to be distributed between parents and other agents? The next section explores this question.

4 Distributing authority for educational paternalism

Education is paternalistic precisely because we educate students, regardless of what they have to say about the matter, *for their own good*. But who has authority to determine what that good is? Controversy surrounds the division of authority between parents and the state to direct the paternalistic education of children, and one frequent target of this controversy concerns the ways in which states aim, through education, to enable students to become autonomous adults.

Schools can facilitate students' development of autonomy both by exposing them to diversity and by fostering the intellectual capacities for judging well among the life courses available to them. Done well, schooling can expose children to ways of life very different than those enacted in their families and communities of origin, and it can provide an environment in which children come to know and care for peers and adults whose values and lives are much different than their own. It can also foster students' development of the skills to choose well among various worthy ways of living, or to determine that ways of living are not to be scrutinized and chosen among but are instead to be accepted humbly out of faith or respect for tradition. In short: schooling can develop students' capacity to live a life that is authentically their own, based on their own judgments and on values they can reflectively endorse.[11]

But the value of autonomy is highly controversial, and autonomy education has plenty of critics. Philosophical criticisms tend to emphasize the rights of parents to determine whether their children should be educated for autonomy.[12] Those who celebrate spiritual heteronomy and believe that the best life is one lived in obedience to God might worry that autonomy will lead to spiritual alienation. Even those who don't regard autonomy as harmful in itself might worry that equipping children to choose autonomously means risking that they will not choose well. If I am convinced I know what kind of life is best for my child, I might sensibly think she is better off non-autonomously living it than autonomously choosing something worse. We might also worry that autonomy-development can threaten traditional and communal ways of life, thereby – assuming these are *good* ways of life – harming the children who are a part of them. In the US Supreme Court case *Wisconsin v. Yoder*, for example, Amish parents argued that educating their children beyond eighth grade jeopardized the children's salvation and the Amish way of life, because it risked alienating Amish children from Amish values. The Court ruled in their favor.[13]

This debate over autonomy education raises broader questions about the distribution of authority to act paternalistically toward children by directing their education: To what degree is the democratic state licensed to speak on behalf of children and to make judgments about which educational regime best serves their interests? To what degree, on the other hand, may *parents* determine which educational experiences best promote their children's interests? If parents and the state disagree as to the value of autonomy, whose judgments should prevail in determining how children will be educated?

Plausibly, autonomy is only one value to be sought among others – including, perhaps, the preservation of community and tradition. Plausibly, too, the state would be reasonable to withhold judgment to some degree on the relative value of autonomy and other goods that schooling might achieve for students. At best, autonomy is *one* good that students have an interest in attaining. But the state need not aim to promote autonomy at all costs in order for this tension to arise. In a democratic society, the authority of the state derives from its claim to be acting on behalf of the citizens. To whatever degree citizens reasonably judge that compulsory education should aim to make students autonomous, and to whatever degree some parents oppose that aim, the tension between parental authority and democratic authority arises.

Notice that compulsory autonomy education might be justifiable even if the state is not authorized to compel it *on paternalistic grounds*. Apart from serving the interests of students themselves, we rely on education to provide certain *public* goods. For example, schools are supposed to equip students to be good citizens – careful, deliberative decision-makers motivated by concern for the public good.[14] When schools perform these tasks well, we all benefit, and it is in part on the basis of this benefit that schooling is justified as a public investment. Although parents generally enjoy some prerogative to direct the lives of their children, this prerogative might legitimately be constrained if constraints are necessary to ensure that crucial public goods are adequately supplied. Plausibly, good citizenship requires a capacity to reflect critically on various social ends and a capacity to engage in mutually respectful deliberation with other citizens who are very different from us (Gutmann 1987). If an autonomy-facilitating education is necessary to develop these capacities, then compulsory autonomy education could be justified nonpaternalistically. Presumably, not every student must be autonomous in order for the good of an educated citizenry to be secure. But opt-outs also jeopardize schools' capacity to facilitate the kind of autonomy-development necessary for citizenship: by diminishing the presence in schools of students whose families espouse values at odds with mainstream schooling, opt-outs reduce ideological diversity in schools, thus diminishing schools' capacity to educate students for citizenship in a diverse democracy.[15]

Must we rely on this public goods justification for restricting parental authority? Or can we build a case for compulsory autonomy education based directly on *students' own interests*? Can we, in other words, establish that in the domain of autonomy education, the state has the authority to dictate the terms of paternalism toward children, even against the wishes of parents?

First consider the prospects for developing an argument for the *instrumental* value of autonomy to the student being educated: even if some children could live perfectly well without autonomously reflecting on their inherited ways of life, others will find the values that they are raised to endorse alienating, at odds with their own experience of who they are. Many such students could live better if they could live a life of their own making, aligned with their deep convictions and self-understandings. For these students, autonomy is crucially instrumentally valuable to finding a good way of life (Brighouse 2006). Moreover, even if some could live well without the capacity for autonomy, we cannot discern in advance who will need autonomy education to live well and who could do without it. This tells in favor of educating *all* students with the skills associated with autonomy.

Lacking the capacity for autonomy can certainly be harmful to those who need it to find a life that they can live well, but there are countervailing considerations: educating students for autonomy might jeopardize communal ways of life – and thus the well-being of the students who could have flourished within them – as we might have worried compulsory schooling beyond eighth grade did for the Amish community – and at least some Amish children – in *Yoder*.

For the instrumental argument for autonomy education, then, we need to consider the number of people who would benefit from it, the number who would be *harmed* by it, and the magnitude of both the benefits and harms; and we need to perform a moral and political calculation. Does the harm of going without autonomy education for those who could benefit from it outweigh the harm of equipping all students with the skills for autonomy given that we cannot discern in advance which students will benefit from having them? In answering this question, we must keep in mind that the students in each group will vary depending on features of the society in question. We must ask, too, whether there is a substantive and not merely technical distinction between equipping students with the skills for autonomy on the one hand, and encouraging them actually to exercise those skills on the other. If so, then we might mitigate the harm of autonomy education by limiting it to the former. There *seems* to be a substantive difference: To equip students with the skills for autonomy – to develop their *capacity* for autonomy – involves developing their skills for critical thought and reflection. To encourage the *exercise* of those skills – to *make* them autonomous – involves disposing them to *value* critical thought.

Would such a case for autonomy education be strong enough to overcome an appropriately weighted presumption in favor of allowing parents to dictate the terms of educational paternalism?[16] Perhaps. I want to suggest, though, that our very reason for finding paternalism toward adults presumptively repugnant might be invoked as a premise in an argument that the capacity for autonomy has value beyond the merely instrumental value of enabling students to exit ways of life that they cannot live well. Recall that we generally find paternalism repugnant because it is insulting to treat an adult like a child – to treat someone who is capable of acting for their own reasons as someone whose own reasons do not carry authoritative moral weight. Still we must ask: *Why* is this insulting? We often do for others what they are perfectly capable of doing for themselves. Why, then, is it insulting to substitute *our judgment about their interests* for *their own judgment about their interests*? The most plausible explanation, I suggest, is that there is value in adults being able to form and pursue their own conception of their good: to act on reasons of their own.[17]

As we have seen, this value is controversial. Schooling policy must take care in invoking controversial values, in part *because* schooling constrains the authority of parents who may disagree about the values at stake. But a weak construal of the value of autonomy will suffice for our purposes: *other things equal*, it is better for us that we *can* act on our own reasons in pursuit of ends that we authentically endorse, even if we ultimately choose *not* to exercise our capacity for autonomy. Invoking the value of the *capacity* for autonomous choice is still controversial, but it is far less so than claiming that all autonomous choices are valuable, or more valuable than they would otherwise be, in virtue of being made autonomously.

How to argue for this value? If we are right to regard paternalism as presumptively bad, then plausibly it is presumptively good for individuals to be able to act for their own reasons. The value that justifies educating students to be capable of autonomy, then, is suggested by the very conviction that paternalism is presumptively problematic and stands in need of justification. Insofar as that conviction is on firm footing, the capacity for autonomy seems vindicated as not only valuable, but also strongly morally important.

Insofar as we justify autonomy education on the basis of students' own interests, it constitutes educational paternalism. If parents prefer, on paternalistic grounds, that their children *not* receive such an education, then we have conflicting claims of authority to act paternalistically toward children. But if our judgments of paternalism as generally repugnant are apt, this suggests that we are on secure footing in judging the *basic capacity* for autonomy to be in students' best interests, whether or not those students ever come to value the substantive exercise of that capacity.[18] Children have an interest in eventually becoming adults, toward whom it is presumptively repugnant to act paternalistically. We can promote that interest, paternalistically, by educating them in ways that equip them to be autonomous. Even if parents enjoy a strong prerogative to speak on behalf of their children's interests, there is good reason to think that prerogative is overridden in the case of autonomy education. If we are right to regard paternalism as presumptively repugnant, this suggests that there is value in adults being (able to be) autonomous. Ensuring that students can develop into autonomous adults serves fundamental interests whose pursuit can overcome a presumption in favor of deference to parents.

Autonomy education is just one case in which we can see the tension between parents' claim and the democratic state's claim to act paternalistically toward children in directing their education. In this case, I think the state would be right to insist that a capacity for autonomy is good for students, and right to require that students be educated to attain it. While authority to act paternalistically toward children will often reside with parents, compulsory education for autonomy falls within the state's legitimate domain of authority.

Paternalism toward children raises distinct philosophical questions. When thinking about *educational* paternalism, two among these questions stand out as particularly urgent. First: When do students transition from children to adults, and correspondingly, when and how do they acquire the status that makes paternalism toward them presumptively repugnant? And second: How is paternalistic educational authority to be distributed between students' parents and the democratic state? I hope these brief remarks begin to elucidate the difficult philosophical work that will be required to answer these questions well.

Related topics

Egalitarian Perspectives on Paternalism; The Concept of Paternalism; Paternalism and Autonomy.

Notes

1 Dworkin (2015) considers variations on this understanding of paternalism. Although I will explore the content of some of these distinctions in Section 1, I do so without invoking Dworkin's terminology (e.g., soft vs. hard paternalism). Variations in terminology across contributions to the literature introduce complexity that is unnecessary for our discussion.
2 Consider Mill's case of an individual about to cross a rotten bridge (1974 [1859]). Is it paternalistic to intervene just to make sure the individual knows the bridge is rotten? Presumably not. Perhaps, then, we might think that the relevant kind of paternalism involves interference with voluntary choices (Feinberg 1986).
3 But see Arneson (2015). For more on the debate over whether paternalism requires acting against or merely without regard to the agent's will, see Gert and Culver (1976) and Bullock (2015).
4 See Groll (2012). If I am right about this case, this provides some support for a motivational or justification-based account of paternalism, according to which the defining feature of paternalism is the distinctly insulting motive or justification of the agent acting paternalistically. See Shiffrin (2000). See Sunstein and Thaler (2003) for a different account of paternalism that does not require the paternalistic action to go against the individual's own preferences.
5 While Dworkin formulates his candidate definitions using biconditionals, for my purposes we can regard the criteria as necessary conditions only. I leave open whether paternalism makes essential reference to the paternalizer's (objective) justification on the one hand, or to their (subjective) motivation on the other (for different takes on this, see de Marneffe (2006), Husak (2003), and Sunstein and Thaler (2003)). I intend for this working definition to be relatively neutral among competing accounts of paternalism; for an account that it apparently does not accommodate, see Shiffrin (2000).
6 In her Kantian account of childhood, Tamar Schapiro puts the point like this: "Paternalism is prima facie wrong because it involves bypassing the will of another person. . . . But if the being whose will is bypassed does not really 'have' a will yet . . . then the objection to paternalism loses its force" (1999: 730–731).
7 See Feinberg (1994) for a gradualist account of children's development into adults.
8 Although see Mullin (2007) for an argument that even very young children have some capacity for autonomy.
9 For a helpful account, see Schapiro (1999).
10 See Rawls (1999: 183) and Gutmann (1980) for discussions of how all-purpose (social primary) goods must guide paternalistic action when the particular preferences of the person toward whom we are acting paternalistically are unknown or unsettled. (For a challenge to the social primary goods metric as applied to children, see Macleod (2010).)
11 For arguments in favor of autonomy education for children's sake, see Raz (1986), Gutmann (1987), and Brighouse (2006).
12 For challenges to autonomy education, see Burtt (1994), Lomasky (1987), and Galston (1995).
13 For philosophical discussion of the *Yoder* decision, see Gutmann (1980, 1995) and Galston (1995).
14 See Gutmann (1993) and Brighouse (2006).
15 Proponents of autonomy education as a component of citizenship education include Gutmann (1987, 1995), Macedo (1995), Costa (2004), and Callan (2002). Even a system that restricts opt-outs may fail to expose students to diversity; our current system is highly segregated even setting aside opt-out policy. But a system that restricts opting out has greater capacity to expose students to diversity than one that allows it. See Anderson (2007) on the importance of integration across ideological differences. See Brighouse (2006) for concerns about curtailing parents' prerogatives.
16 See Brighouse (2006). For further discussion of how children's rights or interests constrain the ways in which their parents may permissibly restrict their development of autonomy, see Gutmann (1980), Feinberg (1994), Clayton (2006, 2011), and Callan (2002).

17 In her survey of recent scholarship on paternalism, Jessica Begon (2016: 357) says that

> it is surely a concern for autonomy, rather than mere negative liberty, that motivates the anti-paternalist stance: the idea that individuals should be enabled to form and pursue their own conception of the good, without outside pressure, coercion, or manipulation.

The value of autonomy is associated with the view that paternalism is justified to prevent nonvoluntary harm to oneself. This view has been criticized from many different angles. See, for examples, Arneson (2005), Begon (2015), Cholbi (2013), Shafer-Landau (2005), de Marneffe (2006), and Ben-Porath (2013). (Much of this criticism notes that interfering with voluntary choice may actually promote autonomy, however, and so is consistent with the claim that autonomy is of value.)

18 Amy Gutmann (1980: 338) argues that

> a liberal justification for paternalism toward children must rest upon . . . the interests of children as potentially rational beings, beings capable of choosing freely among a range of competing conceptions of the good life and of intelligently governing themselves in a democratic society.

References

Anderson, E. (1999) "What Is the Point of Equality?" *Ethics* 109: 287–337.

———. (2007) "Fair Opportunity in Education: A Democratic Equality Perspective," *Ethics* 117: 595–622.

Arneson, R. (2005) "Joel Feinberg and the Justification of Hard Paternalism," *Legal Theory* 11: 259–284.

———. (2015) "Nudge and Shove," *Social Theory and Practice* 41: 668–691.

Begon, J. (2015) "What Are Adaptive Preferences? Exclusion and Disability in the Capability Approach," *Journal of Applied Philosophy* 32: 241–257.

———. (2016) "Paternalism," *Analysis* 76: 355–373.

Ben-Porath, S. (2013) "Paternalism, (School) Choice, and Opportunity," in C. Coons and M. Weber (eds.) *Paternalism: Theory and Practice*, New York: Cambridge University Press, pp. 247–265.

Brighouse, H. (2006) *On Education*, New York: Routledge.

Bullock, E. (2015) "A Normatively Neutral Definition of Paternalism," *The Philosophical Quarterly* 65: 1–21.

Burtt, S. (1994) "Religious Parents, Secular Schools: A Liberal Defence of an Illiberal Education," *Review of Politics* 56: 51–70.

Callan, E. (2002) "Autonomy, Child Rearing, and Good Lives," in C. Macleod and D. Archard (eds.) *The Moral and Political Status of Children*, Oxford: Oxford University Press, pp. 118–141.

Cholbi, M. (2013) "Kantian Paternalism and Suicide Intervention," in C. Coons and M. Weber (eds.) *Paternalism: Theory and Practice*, New York: Cambridge University Press, pp. 115–133.

Clayton, M. (2006) *Justice and Legitimacy in Upbringing*, Oxford: Oxford University Press.

———. (2011) "Debate: The Case Against the Comprehensive Enrollment of Children," *The Journal of Political Philosophy* 20: 353–364.

Costa, M.V. (2004) "Rawlsian Civic Education: Political Not Minimal," *Journal of Applied Philosophy* 21: 1–14.

de Marneffe, P. (2006) "Avoiding Paternalism," *Philosophy & Public Affairs* 34: 68–94.

Dworkin, G. (2014) "Paternalism," in E. N. Zalta (ed.) *The Stanford Encyclopedia of Philosophy*, http://plato.stanford.edu/entries/paternalism/.

———. (2015) "Defining Paternalism," in T. Schramme (ed.) *New Perspectives on Paternalism and Health Care*, New York: Springer, pp. 17–29.

Feinberg, J. (1986) *The Moral Limits of the Criminal Law, Volume 3: Harm to Self*, New York: Oxford University Press.

———. (1994) "The Child's Right to an Open Future," in J. Feinberg (ed.) *Freedom and Fulfillment*, Princeton, NJ: Princeton University Press.

Galston, W. (1995) "Two Concepts of Liberalism," *Ethics* 105: 516–534.

Gert, B. and Culver, C. (1976) "Paternalistic Behavior," *Philosophy & Public Affairs* 6: 45–57.

Groll, D. (2012) "Paternalism, Respect, and the Will," *Ethics* 122: 692–720.

Gutmann, A. (1980) "Children, Paternalism, and Education: A Liberal Argument," *Philosophy & Public Affairs* 9: 338–358.

———. (1987) *Democratic Education*, Princeton, NJ: Princeton University Press.

———. (1993) "Democracy and Democratic Education," *Studies in Philosophy and Education* 12: 1–9.

———. (1995) "Civic Education and Social Diversity," *Ethics* 105: 557–579.

Husak, D. (2003) "Legal Paternalism," in H. LaFolette (ed.) *The Oxford Handbook of Practical Ethics*, New York: Oxford University Press.
Lomasky, L. (1987) *Persons, Rights, and the Moral Community*, Oxford: Oxford University Press.
Macedo, S. (1995) "Liberal Civic Education and Religious Fundamentalism: The Case of God v. John Rawls," *Ethics* 105: 468–496.
Macleod, C. (2010) "Primary Goods, Capabilities, and Children," in H. Brighouse and I. Robeyns (eds.) *Measuring Justice: Primary Goods and Capabilities*, Cambridge: Cambridge University Press, pp. 174–192.
Mill, J. S. (1974 [1859]) *On Liberty*, ed. G. Himmelfarb, London: Penguin.
Mullin, A. (2007) "Children, Autonomy, and Care," *Journal of Social Philosophy* 34: 536–553.
Rawls, J. (1999 [1971]) *A Theory of Justice*, Revised edition, Cambridge, MA: Harvard University Press.
Raz, J. (1986) *The Morality of Freedom*, Oxford: Oxford University Press.
Shafer-Landau, R. (2005) "Liberalism and Paternalism," *Legal Theory* 11: 169–191.
Sunstein, C. and Thaler, R. (2003) "Libertarian Paternalism Is Not an Oxymoron," *The University of Chicago Law Review* 70: 1159–1202.
Schapiro, T. (1999) "What Is a Child?" *Ethics* 109: 715–738.
Shiffrin, S. V. (2000) "Paternalism, Unconscionability Doctrine, and Accommodation," *Philosophy & Public Affairs* 29: 205–250.

27

PATERNALISM AND INTIMATE RELATIONSHIPS

George Tsai

1 Introduction

Does being in an intimate relationship make a normative difference to the moral assessment of paternalistic interference or intervention? In particular, does it make a difference to the moral acceptability (or permissibility) of such interference? Roughly, paternalistic interventions involve intrusions on the other's sphere of agency for their own sake. To treat someone paternalistically is to relate to them in a way that limits their exercise of agency or their options, out of a beneficent concern for their welfare, good, or interests.[1] Might paternalistic interference that would otherwise be morally unjustified be justified, in virtue of one's friendship or loving relationship to the other? Could the fact that one stands in some such relationship to another provide one with more reason to interfere in their affairs and limit their autonomy for their own good? Could an intimate relationship remove or cancel reasons one would otherwise have not to intervene?

Most discussions of the primary factors that justify an otherwise unjustified paternalistic intervention concentrate on the following considerations:

(1) What's at stake? That is, how severe or substantial is the potential welfare loss, injury, harm, or setback to the target in the case of non-intervention?
(2) How rationally or cognitively impaired is the target? That is, do they possess the capacity for voluntary action, or satisfy some sufficient threshold of it?

With regards to consideration (1), it is often thought that the more that is at stake – that is, the greater the likelihood of serious harm to the target of non-interference – the stronger the justification for the paternalistic intervention. The comparative harm of interference is also relevant. That is, in determining the justification of paternalistic interference, the harms (and benefits) faced by the target in the case of non-interference should be weighed against the harms (and benefits) faced by the target in the case of interference.[2]

With regards to (2), it is often thought that the greater the degree of cognitive or deliberative impairment suffered by the target, the greater the justification for paternalistic intervention. The conditions that limit the capacity for voluntariness have been widely theorized. The list of factors that circumscribe the voluntariness of actions is typically taken to include intoxication, ignorance, and serious psychological disorders as paradigm cases. Joel Feinberg has argued that the list should also include "powerful passion[s], e.g. rage, hatred, lust, or a gripping mood, e.g. depression, mania" (1986: 115).[3]

Both (1) and (2) are relationship-independent considerations: they make no mention of anything about the target's relationship to the one who is paternalistically interfering, or who might be in a position to do so. In this chapter, I explore the possibility that the following is also relevant to the justification of paternalism:

(3) What is the nature of the relationship between the target and the paternalistic intervener? That is, who is the person doing the intervening, and specifically what relationship (if any) does she stand in to the paternalized agent?

In particular, I want to explore and uphold the general claim that the closer, more intimate the relationship between the relevant parties, the stronger the overall moral justification for paternalistic intervention. The presence of a relationship may generate additional or stronger reasons for interference. Or put more precisely, perhaps, even if a relationship does not in itself provide one with an additional reason to interfere, it might still have normative significance in defeating or canceling some of the presumptive reasons one would otherwise have *not* to interfere.[4]

In addressing whether one's participation in an intimate relationship can make a normative difference, I begin with general reflections on the nature of intimate relationships (in Section 2), as well as paternalism and its normative significance (in Section 3). I also discuss some important normative differences between paternalism in the institutional and interpersonal contexts (in Section 4). I then argue (in Section 5) that intimate relationships can make a normative difference to paternalism, in virtue of some of the constitutive elements of intimate relationships. These difference-making elements include shared history, mutual knowledge and understanding, joint identification and projects, and reciprocated trust and vulnerability. The presence of these elements can make a normative difference to whether paternalistic interference involves the wrong-making features of paternalism (when it is wrongful). Paternalistic interference that would be wrongful if performed by non-intimates may be morally acceptable (or less morally unacceptable) – and even morally admirable or obligatory – if performed by intimates. This does not mean, however, that intimates can never act objectionably or wrongfully paternalistically. It is still possible, I maintain, for one to treat a friend or lover in a way that is wrongfully or objectionably paternalistic.[5]

2 Intimate relationships

Paradigmatically, *intimate relationships* include close friendships, romantic or committed relationships, and the relationship between parent and adult child. They may also include relationships between siblings, colleagues, and neighbors. Our ordinary usage of the term "intimate relationships" probably also covers the relationship between parent and young child, but in this discussion I set to the side the adult–young child case and focus on intimate relationships between autonomous adult persons. This is because I do not want to overly complicate matters, as the moral consideration of respect for the other's agency and autonomy may be differently relevant or weighty in the deliberative situation (the situation wherein one faces the question whether to paternalistically intervene or not), depending on whether the person one is relating to is a young child or is a fully able-minded, competent adult. In the case of the young child – and much hangs on what is taken to count as a *young child* – the person is typically taken as not fully autonomous, or at least as comparatively less autonomous than the adult.

One useful, rough test for whether two people count as having an intimate relationship in the relevant sense is whether it is apt to say of them not merely that they *stand in some relation* to one another, but that they *have a relationship* with one another. When we say of two people that they *have a relationship*, we typically mean more than that they stand in a relation in the thin, logical sense of relation involved whenever two people satisfy some two-place predicate (Kolodny 2003). T.M. Scanlon suggests that a relationship such as friendship just is "a set of intentions and expectations about our actions and attitudes toward one another that are justified by certain facts about us" (2013: 86). While the constitutive conditions of a friendship certainly include its members' intentions, expectations, and dispositions, Scanlon's account seems to leave out the crucial element of *interaction* and *mutual shaping*.

David Owens does a better job capturing these interactive and reciprocal aspects of relationships when he characterizes relationships he calls "involvements" as involving "a dynamic syndrome of attitudes, of behaviour that expresses (or purports to express) those attitudes and of norms that govern both attitudes and behaviour."[6] He observes that these essential elements of "attitude, behavioral disposition, and applicable norm all evolve in tandem: people who start to keep in touch, begin to want to keep in touch and come to feel they ought to keep in touch, all of a piece" (2012: 98).

Dean Cocking and Jeanette Kennett (1998) argue that friendship is partially constituted by the mutual willingness to be directed and interpreted by the other. A crucial dimension of intimate relationships is the mutual shaping and modification of attitudes, dispositions, and behavior between members in the relationship. What makes the adult relationships I want to focus on "intimate" in the relevant sense is a certain robust shared history between the individuals in the relationship – some sufficient degree of engagement, interaction, regard, and mutual shaping between them over time.

3 Paternalism and our objections to it

My focus is on the sort of *interpersonal* paternalism that occurs between adults in ordinary exchanges and transactions in intimate relationships: the kind of paternalism that occasionally occurs in common life in interactions between friends, family members, loved ones, spouses, and acquaintances. Thus, my discussion leaves to the side often discussed forms of *institutional* paternalism, such as paternalism by the state toward its citizens, and by doctors toward patients. This scope restriction is important, for there are interesting and often overlooked differences in the character of the normative significance between interpersonal and institutional paternalism. (I shall develop this point further in the next section.)

In describing an act as paternalistic in the relevant interpersonal contexts, I am not committing to the *pro tanto* wrongness or impermissibility of the act. That is, I do not presuppose that paternalism is presumptively morally unjustified.[7] My adoption of a normatively neutral conception is mainly for ease of exposition, since my concern is whether actions or activities that interfere with another's agency (or limit their autonomy), performed out of a concern for their welfare, can be morally justified, partly in virtue of the fact that the person doing the interfering is in an intimate relationship with the interfered with person (i.e., their friend or lover). In calling a token interpersonal interaction or transaction *paternalistic*, then, I commit only to its conforming to the broad pattern of interference or limitation of another's agency, performed out of other-regarding concern for the other's welfare. The description of an act as *paternalistic* could be taken as mere shorthand for the act's fitting the broad pattern. The question is whether, and why, interference that conforms to the broad description is sometimes objectionable (or more objectionable) if one is not a friend or lover to the target, but not objectionable (or less objectionable) if one is a friend or lover.

More generally, why is paternalism objectionable, when it is objectionable? What is central to our normative reactions to being treated paternalistically, particularly in cases where we reasonably *resent* it, find it *objectionable*?[8] I believe our objections to paternalism (when it is objectionable) typically concern one of two things (or both):

(1) the distrusting attitude of the paternalist agent manifest in the paternalistic action, and
(2) the autonomy-limiting aim (or intended effect) of the paternalistic action.

Though these two aspects are often linked in practice, they should be separated conceptually.

Consider the attitude of distrust of the paternalist as manifested by their action. Recipients of objectionable interpersonal paternalistic treatment often feel patronized and condescended to by the paternalist, whose behavior betrays the *thought* or *judgment* that they cannot be trusted to effectively advance their own interests in some deliberative domain or situation. More precisely, paternalist distrust by A toward B can consist in A's judgment that B is insufficiently competent to advance B's own interests, or that B is less competent than A to do so. Moreover, A may distrust B's capacity to judge correctly what is in her good, or distrust B's capacity to act effectively to practically implement or secure her good, or distrust both. The person who has acted objectionably paternalistically typically sees herself as better suited to judge or implement that which is in the target's interests (with respect to some deliberative domain or situation) than the target.

An autonomous agent has reason to resent other's distrust of her agency, insofar as it is being undervalued or not respected.[9] It is appropriate to feel insulted when you are treated in a way that conveys that you are insufficiently capable of advancing your own interests. To be clear, what is objectionable is a certain form of *treatment* – being treated as incompetent, related to as someone incapable of looking after your own good. What is objectionable is not simply *feeling insulted*. For it is possible for someone to suffer this kind of *experiential consciousness* due to psychological causes that have nothing to do with how one is being treated objectionably by others.

Autonomous agents also have reason to object to the objectionably paternalistic action to the extent that it aims to limit or diminish (in some way) the legitimate exercise of their agency. That is, the objectionably paternalistic action's attempt to *take over* – to interfere with, intrude on, circumvent, supplant, or replace – some aspect of the autonomous person's sphere of agency is also objectionable (absent special justification). The paternalists in these cases overstep their bounds, arrogating to themselves something that should be properly left to the other's control. Autonomous agents have reason to resent the paternalistic action, insofar as it is intended to preclude them from exercising their agency fully, as a competent person may reasonably expect to do (or be allowed to do) in the situation.

4 Interpersonal and institutional paternalism: some normative differences

In the previous section, I argued that our objections to paternalism are typically directed toward two aspects: (1) to the paternalist's display of distrust and insulting treatment, and (2) to the paternalist's intrusion on our sphere of agency.[10] That is, what is morally distinctive about wrongful paternalism is that it insultingly or disrespectfully conveys that the target is insufficiently capable of advancing her own good, and/or that it inappropriately intrudes on the target's sphere of agency.

This rather broad characterization of our objections to paternalism reflects my sense that different cases of wrongful paternalism need not be always wrongful in exactly the same way. For example, it seems to me inaccurate to say that all forms of (wrongful) paternalism are morally insulting – or morally insulting in the same way. While I find a motive-based, insult-conveying characterization of paternalism compelling in application to interpersonal paternalism, it seems to me to fit less well with paternalism in the larger-scale institutional context, where our concerns with paternalism are primarily to do with its liberty-limiting effects.

There are important differences between our normative reactions to interpersonal and to institutional paternalism. In particular, I believe the primary driving force behind our moral aversion to most paternalistic social policy is not the sense that our capacities and powers as rational agents are deemed to be untrustworthy by social authorities, but rather the sense that our choices over matters that we deeply care about and wish to have control over are being limited. In the case of government paternalism, we commonly think that, as Peter de Marneffe puts it, "paternalism matters because the moral limits to government authority over our choices matter" (2006: 76). The idea is that we care about paternalistic government policies and legal restrictions primarily because they constrain our lives in ways that we do not want to be constrained. This is what is central in our moral objections to paternalistic legal restrictions and (many forms of) paternalistic medical interventions. But it seems untrue to the phenomenology of our moral experience to claim that what is central in our normative reactions to *all* paternalistic government policy (and paternalistic institutional actions and policies, more generally) is the feeling that our rational capacities have been devalued or insufficiently respected by the government or other social institutions.

In contrast to larger-scale institutional cases, the perception that our capacities and powers as rational agents are being undervalued *is* very much at the core of our normative reactions to being treated in a paternalistic manner by friends, loved ones, relatives, colleagues, acquaintances, and so forth. It is toward persons in an interpersonal context (rather than toward institutions in the social and political context) that we tend to get most emotionally exercised in the special way tied to the perception that our ability to judge and to act to implement our own good has been insultingly underestimated in being deemed untrustworthy by another. Anger and resentment over the notion that "the government thinks we are too stupid to run our own lives" – though I do not deny that individuals can sometimes have these responses toward their government (often because they've been stirred up by the rhetoric of politicians and political pundits) – are not normally the primary reactions that individuals have to laws and policies that they do not like, but which they understand were put in place out of a concern for their good.

Our different moral concerns about paternalism in the interpersonal and institutional contexts may partly be explained by the different *means* by which the paternalistic interference is typically realized in the two contexts. That is, paternalistic government policy and paternalistic interpersonal treatment are often different in the way they go about limiting our liberty or choice options. For instance, while the government might ban harmful products such as cigarettes, individuals cannot unilaterally do this. Out of paternalistic concern, a friend might refuse to drive me to the corner store to buy cigarettes or hide my cigarettes. In so doing, my friend may be limiting my choice options, but my friend is not threatening to fine or imprison me.

The *particularity* of the paternalistic interference – to whom is it directed or addressed – may also explain differences between our moral reactions to institutional and interpersonal paternalism. Interpersonal paternalism can seem to be a more insulting form of treatment than paternalistic laws and policies because the former is typically targeted toward a specific person, whereas the latter has a more generalized target.[11] It is one thing for a government official to come out and say, "Americans need to save more money." It is another when your significant other (or in-law) tells you that you need to save more money. These considerations – of *means* and *particularity* – help to explain why concerns about giving the government authority over our choices are more salient in cases of paternalistic government policy, and why concerns about respect for our competence as agents are more salient in cases of interpersonal paternalism.

To be sure, there are some exceptions. Some cases of institutional paternalism are presumptively wrong in part because they devalue the target's rational capacities. Consider a case of medical paternalism: A patient consents to receive a certain medical treatment, but not another. Because the doctor judges that the patient's choice is imprudent, the doctor administers to the patient the second treatment (perhaps while the patient is under general anesthesia, to which she consented) without the patient's knowledge or consent. Here, the patient would be reasonable to object on the grounds that the doctor should have trusted him to make his own decisions. Consider also government policies that provide people aid "in-kind," or restrict the ways in which they can use assistance (e.g., the government offers economically disadvantaged people food stamps, but does not allow them to be used for unhealthy food). Because this policy expresses an objectionable form of distrust – a condescending judgment about recipients' abilities to handle their own affairs – it is reasonably interpreted as insulting treatment.[12]

My general point is that there are important differences between why paternalism matters in different cases where paternalism matters. I believe these differences have been overlooked because standard discussions of paternalism often proceed by first clarifying the concept of paternalism, connecting it with other notions such as "interference," "freedom," and "welfare." This approach usually involves offering up a generalized moral definition of paternalism, a moral definition meant to apply to different forms of institutional paternalism as well as to interpersonal forms of paternalism. There is, of course, nothing in principle wrong with offering a general moral definition of paternalism. Such a definition can be helpful in drawing our attention to what morally relevant features paternalistic social policies and legal restrictions have in common with interpersonal paternalistic behavior and the actions of individuals toward other individuals. At the same time, however, we should not be too quick to assume that there are not important moral differences between paternalism exercised by the state (or any other social institution) and paternalism on the part of individuals toward other individuals in the interpersonal context – differences that a general moral definition may be obscuring or overlooking. Among the things that a general moral definition may lead us to oversimplify are the differences that exist between our normative reactions to interpersonal and to institutional paternalism. The common tendency to offer general moral definitions of paternalism as a starting point to investigations about paternalism in a particular context may explain the failure to notice that in different contexts of paternalism, different strands in our normative reactions to paternalism can be more or less central. A better approach, then, adopts different characterizations of paternalism, depending on the context and the normative issues in question.[13]

5 Why relationship is relevant to paternalism

Consider some familiar cases of paternalism in intimate interpersonal contexts:

- A and B are friends. A hides B's cigarettes out of concern for B's health.
- A and B are romantic partners. A discards the credit card offer addressed to B because A thinks B will subject herself to punishing interest rates.
- A and B are siblings. A ignores B's request to stop by the liquor store because A does not want B to develop an alcohol dependence.

- A and B are married. A replaces the cardigan that B has packed in his suitcase with a blazer just as B is about to leave for the airport, because A believes B is better off wearing the blazer to his job interview.
- A and B are roommates. A decides not to tell B that B's abusive ex-boyfriend has dropped by to see her because A worries that B will get back together with him.[14]

I submit that in each of these cases, the special relationship that obtains between A and B can make a difference to whether A acts *objectionably* toward B. That is, an act that is *pro tanto* wrong on paternalistic grounds may turn out *not* to be objectionable *at all* in virtue of the special relationship between A and B.[15] On the other hand, if A and B do not stand in a special relationship (friendship, marriage, etc.), then A acts in a way that is (*pro tanto*) wrong. If relationships are normatively significant in this way, what is the basis of the normative significance?

The question rests on the assumption that standing in a valuable intimate relationship may justify paternalistic interference. Someone skeptical of the assumption might hold that what makes a normative difference to the overall moral justification of paternalistic intervention depends entirely on *non*-relationship facts. These include facts concerning the *welfare* of the potential target of the paternalistic interference: what aspects of the target's interest are at stake, how likely the target is to suffer a setback, and how substantial the target's losses will be, in the case of non-interference. They might also include psychological facts about the target: that is, the mental condition or deliberative capacity of the paternalized target, and in particular, whether the target is able-minded, is a responsible agent, has the "capacity of voluntariness," or some such. The welfare and psychology of the potential target are non-relationship factors, insofar as they can be understood without reference to the potential intervener's identity (and specifically, their identity with respect to the target: e.g., *wife of* the target, *sister of* the target, *friend of* the target).

In rejecting the assumption that there is something normatively significant about specific kinds of relationships (such as friendship and love) per se, one might maintain that the normative significance of relationships actually resides in the better *epistemic access* intimates have to the relevant facts about paternalized target's welfare and psychological state. True, friends and lovers typically have better information or knowledge about us, but in principle they needn't be the only ones. Imagine, for instance, a mere acquaintance or stranger who somehow had all of the relevant information about you: knowledge that you are likely to suffer a substantial welfare setback or lack the psychological capacity to act voluntarily. Such an acquaintance or stranger could in principle be justified in paternalistically interfering with you, according to this argument. Conversely, a friend or loved one who does not possess all the relevant information or justified beliefs would not be justified in paternalistically interfering with the other person. As I shall argue, however, it is not simply contingent epistemic access that is typically (but not exclusively) available in intimate relationships that makes a normative difference. Rather, what accounts for the normative difference is the set of important constitutive features of intimate relationships.

A different explanation of what provides the justification of paternalistic interference toward intimates appeals to the notion that consent and promises are "normative powers" that alter our "normative relationships."[16] Acts of consenting and promising are normative powers in that they alter how we may permissibly treat each other, or the rights that we may hold against each other. In application to intimate relationships like friendship, the suggestion is that we have, in the course of being friends, consented (tacitly) to our friend's potential interference in our lives. Or we have promised (tacitly) to interfere in the lives of each other: to help the other when they are in need of help, including when help is needed but unwanted by the other.

However, it seems to me untrue to our experience of friendship – or the process through which we become friends with another – to say that we consent or promise to being paternalistically interfered with. For in most intimate relationships (with the exception of marriage), we do not strictly consent to entering into the relationship or promise to be in one, though it is true that in many cases we do enter into relationships of our own choosing, deliberately or voluntarily.

Owens writes that, "One need not intend to become someone's friend but if one does become their friend, one does so intentionally" (2012: 102). To support this, he gives the following example of the role of choice in the emergence of a friendship:

> I find myself taking the bus home from work with a certain colleague. Perhaps this colleague isn't someone I would have singled out from the others for special intimacy; friendship with him is not something I'm aiming for. Still in the course of our conversations and exchanges of small favors, a friendship grows up between us. We're not soul mates but we get along well enough and the mere fact of having spent time together changes things between us. With more or less enthusiasm, I become his friend.... This result could have been avoided; I could have taken a less convenient bus or contrived not to meet him at the bus stop and so forth. In deciding not to do these things, I allowed a relationship to develop which imposes various obligations on us both.
>
> (Owens 2012: 102)

Owens goes on to describe these obligations as "owing various forms of aid and concern," and I would argue that among the forms of aid and concern that we owe to our friends are paternalistic ones (2012: 102). At any rate, the key point is that, if, indeed, we (typically) do not intend to be someone's friend (though we do become someone's friend intentionally), then *a fortiori* we also do not consent or promise to be someone's friend.[17] In other words, we should not stretch the notions of promising and consent and distort the phenomenology of friendship in the service of upholding a philosophical view, by insisting that we perform an act of consenting or promising to enter into a friendship (or other intimate relationship).

I shall now argue that it is not simply consent or promising that gives relationships their normative significance with respect to paternalism. Aspects of intimate relationships that help to constitute it – such as mutual concern, joint identification and shared projects, trust and vulnerability, relationship history and idiosyncratic habits, mutual knowledge and understanding – can make a normative difference to whether paternalistic interference involves the wrong-making features of paternalism (when it is wrongful). In virtue of these constitutive elements of intimate relationships, paternalistic interference that would be wrongful if performed by non-intimates may be morally acceptable (or less morally unacceptable) – and even morally admirable or obligatory – if performed by intimates.

Just as we might appeal to these aspects of friendship to explain why friends are permitted (perhaps obligated in certain circumstances) to ask us personal questions or call us very late in the evening or take liberties with our belongings, so we can also appeal to them to explain why friends are permitted (perhaps obligated in certain circumstances) to treat us paternalistically. In some cases, the presence of a relationship may generate additional or stronger reasons of beneficence for interference that are tied to considerations of love and partiality.[18] We may thus have stronger reasons of beneficence. In other cases, even if a relationship does not in itself provide one with an additional reason to interfere, it might still have normative significance in defeating or canceling some of the presumptive reasons one would otherwise have *not* to interfere.

For most people, participation in intimate relationships such as love and friendship is central to leading a satisfying, fulfilling human life. Participation in these relationships is sometimes described as involving a meshing of selfhood that is bound up with the pursuit of projects and aims that are *shared*, pursued jointly.[19] Examples include raising children and renovating one's home. The successful pursuit of these shared projects and aims are clearly connected to both one's own well-being as well as the welfare of the intimate with whom one shares the projects and aims. Thus, there will be cases where one has project-based reasons to do X (or forbear doing X) that may involve interfering with an intimate's agency (or limiting her choice options). That is, furthering the success of a project that one shares with an intimate, B, may involve acting in a way that circumvents B's agency but does not objectionably intrude on B's sphere of agency. The success of the project may be welfare promoting for both oneself and B. Since the shared project is partly one's own project, one may have project-based reasons to act that others who are not part of the relationship (and so do not partake in the shared project) do not have.[20]

Here one might wonder whether the cases of intervention involving shared projects are best seen as paternalism, since the shared projects partly concern A's interests. For instance, suppose that A and B are co-parents; their shared project is "raising the kids well." A interferes with B's self-harming behavior out of concern that the behavior will lead to the impairment of B's abilities as a parent. Given that A's motivating concern is that B's behavior would lead to A's having to "pick up the slack," A's motive is not purely other-regarding. Moreover, one could argue that the issue in question does not, ultimately, reside wholly within B's sole legitimate sphere of control, but that it resides within A and B's joint sphere. While I acknowledge that this may not be a paradigmatic case of *paternalism*, the crucial point morally speaking is that there is greater overall justification for intervention in virtue of the shared project of raising the children.

Of course, there are cases of justified paternalism in intimate relationships that do not further a shared project with the target. There are also cases of justified paternalism where the target's self-harming behavior falls short of impairing the pursuit of their own project. (An example falling under both categories: intervening to prevent a friend from engaging in foolish and imprudent gambling that is unlikely to result in consequential financial setback for them.) In these cases, the justification of paternalism would have to be grounded in constitutive elements of the relationship other than the fact that there are shared projects, such as vulnerability and trust.

It is virtually a conceptual point that intimate relationships involve vulnerability and trust. Elsewhere, I have argued that valuable intimate relationships (e.g., love and friendship) involve vulnerability essentially, and that the distinctive vulnerability connected with participation in intimate relationships exposes one to harms that can at most be mitigated but not eliminated.[21] The fact that friendship and loving relationships typically entail greater vulnerability is non-accidentally related to the fact that it is often not simply unobjectionable or admirable to paternalistically intervene as a friend but also a requirement of being a good friend or lover. Our friends and loved ones are especially aware of our vulnerabilities and personal insecurities in part because we are more willing to trust them, more open with them about our fears, secrets, and insecurities. This openness is closely connected with the special goods that participation in an intimate relationship makes available, goods such as: care, affection, intimacy, mutual understanding, sense of connection, shared feelings and experiences, shared purposes, and joint activities. Our friends and loved ones are often best placed to relate well to our insecurities and fears – to do things for us that promote our own interests, aims, and goals given our weaknesses.

Part of what we value in valuing intimate relationships is being in a trusting relationship: a relationship in which the participants are mutually vulnerable to one another in part because they have placed their trust in one another. While we do not value placing ourselves in just anyone or everyone's hands, we do value placing ourselves in certain people's hands. We value intimate relationships partly because such relationships enable us to relax the default self-protective strategies that we usually have in life. With friends and loved ones, we are more at ease to reveal our helpless side and more freely able to acknowledge the fact that we have less control than we like over much of what we care about. Thus, when intimates treat us paternalistically – that is, are moved out of beneficent concern to limit our agency, seeing that we need help, that we are not self-sufficient, that we are vulnerable and fallible in the relevant deliberative situation – their motivating concerns about our ability to adequately help ourselves on our own are often not experienced as insulting or disrespectful. One may call motivating concerns of this kind a sort of distrust or lack of faith, if one wishes. But since such distrust or lack of faith is based on our self-presentation to them as the vulnerable persons we actually are (a self-presentation we do not provide to just anyone but only to special others), the paternalistic motive is not typically disrespectful or insulting, and so the paternalistic intervention is not insulting treatment.

The fact that we are more vulnerable (emotionally and otherwise) to our friends and loved ones – and often we have *made* ourselves more vulnerable to them – is also importantly connected to the general point that the expressive meaning of an act of kindness depends on *who* is performing the act, and in particular, what relationship the person performing the act bears to the person acted on. The act of inquiring about the health of someone may be inappropriate and received as creepy and objectionably intrusive if done by a mere acquaintance, but it may be *an act of love or friendship* – expressive of the kind of concern only friends and loved ones can express toward each other – and received warmly and with the effect of cheering the person up if done by an intimate. Acts (such as a telephone call or hospital visit to a sick person) have a different meaning or significance for the recipient – they are valued or cherished differently by the recipient, and generate in them different reactive attitudes, such as resentment or gratitude – depending on the identity of the agent, and specifically his or her relationship to the person toward whom the act is directed. The same action, then, can have different meaning or significance for different people depending on their relationship to the agent. For whether the act can actually indicate, reveal, or convey a certain kind of welcome concern or intimacy is not something that is equally available to everyone, but only to special others to whom one has made oneself vulnerable. Paternalistic treatment typical of friendships and loving relationships is often just the expression of the kind of intimacy that we cherish.

In previous work, I considered several factors that make a difference to whether the provision of reasons (for example, in offering advice) is disrespectful, intrusive, or insulting.[22] Here, I want to focus on one factor: the nature of the relationship or personal history between the person offering the reason and the person receiving it. Applied to the concerns of this discussion, the question is how the nature of the relationship or personal history between the interfering and interfered with persons makes a difference to whether the interference is morally acceptable. My claim is that whether paternalistic interference is intrusive or inappropriate or insulting can depend on whether the relevant parties stand in the right kind of relationship, having cultivated a relationship that involves intruding on one another's sphere of agency in ways that are understood by the participants as permissible (even welcome) or obligatory.[23]

Take the case of offering someone advice or providing them with a reason for action, which I have argued is sometimes objectionably paternalistic (Tsai 2014). Whether offering someone advice is objectionably paternalistic or unobjectionably paternalistic depends on the relationship one stands in to the other. For close interpersonal relationships are partially constituted by the willingness of their members to guide and help each other out, as well as normative and psychological expectations that they do so, which are generated by patterns of past behavior. Such guidance and assistance to a friend or family member can often take the form of offering advice, reminders, even warnings.

But although we all want to be cared for to some degree or extent – indeed, care is one of the special goods that intimate relationships make available – there are limits to the care we want to receive or receive from certain people, even intimates. For example, as we come nearer to adulthood (and especially in the years of young adulthood), mere attempts by parents to benefit us out of love can be experienced as a threat to one's independence. But more generally, not only do we not always want to be looked after by those we love, we may not always want them or others to see us as in need of being looked after. This is because, though we value care, we also value autonomy and a sense of *dignity*, which may be compromised if we are seen as or become in fact overly dependent. The point might also be put in terms of *equality*. Sometimes when one person benefits another, they may, as a result of the benefitting, no longer be equals: there is a subtle alteration of status. The inequality is generated out of the fact that with respect to B's welfare, A has become a surrogate for B's agency, providing what B was not in a position to provide for herself. B has become dependent (albeit in a limited way). And that might be objectionable to B.[24]

On the other hand, given that in paradigmatic intimate relationships, such as close friendship and marriage, each party will be dependent on the other, there may be less significant concerns about the inequality generated by a token intervention. Thus, one might think it would therefore be easier to justify inequality-generating interventions in these relationships. The inequality-generating interventions would indeed be easier to justify if the following is true: though token interventions generate some inequality and dependency for one party in relation to the other, when *the relationship as a whole* is considered, it is not the case that the interventions are one-sided or leading to a relationship that is structurally unequal, conducted on terms of inequality. Suffice it to say, our interest in holding onto some measure of autonomy and preserving a sense of equality, while balancing this interest with the characteristic dependency and vulnerability in our intimate relationships, is a complicated matter.[25]

More generally, how exactly are the reasons of beneficence that point toward interference and the reasons of respect that point toward non-interference to be weighed when opportunities for paternalistic interference arise in intimate relationships? Differences in the goals and means of the paternalistic intervention further complicate the question of its overall moral justification. Moreover, unique understandings developed within relationships may modify the standard expectations of participants in such relationships, such that what paternalistic interventions count as permissible, impermissible, or even obligatory can diverge depending on the particular relationship and the personalities of the members of the relationship.

In light of these nuances, complexities, and qualifications, one might wonder how one is supposed to discern in a given situation whether it would be appropriate or respectful to paternalistically intervene. Perhaps we can say that paternalistic intervention is morally justified (overall) when it strikes an appropriate balance between beneficence and respect. But this is probably not going to be terribly helpful in many cases. Yet, we are also not completely at sea, for to understand the nature of respect for other people *just is*, in part, to have some grasp of the kinds of circumstances wherein certain paternalistic interventions would be appropriate. Put differently, judging well whether and how one can justifiably paternalistically intervene in an intimate's life is an art – an aspect of practical wisdom that one can cultivate. But, then again, this is true of treating people with respect and consideration in general.

To summarize: I have identified some constitutive elements of intimate relationships and argued that they help to explain the intuition that paternalism towards intimates is sometimes justified. These elements include joint identification and shared projects, trust and vulnerability, mutual understanding, and shared history. I have argued that paternalistic treatment that would otherwise be morally objectionable (because it constitutes insulting treatment or an unwarranted intrusion into the target's sphere of agency) may be justified in virtue of these constitutive elements. The fact that A and B are in an intimate relationship can mean that paternalistic treatment by A towards B needn't be understood as involving an objectionable motive of distrust in B's competence, as disrespectfully conveying that B lacks competence, or as limiting B's legitimate exercise of agency. Sometimes, the presence of a relationship generates additional or stronger reasons of beneficence to interfere. In others, the relationship weakens or cancels some of the presumptive reasons of respect one would otherwise have *not* to interfere.[26]

Related topics

Paternalism and Sentimentalism; Perfectionism and Paternalism.

Notes

1 I shall use the notions of a person's welfare, interests, and good interchangeably in this discussion, though I acknowledge that they are not exactly equivalent in ordinary usage.
2 Thanks to Kalle Grill for pressing me to clarify this point.
3 Donald VanDeVeer offers a similar list, including such factors as "disease, injury, fainting, drunkenness, drug usage, embarrassment, fear, and so on" (1986: 347).
4 Let us suppose that paternalism is morally justified when it strikes an appropriate balance between beneficence and respect. The suggestion is that when A stands in a close relationship to B, the reasons of beneficence for A to intervene may be no different, but the reasons of respect not to intervene may be weaker (or maybe stronger). Thanks to Jason Hanna for this suggestion.
5 Throughout this discussion, I shall use the expressions "objectionable paternalism" and "wrongful paternalism" (and their cognates, e.g., "objectionably paternalistic," "wrongfully paternalistic") interchangeably.
6 What Owens means by involvements does not exactly coincide with what I mean by intimate relationships. For Owens, "involvements" are "valuable forms of human relationship" that are marked by two features: they are "in some sense chosen" and "entail obligation" (2012: 96). Involvements include relationships between neighbors, acquaintances, guest and host, conversational participants, and friends. But they do not include such relationships as between parent and child, family members, and fellow citizens, insofar as these relationships are not chosen.
7 Many discussions of paternalism adopt a moralized or normatively loaded definition of paternalism. Indeed, I adopted such usage (in Tsai 2014).
8 This section draws on points advanced in Tsai (2014).
9 An interesting question is whether a person's agency is being undervalued if another distrusts her because she really is likely to choose imprudently. See Enoch (2016).
10 In Tsai (2014), I argue that there is an important link between the two wrong-making dimensions of paternalism and the motive of the paternalist agent. On the distinction between motive-centered and effect-centered characterizations of paternalism, see Shiffrin (2000: 211–220). On the role of motive in paternalism more generally, see Quong (2011).
11 Thanks to Jason Hanna for this suggestion.
12 Thanks to Jason Hanna for the examples in this paragraph.
13 If this is right, then there is reason to doubt that either a strictly motive-centered or effect-centered account of paternalism can accommodate both interpersonal and institutional forms of paternalism.
14 These examples are presented in Tsai (2014).
15 I leave open the possibility that some of these cases may also be objectionable on other nonpaternalistic grounds.
16 See Shiffrin (2008), Owens (2012), and Dougherty (2015).
17 One might argue that if someone does something deliberately and voluntarily, and there is no background pressure or difficult circumstances, then that person automatically consents to the expected consequences. Even if this is right, there is the question of how to understand the relevant "difficult circumstances" in the case of becoming friends with someone. Another important related issue is whether consent requires communication. See Dougherty (2015).
18 Many accept that we have stronger reason to care for and promote the interests of those with whom we stand in special relationships. See Scheffler (1997) and Keller (2013).

19 In Tsai (forthcoming), I consider the normative importance of shared projects that arise out of the support we offer our intimates pursuing their personal projects. There I write, "When I support you in your projects, your projects become our projects through my investment in your projects."
20 Conversely, non-intimates can also share projects, and these shared projects could be the basis of justified paternalism. Consider an application of the shared projects idea to the case of medical paternalism: a cancer patient's health and physical well-being could be viewed as a project shared with her oncologist. It is an intriguing notion that some forms of medical paternalism might be justified by appeal to the notion of a shared project. Thanks to Jason Hanna for this suggestion.
21 Tsai (2016).
22 In Tsai (2014), I discuss five relevant factors: subject matter, mode of presentation, timing, relationship, and epistemic access.
23 Or put somewhat differently, if it is not insulting for the paternalist agent so to act – say, because the project is shared in some sense – then maybe the act is not actually an incursion into another's sphere of agency. (To say that something is within my sphere of agency may suggest that I ought to have exclusive rights to determine what to do, or how to resolve the issue.) The upshot of seeing matters in this way would be normatively equivalent to the construal in the main text, insofar that the act of benefitting in question would be unobjectionable (or not objectionable in the way the act would be objectionable if performed by a non-intimate). Thanks to Jason Hanna for suggesting this alternative construal of the point.
24 It may be that the risk of making the relationship less equal as a result of the benefitting is greater in the case where the beneficiary has invited or even requested the benefits from the benefactor. Nonetheless, I think a similar risk of generating inequality in the relationship also exists in the case of uninvited, unrequested paternalistic benefitting (even if the risk is perhaps not as great).
25 Thanks to Jason Hanna for raising the concerns in this paragraph.
26 I would like to thank Kalle Grill and Jason Hanna for their very helpful written comments on an earlier draft of this chapter.

References

Cocking, D. and Kennett, J. (1998) "Friendship and the Self," *Ethics* 108: 502–527.
de Marneffe, P. (2006) "Avoiding Paternalism," *Philosophy & Public Affairs* 34: 68–94.
Dougherty, T. (2015) "Yes Means Yes: Consent as Communication," *Philosophy & Public Affairs* 43: 224–253.
Enoch, D. (2016) "What's Wrong With Paternalism: Autonomy, Belief, and Action," *Proceedings of the Aristotelian Society* 116(1): 21–48.
Feinberg, J. (1986) *The Moral Limits of the Criminal Law, Volume 3: Harm to Self*, Oxford: Oxford University Press.
Keller, S. (2013) *Partiality*, Princeton, NJ: Princeton University Press.
Kolodny, N. (2003) "Love as Valuing a Relationship," *Philosophical Review* 112: 135–189.
Owens, D. (2012) *Shaping the Normative Landscape*, Oxford: Oxford University Press.
Quong, J. (2011) *Liberalism Without Perfection*, Oxford: Oxford University Press.
Scanlon, T. M. (2013) "Interpreting Blame," in D. Coates and N. Tognazzini (eds.) *Blame: Its Nature and Norms*, Oxford: Oxford University Press.
Scheffler, S. (1997) "Relationships and Responsibilities," *Philosophy & Public Affairs* 26: 189–209.
Shiffrin, S. (2000) "Paternalism, Unconscionability Doctrine, and Accommodation," *Philosophy & Public Affairs* 29: 205–250.
———. (2008) "Promising, Intimate Relationships, and Conventionalism," *Philosophical Review* 117: 481–524.
Tsai, G. (2014) "Rational Persuasion as Paternalism," *Philosophy & Public Affairs* 42: 78–112.
———. (2016) "Vulnerability in Intimate Relationships," *Southern Journal of Philosophy* 54(Spindel Supplement): 166–182.
———. (forthcoming) "The Virtue of Being Supportive," *Pacific Philosophical Quarterly*.
VanDeVeer, D. (1986) *Paternalistic Intervention: The Moral Bounds on Benevolence*, Princeton, NJ: Princeton University Press.

INDEX